# SPATIAL PATTERN IN PLANKTON COMMUNITIES

## NATO CONFERENCE SERIES

- I Ecology
- II Systems Science
- III Human Factors
- IV Marine Sciences
- V Air–Sea Interactions
- VI Materials Science

## IV MARINE SCIENCES

*Volume 1* Marine Natural Products Chemistry
edited by D. J. Faulkner and W. H. Fenical

*Volume 2* Marine Organisms: Genetics, Ecology, and Evolution
edited by Bruno Battaglia and John A. Beardmore

*Volume 3* Spatial Pattern in Plankton Communities
edited by John H. Steele

Nato conference on Marine Biology)
= Erice, Italy, 1977.

# SPATIAL PATTERN IN PLANKTON COMMUNITIES

Edited by
**John H. Steele**
*Woods Hole Oceanographic Institution*
*Woods Hole, Massachusetts*

Published in coordination with NATO Scientific Affairs Division
PLENUM PRESS · NEW YORK AND LONDON

Library of Congress Cataloging in Publication Data

Nato Conference on Marine Biology, Erice, Italy, 1977.
Spatial pattern in plankton communities.

(NATO conference series: IV, Marine sciences; 3)
Includes index.
1. Marine plankton — Congresses. 2. Vertical distribution (Aquatic biology) — Congresses. 3. Marine ecology — Congresses. I. Steele, John H. II. Nato Special Program Panel on Marine Science. III. Title. IV. Series.
QH91.8 P5N37 1977 574.9'2 78-12017
ISBN 0-306-40057-X



Proceedings of the NATO Conference on Marine Biology
held at Erice, Italy, November 13—21, 1977, and sponsored by the NATO
Special Program Panel on Marine Science


A Division of Plenum Publishing Corporation
227 West 17th Street, New York, N.Y. 10011



Printed in the United States of America

# FOREWORD

The planning for the conference held at Erice, Sicily, in November 1977, began with discussions among oceanographers from several countries on the need to consider the special problems and the recent results in the study of plankton "patchiness." An approach to the Marine Sciences Panel of the NATO Science Committee resulted in a planning grant to determine the probable content and participation in such a meeting.

The planning group consisted of B. Battaglia (Padua), G.E.B. Kullenberg (Copenhagen), A. Okubo (New York), T. Platt (Halifax, Nova Scotia) and J. H. Steele (Aberdeen). The group met in Aberdeen, Scotland, in September 1976.

The proposal for a NATO School on the subject of "Spatial Pattern in Plankton Communities" was accepted by the Marine Science Panel and it was agreed that it be held at the Ettore Majorana Centre for Scientific Culture in Erice.

The Centre began in 1963 with an International School of Subnuclear Physics and has since developed to include courses in many other subjects which cover various fields of basic and applied research. The original aim of the Centre was to create, in Italy, a cultural forum of high scientific standard which would allow young research workers to appreciate problems currently of major interest in various fields of research. This is now attained by encouraging and promoting national and international meetings among specialists actively engaged in advanced disciplines, thus offering the younger scientists a unique opportunity to discuss with their more qualified colleagues the results of their investigations. Since these aims corresponded closely to those of the NATO School, we were happy to accept the kind invitation of Professor Zichichi, Director of the Centre. We wish to thank him and his colleagues at Erice for the hospitality enjoyed by all the participants.

The Centre is sponsored by the Sicilian Regional Government, the Italian Ministry of Public Education, and the Ministry of Scientific and Technological Research. The various schools take place under the sponsorship and with some financial aid from these national agencies.

The Centre has its own journal: "Progress in Scientific Culture" whose articles have an interdisciplinary character. The Journal has published summaries of the lectures delivered during this NATO School.

The meeting at Erice provided valuable opportunities for interdisciplinary consideration of new and important questions in marine science. We hope this volume reflects the nature of these questions and the recent advances in this field. We are grateful to the NATO Science Committee for its support of the School.

Bruno Battaglia

John H. Steele

CONTENTS

## PARTICIPANTS

G. Andreoli, Italy
M. Azzali, Italy
B. Battaglia, Italy
R. Benzi, Italy
C. M. Boyd, Canada
P. Chryssovergi, Greece
W. C. Clark, Canada
J. M. Colebrook, U.K.
B. Coste, France
M. E. Cunha, Portugal
K. L. Denman, Canada
D. M. Dubois, Belgium
R. C. Dugdale, U.S.A.
M. Van Der Eijk, Netherlands
M. Estrada, Spain
G. T. Evans, U.K.
M.J.R. Fasham, U.K.
P. Franco, Italy
S. Gaiardi, Italy
L. Guglielmo, Italy
G. Harris, Canada
P. M. Holligan, U.K.
J. W. Horwood, U.K.
L. Ignatiades, Greece
G.E.B. Kullenberg, Germany
R. Lasker, U.S.A.
S. A. Levin, U.S.A.
A. R. Longhurst, U.K.
D. Mackas, Canada
G. Magazzu, Italy
S. Mater, Turkey
J. A. McGowan, U.S.A.
M. Moraitou-Apostolopoulou, Greece
P. Nival, France
A. Okubo, U.S.A.
T. Ongan, Turkey
I. Ozel, Turkey
V. Parisi, Italy
R. D. Pingree, U.K.
T. Powell, U.S.A.
P. R. Pugh, U.K.
G. Radach, Germany
P. J. Richerson, U.S.A.
S. E. Salusti, Italy
A. P. dos Santos Matos, Portugal
F. Schott, Germany
W. Silvert, Canada
N. Staresinic, U.S.A.
J. H. Steele, U.S.A.
K. H. Szekielda, Germany
S. Tilseth, Norway
G. M. de Vasconcelos, Portugal
J. J. Walsh, U.S.A.
G. White III, Canada
J. D. Woods, Germany
J. S. Wroblewski, Canada
C. S. Yentsch, U.S.A.
D. Zafiropoulos, Greece
M. Zarull, Canada
B. Zeitzschel, Germany

# SOME COMMENTS ON PLANKTON PATCHES

John H. Steele

Woods Hole Oceanographic Institution
Woods Hole, Massachusetts 02543

## INTRODUCTION

The title of this volume mentions only the spatial patterns of plankton, but the populations we are concerned with have other types of variation which cannot be ignored. They have different life spans which must have an effect on their spatial distributions. Also, we must consider the food chain relations between organisms as a factor affecting spatial structure and this involves aspects such as the size distribution of the plants and animals and the numbers of different species within each trophic grouping.

The interactions of these various space and time scales with species diversity form the general problem of variability in the marine ecosystem. The concepts we shall consider must concern all those aspects. Similarly, for the problems of sampling this environment, we have to take account of these different dimensions. In practice, however, for the pelagic environment we are forced to concentrate on spatial variations. The nature of water movements and dispersion make it impossible to obtain a pure time series of events in the open sea, although some approximation to a time sequence may be possible in special circumstances such as Gulf Stream Rings (Wiebe *et al.*, 1976). Nor do we yet have instrumentation for biological parameters that can be left moored in the sea. Strictly, it is also true that we cannot make instantaneous measurements in the horizontal dimension, but with present shipboard instruments we can make effectively continuous measurements over distances of 10-100 km in a few hours. Such measurements can be considered synoptic in relation to events which we expect to change on a time scale of days or weeks. The basis for this assumption will be considered in detail in further papers.

Thus the title of this volume indicates the nature of the data from which we must formulate our ideas, rather than a limitation on the concepts needed to explain the observations. Within this initial concentration on spatial patterns, there are further limitations. In the vertical we shall be concerned mainly with events in the upper 100 m. This is the region to which phytoplankton growth is confined by the limited penetration of light. The measurement of the flucrescence of the pigments in phytoplankton is one of the major methods available at present to describe the variability of an important component of the pelagic ecosystem. By towing a fluorometer through the water or by pumping water through a fluorometer on a ship we can obtain profiles indicating changes in phytoplankton concentration. For vertical profiles resolution to about 10 cm is obtainable. In the horizontal a resolution of less than a meter would be possible from a very slowly moving or drifting ship and some measurements near this range have been obtained. However, there is some reason to believe that horizontal variations at these small scales are dominated by physical changes (Platt, 1972; Steele, 1976) and it is at larger scales that biological interactions are most likely to occur. For this reason we have more data relating to scales from 10 m to 100 km collected from faster moving ships. Thus these scales will form the core for much of the later discussion.

The fluorometer technique has provided most of the recent information on biological variability in the upper layers of the sea. Similar continuous measurements can be made for some physical and chemical parameters such as temperature and salinity which will partly define the physical environment, and for nutrients such as nitrate, phosphate and silicate which provide a link between the physical conditions and the growth and abundance of the phytoplankton. Figure 1 provides an example of such data collected at the start of the spring outburst in the North Sea when zooplankton populations were negligible, giving a good relation of phytoplankton growth to nitrate uptake (Steele and Henderson, 1977).

Many of the available data, showing comparable variability, are from regions of the continental shelf. In the North Atlantic between the Scottish shelf and Rockall Bank considerable variability in chlorophyll and nitrate has been observed (Figure 2). In this case there was no obvious correlation between chlorophyll and nitrate. In contrast, Lorenzen (1971) found that for large areas of the Pacific Ocean the chlorophyll levels were approximately uniform.

For the herbivorous copepods which graze on the phytoplankton we are not so fortunate in terms of methodology. The diurnal vertical migrations of these animals pose sampling problems which require, either scanning of the whole water column for population concentration (Longhurst *et al.*, 1966) or particle numbers (Boyd,

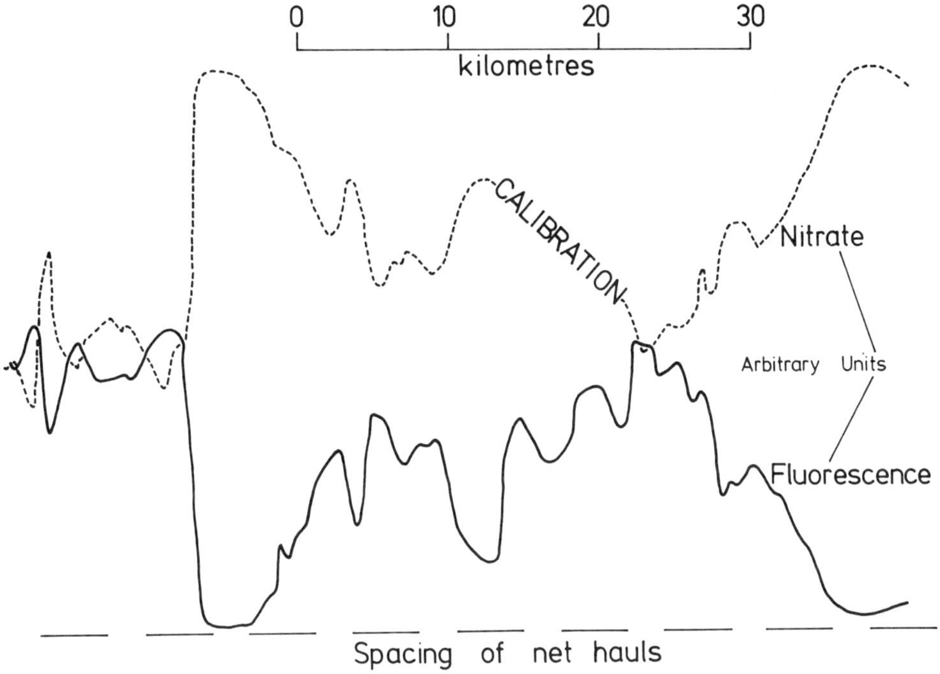


Figure 1. An example of observed traces at 3 m depth from the northern North Sea on 10 April 1974. The lines along the bottom of the figure indicate the distance covered by oblique zooplankton net hauls.

1973; Haury *et al.*, 1976); or the use of vertically or obliquely hauled nets which integrate over the water column. The former has still to be developed as a routine method (Haury, 1973); the use of nets towed from a moving ship gives samples with a spacing of about 5 km which is one or two orders of magnitude greater than that of the phytoplankton measurement (see Figure 1). For the higher trophic levels of invertebrate predators and pelagic fish we have very little information on smaller-scale variability although the use of echo-integrating techniques may help with the latter.

These brief comments on sampling problems indicate present inadequacies but also the potential from new methods and these will be dealt with in more detail later in this volume. At present, fluorometer data provide a small window through which we observe a part of the pelagic ecosystem and from which we may infer possible features of the other components which interact with the phytoplankton. These deficiencies must be remembered in the following discussion of some of the general problems.

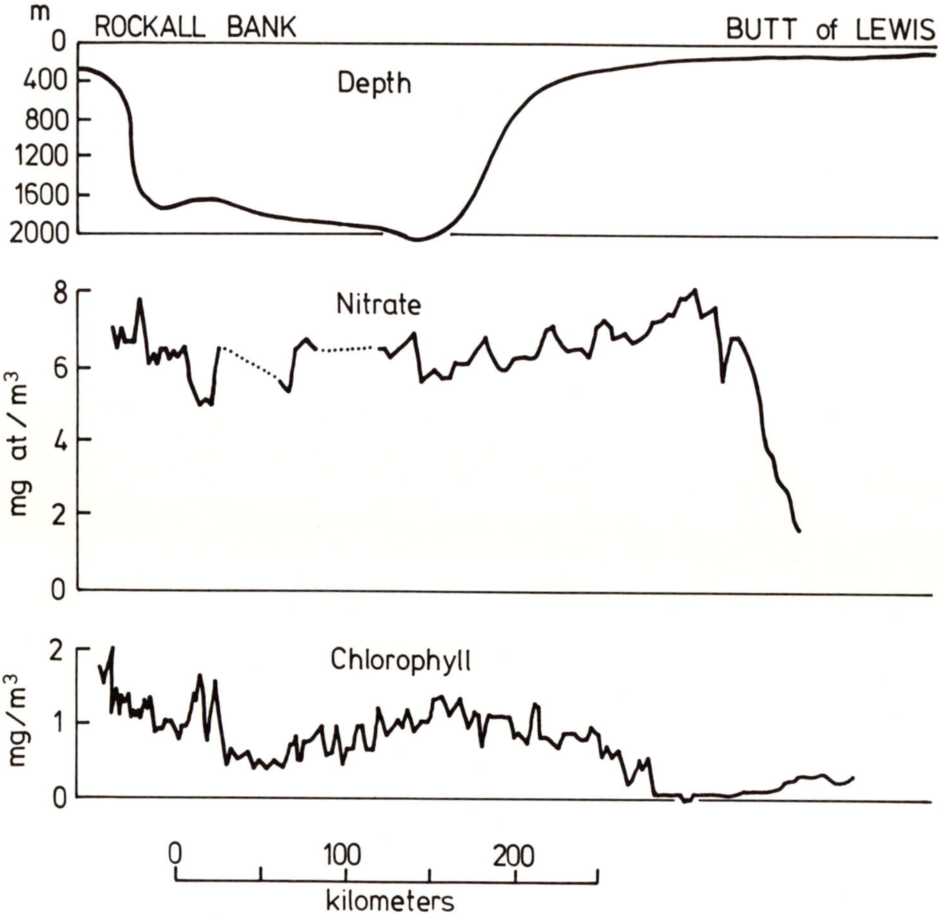


Figure 2. Nitrate and chlorophyll data from 3 m on a section in the Northeast Atlantic on 11-12 May 1968.

## SPACE-TIME INTERACTIONS

Some idea of the conceptual and logistic problems can be obtained from a very simplified representation of possible characteristic scales for three main trophic components (Figure 3) in the context of the North Sea food chain. The life spans of phytoplankton, copepods and pelagic fish such as herring are of the order of 1-10, 30-100, 600-2000 days. It has been proposed (e.g. Steele, 1976) that the minimum scales at which phytoplankton growth can maintain patch structure in face of horizontal turbulent dispersion is of the order of a few kilometers. Scales of zooplankton patches appear to occur at tens of kilometers (e.g. Cushing and Tungate, 1963). For pelagic fish I have taken the diameter of the

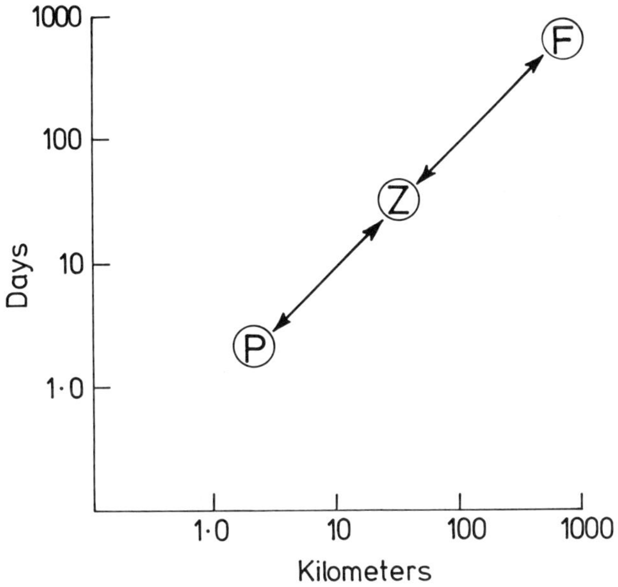


Figure 3. A very simplified representation of typical time and space scales associated with plants (P), herbivorous zooplankton (Z) and pelagic fish (F).

annual migration cycle as a scale appropriate to the life span. For the North Sea (Harden-Jones, 1968) this is of the order of 1000 km. To understand the dynamics of these populations, sampling programs and theoretical studies should be concerned with the interactions along the diagonal of Figure 2. The former is obviously impossible logistically and we have to content ourselves with obtaining data on some chosen horizontal (spatial) scales and repeating these as frequently as possible. The classical survey used for fish stock assessment would sample at stations with a 50-100 km grid and repeat the program, if possible, 3-6 times per year over several years to give a 50-100 day spacing. It can be seen (Figure 4) that such sampling, with minimum frequencies corresponding to expected herbivore variability, is almost perfectly designed to miss this component producing apparently random data (Steele and Henderson, 1977).

Using the fluorometer and other techniques, sampling is now concerned with much smaller scales. By a combination of stationary and moving ships, a recent international effort (FLEX) has tried to study a 100 km box for 100 days (Figure 4). The large effort required, about 20 ships were involved, is an indication of the effort needed to tackle even one part of the scales indicated in Figure 3.

The conceptual problems are equally difficult to analyze. The variability of the physical environment extends over the range of scales given in Figure 3, and beyond. Mixing will occur at the smallest scales we can measure and extends to the effects of current systems on large-scale migrations of pelagic fish (Harden-Jones, 1968). Again, there are general relations between time and space scales. Okubo (1971) from an analysis of experiments

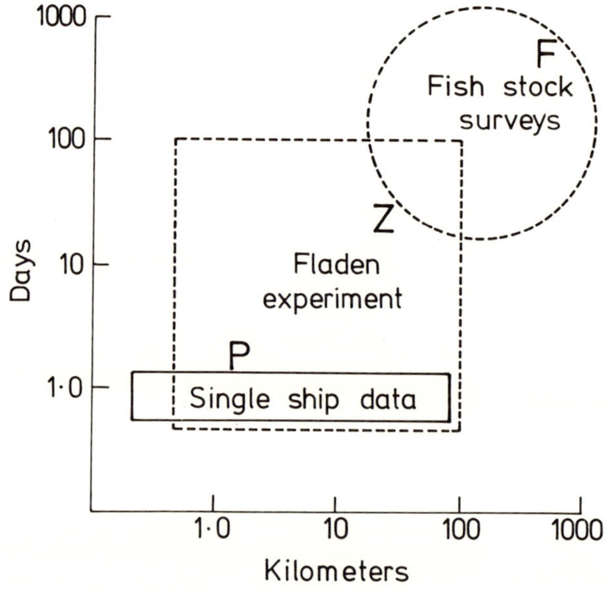


Figure 4. An indication of the space and time scales covered by various types of sampling programs.

measuring the dispersion of dyes from "point source" injections, derives the empirical relation between the standard deviation, $\sigma$, (km) and time, $t$, (days)

$$\sigma = 0.6\, t^{1.17}.$$

By comparison with Figure 3 we can see in Figure 5 that there is a rough equivalence in the scales of physical dispersion and the biological scales associated with growth of organism. On this basis patchiness might be seen as a deviation of the biological spectrum from the physical spectrum of turbulence rather than as increases or decreases from a uniform spatial distribution. Such deviations have been observed in phytoplankton and linked to growth rates (Denman and Platt, 1976).

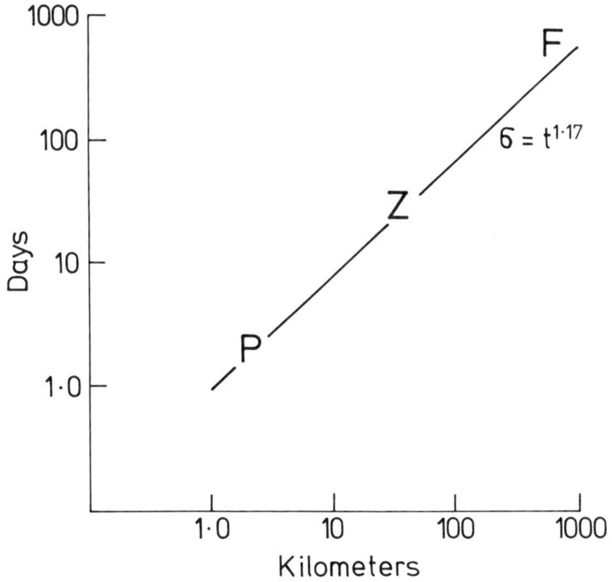


Figure 5. A comparison of mixing scales with the trophic scales of Figure 3.

## TROPHIC LEVEL INTERACTIONS

The underlying assumption in Figure 3 is that biomass created by phytoplankton is passed up the diagonal by a set of prey-predator interactions. These interactions depend on another length scale -- the size of the organisms themselves. There is a general relation (Sheldon *et al.*, 1972) between equivalent diameter and length of life (Figure 6). This figure shows the range of sizes and life spans associated with each of the main groups. It includes a group of invertebrates which contain carnivores such as *Sagitta* and omnivores such as the Euphausids. Also, the fish group contains anchovy which can be herbivorous. In general, however, increasing size and life span corresponds to an increase in the trophic level. Further, as Sheldon *et al.* point out, a 10% efficiency of yield in terms of food intake, combined with a factor of 10 in the ratio of life span for fish (herbivores, phytoplankton) implies a rough equality in biomass at the various trophic levels. Thus, at all levels, there is a similar problem in obtaining food from a given

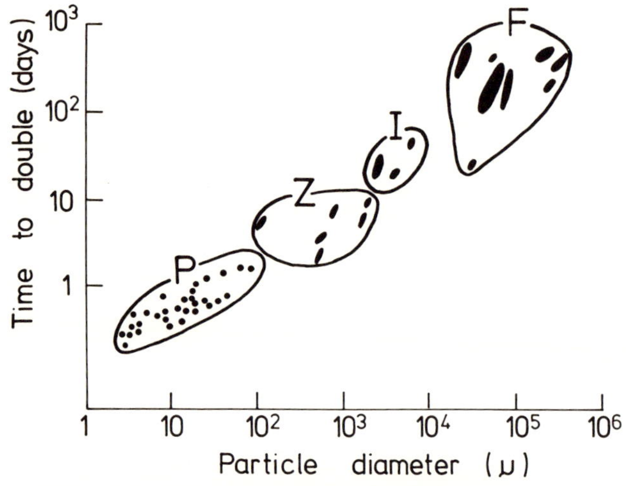


Figure 6. The relationship between doubling time and particle size from data for phytoplankton (P); herbivores (Z); invertebrate carnivores or omnivores (I); and fish (F) (from Sheldon *et al.*, 1972).

volume of water, and variations in food concentration will be significant for rate of feeding.

There are other features of the life cycle which must be considered. The animals span a significant range of sizes (and so feeding habits) as they pass from egg to adult. Larvae of fish, such as herring, when they start feeding, have an equivalent diameter of $10^3$ μm comparable to the herbivores. They can feed only on the smallest copepod stages and are themselves liable to predation by older fish. In consequence, although energy flows upwards along the diagonals in Figures 3 and 6, predation and reproduction at higher trophic levels can affect the size composition at lower levels and thus smaller sizes (Gamble *et al.*, 1977; Steele and Frost, 1977).

These aspects form the background to questions of interaction in relation to spatial structure. The increasing size and life span of organisms can permit greater independence of water movements and mixing. Thus phytoplankton (apart from possible slow sinking rates) are assumed to be carried passively by both horizontal and vertical motions. Zooplankton have little directed movement at horizontal scales above a few meters but can move vertically over tens of meters in response to diurnal changes in light and also, possibly, in relation to vertical variations in food concentration. Pelagic fish, as well as exhibiting shoaling behavior, carry out horizontal migrations of hundreds of kilometers using, to some extent, regular features of ocean currents. For these reasons there will be a greater independence of physical variability at higher trophic levels and so one would expect the effects of such variability to be passed from lower to higher levels. On the other hand, the greater population interactions and greater mobility at higher trophic levels can mean that some spatial features could be passed down the trophic web.

Thus factors for each group may influence not only their own distribution in space but also those of their food supply -- or of their predators. For these reasons we could expect spatial patterns to be transmitted both up and down the diagonal of Figure 3.

## SPATIAL INTERACTIONS

Early data on spatial distributions gave the impression of actual patches of plankton observed in the open sea at scales of tens of kilometers (Bainbridge, 1957; Cushing and Tungate, 1963; Steele, 1974). It might be supposed that results based on spatially discrete observations, at intervals of a few kilometers, missed the finer structure which can now be detected with instrumentation such as autoanalyzers or fluorometers. Much of the fluorometer data supports this assumption (for example, Figure 1, but see also Platt, this volume).

The question arising is how spectral distributions would be affected by biological interactions, especially the effects of grazing on the phytoplankton. The simplest theoretical relations of horizontal mixing and phytoplankton population growth suggest that the smallest patches should be inversely proportional to $(\text{growth})^{\frac{1}{2}}$. If population growth is taken as a balance between cell division and grazing rates, then increasing herbivore populations should lead to increased critical scales at which patches can occur due to biological factors. Finally, if the two rates, production and grazing, are in balance then phytoplankton variability would depend only on physical factors. This, however, depends on the assumption of spatial uniform herbivore distribution.

From data collected during the FLEX cruises, the opposite appears to occur. Figure 7 shows the repetition of a transect over three days with several significant features. (The water was sampled at 3 m and the data logged at one minute intervals corresponding to a horizontal distance of approximately 200 m.) The marked change at the northern end corresponds to a thermal front (Figure 8), but none of the other features show any obvious relation to temperature or salinity (and this is true for the rest of these surveys). Except for the last run (Figure 7), there is no evidence of fine structure comparable to data collected earlier in the season (see Figure 1). Instead, there are definite features with dimensions in the range 10-20 km. Although the samples were taken at 3 m, vertical profiles showed a fairly uniform layer down to 40-50 m depth. Thus the variability is not associated with transient near-surface events. This is confirmed by the persistence of the main features over the first 24 hours. The last run (Figure 7) two days after the start showed some breakdown of the main patches and also increase in variations at the smaller scales. On this and other occasions the increase in variability at smaller scales appeared to be associated with increasing wind force.

These, and other transects, give the impression of sections through patches or fronts, and a smaller scale grid a few days later gives an indication of the two-dimensional near-surface structure found in this area (Figure 9). Because of the size of the box, some physical features seem to have the same scales but no two of the parameters show an obvious similarity so that there are no simple explanations for the observed features.

These data confirm the earlier conclusions that patches can be observed as events in space, as well as patchiness occurring as variations in spectral density. The major difference between the two environments illustrated here is that in the earlier case

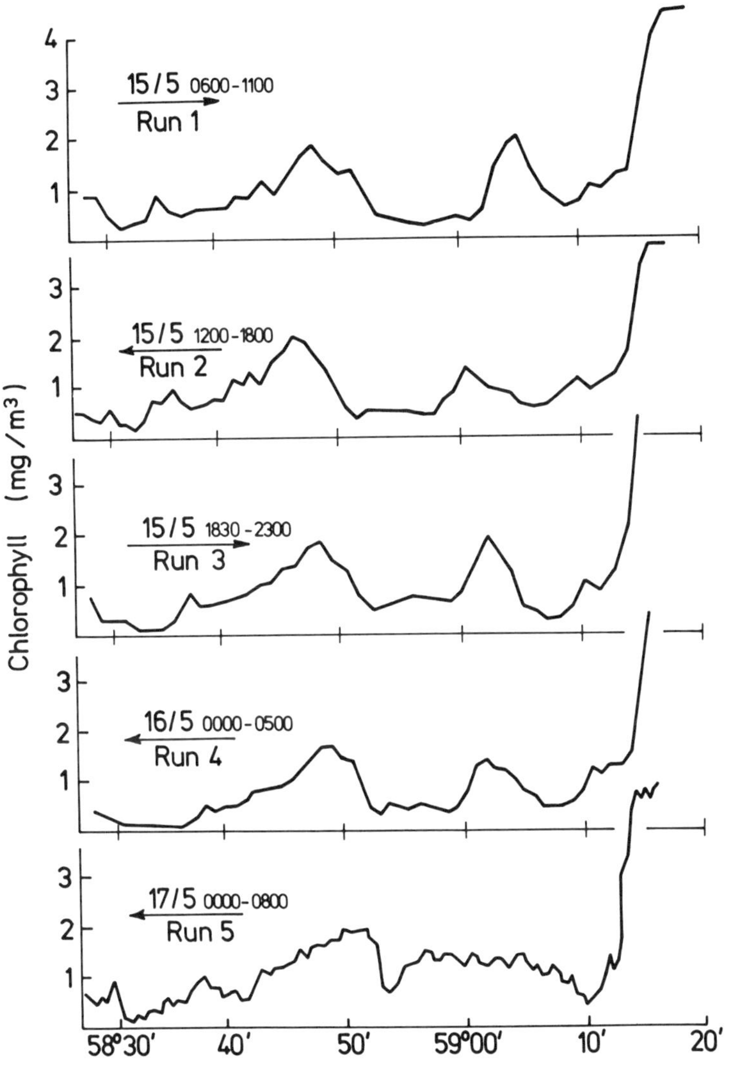


Figure 7. Repeated measurements of chlorophyll *a* at 3 m depth in the northern North Sea during May 1976.

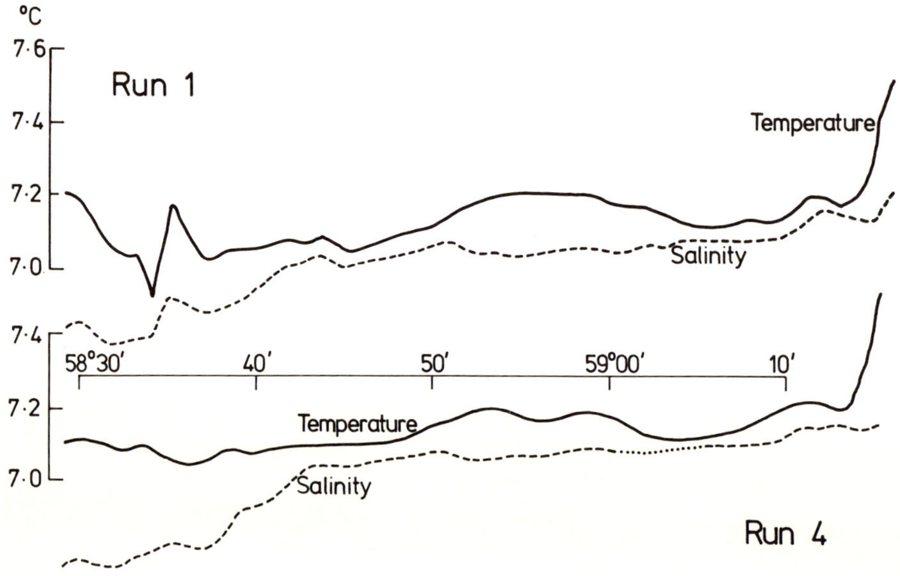


Figure 8. Temperature and salinity data from the first and fourth runs in Figure 6.

(Figure 1) the herbivorous zooplankton were few in number whereas during the period of FLEX shown here there were generally large populations of *Calanus* and other copepods. This suggests circumstantially that the change in structure might result from the interaction of plants and herbivores through some mechanisms other than a spatially uniform balance between growth and grazing. Data on zooplankton numbers obtained with a particle counter by Mackas (1976) show high herbivore densities and great spatial variability in samples taken at 3 m during the night.

It is apparent from Figures 1 and 7 that sharp (<1 km) changes in chlorophyll can occur and may be associated with physical changes usually referred to as "fronts." It is likely that some of the other variations in chlorophyll are related to physical features not observable in the near-surface water. These physical features could be due to the bottom topography but there does not seem to be any direct association and variability also occurs over deep water (Figure 2).

However, irregular bottom topography, among other factors, could generate irregularities in the flow resulting in variations in the depth of density changes. These in turn could provide an energy source for subsurface variations at the scale of the Rossby radius of deformation, defined as

$$R = \sqrt{g'h/f}$$

when $h$ = depth (approx. 120 m in the Fladen Ground),

$$g' = \frac{\Delta\rho}{\rho}g \text{ (approx. 0.3 cm/sec}^2\text{)},$$

when ρ is the density and Δρ is the density difference across the thermocline

$f$ = Coriolis force due to the earth's rotation ($10^{-4}$ sec$^{-1}$).

This gives a value of $R$ = 5 km, approximately, and this is of the right scale in relation to the observed patches.

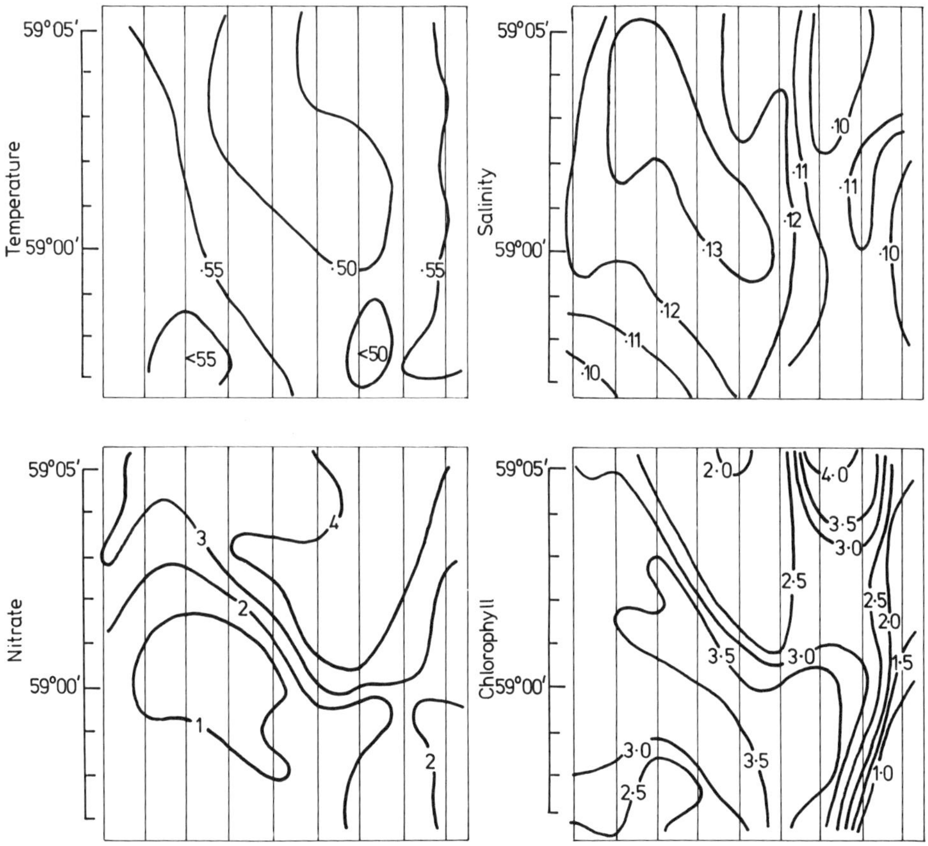


Figure 9. Data at 3 m from a small area of the northern North Sea during May 1976. The vertical lines indicate continuous sampling transects.

It is likely that there will be internal motions at other scales which could affect the distribution or growth rate of organisms. For example, Schott (1970) has deduced a scale of 50 km for inertial period motions in the northern North Sea and fine structure of plankton such as that observed in the last run of Figure 7 could be related to the intermittency of turbulence described by Woods (1974).

The existence of an appropriate scale of physical structure does not necessarily lead to the existence of biological phenomena with this scale since there must be significant coupling with the plant population growth rate or with animal behavior to produce patches with 3-4 times the biomass of populations in the surrounding water. Nor need the changes be considered purely in a space.

There is the further possibility that the observed spatial variations are the result of temporal differences. Cycles of production and grazing may start at different times in different parts of the area so that simultaneous observations may correspond to different phases of a growth cycle such as the phytoplankton spring outburst. From another point of view the rates at which this cycle proceeds in different locations may differ due to physical factors. The patches or plumes observed in an upwelling area can be regarded in this way, where the size of the plume is a joint function of the rate of upwelling and the rate of growth, grazing and nutrient regeneration (Walsh, 1977).

In the North Sea during FLEX and in 1974 (Steele and Henderson, 1977) there is some evidence that cycles of production differed along a 100 km line with production occurring earlier, or faster, at the southern end. Data on the zooplankton show comparable differences leading to populations with different age structure.

## VERTICAL STRUCTURE

These data and others (Fasham and Pugh, 1976; Herman, 1975) show that horizontal variations in phytoplankton are unlikely to be explained solely in terms of lateral turbulent diffusion in an upper mixed layer. Variations in the vertical must be considered for both the physical and biological factors which can affect phytoplankton concentration.

Details of vertical structure of phytoplankton and of temperature can now be obtained from towed undulating devices such as the "batfish" (Herman, 1975). These show the complexities in distribution which occur within the thermocline. Such fine structure can be expected to interact with the upper layers. This is most obvious when the thermocline outcrops at the surface to form a thermal front

(Pingree *et al.*, 1977). Very high phytoplankton concentrations occur near the front. In terms of nitrogen content, these concentrations represent more than the original concentration of nutrient in the water implying movement of cells through the water by sinking or, for cells with flagellae, by "swimming" in response to changing light conditions induced by vertical motions of the water. Such "red tides" might be considered as the most dramatic example of the effects of alteration in the thermocline depth which, generally, are not simply observable by near surface temperature or salinity. The importance of these vertical variations for the rest of the food chain is discussed later (by Lasker). The relevant point, at present, is the intrinsically three-dimensional nature of the physical interactions which make it difficult, if not impossible, to separate horizontal and vertical variability.

These interactions between vertical and horizontal may be considered to have deterministic and random components, the latter typified by the intermittency associated with the breaking of internal waves (Woods, 1974). For the former, the effects of velocity shear between horizontal layers, combined with vertical exchange between these layers, provides a mechanism for lateral dispersal (Bowden, 1965; Kullenberg, 1972) of substances in the water such as dye, nutrients or phytoplankton. When these physical mechanisms are combined with the vertical migrations of the herbivores, there is a means for a further component of horizontal motion of the animals which can then differ from that of the phytoplankton. Differences in rates of horizontal dispersion of the two components, plants and herbivores, can then affect the production of patches (Segel and Jackson, 1972; Okubo, 1974; Evans *et al.*, this volume).

## POPULATION STABILITY

Much of the general interest in patchiness of organisms in their environment derives from the effect this may have on their persistence within the larger community or ecosystem. The spatial heterogeneity of the physical conditions may provide a range of environments which will enhance the growth, or even determine the survival, of a number of species.

The term "stability" can be used in several ways (May, 1973) but in this context I shall be concerned with the ability of a system to recover from perturbations. The greater the perturbations which can be absorbed, the more robust is the system. The stabilizing effect of horizontal dispersion has been discussed (Steele, 1974 ; Platt and Denman, 1975) and will be considered further in this volume. The possibility of destabilizing consequences of different dispersion rates for plants and herbivores has also been considered (Okubo, 1974).

At an experimental level there could be a comparison with concepts in terrestrial ecology derived from the classical work on mites (Huffaker, 1958) where a complicated and heterogeneous experimental environment was needed to prevent rapid collapse of a prey-predator system. It might be considered that horizontal dispersion plays the same role in the sea, and that without it the structure of plankton communities would rapidly collapse. The appropriate aquatic experiment is the opposite of Huffaker's and requires the isolation of a column of water so that motion of water, plants and animals is in one, vertical dimension only, at least in relation to the horizontal scales considered here. Such experiments have been conducted in large plastic enclosures in Canada (Menzel, 1977) and Scotland (Gamble *et al.*, 1977). The results from these, particularly the latter, indicate that interacting communities of plants, herbivores and carnivores do not collapse rapidly after their separation from the large water mass. In these experiments not only is horizontal mixing removed but the nature and rates of vertical mixing are changed (Steele *et al.*, 1977). This indicates a marked degree of robustness in such three-level trophic ecosystems for time scales of the order of 30-60 days. By reference to Figure 3 this would suggest that horizontal mixing on scales less than 30 km is not an essential factor for community persistence at the plant-herbivore levels. On this provisional basis, theoretical instabilities in a plant-herbivore system, without diffusion, could not be used as a mechanism for generating patches in a dispersive system.

In the Scottish experiments, however, there was a gradual divergence quantitatively and quanlitatively in plants and herbivores in two bags that were initially indistinguishable within the levels of sampling error. We think this was caused by slight initial differences in invertebrate predators whose subsequent different population sizes gradually affected herbivores and then, in turn, plant populations. This divergence between two initially similar systems indicates the role that predators could play in generating spatial heterogeneity in open sea systems.

Invertebrate predators such as ctenophores can reproduce as rapidly as herbivores and may occur in local high concentrations. Pelagic fish commonly move in shoals and although these may disperse at night to feed, this feeding is likely to be very uneven horizontally. Patterns imposed from this trophic level may then appear in both the quantity and species composition of the herbivores and plants (Steele and Frost, 1977). The quantification of data from echosounders may provide spatial distributions comparable to that for the lower trophic levels (Cushing, this volume).

At the same time the nature of the predation in terms of the preferred size structure of the food and response time of the predators to changes in food concentration must affect the temporal pattern of development of the herbivores and so their population stability. For pelagic fish we are then concerned with periods of years and correspondingly large spatial scales; and with the further question of how such large period variations might have consequences at smaller space and time scales.

## CONCLUSION

The deliberate intention of these introductory remarks has been to indicate the complexity of the interactions with which we are concerned. Discussion of the regularity, or conversely, the variability of plankton communities is not new, going back to Hensen (1911). However, recent technology gives us the ability to measure the variability and to quantify its spatial structure. We are becoming more conscious of its importance to the organisms themselves (Lasker, 1975). Especially, it seems likely that the statistical nature of this variance may display more regular or predictable features than the mean levels of populations (Platt and Denman, 1975). This situation would be comparable to that in meteorology where the development and testing of statistical theories of meteorological distributions provides a basis for determining the limits on weather prediction. If through mathematical or computer simulations we could explore the ways in which our observed smaller-scale variances may be caused, this could provide a firmer basis for ecological theory than attempts to determine long-term or large-scale mean conditions. Similar considerations are being explored for terrestrial ecosystems (Taylor and Taylor, 1977) and human geography (Cliff and Ord, 1976). These could provide a basis for comparison which is required between basically comparable systems.

## SUMMARY

The study of plankton patchiness can provide a basis not only for an understanding of variability but also for the investigation of the relative importance of physical and biological factors in the control of pelagic ecosystem structure. Estimates of the space and time scales of this biological variance may provide the soundest theoretical foundation for comparison with theory and also for practical predictions.

## REFERENCES

Bainbridge, R., 1957: The size and shape and density of marine phytoplankton concentrations. Biol. Rev., 32, 91-115.

Bowden, K. F., 1965: Horizontal mixing in the sea due to a shearing current. J. Fluid Mech., 21, 83-95.

Boyd, C. M., 1973: Small scale spatial patterns of marine zooplankton examined by an electronic in situ zooplankton detecting device. Neth. J. Sea Res., 7(4), 555-666.

Cliff, A. D. and J. K. Ord, 1976: Model building and the analysis of spatial pattern in human geography. J. Roy. Stat. Soc., 37(3), 297-348.

Cushing, D. H. and D. S. Tungate, 1963: Studies on a Calanus patch. I. The identification of a Calanus patcn. J. Mar. Biol. Ass. U.K., 43, 327-337.

Denman, K. L. and T. Platt, 1976: The variance spectrum of phytoplankton in a turbulent ocean. J. Mar. Res., 34(4), 593-601.

Fasham, M.J.R. and P. R. Pugh, 1976: Observations on the horizontal coherence of chlorophyll a and temperature. Deep-Sea Res., 23, 527-538.

Gamble, J. k., J. M. Davies and J. H. Steele, 1977: Loch Ewe bag experiment, 1974. Bull. Mar. Sci., 27(1), 146-175.

Harden Jones, F. R., 1968: Fish Migration. Arnold, London, 325 pp.

Haury, L. B., 1973: Sampling bias of a Longhurst-Hardy plankton recorder. Limnol. Oceanogr., 18, 500-506.

Haury, L. R., P. H. Wiebe and S. H. Boyd, 1976: Longhurst-Hardy plankton recorder systems: Design and use to minimize bias. Deep-Sea Res., 23, 1217-1229.

Hensen, V., 1911: Das Leben im Ozean nach zahlungen seiner Bewohner. Ubersicht und Resultaten der quantitativen Untersuchungen. Ergebn. Plankton Expdn. der Humboldt Stiftung V., 406 pp.

Herman, A. W., 1975: Chlorophyll and dye detection with the variosens fluorometer. Bedford Inst. of Oceanography Report Series B1-R.75-2.

Huffaker, C. B., 1958: Experimental studies on predation: dispersion factors and predator-prey oscillations. Hilgardia, 27(14), 343-383.

Kullenberg, G., 1972: Apparent horizontal diffusion in stratified vertical shear flow. Tellus, 24, 17-28.

Lasker, R., 1975: Field criteria for survival of anchovy larvae: the relation between inshore chlorophyll maximum layers and successful first-feeding. Fish. Bull., 73, 453-462.

Longhurst, A. R., A. D. Reith, R. E. Bower and D.L.R. Seibert, 1966: A new system for the collection of multiple serial plankton samples. Deep-Sea Res., 13, 213-222.

Lorenzen, C. J., 1971: Continuity in the distribution of surface chlorophyll. J. Cons. Int. Explor. Mer., 34, 18-23.

Mackas, D., 1976: Horizontal spatial heterogeneity of zooplankton on the Fladen Ground. I.C.E.S. C.M.1976/L:20.

May, R. M., 1973: Stability and complexity in model ecosystems. Princeton University Press, Princeton, N.J., 235 pp.

Menzel, D. W., 1977: Summary of experimental results: controlled ecosystems pollution experiment. Bull. Mar. Sci., 27, 142-145.

Okubo, A., 1971: Oceanic diffusion diagrams. Deep-Sea Res., 18, 789-802.

Okubo, A., 1974: Diffusion-induced instability in model ecosystems: another possible explanation of patchiness. Tech. Report 86, Chesapeake Bay Inst., Johns Hopkins Univ., Baltimore.

Pingree, R. D., P. M. Holligan and R. N. Head, 1977: Survival of dinoflagellate blooms in the western English Channel. Nature, 265, 266-269.

Platt, T., 1972: Local phytoplankton abundance and turbulence. Deep-Sea Res., 19, 183-188.

Platt, T. and K. L. Denman, 1975: Spectral analysis in ecology. Ann. Rev. Ecol. Syst., 6, 189-210.

Schott, F., 1971: Spatial structure of inertial-period motions in a two-layered sea, based on observations. Sears Foundation: J. Mar. Res., 29, 85-102.

Segel, L. A. and J. Jackson, 1972: Dissipative structure: an explanation and an ecological example. J. Theor. Biol., 37, 545-559.

Sheldon, R. W., A. Prakash and W. H. Sutcliffe, Jr., 1972: The size distribution of particles in the ocean. Limnol. Oceanogr., 17, 327-340.

Steele, J. H., 1974: The Structure of Marine Ecosystems. Harvard University Press.

Steele, J. H., 1976: Patchiness. In Ecology of the Sea, D. H. Cushing and J. J. Walsh (eds.), pp. 98-115, Blackwell.

Steele, J. H., D. M. Farmer and E. W. Henderson, 1977: Mixing in large enclosures. J. Fish. Res. Bd. Can., 34, 1095-1104.

Steele, J. H. and B. W. Frost, 1977: The structure of plankton communities. Phil. Trans. Roy. Soc. Lond., 280, 485-533.

Steele, J. H. and E. W. Henderson, 1977: Plankton patches in the northern North Sea. In Fisheries Mathematics, J. H. Steele (ed.), pp. 1-19, Academic Press, London.

Taylor, L. R. and R.A.J. Taylor, 1977: Aggregation, migration and population mechanics. Nature, 265, 415-421.

Walsh, J. J., 1977: A biological sketchbook for an eastern boundary current. In The Sea, E. D. Goldberg, I. N. McCave, J. J. O'Brien and J. H. Steele (eds.), Vol. 6, 923-968, Wiley-Interscience, New York.

Wiebe, P. H., E. M. Hulburt, E. J. Carpenter, A. E. Jahn, G. P. Knapp III, S. H. Boyd, P. B. Ortner and J. L. Cox, 1976: Gulf Stream cold core rings: large scale interaction sites for open ocean planktonic communities. Deep-Sea Res., 23, 695-710.

Woods, J. D., 1974: Space-time characteristics of turbulence in the seasonal thermocline. Mem. Soc. Royale des Sciences de Liege, 6e serie, VI, 109-130.

# HORIZONTAL DISPERSION AND CRITICAL SCALES FOR PHYTOPLANKTON PATCHES

Akira Okubo

Marine Sciences Research Center
State University of New York
Stony Brook, New York 11794

## INTRODUCTION: HORIZONTAL DISPERSION IN THE SEA

Planktonic organisms in the sea are subject to physical processes in addition to biological processes. It can be seen that physical effects become important in the distribution of organisms when the time scale of the process is comparable to (or smaller than) that of biological factors such as the time scale of organism reproduction.

There are two kinds of physical processes contributing to mixing: advection and diffusion. Roughly speaking, in advection, large-scale movements of water occur and the imbedded organisms are carried with the water; whereas with diffusion, a spatial exchange of organisms takes place without any overall transport of water. As a result of the two processes, physical mixing of organisms occurs. The term "diffusion" is sometimes used synonymously with "mixing" or "dispersion". Strictly speaking, "diffusion" is equivalent to "dispersion" only when there is no spatial variation of mean velocity in the field. In a loose sense, however, we shall use "diffusion" as the result of mixing of some property; when we wish to stress the process of mixing, however, we shall use the term "diffusive process", instead of simply "diffusion".

The advective process includes the horizontal and vertical movements of water associated with ocean currents, and the diffusive process implies turbulent diffusion due to a wide range of random eddies or turbulent motions in three dimensions. It is, however, a little arbitrary to divide the oceanic motion into a regular pattern of flow (currents) and an irregular random motion (turbulence).

Since motion in the sea consists of a nearly continuous spectrum containing scales which range from the microscale of turbulence up to the ocean wide general circulation, the division between one part of the motion which is assigned to the advective processes and the other part of the motion which is assigned to the diffusive processes depends not only upon the spectrum but also on the scale of the phenomenon in which we are interested.

In addition to the two physical processes referred to above, molecular diffusion or random dispersion due to Brownian motion also contributes to the dispersion of dissolved or microscopic organisms in the ocean. However, an averaging procedure is normally employed in order to formulate the mixing problem in tractable mathematical form. The molecular diffusion terms have been shown to be negligible compared to the turbulent diffusion terms which result from this averaging process. This does not mean that molecular diffusion plays an insignificant role in mixing. A complete physical understanding of mixing could not be accomplished without taking molecular diffusion into account.

Although the actual mechanism of oceanic mixing is extremely complex, the essential features of the advective and diffusive processes in the ocean are as follows. We shall consider a diffusing patch of phytoplankton in the sea. Eddies much larger than the patch size advect the patch as a whole, while eddies much smaller than the patch size produce the dispersion of phytoplankton about its center of mass by turbulent diffusion. However, eddies of the order of size of the patch deform the patch as well as contributing to diffusion. These eddies produce shears or spatial variations in the velocity field; the velocity inhomogeneity tends to advect one part of the patch relative to another, resulting in the deformation of the patch. Such a deformed patch will have larger mean concentration gradients than a nondeformed patch. Consequently, turbulent eddies whose average scale is smaller than the patch size can effectively produce diffusion of the deformed patch.

As the patch size increases, the boundary between the eddies contributing to advection, and those responsible for diffusion, shifts toward larger scale; or stated another way, toward smaller wave number in the kinetic energy spectrum of oceanic motion. The distribution of the kinetic energy of water movements between the two processes depends on the structure of the spectrum. As a result, the apparent power of mixing which acts on a diffusing patch increases with patch size. It is apparent, therefore, that a model based on Fickian diffusion using a constant diffusivity is inappropriate for describing the mixing of a patch of phytoplankton.

Tracer diffusion data in the sea support the non-Fickian nature of oceanic horizontal dispersion. Thus, an apparent horizontal diffusivity increases with the scale of phenomenon, as shown in

Figure 1 (Okubo, 1971). It should be noted that the diffusivity may vary, probably depending on the environmental conditions, by an order of magnitude for the same scale of phenomenon.

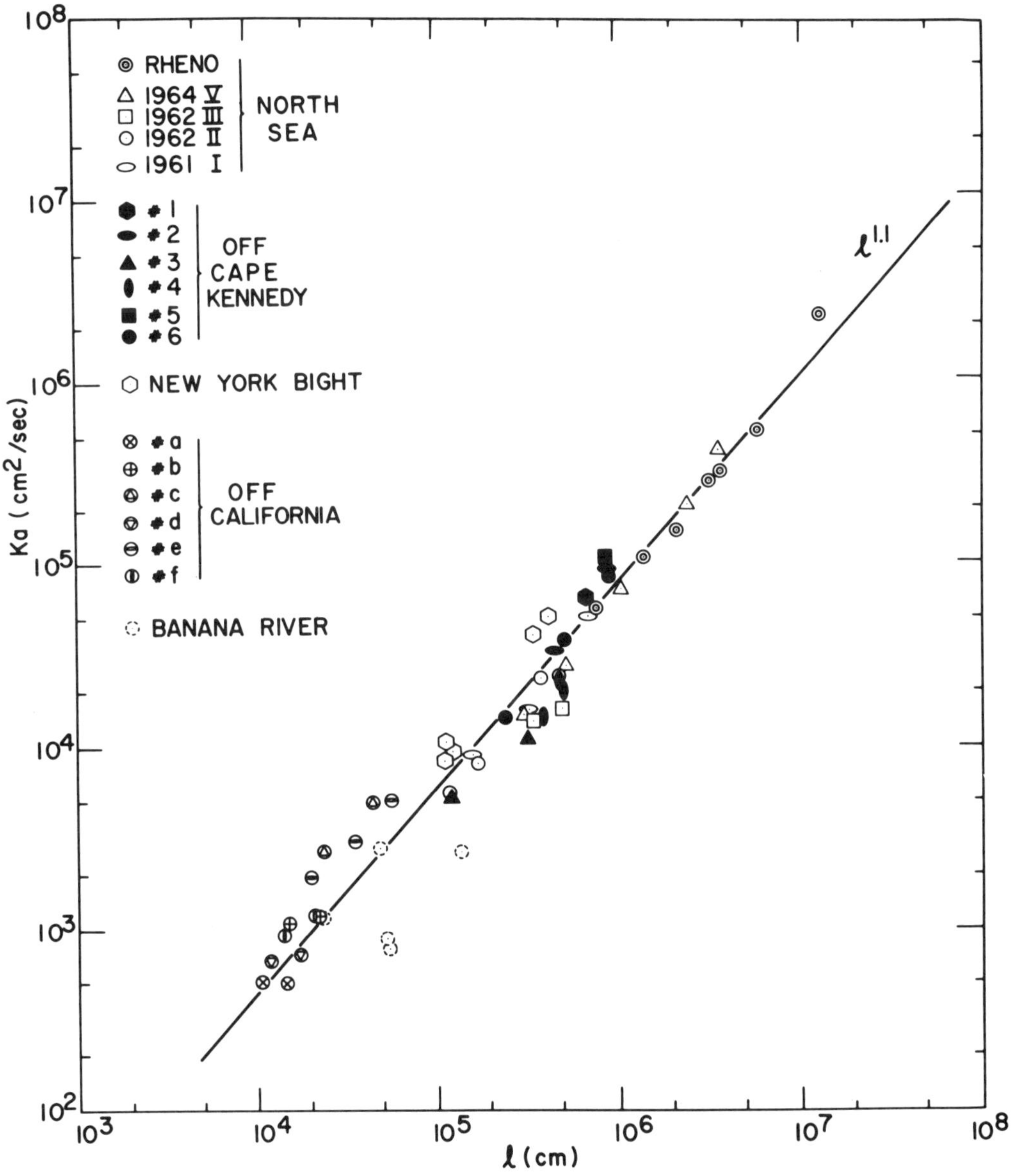


Figure 1. Apparent horizontal diffusivity, Ka, vs. scale of diffusion, ℓ (from Okubo, 1971).

The above discussion is concerned primarily with the horizontal spread of plankton patches by horizontal mixing. When significant vertical movement or mixing of organisms occurs, the combined action of vertical shear in the horizontal currents and vertical diffusion can produce an effective dispersion of organisms in the horizontal direction (Bowden, 1965). This "shear effect" means that the vertical and horizontal processes of mixing cannot be treated separately, and that the advection-diffusion in the vertical direction is responsible for the dispersion of plankton in the horizontal direction.

An important role for the shear effect in contributing to horizontal dispersion is now well recognized (Kullenberg, 1971, 1972; Evans, 1977; Oonishi and Kunishi, 1977). Steele (1976), Evans, Steele and Kullenberg (1977), Kullenberg (this volume), and Evans (this volume) apply the vertical shear diffusion theory to the problem of plankton patchiness. The main purpose of this contribution is to investigate the role of horizontal dispersion *per se* in the patchy distribution of plankton in the sea, without paying attention to the precise mechanism of the shear diffusion process. In this sense, we understand the pertinent variables, such as the concentration of phytoplankton, to be depth averaged quantities. Thus, normally, the plankton concentration would be averaged over the depth of the euphotic layer.

## DISPERSION OF PLANKTONIC ORGANISMS: BIODIFFUSION

Wolfenbarger (1975) introduces the following definition: when movement is supplied by energy from within the organism, the dispersion is termed active; and when it is supplied by energy from outside the body, the dispersion is termed passive. Thus, many phytoplankters diffuse approximately passively, while many zooplankton, in swimming, undergo varying proportions of passive and active diffusion, according to the degree of movement of the ambient water.

Theories of oceanic diffusion are expected to hold approximately for phytoplankton. In this case, however, attention must be paid to the fact that, depending upon the size of the organism, certain components, i.e. small-scale motions of the environmental turbulence, will not be effective agents of the dispersion.

A degree of success has been achieved in the analysis of dispersal of animals by starting with a direct analogy to the random walk or physical diffusion. The work of Skellam (1951) has provided a profound and lasting contribution to this approach. However, animal movements are often anti-diffusive (Clark *et al.*, this volume). A more realistic model of animal diffusion must be built by taking into account interactions between individual animals, effects of stimuli, and animal response to environment and food organisms (Skellam, 1972, 1973).

According to Strickler (1977), the lack of knowledge in autecology (which includes such aspects as the mode and energy expenditure of locomotion, the randomness of the path in search of food resources and the speed, direction and endurance of the animal during an escape from its predators) for pelagic zooplankton is astonishing when compared with the detailed information on air-borne insects. However, recent studies in zooplankton autecology by Strickler (1975, 1977), Gerritsen and Strickler (1977), Lehman (1977), Vlymen (1970), and Haury and Weihs (1976) have proved encouraging.

Diffusion models for zooplankton that take into account the above-mentioned behavioral aspects are nearly nonexistent (Wiebe, personal communication). An intuitive approach suggests that the autecological aspect should be involved in both the advection and diffusion terms for zooplankton, the advection velocity and virtual diffusivity for a species should be functions of population densities of the same species and other interacting species, as well as of time and space. Thus, the advective and diffusive processes of animals become essentially nonlinear, and one must expect to encounter considerable mathematical difficulty in solving the problem.

In this context, models for chemotaxis of microscopic organisms should be worthy of serious attention in relation to zooplankton behavior (Keller and Segel, 1971 a,b). It is suggested that the swarming behavior of zooplankton should play an important role in producing patchy distributions of phytoplankton as they undergo grazing (Wiebe, personal communication).

To illustrate the general interactions we shall present a set of advection-diffusion-reaction equations for multispecies marine populations (for details see Platt, Denman and Jassby, 1977; Dubois, 1975 a,b):

$$\frac{\partial S_i}{\partial t} = - \underset{\sim}{\nabla}_1 \, (\underset{\sim}{U}_i S_i) + \underset{\sim}{\nabla}_1 \, (K_i \underset{\sim}{\nabla}_1 S_i) + F_i \, (S_j, e_k, \underset{\sim}{x}, t) \tag{1}$$

$$i, \; j = 1, \; 2, \; . \; . \; . \; . \; n$$

$$k = 1, \; 2, \; . \; . \; . \; . \; m$$

Here we define

$\underset{\sim}{\nabla}_1 = \underset{\sim}{i}_x \frac{\partial}{\partial x} + \underset{\sim}{i}_y \frac{\partial}{\partial y}$ ($\underset{\sim}{i}_x$ and $\underset{\sim}{i}_y$ are unit vectors in the respective directions), $\underset{\sim}{U}_i = (U_i, V_i)$: advection of ith species, $S_i$: concentration of ith species, $K_i$: horizontal diffusivity of ith species, $F_i$: inter-, intra-specific interactions of ith species, $e_k$: environmental parameters such as light intensity and temperature. In general, $\underset{\sim}{U}_i$ and $K_i$ are functions of $S_j$ as well as of $e_k$, $\underset{\sim}{x}$, and $t$. Detailed formulations of the $F_i$ terms in marine ecosystems are discussed by Steele and Mullin (1977) and Platt, Denman and Jassby (1977).

## CRITICAL SIZE OF PHYTOPLANKTON PATCHES

The mechanism by which plankton patchiness arises is still not well understood. Patchiness of marine organisms can originate from various mechanisms and processes under various conditions. Yet, common to many cases, a single process acts as an "anti-patchiness" agent: diffusion, due to turbulence in waters or to random movements of organisms. Generally speaking, the diffusive process tends to destroy aggregations of organisms and to give rise to more uniform distributions.

The entire story is not so simple, however. When, for instance, nonlinearity in predator-prey interactions is combined with diffusion, a new instability may arise which leads to "patchiness" or enhances the existing aggregations. (This aspect will be discussed in the following section.) Remembering this instability, we here focus our attention on the balance between the aggregative process of organism growth or reproduction and the anti-aggregative process of diffusion.

We consider a water mass within which phytoplankton are blooming. This water is surrounded by ambient water, physiologically unsuitable for the growth of plankton. In reality, this would be the situation when a nutrient-rich water mass is introduced, say by upwelling, into a nutrient-poor area (Wang and Walsh, 1975). Since a part of the population is continuously lost to the surrounding water owing to diffusion, this plankton patch would cease to exist unless the reproduction of organisms within it counterbalances the diffusion loss.

Now the loss of organisms due to diffusion takes place through the boundary of the patch of two dimensions and hence its rate is proportional to the circumference, while the reproduction takes place locally within the patch, and hence its rate is proportional to the surface area. Since the ratio of circumference to surface area is inversely proportional to the radius of the patch, a larger patch carries less circumference relative to its surface area in comparison to a smaller patch. As the size of a water mass decreases, diffusion plays a relatively important role and eventually a limit is reached beyond which reproduction can no longer compensate for the loss due to diffusion.

Thus there exists a minimum horizontal scale of water mass, i.e., critical size, below which no increase of plankton population is possible. We now estimate the critical size first on the basis of simple dimensional analysis.

Let $D$ and $\alpha$ be diffusivity and the rate of growth of plankton, respectively. Suppose that the critical size, $L_c$, can be determined by the two parameters: $L_c = f(D,\alpha)$. Dimensionally, $D = [L^2T^{-1}]$, $\alpha = [T^{-1}]$ and $L_c = [L]$, so that dimensional analysis leads to

$$L_c = b(D/\alpha)^{1/2} \tag{2}$$

where $b$ is a nondimensional constant of order unity.

A more rigorous treatment for the critical size depends upon a diffusion-reaction equation:

$$\frac{\partial P}{\partial t} = D\nabla_1^2 P + \alpha P \tag{3}$$

where $P$ is the concentration of plankton. In fact, Kierstead-Slobodkin (1953) and Skellam (1951) independently showed from this model equation (hereafter abbreviated as the KISS model) that

$$L_c = \pi(D/\alpha)^{1/2} \tag{4}$$

in one-dimensional space, and

$$R_c = 2.4048\ (D/\alpha)^{1/2} \tag{5}$$

for the critical radius $R_c$ of a circular patch in two-dimensional space.

The critical size problem has been extended to a population growing logistically (e.g., Landahl, 1957, 1959; Wilhelm, 1972) and to asocial populations (Bradford and Philip, 1970 a,b).

Wroblewski, O'Brien and Platt (1975), and Platt and Denman (1975) have included the effect of herbivore grazing in the diffusion-reaction equation for phytoplankton. Upon introducing an Ivlev type grazing function, they calculate the critical size that results in one-dimensional space,

$$L_c = \pi \left( \frac{D}{\alpha - R_m \lambda} \right)^{1/2}, \text{ for } \alpha > R_m \lambda \qquad (6)$$

where $R_m$ is the maximum grazing ration of the herbivores and is the Ivlev constant. If $\alpha \leq R_m \lambda$, all patches must vanish as time progresses. Thus, the effect of the herbivore grazing is to reduce the net growth rate of the phytoplankton and so requires a larger critical size than given by the KISS model.

As mentioned in the previous section, another way of refining the KISS model is by incorporating the scale dependence of the diffusivity. The combination of scale-dependent diffusion and simple growth gives rise to the following mathematical model for phytoplankton concentration, $P$:

$$\frac{\partial P}{\partial t} = \frac{c}{r} \frac{\partial}{\partial r} \left( r^{m+1} \frac{\partial P}{\partial r} \right) + \alpha P, \; c > 0, \; m \geq 0, \qquad (7)$$

in two-dimensional space. In equation (7), $c$ and $m$ are constants. Scale-dependent diffusivity, $K(r) = cr^m$, is also used (Bowden, 1975). As a special case with $m=0$, equation (7) includes the KISS model.

From equation (7) we can calculate the minimum critical radius, $R_c$, of a circular patch,

$$R_c = a_1^{\nu} \nu^{-2} (c/\alpha)^{\nu/2}$$

where $\nu = 2/(2-m)$ and $a_1$ is the smallest positive root of the equation:

$$J_\nu(a_j) = 0.$$

$J_\nu$ is a Bessel function of the first kind and order $\nu$.

For various models of diffusion, formulations of the critical size in two dimensions are:

i) $m = 0$, $c = D$: constant diffusion (KISS model)

$$R_c = 2.408 \, (D/\alpha)^{1/2}$$

ii) $m = 1$, $c = p$ (diffusion velocity): Joseph-Sendner (1958) diffusion model

$$R_c = 3.67 \, (p/\alpha)$$

iii) $m = 4/3$, $c = \frac{9}{4}\gamma$ ($\gamma$: energy dissipation parameter): Ozmidov (1958) diffusion model

$$R_c = 16.9\ (\gamma/\alpha)^{3/2}$$

iv) $m = 1.15$, empirical relation, Okubo (1971)

$$R_c = 4.13\ (c/\alpha)^{1.176}$$

Table 1 shows the estimated values for $R_c$ for various models of scale-dependent diffusion, using $\alpha$ = one division per day as maximum and $\alpha$ = one division per ten days as minimum (Steele, 1974) and $p = 0.5$ cm/sec, $\gamma = 5 \times 10^{-3} (\text{cm})^{1.5}\ \text{sec}^{-1}$ and $c = 0.0687\ \text{cm}^{0.85}\ \text{sec}^{-1}$ (Okubo, 1976). By and large, the critical size is 1 to 2 km for $\alpha$ = one division per day and 20 to 50 km for $\alpha$ = one division per ten days. Steele (1976) evaluated the critical size on the basis of the KISS model in combination with Okubo's (1971) diffusion diagram to obtain a critical size ranging from 2 km to 50 km for two values of $\alpha$.

These theoretical models support the general observations that plankton patches in the open sea appear to occur at scales of the order of 10-100 km (Steele, 1976). However, the theory is based on the initial dynamics of the patches; i.e., it assumes that the concentration of phytoplankton, $P$, is infinitesimally small in the beginning. Also, the boundary condition that the patch is surrounded by unsuitable water is rather arbitrary. To achieve realistic modelling the necessity of including nutrient limitation of phytoplankton growth or of coupling nutrient dynamics with phytoplankton dynamics is obvious.

Table 1

Estimated Critical Size of Phytoplankton for Various Oceanic Diffusion Models

| | Diffusion Model | | | |
|---|---|---|---|---|
| $R_c$ | Joseph-Sendner | Ozmidov | Empirical | KISS |
| $\alpha$ = 1/day | 1.6 km | 1.5 km | 1.1 km | 2 km |
| $\alpha$ = 1/10 day | 16 km | 48 km | 17 km | 50 km |

The theory should also deal with interacting populations, such as phyto- and zooplankton, by means of a set of diffusion-reaction equations. In this context, Dubois (1975 a,b) combined Lotka-Volterra population dynamics with diffusion and determined the critical patch scale. Thus, for two competing species, a critical size was given by

$$L_c = \pi\{(D_1/\alpha_1)^{1/2}\ (D_2/\alpha_2)^{1/2}\}^{1/2} \tag{8}$$

where $D_1$ and $D_2$ are the diffusivities and $\alpha_1$ and $\alpha_2$ are the growth rates of the two competing species, respectively. It can be seen that the overall effect of two competing species on the actual size is a geometrical average of the KISS type relations.

For two predator-prey populations, a characteristic size was deduced:

$$\lambda_c = \pi\ (2D/\omega)^{1/2} \tag{9}$$

$$\omega \equiv (a_1 a_2)^{1/2}$$

where $D$ is the common diffusivity of the two populations, $a_1$ is the net growth rate of prey (phytoplankton), and $a_2$ is the net death rate of predator (zooplankton). Note that $\omega$ is the frequency of the oscillations of predator and prey populations in the neighborhood of their equilibrium point. The characteristic size is determined by diffusivity and the frequency parameter. Dubois (1975) estimated the characteristic size for the predator-prey system in the sea to be 7.5 km.

It is interesting to note that no matter what the model may be, the essential feature is preserved: the critical size is determined by the balance of diffusion rate and reaction rate, and the estimated value of the critical size lies in the range of 10-100 km.

The critical size may be rather different when phytoplankton patches are entrained in a field of flow convergence or divergence. Elongated, ribbon-like patches of plankton are often found in frontal zones (Bainbridge, 1957) where there are regions of convergence associated with relatively strong vertical motions of water. In such a case the KISS model can be modified to include the advection term which arises from the presence of convergence or divergence. As a result, it has been shown (Okubo, 1972) that convergence has a general tendency to decrease the critical size, as may be expected, while divergence has a rather dramatic effect on patches; even a weak horizontal divergence of a few centimeters per second is capable of eliminating a very strong phytoplankton patch with $\alpha$ = one division per day.

## DIFFUSION-INDUCED INSTABILITY

Another interesting aspect of diffusion is its role in destabilizing ecosystems that are basically stable without diffusion. Since diffusion is a process that tends to give rise to a uniform density of population in space, it might be expected that diffusion, when it occurs, plays a general role of increasing stability by decreasing the patchiness in marine ecosystems of mixed populations and resources.

However, there is an important exception termed "diffusion-induced instability" or "diffusive instability". This exception might not be rare in marine systems. Under certain conditions, diffusion can indeed destabilize otherwise stable interacting populations to produce spatial and temporal patterns. The concept was originally proposed by Turing (1952) as a physico-chemical model for morphogenesis. Actually, one could extend this concept to any set of diffusion and reaction (interaction) kinetics.

Whether an instability of this sort can occur in an ecosystem depends upon the diffusion rates and interactions among species both biotic and abiotic. Thus, a system of two interacting populations which would coexist in a spatially uniform distribution with no patterns may exhibit a diffusive instability.

The mathematical basis of this diffusive instability in ecology was presented in detail by Segel and Jackson (1972) and also by Levin (this volume). The concept of diffusion-induced instability may be expressed in words as follows: A basically stable ecosystem that is spatially homogeneous may continue indefinitely to be homogeneous; however, irregularities and statistical spatial fluctuations in the abundance of the populations, or in the environmental parameters, are present. If every species of the system diffuses at the same rate, the fluctuations tend to be smoothed out by the process of diffusion without disturbing the balance among the interacting species; the stability of the system remains. On the other hand, if the diffusion rate differs among the species, the balance may be upset in the presence of fluctuations.

The event leading to instability occurs in the following manner. Suppose a regulatory species (e.g. a predator species) diffuses faster than a prey species (Figure 2). Then, in a relatively populated region (i.e. a positive part of the fluctuation), the prey number starts to increase with the relative lack of predation, while in a relatively less populated region (i.e. a negative part of the fluctuation), the prey number starts to decrease with the increase of predation. This situation is followed by a feedback process. That is, the growth of prey favors the

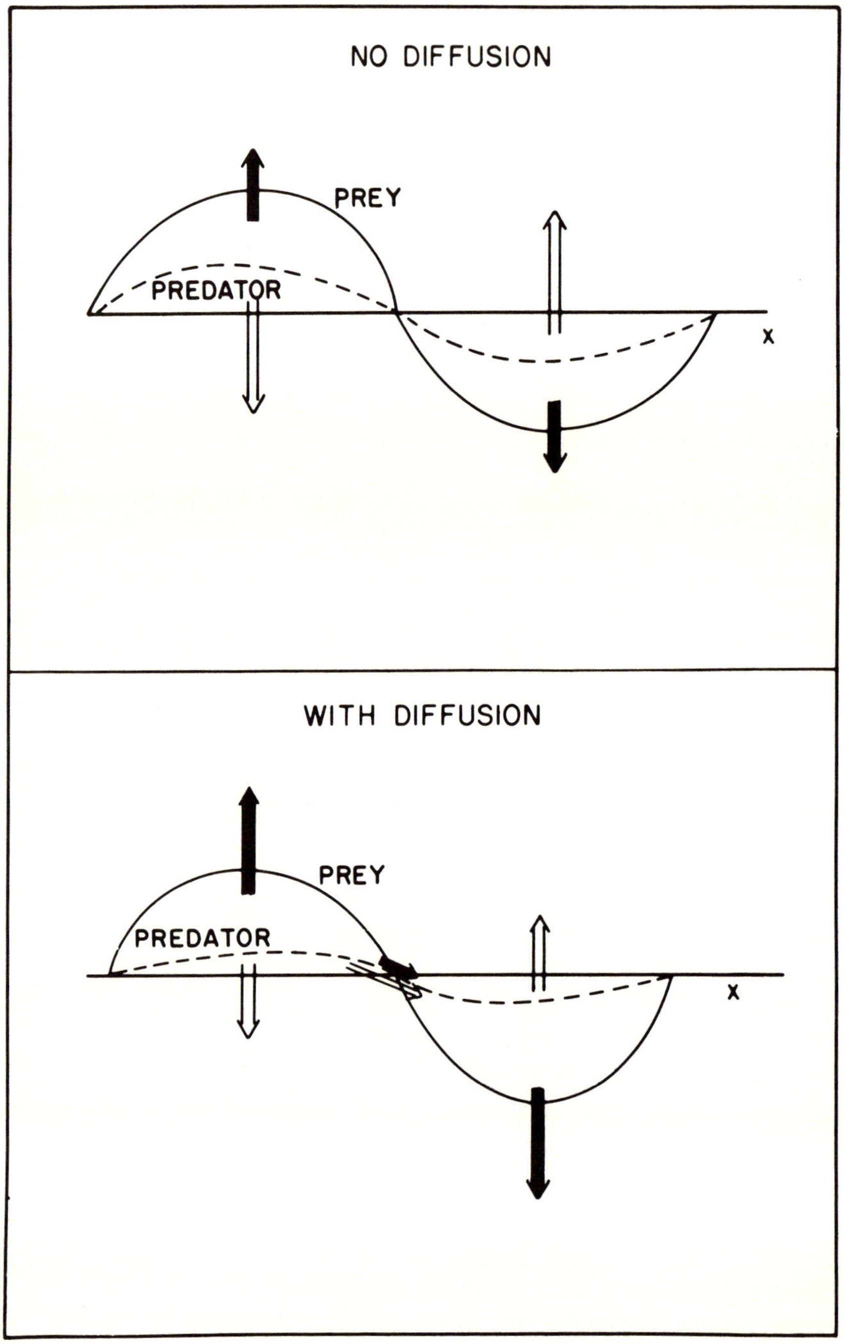


Figure 2. Diffusive instability of a predator-prey system.

growth of predator in one part of the region while the diminishing of prey reduces the number of predators in the other part.

Of course, the inherent, homogenizing process of diffusion always exists so that perturbations of relatively short spatial period (wave length) may be effaced by diffusion at a faster rate than the diffusive instability operates, and hence these perturbations never grow in time. For perturbations of relatively long spatial period, on the other hand, the regulatory species takes too long in dispersing from one part of the perturbation to the other to produce its destabilizing effect on the second species (refer to the perturbation theorem by Levin, 1974). Thus, only fluctuations in a certain wave number range may exhibit diffusion-driven instability.

We shall demonstrate diffusive instability using some specific models of plankton dynamics.

## Phytoplankton-Herbivore System (Levin and Segel, 1976)

Let $P$ and $H$ be the phytoplankton density and herbivore density, respectively. The diffusion-reaction equations for the two species are

$$\frac{\partial P}{\partial t} = D_1 \frac{\partial^2 P}{\partial x^2} + aP + eP^2 - bPH$$

$$\frac{\partial H}{\partial t} = D_2 \frac{\partial^2 H}{\partial x^2} + cPH - dH^2$$

where $a$, $b$, $c$, $d$, and $e$ are all assumed positive constants, and $D_1$ and $D_2$ are constant diffusivities.

The system is seen to have a spatially uniform stable equilibrium without diffusion:

$$P^* = ad/(bc-ed)$$

$$H^* = ac/(bc-ed)$$

provided $bc - ed > 0$, $c > e$.

Levin and Segel (1976) show that small disturbances on $P^*$ and $H^*$ tend to grow if

$$D_2/D_1 > \{b/a - (b/d - e/c)^{1/2}\}^{-1} \equiv \sigma_{cr} \qquad (10)$$

When condition (10) is met, the local activation effect due to the autocatalytic term ($eP^2$) interacts with the long range inhibition to destabilize the uniform state.

In a nonlinear stability analysis, Segel and Levin (1976) have shown that for $D_2/D_1$ slightly greater than $\sigma_{cr}$, the uniform state is replaced by a new steady state in which phytoplankton and zooplankton are more concentrated in certain regions: giving patchiness.

## Nutrient-Phytoplankton System (Okubo, 1974)

Let $N$ and $P$ be the concentration of nutrient and phytoplankton density, respectively. The diffusion-reaction equations for the two species are

$$\frac{\partial N}{\partial t} = K \frac{\partial^2 N}{\partial x^2} + Q - \alpha(N)P$$

$$\frac{\partial P}{\partial t} = D \frac{\partial^2 P}{\partial x^2} + \beta(N)P - Z$$

where $K$ and $D$ are diffusivities, $Q$ and $Z$ are, respectively, constant rates of nutrient supply and of grazing, and $\alpha$ and $\beta$ are arbitrary functions of $N$ only.

This model is concerned with a well-mixed upper layer of the ocean, $Q$ being the supply rate from the lower layer. The grazing generally involves both phytoplankton density and herbivore density. For simplicity we regard this grazing term as a constant sink; this implies that a great number of phytoplankton exist and that a constant population of herbivores graze them. More rigorous treatment would include the herbivores in the system (3-species system).

It can be shown that for the diffusive instability to occur we require, first,

$$\frac{Q/N^*}{Z/P^*} > \frac{\alpha(N^*)/N^*}{\alpha'(N^*)} > \frac{\beta(N^*)/N^*}{\beta'(N^*)}$$

where $N^*$ and $P^*$ are spatially uniform, steady state concentrations of nutrient and phytoplankton, respectively. In other words, the ratio of the nutrient supply rate, per nutrient concentration to the grazing rate, per phytoplankton density must be greater than the ratio of the mean slope of the uptake coefficient to the gradient of the coefficient at the reference state and, simultaneously, the latter ratio must be greater than the ratio of the mean slope of the growth coefficient to the gradient of the coefficient at the reference state.

We then require that the diffusivity of nutrient must be greater than that of phytoplankton to such a degree that

$$\frac{K}{D} > \frac{\alpha' Q}{\alpha\beta} + \frac{2(\beta' Q - \alpha' Z)}{\beta^2} - \{ \frac{4(\beta' Q - \alpha' Z)^2}{\beta^4} + \frac{\alpha' Q}{\alpha\beta^3} (\beta' Q - \alpha' Z)\}^{1/2}$$

where $\alpha$, $\alpha'$, $\beta$, and $\beta'$ are referred to the values at the reference state.

Measurements of horizontal diffusivity of planktonic organisms are very few (Talbot, 1974). We intuitively assume that horizontal diffusivities of soluble nutrients and of phytoplankton should be the same. However, the shear effect on the horizontal dispersion suggests that this is not necessarily true. If phytoplankton concentrate in the vicinity of the water surface while nutrients are distributed more or less uniformly in the upper mixed layer, the shear effect on phytoplankton dispersion may be smaller, in the order of magnitude, than that of the soluble substance (Okubo, 1968). In this connection, a model that takes into account the vertical distribution and shear in currents is highly desirable (Kullenberg and Evans, this volume).

These specific examples are only a few among many conceivable models. The conditions for diffusive instability to occur in plankton communities seem rather stringent. Thus, interacting species-systems of a simple Lotka-Volterra type do not exhibit diffusive instability. Yet the preceding two examples suggest that the instability can occur in marine ecosystems.

The diffusion-induced instability, if it occurs, results in a form of "biological and chemical waves" in space and time, thus leading to patchiness. A critical size of the patchiness in Case 2 is given by

$$L_c = \left(\frac{a_2}{2D} - \frac{a_1}{2K}\right)^{-1/2}$$

where $a_1 \equiv \alpha'(N^*)P^*$ and $a_2 \equiv \beta(N^*) = Z/P^*$; especially if $K \gg D(a_1/a_2)$,

$$L_c = \pi\, (2D/a_2)^{1/2}.$$

Thus, the critical size is the same order as that of the KISS model.

The case of three or more species can be treated in a similar way, doubtless with increasing complexity in mathematical analysis. Each additional species will produce a richer range of possibilities for diffusive instability. However, it seems premature to conclude that the diffusion-induced instability does play a role in plankton patchiness. Until further critical studies are made, the concept of diffusive instability remains no more than another possibility to explain patchiness.

## SUMMARY

Horizontal dispersion of a physical quantity in the sea may be parameterized by an effective diffusivity, the value of which increases with the scale of the phenomenon. Phytoplankton are expected to diffuse approximately in a passive manner, whereas zooplankton are expected to diffuse to a large extent in an active manner. Thus, the swarming behavior of zooplankton may play an important role in their spatial distribution and in their interactions with phytoplankton; however, the modelling of the dispersion of zooplankton is very limited at present. The critical scale of phytoplankton patches was first evaluated by Kierstead and Slobodkin (1953) and, independently, by Skellam (1951) in a more general context. The critical size problem is extended to populations of logistic growth and with species interactions, and to phytoplankton populations subject to grazing. It can also be extended to scale-dependent diffusion. However, most of these cases tend to provide us with the same critical patch size of from 2 to 50 km, as given by the Kierstead-Slobodkin-Skellam model.

Diffusion-driven instability of interacting populations is another possible mechanism to account for phytoplankton patchiness.

## ACKNOWLEDGMENT

Contribution No. 209 from the Marine Sciences Research Center, State University of New York, Stony Brook.

REFERENCES

Bainbridge, R., 1957: The size, shape and density of marine phytoplankton concentrations. Cambridge Philosophical Soc., Biol. Rev., 32, 91-115.

Bowden, K. F., 1965: Horizontal mixing in the sea due to a shearing current. J. Fluid Mech., 21, 83-95.

Bowden, K. F., 1975: Oceanic and estuarine mixing processes. In Chemical Oceanography, J. P. Riley and G. Skirrow (eds.), Vol. 1 (2nd ed.), pp. 1-41, Academic Press.

Bradford, E. and J. R. Philip, 1970a: Stability of steady distributions of asocial populations dispersing in one dimension. J. Theor. Biol., 29, 13-26.

Bradford, E. and J. R. Philip, 1970b: Note on asocial populations dispersing in two dimensions. J. Theor. Biol., 29, 27-33.

Dubois, D. M., 1975a: Simulation of the spatial structuration of a patch of prey-predator plankton populations in the southern bight of the North Sea. Mem. Soc. Roy. des Sci. Liege, 6e seire, tome VII, 75-82.

Dubois, D. M., 1975b: A model of patchiness for prey-predator plankton populations. Ecol. Modelling, 1, 67-80.

Evans, G. T., 1977: A two-layer shear diffusion model. Deep-Sea Res., 24, 931-936.

Evans, G. T., J. H. Steele and G.E.B. Kullenberg, 1977: A preliminary model of shear diffusion and plankton populations. Scot. Fish. Res. Rep. No. 9, 20 pp.

Gerritsen, J. and J. R. Strickler, 1977: Encounter probabilities and community structure in zooplankton: a mathematical model. J. Fish. Res. Bd. Canada, 34, 73-82.

Haury, L. and D. Weihs, 1976: Energetically efficient swimming behavior of negatively buoyant zooplankton. Limnol. Oceanogr., 21, 797-803.

Joseph, J. and H. Sendner, 1958: Über die horizontale Diffusion im Meere. Deut. Hydrogr. Z., 11, 49-77.

Keller, E. F. and L. A. Segel, 1971a: Model for chemotaxis. J. Theor. Biol., 30, 225-234.

Keller, E. F. and L. A. Segel, 1971b: Traveling bands of chemotactic bacteria: a theoretical analysis. J. Theor. Biol., 30, 235-248.

Kierstead, H. and L. B. Slobodkin, 1953: The size of water masses containing plankton bloom. J. Mar. Res., 12, 141-147.

Kullenberg, G., 1971: Vertical diffusion in shallow waters. Tellus, 23, 129-135.

Kullenberg, G., 1972: Apparent horizontal diffusion in stratified vertical shear flow. Tellus, 24, 17-28.

Landahl, H. D., 1957: Population growth under the influence of random dispersal. Bull. Math. Biophys., 19, 171-186.

Landahl, H. D., 1959: A note on population growth under random dispersal. Bull. Math. Biophys., 21, 153-159.

Lehman, J. T., 1977: On calculating drag characteristics for decelerating zooplankton. Limnol. Oceanogr., 22, 170-172.

Levin, S. A., 1974: Dispersion and population interactions. Am. Nat., 108, 207-228.

Levin, S. A. and L. A. Segel, 1976: Hypothesis for origin of plankton patchiness. Nature, 259, 659.

Okubo, A., 1968: Some remarks on the importance of the "shear effect" on horizontal diffusion. J. Oceanogr. Soc. Japan, 24, 60-69.

Okubo, A. 1971: Oceanic diffusion diagrams. Deep-Sea Res., 18, 789-802.

Okubo, A., 1972: A note on small organism diffusion around an attractive center: a mathematical model. J. Oceanogr. Soc. Japan, 28, 1-7.

Okubo, A., 1974: Diffusion-induced instability in model ecosystems. Chesapeake Bay Inst., The Johns Hopkins Univ. Tech. Rep. 86, Ref. 74-3, 17 pp.

Okubo, A., 1976: Remarks on the use of "diffusion diagrams" in modeling scale-dependent diffusion. Deep-Sea Res., 23, 1213-1214.

Oonishi, Y. and H. Kunishi, 1977: On the coefficient of shear diffusion - its time-dependence. J. Oceanogr. Soc. Japan, 33, 165-167.

Ozmidov, R. V., 1958: On the calculation of horizontal turbulent diffusion of the pollutant patches in the sea. Doklady Akad. Nauk, SSSR, 120, 761-763.

Platt, T. and K. L. Denman, 1975: A general equation for the meso-scale distribution of phytoplankton in the sea. Mem. Soc. Roy. des Sci. Liege, 6e serie, tome VII, 31-42.

Platt, T., K. L. Denman and A. K. Jassby, 1977: Modeling the productivity of phytoplankton. In The Sea, Vol. 6, pp. 807-856, Wiley.

Segel, L. A. and J. L. Jackson, 1972: Dissipative structure: an explanation and an ecological example. J. Theor. Biol., 37, 545-559.

Segel, L. A. and S. A. Levin, 1976: Application of nonlinear stability theory to the study of the effects of diffusion on predator-prey interactions. In Topics in Statistical Mechanics and Biophysics; a memorial to Julius L. Jackson, AIP Conf. Proc. No. 27, R. A. Piccirelli (ed.), pp. 123-152, Amer. Inst. Phys., New York.

Skellam, J. G., 1951: Random dispersal in theoretical populations. Biometrika, 38, 196-218.

Skellam, J. G., 1972: Some philosophical aspects of mathematical modelling in empirical science with special reference to ecology. In Mathematical Models in Ecology, J.N.R. Jeffers (ed.), pp. 13-28, Blackwell Sci. Publ., London.

Skellam, J. G., 1973: The formulation and interpretation of mathematical models of diffusionary processes in population biology. In The Mathematical Theory of the Dynamics of Biological Populations, M. S. Barlett and R. W. Hiorns (eds.), pp. 63-85, Acad. Press.

Steele, J. H., 1974: The Structure of Marine Ecosystems. Harvard University Press, Cambridge, 128 pp.

Steele, J. H., 1976: Patchiness. In The Ecology of the Seas, D. H. Cushing and J. J. Walsh (eds.), pp. 98-115, W. B. Saunders Co., Philadelphia.

Steele, J. H. and M. M. Mullin, 1977: Zooplankton dynamics. In The Sea, Vol. 6, pp. 857-890, Wiley.

Strickler, J. R., 1975: Swimming of planktonic Cyclops species (Copepoda, Crustacea): pattern, movements and their control. In Swimming and Flying in Nature, T.Y.T. Wu et al. (eds.), Vol. 2, pp. 599-613, Plenum.

Strickler, J. R., 1977: Observation of swimming performances of planktonic copepods. Limnol. Oceanogr., 22, 165-170.

Talbot, J. W., 1974: Diffusion studies in fisheries biology. In Sea Fisheries Research, F. R. Harden Jones (ed.), pp. 31-54, Wiley.

Turing, A. M., 1952: The chemical basis of morphogenesis. Phil. Trans. Roy. Soc. Lond., B237, 37-72.

Vlymen, W. J., 1970: Energy expenditure of swimming copepods. Limnol. Oceanogr., 15, 348-356.

Wang, D. P. and J. J. Walsh, 1975: Objective analysis of the upwelling ecosystem off Baja California. J. Mar. Res., 34, 43-60.

Wilhelm, H. E., 1972: Nonlinear initial-boundary-value problem for convection, diffusion, ionization, and recombination processes. J. Math. Phys., 13, 252-256.

Wolfenbarger, D. O., 1975: Factors affecting dispersal distances of small organisms. Exposition Press, Hicksville, New York, 230 pp.

Wroblewski, J. S., J. J. O'Brien and T. Platt, 1975: On the physical and biological scales of phytoplankton patchiness in the ocean. Mem. Soc. Roy. des Sci. Liege, 6e serie, tome VII, 43-57.

## APPENDIX

$a_1$: The net growth rate of phytoplankton

$a_2$: The net death rate of zooplankton

$D$ : Diffusivity of phytoplankton

$D_1, D_2$: Diffusivities of two competing species

$e_k$: Environmental parameters

$F_i$: Inter- and intra-specific interactions for ith species

$H$ : Herbivore population density

$H^*$: Equilibrium population density of herbivores

$K$ : Diffusivity of nutrient

$K_i$: Diffusivity of ith species

$L_c$: The critical size of phytoplankton patch; i.e. the minimum horizontal scale of water mass below which no increase of plankton population is possible

$N$ : Nutrient concentration

$N^*$: Equilibrium nutrient concentration

$p$ : Diffusion velocity

$P$ : Phytoplankton population density

$P^*$: Equilibrium phytoplankton population density

$Q$ : The rate of nutrient supply

$r$ : Horizontal radial coordinate, $r = (x^2 + y^2)^{\frac{1}{2}}$

$R_c$: The critical radius of a circular patch below which no increase of plankton population is possible

$R_m$: The maximum grazing rate

$S_i$: The concentration of ith species

$t$ : Time

$\underset{\sim}{u}_i$: Advection velocity of ith species

$u_i$: The x-component of advection velocity of ith species

$v_i$: The y-component of advection velocity of ith species

$x, y$: Horizontal coordinates

$Z$ : The rate of grazing

$\alpha$ : The rate of growth of plankton

$\alpha(N)$: The rate of nutrient uptake per unit phytoplankton density

$\alpha'(N^*)$: The gradient of the rate of nutrient uptake at the equilibrium concentration of nutrient

$\beta(N)$: The rate of growth of phytoplankton as a function of nutrient concentration

$\beta'(N^*)$: The gradient of the rate of growth of phytoplankton at the equilibrium concentration of nutrient

$\gamma$ : Energy dissipation parameter

$\lambda$ : The Ivlev constant

$\lambda_c$: The characteristic size of patches for predator-prey populations

$\omega$ : The frequency of the oscillation of predator-prey populations

$\underset{\sim}{\nabla}$ : Horizontal gradient vector

$\nabla_1^2$: Horizontal Laplacian = $\partial^2/\partial x^2 + \partial^2/\partial y^2$

# VERTICAL PROCESSES AND THE VERTICAL-HORIZONTAL COUPLING

G.E.B. Kullenberg

Institute of Physical Oceanography
University of Copenhagen
DK-2200 Copenhagen N, Denmark

## INTRODUCTION

In this chapter some aspects of vertical processes and vertical mixing will be discussed, as well as the coupling between vertical processes and horizontal ones.

Several recent papers combine observations and modelling studies to show that interactions between the vertical and the horizontal dimensions, on at least some scales, cannot be disregarded. Many experimental investigations of the top 100 m of the ocean also clearly show the importance of the vertical shear effect on horizontal mixing (Pritchard *et al.*, 1966; Bowden, 1965, Okubo, 1967, 1968; Talbot and Talbot, 1974; Kullenberg, 1972, 1974). By shear effect is meant the apparent increase of the mixing in one plane due to the combined action of diffusion and velocity variations in a perpendicular plane. In particular, the vertical shear effect, in which vertical diffusion and vertical shear combine in generating an apparent horizontal mixing, is considered here. Since the horizontal currents in the surface layer of the sea in general have a considerable vertical velocity gradient, it is conceivable that this mixing mechanism can be important on intermediate and small scales (i.e., in the range 0.1-10 km). Due to the vertical migration of some zooplankton, it is also conceivable that this mechanism can be of profound biological significance. Purely horizontal diffusion was discussed by Okubo (this volume) and will not be further considered here.

## VERTICAL MIXING

Direct atmospheric influence is of major importance in determining the conditions in the upper layer of the ocean. Several thermocline models have been formulated to predict the temperature profile and the depth of the mixed layer as a function of time, wind speed and insolation. Many aspects of this problem are reviewed in Kraus (1977).

Niiler and Kraus (1977) and Kitaigorodsky (1978) present extensive discussions of one-dimensional models which include only the vertical coordinate and time. Thereby, the conditions in the upper layer are treated as horizontally homogeneous. In order to solve the model equations, expressions for the turbulent fluxes must be found and techniques for this include the use of Reynolds fluxes, mixing length theories and eddy coefficients. However, these formulations are not very satisfactory since it is necessary at some stage to make use of rather arbitrary assumptions which are difficult to test and often are not physically well-founded. In this respect, the mixed layer approach is more attractive. This is based on the observation that scalar distributions in the upper layer very often are also approximately vertically homogeneous. Assuming vertical homogeneity, the equations may be vertically integrated from the mixed layer base to the surface to yield expressions for the turbulent fluxes as functions of mean quantities and external inputs.

The essential parameters in the mixed layer models are the wind stress $\tau_o$, defining the friction velocity $u_*$, the stratification, as given by $N^2$, the net heat flow and the thickness $h$ of the mixed layer (see Appendix for list of symbols). The turbulent energy equation often forms the basis for the model. De Szoeke and Rhines (1976) summarized the different time regimes for the rate of deepening of the mixed layer and showed that the models proposed by Niiler (1975), Pollard, Rhines, and Thompson (1973), Kraus and Turner (1967) were special cases of the same general formulation. Six regimes were identified for the rate of deepening. Deepening by erosion, proportional to $t^{1/3}$, occurs first for very small time scales, on the order of minutes, and reestablishes itself at the last stage of the sequence, on the order of inertial period. At intermediate time scales the shear production at the base of the mixed layer and the rotational effect, gradually veering the current direction from the wind direction, dominates the rate of deepening, which is then proportional to $t^{1/2}$ and $t^{1/4}$, respectively.

The depth of the mixed layer was found by Niiler (1975) to be

$$h \sim (12\, m_o)^{1/3} \cdot \left(\frac{\tau_o}{\rho_o N^2}\right)^{1/2} \cdot (tN)^{1/3} \tag{1}$$

where $m_o$ is the fraction of turbulent energy generated in the wave zone which is available for mixing. Pollard, Rhines and Thompson (1973) disregarded this energy source and found

$$h \sim 2^{1/2} \left(\frac{\tau_o}{\rho_o N^2}\right)^{1/2} \cdot (tN)^{1/2} \qquad (2)$$

The thickness of the mixed layer may range from meters to several tens of meters.

An important problem in relation to conditions in the upper layer is the distribution and magnitude of the wind energy transferred to the sea. Experimental studies over time scales from hours to months have shown that only about $1^o/oo$ of the available wind energy is consumed for vertical mixing in the water (e.g. Turner, 1969; Denman and Miyake, 1973; Kullenberg, 1976). These results are supported by laboratory experiments using shear flows (e.g. Kato and Phillips, 1969; Kantha and Phillips, 1976). The transfer of energy from the wind to the water is coupled both to the wind generated waves and the current. Studies by Stewart and Grant (1962) and Dobroklonskiy and Kontoboytseva (1973) suggest that the turbulence generated by breaking waves does not penetrate to any great depth and that the surface wave energy dissipation occurs in a very shallow zone. This would imply a considerable vertical shear in that zone.

A basic process in the mixed layer models relates to the deepening of the layer. Assuming that the mechanical energy for the mixing is supplied by the wind stress, the mixed layer will deepen by the process of entrainment or erosion. Thereby, fluid from the weakly (or non-) turbulent layer below is entrained upwards into the mixed layer implying a net volume transport (e.g. Townsend, 1956, 1976; Phillips, 1966, 1977; Turner, 1973). The rate of deepening is expressed by the entrainment velocity $w_e$ and dimensional arguments yield

$$w_e = u_* \cdot f(Ri_*) \qquad (3)$$

where $Ri_*$ is the overall Richardson number defined by

$$Ri_* = \frac{g\Delta\rho h}{\rho_o u_*^2}$$

where $\Delta\rho$ is the density difference between the layers. Laboratory experiments suggest that the function $f(Ri_*)$ is a curve with asymptotic behavior when $Ri_* \to o$ and $\infty$ but with a form $f(Ri_*) \sim Ri_*^{-1}$ at intermediate values of $Ri_*$ (Kato and Phillips, 1969;

Kantha, 1976; Kantha and Phillips, 1976). These results (Figure 1a) are to a limited extent corroborated by field data (Ottosen-Hansen, 1975; Kullenberg, 1977). The implication of the $Ri_*^{-1}$ law is that the rate of change of potential energy due to the entrainment is proportional to the energy input from the wind or $\rho u_*^3$.

The mixed layer models assume homogeneity both horizontally and vertically. The latter assumption may be reasonable for some applications and some observations indicate that the mixed layer may move as a slab for a period of time after the passage of a storm (e.g. Halpern, 1974). In this case the current shear is concentrated across the thermocline layer. However, there are also observations showing the presence of considerable shear close to the surface (e.g. Thorpe, 1977). In some cases, as for instance in the present context, the shear in the top layer is a very essential factor and cannot be disregarded. It is also quite clear that normally there is vertical shear over the whole surface layer due to wind and wave action and related both to the magnitude and directional variations of the horizontal motions. Without the assumption of homogeneity the problem becomes more complex. No satisfactory way of handling it has been found although various closure formulations for the equations have been proposed. However, a parameterization of the mixing can be carried through based on observable parameters and by relating the effective mixing to local conditions.

Two basic parameters for mixing in stratified conditions are the Richardson number $Ri$ and the flux Richardson number $Rf$, defined as follows

$$Ri = N^2 / \left(\frac{dU}{dz}\right)^2, \quad Rf = \frac{g\,\dfrac{\overline{w'\rho'}}{\rho_o}}{-\overline{u'w'}\,\dfrac{dU}{dz}} \text{ ; } z \text{ positive upwards.}$$

The flux number expresses the ratio of the energy consumed for mixing against buoyancy forces and the total turbulent energy. Introducing the vertical turbulent transfer coefficients for mass $K$ and momentum $K_m$ we find

$$\frac{K}{K_m} = \frac{Rf}{Ri} \tag{4}$$

The mean local buoyancy flux is

$$B(z) = -\frac{g\overline{w'\rho'}}{\rho_o}$$

At the surface this flux is governed by evaporation, precipitation and heat conduction. In the near surface layer the governing parameters are $B(o)$ and $u_*$, which when combined yield the Monin-Obukhov length scale (e.g. Phillips, 1966).

$$L = u_x^3 / \kappa_o B(o)$$

which represents the depth where production of turbulent energy equals the buoyancy flux ($\kappa_o \sim 0.4$ is the von Karman constant). For depths $z \ll |L|$ the buoyancy flux plays a minor role, whereas it is dominating for $z > |L|$. In the limit of strong stability, the Monin-Obukhov theory implies that $Rf \to Rf_c$, the constant critical flux Richardson number. This is inherently much less than one (Stewart, 1959). In the near surface layer it may be argued that the stress $\tau$ is locally independent of depth and determined by the wind stress $\tau_o$ (Phillips, 1966). For this case, Kullenberg (1971, 1974) proposed an expression for $K$ based on results from dye experiments,

$$K = c_1 \cdot \frac{W_{10}^2}{N^2} \left|\frac{dq}{dz}\right|, \quad W_{10} > 5 \ m/s \tag{5}$$

where $c_1$ is a constant of the order of $10^{-7}$.

The momentum transfer coefficient for this case may also be related to the wind by using an energy argument (Kullenberg, 1976). The energy input $E$ from the wind to the water and the wind generated surface current $V_o$, respectively, are generally expressed in the form

$$E = k \cdot \tau_o \cdot W_{10} = V_o \tau_o, \quad V_o = k W_{10}$$

Combining this with the Ekman expression for the surface current, namely

$$V_o = \frac{\tau_o}{\rho_o \sqrt{K_m \cdot f}} = k \cdot W_{10}, \tag{6a,b}$$

it is found that

$$K_m = \frac{1}{f} \left(\frac{\rho_a c_d}{\rho_o k}\right)^2 W_{10}^2, \tag{7}$$

which relates momentum transfer to wind speed and the Coriolis effect.

The ratio

$$\frac{K}{K_m} = c_1 \cdot f \cdot \left(\frac{\rho_o k}{N\rho_a c_d}\right)^2 \cdot \left|\frac{dq}{dz}\right| \tag{8}$$

can be calculated for given conditions and Kullenberg (1977) found it to conform with the result $K/K_m \propto Ri^{-1}$ for $Ri > 0.1$ (Figure 1b).

This parameterization of the mixing can be expected to hold only for the surface layer in cases when the wind stress is the mechanical energy source. The direct influence of the surface is not expected to reach much deeper than the Monin-Obukhov length (e.g. Turner, 1973). For deeper layers another parameterization should be attempted. It is therefore important to define an appropriate vertical length scale which can form a basis for the desired observational resolution. This aspect has been studied by Woods (1973) and Thorpe (1977). A length scale which defines the vertical interval over which the turbulence can produce overturning in stratified conditions was proposed by Ozmidov (1965) and Phillips (1966, p. 232, also citing Lumley, 1964), and further discussed by Woods (1973), Kullenberg (1974) and Thorpe (1977). The scale is based on the rate of energy dissipation per unit mass for locally isotropic turbulence which is given by

$$\varepsilon = const. \frac{(\Delta u)^3}{\ell} \tag{9}$$

where $\Delta u$ is the velocity variation over the distance $\ell$. The average velocity gradient may then be estimated as (Ozmidov, 1965)

$$\frac{dU}{d\ell} = C_1 \cdot \varepsilon^{1/3} \cdot \ell^{-2/3} \tag{10}$$

Ozmidov used this expression to obtain the critical length scale $\ell_{cr}$ for which the buoyancy influence will cease to be important. The Richardson number is

$$\frac{N^2}{C_1^{\,2}\, \varepsilon^{2/3}\, \ell^{-4/3}} = a$$

yielding

$$\ell_{cr} = (a\, C_1^{\,2})^{3/4} \cdot \left(\frac{\varepsilon}{N^3}\right)^{1/2} \tag{11}$$

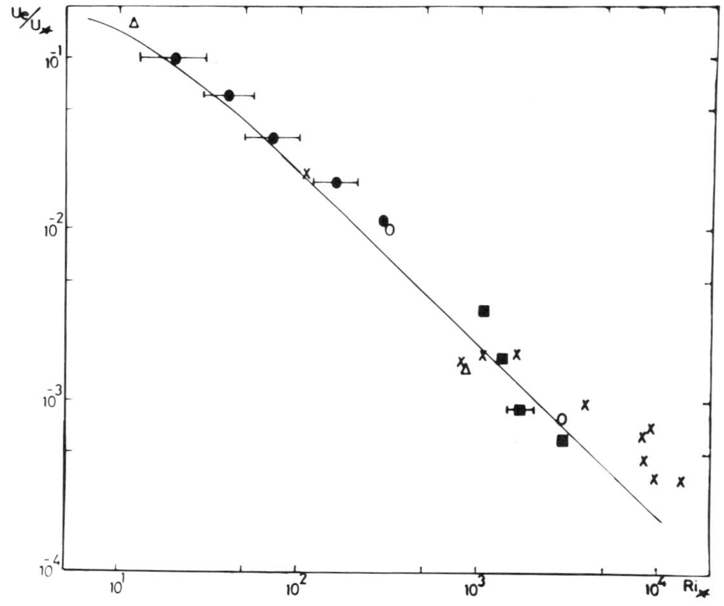


Figure 1a. Ratio of entrainment velocity ($u_e \equiv w_e$) to friction velocity $u_*$ vs. overall Richardson number $Ri_*$. Experimental values from Kato and Phillips (1969) ■; Ottesen Hansen (1975) O,●; Kullenberg (1977) X, Δ.

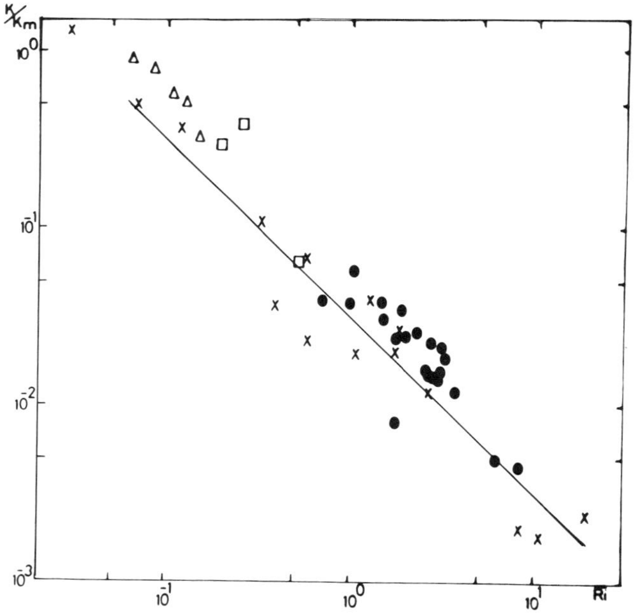


Figure 1b. Ratio of vertical mixing coefficient $K$ to vertical momentum transfer coefficient $K_m$ vs. mean Richardson number $\overline{Ri}$. Experimental values from Lofquist (1960) ●; Bowden (1973) □; Thorpe (1973) Δ; Kullenberg (1974) X.

which separates the scales where buoyancy is important from those where it is not and where the turbulence can be expected to be locally isotropic (e.g. Phillips, 1966). The Ozmidov scale can vary from centimeters to meters, the latter in very weak stratification. The maximum turbulent mixing coefficient is given by (Ozmidov, 1965)

$$K_{max} = C_2 \cdot \varepsilon^{1/3} \cdot \ell_{cr}^{4/3} = const. \ \varepsilon \cdot N^{-2} \tag{12a,b}$$

or alternatively (Thorpe, 1977)

$$K_{max} = \frac{C_2}{C_1 \cdot a^{1/2}} \cdot N \cdot \ell_{cr}^{2}. \tag{13}$$

In the case of internally shear-generated turbulence the vertical mixing coefficient may be related to the shear, the stress and the stratification (e.g. Kullenberg, 1974)

$$K = Rf \cdot u_*^{2} \cdot \left|\frac{dq}{dz}\right| \cdot N^{-2}. \tag{14}$$

Equating (12a) and (14) yields

$$\frac{N^2}{u_*^{2}/\ell_{cr}^{2}} = \frac{Rf \cdot C_1}{C_2} = constant \tag{15}$$

which may be interpreted as a critical Richardson number for instability to occur. Theory as well as experimental evidence suggests this to be of the order of 0.25 (e.g. Miles and Howard, 1964).

Both formulations of $K$ contain quantities which are difficult to determine experimentally. Thorpe (1977) determined the length scale $\ell_{cr}$ from detailed profile measurements yielding not only $N$ and $dq/dz$ but also the vertical displacement of fluid particles in profiles with temperature inversions. Thorpe identified the displacement with the critical length scale $\ell_{cr}$, which made it possible to calculate $\varepsilon$ as well as $K_{max}$.

The rate of energy dissipation per unit mass $\varepsilon$ is normally found to decrease with depth (e.g. Phillips, 1966, p. 231; Okubo, 1971). This is in accordance with the general decrease of the fluctuating kinetic energy with depth (e.g. Webster, 1969; Fofonoff and Webster, 1971). The internal motion is highly intermittent and the mixing can also be expected to be intermittent, as discussed by Woods (1973) and others. In a field of motion dominated by internal waves, the fluctuating kinetic energy is proportional to $N$

(e.g. Turner, 1973, p. 332); this has been observed in the open ocean (Webster, 1969; Fofonoff and Webster, 1971) and in enclosed seas like the Baltic (Kielman *et al.*, 1969). The mixing is then related to the breaking of waves. Several different types of instabilities can occur, such as Kelvin-Helmholtz, Holmboe, Orlanski-Bryan (e.g. Woods and Wiley, 1972; Turner, 1973; Thorpe, 1977).

Experimental indications suggest that the Kelvin-Helmholtz instability is important (Woods, 1968; Woods and Wiley, 1972; Kullenberg, 1974a; Thorpe, 1977). Although these aspects need considerable more study, it seems clear that internal mixing in stratified conditions occurs intermittently on small scales often coupled to shear type instabilities generated both by internal wave shear and shear in the horizontal currents. An effective vertical mixing coefficient may be used to quantify the mixing for given time and space scales. Quite clearly such a mixing coefficient will depend not only on these scales but also on the local conditions.

Recent developments in the field of large- and meso-scale two- or three-dimensional modelling in the upper ocean are presented in Kraus (1977). In relation to the present subject, the coastal upwelling studies are of particular interest (O'Brien *et al.*, 1977). Most numerical models of coastal upwelling are two-dimensional with two homogeneous layers separated by an interface. Vertical mixing across the interface in the form of entrainment both up- and downwards may be included in these models. Observations suggest that vertical mixing becomes important, relative to advection, several days after the start of an upwelling event, and the models which include mixing can reproduce several observed features. The role of vertical mixing in relation to upwelling patchiness has been investigated very little. The most advanced ecosystem model which includes vertical and horizontal mixing, as well as advection, suggests that advective processes are of dominating importance at least in the Oregon upwelling system (Wroblewski, 1977).

## REVIEW OF SOME SHEAR DIFFUSION MODELS

Taylor (1953, 1954) introduced the concept of increased dispersion due to the combined action of diffusion and velocity variations over the cross-section in studying the longitudinal dispersion of soluble matter in flows through tubes. Both laminar and turbulent conditions were investigated, and the effect was shown to be of central importance. Bowles *et al.* (1958) were the first to use the idea in interpreting observations of horizontal dispersion in the surface layers of shallow waters.

Novikov (1958) developed a rather general theory which describes the dispersion from an instantaneous point source in a horizontal stream with vertical shears. The velocity components were written in the form

$$u = u'(t) + A(t) \cdot z$$
$$v = v'(t) + B(t) \cdot z \qquad (16a,b)$$

Horizontal and vertical turbulent diffusion is included by means of time-dependent diffusion coefficients ($A(t)$ & $B(t)$). The coordinate system is moving with the stream. Novikov transformed the diffusion equation by a Laplace transformation and obtained a general solution. The special case of a steady stream with diffusion coefficients independent of time was considered in detail. In this case it is always possible, Novikov pointed out, to define a coordinate system such that $B = 0$. The result shows that (1) after an initial period, the shear effect will dominate over the purely horizontal diffusion, (2) the peak concentration will then decrease as $t^{-2.5}$ instead of $t^{-1.5}$ as in the absence of the shear effect, and (3) the concentration isolines in the $x$-$z$ plane are ellipses strongly elongated in the direction of the shear. It is noted that no transverse shear effect can be obtained in the steady state case.

Elder (1959) discussed the effect of lateral velocity variations on the dispersion in steady open channel flow, by extending the analysis of Taylor (1954). Elder showed that the longitudinal dispersion is determined by the combined action of lateral diffusion and advection by the mean flow. This implies that a patch of dye injected in the channel will become elongated in the direction of the mean flow. It was further shown by experiments that the concentration distributions are closely Gaussian, so that it is reasonable to model the mixing by effective diffusion coefficients.

Bowden (1965) investigated estuaries and coastal regions and emphasized the case of a vertically bounded flow with a depth dependent mean current velocity. The model was formulated for both an oscillating and a steady current in the longitudinal direction. The effective dispersion was in this case found to be inversely proportional to the vertical diffusion coefficient. The oscillating current component was specified as a single harmonic constituent with negligible variations in phase with depth. Average conditions over one or several tidal cycles were investigated and the significance of the shear effect was clearly demonstrated. Evidently, two very different regimes of the shear effect exist, depending upon whether the vertically bounded or unbounded case is considered.

Carter and Okubo (1965) formulated a model with the velocity spectrum consisting of two essentially separated parts, corresponding to large-scale and small-scale motions. The large-scale motions are responsible for the non-uniformities of the mean velocity field, appearing as steady shears in both vertical and horizontal directions. The case with a stream in one direction only was studied. The velocity was given as a sum of a purely time-dependent part and a transverse and vertical shear component, i.e.,

$$u = U_o(t) - y \cdot \frac{\partial U}{\partial y} - z \cdot \frac{\partial U}{\partial z}$$

$$v = w = o. \qquad (17a,b)$$

The component $U_O(t)$ is the time-dependent motion in the plane $z=0$. The coordinate system is taken as moving with the mean velocity at the plane $z=0$.

The small-scale motions were included in the model by constant diffusion coefficients $K_x$, $K_y$, and $K$ in the $x$, $y$, and $z$ directions, respectively. The scale separating advection and diffusion was not specified since it depends upon the specific problem considered. In general, the division of the spectrum in two separate parts does not conform with reality. But, considering the efficiency of different scales in causing dispersion, the division may be appropriate.

The authors formulated the diffusion equation and solved it by means of Fourier transformation technique for an instantaneous point source. It was assumed that the necessary convergence requirements were fulfilled. The isolines of the theoretical distribution form ellipsoids. The degree of elongation is a function of time, the shears and the diffusion coefficients. During an initial period the shear effect is not very important, but for large times it clearly dominates the dispersion.

The theoretical form of the longitudinal variance is the most specific result or prediction of the theory,

$$\sigma_x^2 = 2K_x \cdot t + \frac{2}{3} K_y \cdot \left(\frac{\partial U}{\partial y}\right)^2 \cdot t^3 + \frac{1}{6} K \cdot \left(\frac{\partial U}{\partial z}\right)^2 \cdot t^3. \qquad (18)$$

No transverse, or lateral, dispersion due to the shear effect can be obtained by this model, and so the lateral variance is

$$\sigma_y^2 = 2K_y \cdot t \qquad (19)$$

The dispersion is obviously governed by the combined effect of diffusion due to small-scale motions and advection due to large-scale motions. It is noted that the shears enter squared, and that the variance increases as $t^3$. The rotationally symmetrical variance increases as $t^2$, in agreement with the prediction of the Joseph and Sendner theory.

The predictions were used to interpret observations of the diffusion of the surface waters in the coastal zone of the Cape Kennedy area, and the plausibility of the model was demonstrated. Pritchard *et al.* (1966) applied the model to a number of surface layer observations. They found that, by and large, the shear model is able to describe the observed diffusion pattern. Both the observed decrease of the peak concentration and increase of the rotationally symmetrical variance conformed with the theoretical predictions.

Okubo (1967) considered the case of an oscillatory current $U(t,z)$ with a single harmonic constituent having a phase independent of depth and superimposed on a steady current $U(z)$ in the same direction. No transverse effects were included, and a purely two-dimensional model in the $x$-$z$ plane with vertical and horizontal eddy diffusion was investigated. The current field was specified as

$$U(t,z) = V_o \cdot (1-z/d') . \sin \omega t$$

$$U(z) = U_o \cdot (1-z/d) \qquad (20a,b)$$

where $U_o$ is the steady current at $z=0$; $V_o$ is the amplitude of the oscillating component at $z=0$; and $d,d'$ are length scales associated with the respective motions. Okubo formulated the diffusion equation and investigated the dispersion in both the vertically bounded and unbounded case.

In the case of an unbounded sea, the equation was transformed by introducing a moment-generating function, and the solution yielded an equation for the variance in the direction of the current. The initial distribution used by Okubo was an instantaneous point source. The variance is determined by the sum of the purely horizontal diffusion and the interaction between the shears and the vertical diffusion. The latter part was found to be proportional to the vertical diffusion coefficient, and consists of the effects of the steady and periodic flows plus a combined effect of the two parts of the current. For diffusion times, large compared to the period of the oscillating current component, the influence of the steady shear is dominating.

For the semi-bounded case a similar form for the variance would be obtained, according to Okubo. The variance would be reduced by a certain numerical factor and the form would not be symmetrical. Okubo stated that, as long as an appreciable amount of the diffusing material remains within a layer thin in comparison to the depth of the water, the results for the unbounded case can be applied. The criterion for an unbounded layer may then be expressed as $D^2 >> \pi^2 \cdot K \cdot t$, with $D$ the water depth.

Okubo proceeded to treat the bounded problem, and two separate cases were distinguished. When the time needed for the contaminant to become uniformly distributed throughout the water column is much larger than the period of the oscillating current component, the shear effect is dominated by the steady current. When the vertical mixing, on the other hand, is essentially completed within one period of the oscillating current, the oscillating shear effect is of equal importance. In both cases the horizontal variance was found to be inversely proportional to the vertical diffusion coefficient. The results agree with those of Bowden (1965). Many areas are characterized by large amplitude tidal oscillations and there the dispersion is often dominated by the shear effect due to the tidal flows (Bowden, 1965; Talbot and Talbot, 1974).

Okubo (1967) furthermore showed that in a bounded sea the horizontal dispersion due to the shear effect is asymptotically governed by a constant effective diffusion coefficient, whereas in the unbounded case the diffusivity increases with time. He concluded that the shorter the period of an oscillating current, the less effect it will have on the asymptotic dispersion. He also concluded that the shear effect due to the shear in the steady flow is most effective. These conclusions conform with the general picture of turbulent dispersion given above. Some indication of an asymptotically constant diffusivity in a bounded sea has been given by Talbot (1974), but this aspect needs more investigation.

Okubo (1968) compared the theoretical predictions of the shear diffusion models with those given by the rotationally symmetrical theories. It was suggested that in the unbounded case the two approaches yield similar results. It should be noted that the rotationally symmetrical theories predict the time dependence of the variance as $\sigma_r^2 \propto t^3$ or $\sigma_r^2 \propto t^2$ depending upon the fundamental assumptions made.

It may be more appropriate to compare the shear models for the vertically bounded case with the rotationally symmetrical theories since these presuppose the contaminant evenly distributed in a layer which it does not leave. Strictly speaking, the comparison is not adequate since the rotationally symmetrical approach is two-dimensional whereas the shear diffusion is basically three-dimensional.

It should be pointed out that the shear diffusion also is of importance in atmospheric dispersion. Pasquill (1962) made observations of the dispersion in the atmospheric boundary layer and found evidence of the shear effect for the bounded case; i.e., the dispersion seemed to be inversely proportional to the vertical mixing. Saffman (1962) investigated theoretically the effect of wind shear on the dispersion from an instantaneous ground source considering both the bounded and unbounded case. He found that in the former case the dispersion was inversely proportional to the vertical mixing coefficient, and that for large diffusion times the dispersion could be described by a constant diffusion coefficient. In the unbounded case this was not possible; dispersion was found to be proportional to the vertical mixing coefficient and the longitudinal variance increased as $t^3$. It is interesting to note that, except for a numerical factor, the expression for the longitudinal variance obtained by Carter and Okubo (1965) is the same as that found by Saffman (1962). This implies that the presence of the ground (or the sea surface) does not influence the functional form of the dispersion relation.

All the models discussed so far treat only the longitudinal or the steady state shear diffusion in detail. Kullenberg (1972, 1974) formulated a time-dependent model for the unbounded case including lateral as well as longitudinal effects. The horizontal current was specified as consisting of a mean current varying with depth only and a superimposed oscillating current component. The amplitudes vary with depth but both phase and period are considered independent of depth.

$$u = U_o(z) + u_o(z) \cos \omega t$$

$$v = V_o + v_o(z) \sin \omega t \qquad \text{(21a,b)}$$

The term $V_o$ is independent of depth since for steady shears the coordinate system can be rotated so that the depth variation is contained in one component. The $x$-axis is positive in the direction of the mean shear and $z$ is positive upwards. The plane $z=0$ is placed in the center of the layer of interest and the system is moving with the mean velocity at $z=0$. Since the goal was to investigate the significance of the vertical shear-effect using a number of submerged tracer experiments in different areas, Kullenberg neglected purely horizontal diffusion and included all the vertical transports in a vertical mixing coefficient, considered constant in the layer. Thus

$$\frac{\partial c}{\partial t} + u \frac{\partial c}{\partial x} + v \frac{\partial c}{\partial y} = K \frac{\partial^2 c}{\partial z^2} \qquad (22)$$

This equation was first transformed using the technique of Okubo (1967) and solved for a short line source parallel to the $y$-axis at $x=0$ and $z=0$. The solution yields expressions for the variances of the concentration distribution, which include several oscillating terms. Neglecting these, the variances are

$$\sigma_x^2 = \frac{2}{3} K \left(\frac{dU_o}{dz}\right)^2 t^3 + \frac{Kt}{\omega^2} \left(\frac{du_o}{dz}\right)^2$$

$$\sigma_y^2 = \frac{9}{8} \frac{Kt}{\omega^2} \left(\frac{dv_o}{dz}\right)^2 \tag{23a,b,c}$$

$$\sigma_z^2 = 2Kt$$

The lateral variance is determined by the oscillating current component, and large periods become increasingly important with increasing diffusion time.

It is possible to develop the model further by including several oscillating components but this will not be pursued here.

For $K=0$ the variances become zero when the source has no vertical extension. However, most sources have a finite thickness $H_O$ and, due to the stretching influence of the shear, this causes an apparent horizontal diffusion. Kullenberg (1972) investigated this aspect and found that for $K=0$

$$\sigma_x^2 = \frac{1}{12} H_o^2 \left(\frac{dU_o}{dz}\right)^2 \cdot t^2. \tag{24}$$

Kullenberg (1972, 1974) compared the predictions of the model with observed variances determined from dye tracer experiments. Current profile measurements in conjunction with the tracing made it possible to determine the mean vertical shear due to the steady current, as well as the periods of the dominating oscillating components and their vertical shears. The vertical mixing coefficient $K$ was also obtained from the dye experiments. The comparison between observation and theory (Figure 2) suggests that the shear effect is an important mixing mechanism in many coastal areas for scales in the range 0.1 to 30 km. Clearly, other mechanisms are also effective and in some areas the vertical shear effect may be insignificant. One requirement for it to be important is, of course, the presence of vertical current shear. The results emphasize the need to include current profile observations in field studies as well as information on the current spectrum.

The problem of whether to consider the upper layer of the sea bounded or unbounded and the transition between these conditions

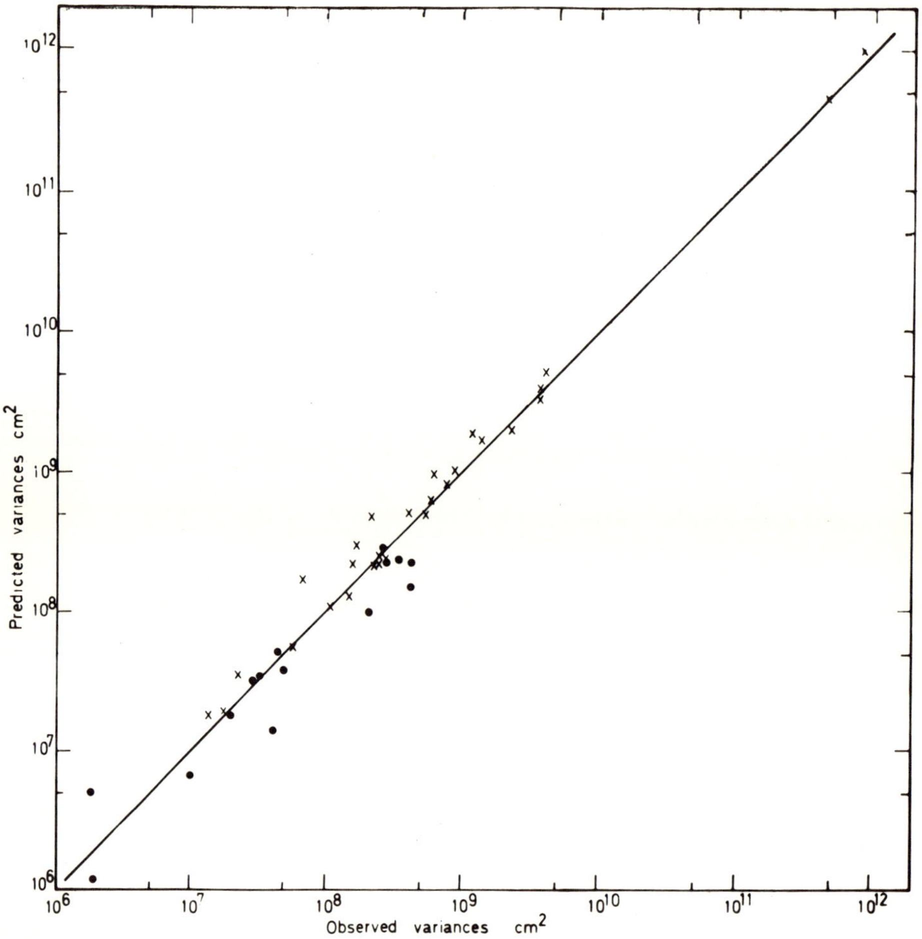


Figure 2. Predicted vs. observed horizontal variances (from Kullenberg, 1974); single value Δ from FLEX 1976.

has recently been studied by Evans (1977) who solved the special case of two well-mixed layers with a constant shear.

Similarity predictions for the horizontal mixing in the sea have been experimentally supported over a range of scales, and therefore it is of interest to consider briefly if the shear diffusion models conform with these similarity predictions (Okubo, this volume). For intermediate diffusion times the shear diffusion model for the unbounded case yields

$$\sigma_x^{\ 2} \propto K \cdot a_o^{\ 2}\ t^3, \tag{25}$$

where $K$ is the vertical diffusion coefficient and $a_o$ is the mean vertical shear in the internal layer under consideration. The effective horizontal diffusion coefficient is then, assuming that $K$ and $a_o$ are constant,

$$K_x \propto \frac{d\sigma_x^2}{dt} \propto K \cdot a_o^2 \; t^2. \tag{26}$$

Elimination of the time yields

$$K_x \propto \sigma_x^{4/3} \cdot (Ka_o^2)^{1/3} \tag{27}$$

in agreement with the Richardson law. For the relationship to conform with the similarity prediction it is necessary that

$$K \cdot a_o^2 \propto \varepsilon, \text{ where } \varepsilon \text{ is the energy dissipation.}$$

This can be made plausible by using the result (see 12 and 14):

$$K = const. \cdot \frac{\tau}{\rho} \cdot \left|\frac{dq}{dz}\right| \cdot \overline{N^{-2}} = const. \cdot \varepsilon \cdot \overline{N^{-2}}$$

$$\text{with } \overline{N^2} = -\frac{g}{\overline{\rho}}\frac{d\overline{\rho}}{dz} \tag{28}$$

We may now introduce the mean, overall Richardson number

$$\overline{Ri} = \overline{N^2}/a_o^2$$

yielding

$$\overline{KN^2} = K \cdot a_o^2 \cdot \overline{Ri} = const. \cdot \varepsilon \tag{29}$$

in agreement with the similarity prediction.

Attempts have also been made to use the shear diffusion models for calculating the shear from observed concentration distributions (Ebbesmeyer, 1975) and to determine variance relations between radially symmetric diffusion models and shear diffusion models (Ebbesmeyer and Okubo, 1975).

Okubo and Platt (this volume) discussed the concept of critical scales in patchiness generation in relation to purely horizontal processes, and it may be of interest to introduce results from the shear diffusion models into the estimation of critical scales. This will also indicate how the critical scale depends upon certain environmental conditions. Most formulations of critical scale predict that the scale depends upon the square root of a horizontal diffusion coefficient $K_x$, i.e., $l_{cx} \simeq K_x^{\frac{1}{2}}$. We may then investigate how $K_x$ is predicted from the shear diffusion models, and we will use the Kierstead and Slobodkin formulation of $l_{cx}$ to exemplify the result, i.e., $l_{cx} = \pi(K_x/r)^{\frac{1}{2}}$ where $r$ is the division rate. For intermediate time scales the longitudinal variance predicted by various shear diffusion models is

$$\sigma_x^2 = \frac{2}{3} K a_o^2 t^3. \tag{30}$$

We may define $K_x$ as

$$K_x = \frac{d\sigma_x^2}{dt} = 2 K a_o^2 t^2. \tag{31}$$

From observations in the surface layer for winds above 5 m/s, Kullenberg (1971, 1974) found that

$$K = 9 \cdot 10^{-8} \frac{W_{10}^2}{N^2} \left|\frac{dq}{dz}\right|$$

where $W_{10}$ is the wind velocity and $N^2 = \frac{q}{\rho_o}\frac{\partial\rho}{\partial z}$ and $q(z)$ is the current vector.

This yields

$$K_x = 54 \cdot 10^{-8} \frac{W_{10}^2}{N^2} a_o^3 t^2, \text{ assuming } \frac{dq}{dz} \approx 3a_o. \tag{32}$$

The critical length scale becomes

$$\ell_{cx} = 2.3 \cdot 10^{-3} \frac{a_o t W_{10}}{N} \sqrt{\frac{a_o}{r}} \tag{33}$$

showing the influence of the shear, the wind and the stratification. This is one example on how the dependence of the critical scale on the environmental conditions can be brought out explicitly by relating the mixing, as originally given by $K_x$, to these conditions through a set of semi-empirical results. It is noted that the time $t$ also enters explicitly. This is due to the fact that horizontal mixing in the sea is scale dependent (see Okubo, this volume). It should be remembered that we have considered a case covering only intermediate time scales. Let us next consider the rotationally symmetrical case with

$$\sigma_r^2 = \sigma_x \cdot \sigma_y, \quad K_r = \frac{d\sigma_r^2}{dt} \tag{34}$$

The shear diffusion model of Kullenberg (1972) then yields

$$\sigma_r^2 = \sigma_x \cdot \sigma_y \approx \left(\frac{2}{3} K a_o^2 t^3 \cdot \frac{b^2}{\omega^2} Kt\right)^{\frac{1}{2}}$$

where $b$ is the transverse oscillating shear.

$$\sigma_r^2 \simeq 0.8 \cdot \frac{Ka_o b t^2}{\omega} = 22 \cdot 10^{-8} \frac{W_{10}^2 a_o^2 bt^2}{N^2 \omega}$$

The critical length scale in this two-dimensional case then becomes

$$\ell_r = 2.4 \sqrt{\frac{K_r}{r}} = 1.6 \cdot 10^{-3} \frac{W_{10} \quad a_o}{N} \sqrt{\frac{bt}{\omega r}}$$

which shows the influence of the steady and oscillating shear as well as the wind and the stratification.

Denman and Platt (1976) gave an alternative formulation of the phytoplankton distribution in the form of a variance spectrum using the turbulence similarity approach. This resulted in a balance between phytoplankton growth and turbulent energy dissipation yielding a critical length scale $l_c \equiv (\varepsilon/r^3)^{\frac{1}{2}}$ where $\varepsilon$ is the energy dissipation and $r$ the growth rate parameter. Turbulence similarity predictions yield $K_x \propto \sigma_x^{4/3} \cdot \varepsilon^{1/3}$ which, using $K_x = \sigma_x^2/t$, is equivalent to $K_x \propto \varepsilon \cdot t^2$. This implies that this approach also predicts the critical scale as proportional to $(K_x)^{\frac{1}{2}}$.

## Application to FLEX Data

As part of the Fladen Ground Experiment 1976 (FLEX), a series of small scale thermocline dye experiments were performed during the second half of May. These experiments were conducted using the *in situ* release and tracing technique developed by Kullenberg (1969, 1972, 1974). From the observed concentration distributions, vertical and horizontal mixing parameters have been calculated (Table 1). The vertical mixing coefficients were determined using the method given by Kullenberg (1971, 1974) and the horizontal

Table 1

Mixing Parameters from Dye Experiments During FLEX '76

$\overline{K}$ mean vertical mixing coefficient, $\overline{P}$ mean horizontal diffusion velocity, $\sigma_r^2$ symmetrical variance, $\sigma_x^2$ longitudinal variance, $\overline{W_{10}}$ mean wind speed during tracing, $\overline{N^2}$ mean stratification parameter, $|\overline{dq/dz}|$ mean shear.

| Date May | Depth m | Diffusion Time h | $\overline{K}$ $cm^2/s$ | $\overline{P}$ cm/s | $\sigma_r^2$ $cm^2$ | $\sigma_x^2$ $cm^2$ | $\overline{W_{10}}$ m/s | $\overline{N^2}$ $s^{-2}$ | $\lvert\overline{\frac{dq}{dz}}\rvert$ $s^{-1}$ |
|---|---|---|---|---|---|---|---|---|---|
| 15 | 20-25 | 7 | 1.3 | 0.5 | - | - | 10 | $5\cdot10^{-5}$ | $7\cdot10^{-4}$ |
| 21 | 10-30 | 5 | - | 0.6 | - | - | 7 | $2.5\cdot10^{-5}$ | - |
| 23-24 | 10-15 | 18 | 1.1 | 0.3 | $0.9\cdot10^9$ | $1.1\cdot10^9$ | 7 | $6.7\cdot10^{-5}$ | $1.7\cdot10^{-3}$ |

diffusion velocity was determined by applying the model of Joseph and Sendner (1958). The variances were determined by integrating the horizontal distributions. Examples of vertical profiles through the dye layer together with the temperature profile are given in Figure 3. For comparison, a profile of chlorophyll observed in a thermocline layer by Pingree *et al.* (1975) is included. It is interesting to note how similar these profiles look although the vertical scales are slightly different. It is not unreasonable to assume that similar processes have influenced the shape of these profiles. The shears given in Table 1 have been obtained by applying equation (5). The values are reasonable in comparison with the shears which can be estimated from the current measurements during FLEX.

Let us now apply these data to (33) for the critical scale, using $r = 0.3/day = 3.5 \cdot 10^{-6}\ s^{-1}$ and $t = 3$ days. This yields, for 15 May, $l_{cx} = 1.0\ km$ and for 23 May $l_{cx} = 2.5\ km$. These values are, of course, only orders of magnitude but they do indicate the possible influence of vertical processes on these scales. There exist several indications from observations in the North Sea of a critical scale of the order of a few kilometers on time scales of a few days (e.g. Steele and Henderson, 1978). The present results would then suggest that vertical processes may be important on these scales.

Using the following mean values for parts of the North Sea:

$$N = 7 \cdot 10^{-3}\ s^{-1},\ W_{10} = 10\ m/s,\ a_o = 5 \cdot 10^{-4}\ s^{-1},\ r = 3.5 \cdot 10^{-6}\ s^{-1}$$

we find critical scales of the order of 110 km for large time scales of the order of 100 days. Two patches given by Cushing and Tungate (1963) can be used to calculate a length scale defined by $\ell = 3\sigma_x$ which may be representative of large time scales for the central North Sea. The scales found in this way are 110 to 150 km, which is of the same order of magnitude as the critical scale estimated for large time scales. Again, it should be stressed that this is only an illustration of how the average environmental conditions may affect the scales. A time scale of 100 days may be outside the range for which the present formulation is valid, although a reasonable estimate of the corresponding length scale is obtained. The time scales used in these examples appear relevant both from a biological and physical point of view. Time scales typical for biological interactions as well as events such as passages of storms will be of particular interest, and these scales often fall within the range appropriate for the present formulation.

We have used the Kierstead-Slobodkin formulation of the critical length scale. This does not include grazing, which will increase the size of the critical scales (Platt and Denman, 1974).

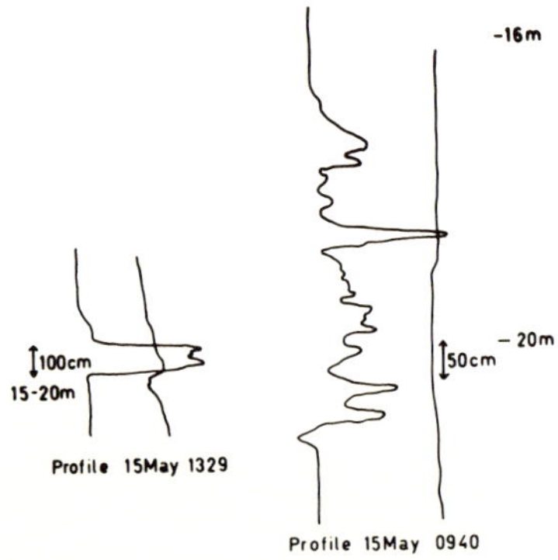


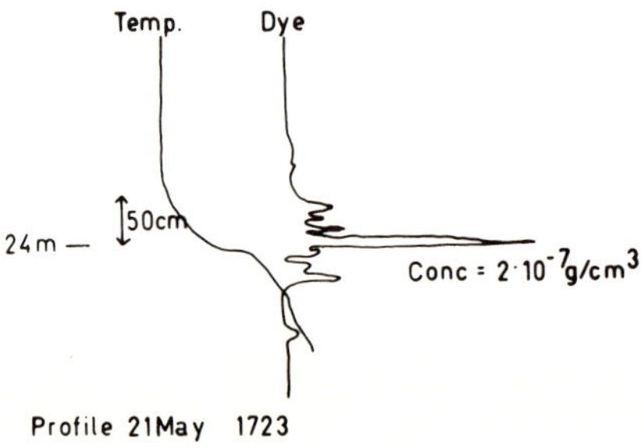


Figure 3a Figure 3b

Dye profiles in different temperature structures observed during FLEX 1976.

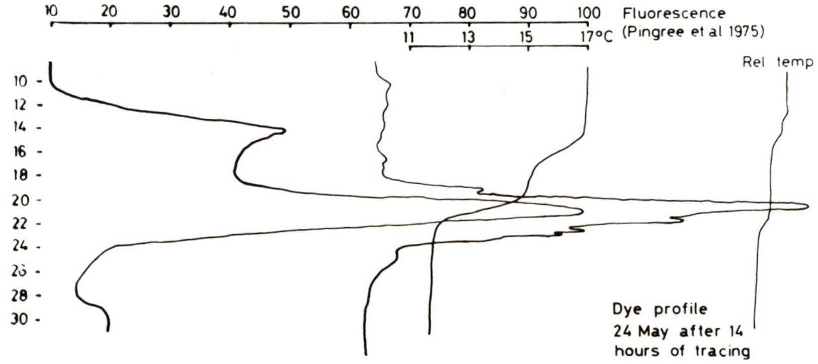


Figure 3c

Dye profile in weak thermocline observed during FLEX 1976 and fluorescence (chlorophyll) profile observed in thermocline layer by Pingree *et al.* (1975).

These latter authors find a critical length scale

$$\ell_c \sim \left( \frac{Kx}{r-R_m\lambda} \right)^{\frac{1}{2}}$$

where $\lambda$ is a constant used in the grazing equation and $R_m$ represents the maximum ration required by the zooplankton population. The product $R_m\lambda$ can vary in the range $(0.1-0.9)\cdot r$, according to Platt and Denman (1974). The horizontal mixing coefficient enters as $K_x^{\frac{1}{2}}$. Hence, the transformation from horizontal to vertical processes as the basic physical approach will yield the same results.

It seems reasonable to make the following conclusions: 1) vertical processes, in combination with the vertical shear of horizontal processes, can be important in generating and maintaining horizontal patchiness and are worth studying in an exploratory model, 2) a simple parameterization of vertical mixing, combined with a shear diffusion model which includes advection and an oscillating current, may be used for that purpose.

The outline of such a model was described by Evans, Steele and Kullenberg (1977) and is further discussed in the contribution by Evans in this volume.

## SUMMARY

Average vertical mixing in the top 100 meters of the sea are discussed and various attempts to parameterize and quantify the mixing rates are presented. The significance on the energy input from the wind is shown and that the vertical entrainment velocity can be related to the overall Richardson number and the wind-generated friction velocity for winds above about 5 m/s.

The vertical shear effect, i.e., the combined action of vertical shear of the horizontal currents and the vertical mixing generating apparent horizontal dispersion is then described and a brief review of vertical shear models is given. Comparisons between model predictions and observations of mixing including relevant environmental parameters, such as average vertical shear and current fluctuations, suggest that the shear effect is a dominating mechanism for horizontal dispersion on scales up to about 10 km. This would then suggest that the shear effect also can influence the biological distributions. In order to appreciate the possible significance of this mechanism, various formulations of characteristic scales, which all depend upon the horizontal mixing in a similar way, are transformed so that the vertical mixing and shear is introduced. The important physical environmental parameters are the wind, the stratification and the shear. The results from field observations in the North Sea are used to show that the shear effect indeed gives the right order of magnitude scales.

ACKNOWLEDGEMENT

Financial support for the FLEX field observations from the Danish Natural Science Research Council and from the Sonderforschungsbereich 94 is acknowledged.

REFERENCES

Bowden, K. F., 1965: Horizontal mixing in the sea due to a shearing current. F. Fluid Mech., 21, 83-95.

Bowles, P., R. Burns, F. Hudswell and R. Whipple, 1958: Sea disposal of low activity effluent. Proc. 2nd Conf. Peaceful Uses of Atomic Energy, U.N., Geneva, 18, 376-389.

Carter, H. H. and A. Okubo, 1965: A study of the physical processes and movement and dispersion in the Cape Kennedy area. Chesapeake Bay Inst., The Johns Hopkins Univ., Ref. 65-2, 162 pp.

Corsin, S., 1960: Theories of turbulent dispersion. In Mecanique de la Turbulence, Marseille 1961, 27-52.

Denman, K. L. and M. Miyake, 1973: Upper layer modification at ocean station Papa: observations and simulations. J. Phys. Ocean., 3, 185-196.

Denman, K. L. and T. Platt, 1976: The variance spectrum of phytoplankton in a turbulent ocean. J. Mar. Res., 34, 593-601.

de Szoeke, R. A. and P. B. Rhines, 1976: Assymptotic regimes in mixed-layer deepening. J. Mar. Res., 34, 111-116.

Debroklonskiy, S. V. and N. V. Kontoboytseva, 1973: Empirical relationships between steady-state waves and wind sppeds for a wide range of fetches. Oceanology, 13, 640-645.

Ebbesmeyer, C. C., 1975: Computing current shear from dye observations using one of Okubo's advective-diffusive models. J. Phys. Ocean., 5, 191-193.

Ebbesmeyer, C. C. and A. Okubo, 1975: Variance relations between variable shear-diffusion and radially symmetric turbulent diffusion. Deep-Sea Res., 22, 441-445.

Elder, J. W., 1959: The dispersion of marked fluid in turbulent shear flow. J. Fluid Mech., 5, 544-560.

Evans, G. T., 1977: A two layer shear diffusion model. Deep-Sea Res., 24, 931-936.

Evans, G. T., J. H. Steele and G.E.B. Kullenberg, 1977: A preliminary model of shear diffusion and plankton populations. Scot. Fish. Res. Rept., 9, 20 pp.

Halpern, D., 1974: Observations of the deepening of the wind-mixed layer in the northeast Pacific Ocean. J. Phys. Oceanogr., 4, 454-466.

Joseph, J. and H. Sendner, 1958: Uber die horizontale Diffusion im Meere. D.H.Z., 11, 49-77.

Kantha, L. H., 1975: Turbulent entrainment at the density interface of a two-layer stably stratified fluid system. Geophy. Fluid Dyn. Lab. Tech. Rept. 75-1, Johns Hopkins Univ.

Kantha, L. and O. M. Phillips, 1976: On turbulent entrainment at a stable density interface. J. Fluid Mech.

Kato, H. and O. M. Phillips, 1969: On the penetration of a turbulent layer into stratified fluid. J. Fluid Mech., 37, 643-655.

Kielmann, J. W. Krauss and L. Magaard, 1969: Uber die Verteilung der kinetischen Energie im Bereich der Tragheits - und Seichesfrequenzen der Ostsee im August 1964. Kieler Meeresf., 25(2), 245-254.

Kitaigorodsky, S. A., 1978: Oceanic surface boundary layer. Paper presented at SCOR/IOC Climate Workshop, Helsinki, 1977.

Kraus, E. B., editor, 1977: Modelling and prediction of the upper layers of the ocean. Pergamon Press, 325 pp.

Kraus, E. B. and J. S. Turner, 1967: A one-dimensional model of the seasonal thermocline. II. The general theory and its consequences. Tellus, 19, 98-106.

Kullenberg, G., 1971: Vertical diffusion in shallow waters. Tellus, 23, 129-135.

Kullenberg, G., 1972: Apparent horizontal diffusion in stratified vertical shear flow. Tellus, 24, 17-28.

Kullenberg, G., 1974: An experimental and theoretical investigation of the turbulent diffusion in the upper layer of the sea. Inst. Phys. Oceanogr., Copenhagen Rept. No. 25, 212 pp.

Kullenberg, G., 1974a: Investigation of small-scale vertical mixing in relation to the temperature structure in stably stratified waters. Adv. Geophys., 18A, 339-351.

Kullenberg, G.E.B., 1976: On vertical mixing and the energy transfer from the wind to the water. Tellus, 28, 159-165.

Kullenberg, G., 1977: Entrainment velocity in natural stratified vertical shear flow. Est. Coast. Mer. Sc., 5, 329-338.

Lumley, J. L., 1964: The spectrum of nearly inertial turbulence in a stably stratified fluid. J. Atmos. Sci., 21, 99-102.

Miles, J. W. and L. N. Howard, 1964: Note on heterogeneous shear flow. J. Fluid Mech., 20(2), 331-336.

Niiler, P. P., 1975: Deepening of the wind-mixed layer. J. Mar. Res., 33, 405-422.

Niiler, P. P. and E. B. Kraus, 1977: One-dimensional models of the upper ocean. In Modelling and Prediction of the Upper Layers of the Ocean, E. B. Kraus (ed.), pp. 325, Pergamon Press.

Novikov, E. A., 1958: Concerning a turbulent diffusion in a stream with a transverse gradient of velocity. A. Appl. Math. and Mech., 22, 576-579.

O'Brien, J. J., et al., 1977: Upwelling in the ocean: two- and three-dimensional models of the upper ocean dynamics and variability. In Modelling and Prediction of the Upper Layers of the Ocean, E. B. Kraus (ed.), pp. 325, Pergamon Press.

Okubo, A., 1967: The effect of shear in an oscillatory current on horizontal diffusion from an instantaneous source. Int. J. Oceanol. Limnol., 1, 194-204.

Okubo, A., 1968: Some remarks on the importance of the "shear effect" on horizontal diffusion. J. Oc. Soc. Jap., 24, 20-29.

Okubo, A., 1971: Horizontal and vertical mixing in the sea. In Impingement of Man on the Oceans, D. Hood (ed.), pp. 738, Wiley-Interscience.

Okubo, A. and R. V. Ozmidov, 1970: Empirical dependence of the coefficient of horizontal turbulent diffusion in the ocean on the scale of the phenomenon in question. Izv. Atm. Ocean. Phys., 6, 534-536.

Ottesen Hansen, N-E., 1975: Entrainment in two-layered flows. Inst. Hydr. and Hydraulic Eng. Tech. Univ. Denmark, ser. paper 7, 99 pp.

Ozmidov, R. V., 1965: On the turbulent exchange in a stably stratified ocean. Izv. Atm. Ocean. Phys. Ser., 1, 493-497.

Ozmidov, R. V., 1965a: Energy distribution between oceanic motions of different scales. Izv. Atm. Ocean. Phys., 1, 257-261.

Pasquill, F., 1962: Some observed properties of medium scale diffusion in the atmosphere. Quart. J. Roy. Me. Soc., 88, 70-79.

Phillips, O. M., 1966: The Dynamics of the Upper Ocean. Cambridge Univ. Press, Cambridge, 26 pp.

Phillips, O. M., 1977: Entrainment. In Modelling and Prediction of the Upper Layers of the Ocean, E. B. Kraus (ed.), pp. 325, Pergamon Press.

Pingree, R. D., P. R. Pugh, P. M. Holligan and G. F. Foster, 1975: Summer phytoplankton blooms and red tides along tidal fronts in the approaches to the English Channel. Nature, 258, 672-677.

Pollard, R. T., P. B. Rhines and R.O.R.Y. Thompson, 1973: The deepening of the wind mixed layer. Geophys. Fluid Dyn., 3, 381-404.

Pollard, R. T. and R. C. Millard, Jr., 1970: Comparison between observed and simulated wind-generated inertial oscillations. Deep-Sea Res., 17, 813-821.

Steele, J. H. and E. W. Henderson, 1977: Plankton patchiness in the northern North Sea. Symp. Inst. Maths. Applications Fish. Maths., Aberdeen, November 24-26, 1975.

Pritchard, D. W., A. Okubo and H. H. Carter, 1966: Observations and theory of eddy movement and diffusion of an introduced tracer material in the surface layers of the sea. In Disposal of Radioactive Wastes into Seas, Oceans and Surface Waters. I.A.E.A. Vienna, 397-424.

Saffman, P. G., 1962: The effects of wind shear on horizontal spread from an instantaneous ground source. Quart. J. Roy. Met. Soc., 88, 382-393.

Stewart, R. W., 1959: The problem of diffusion in stratified fluids. Adv. Geoph., 6, 303-311.

Stewart, R. W. and H. L. Grant, 1962: Determination of the rate of dissipation of turbulent energy near the sea surface in the presence of waves. J. Geophys. Res., 67, 3177-3180.

Talbot, J. W. and G. A. Talbot, 1974: Diffusion in shallow seas and in English coastal and estuarine waters. Rapp. Proces-Verbaus, 167, 93-110.

Taylor, G. I., 1953: Dispersion of soluble matter in solvent flowing slowly through a tube. Proc. Roy. Soc. Lond., A219, 186-203.

Taylor, G. I., 1954: The dispersion of matter in turbulent flow through a pipe. Proc. Roy. Soc. Lond., A223, 446-468.

Thorpe, S. A., 1977: Turbulence and mixing in a Scottish Loch. Phil. Trans. Roy. Soc. Lond., A286, 125-181.

Townsend, A. A., 1956: The Structure of Turbulent Shear Flow. Cambridge University Press, 2nd edition.

Turner, J. S., 1969: A note on wind mixing at the seasonal thermocline. Deep-Sea Res., 16, Suppl., 297-300.

Webster, F., 1969: Vertical profiles of horizontal ocean currents. Deep-Sea Res., 16, 85-98.

Woods, J. D., 1968: Wave-induced shear instability in the summer thermocline. J. Fluid Mech., 32, 791-800.

Woods, J. D., 1974: Space-time characteristics of turbulence in the seasonal thermocline. Min. Soc. Royale des Sciences de Liege 6e Ser., 6, 109-130.

Woods, J. D. and R. L. Wiley, 1972: Billow turbulence and ocean microstructure. Deep-Sea Res., 19, 87-121.

Wroblewski, J. S., 1977: A model of phytoplankton plume formation during variable Oregon upwelling. J. Mar. Res., 35, 357-394.

# APPENDIX

## List of Symbols

| | |
|---|---|
| $a_o$ | mean vertical shear |
| $c$ | scalar concentration |
| $f$ | Coriolis parameter |
| $g$ | acceleration of gravity |
| $H_o$ | initial thickness of contaminated layer |
| $h$ | mixed layer thickness |
| $k$ | wind factor |
| $K, K_m$ | vertical mixing coefficients for mass (heat, salt) and momentum, respectively |
| $K_x, K_y$ | diffusion coefficients in $x$ and $y$ directions, respectively |
| $N$ | Brunt-Väisälä frequency; $N^2 = -\frac{g}{\rho o}\frac{d\rho}{dz}$ |
| $P$ | horizontal diffusion velocity |
| $q(z)$ | horizontal current vector as a function of depth |
| $r$ | (net) growth rate parameter |
| $Ri, Rf$ | gradient and flux Richardson numbers, respectively |
| $Ri_*$ | overall Richardson number |
| $t$ | time |
| $u_*$ | friction velocity |
| $U(z)$ | mean horizontal velocity |
| $u, v, w$ | velocities in the $x, y, z$-directions, respectively |
| $u', w'$ | fluctuating velocities |
| $V_o$ | surface current |
| $W_{10}$ | wind velocity at 10 m level |
| $w_e$ | entrainment velocity |
| $\varepsilon$ | rate of energy dissipation per unit mass |
| $\rho_a, \rho, \rho_o$ | air, water density and mean surface layer density, respectively |
| $\tau_o$ | wind stress |
| $\sigma_r^2$ | horizontal symmetrical variance |
| $\sigma_x^2, \sigma_y^2$ | variances in $x$ and $y$ directions, respectively |
| $\omega$ | frequency of oscillating current component |
| $\kappa_o$ | von Karman constant |

# SPECTRAL ANALYSIS OF SPATIAL STRUCTURE IN PHYTOPLANKTON POPULATIONS

Trevor Platt

Marine Ecology Laboratory
Bedford Institute of Oceanography
Dartmouth, Nova Scotia, Canada

## INTRODUCTION

Our ability to study spatial structure in phytoplankton populations by the methods of spectral analysis depends on the development of a method for continuous measurement of chlorophyll concentration from a moving ship (Lorezen, 1966). Chlorophyll is used as the relative index of phytoplankton abundance, and it can be assayed continuously by Lorenzen's *in vivo* fluorescence technique (see Platt and Denman, in press). Adaptation of this methodology to the particular requirement of phytoplankton patchiness research is discussed in Platt (1972) and in Herman and Denman (1977).

Given this facility, it may be deployed in one of two ways to accumulate the kind of data required for a spectral analysis. The sampling apparatus may be held fixed at a particular depth below the sea surface, and fluctuations in chlorophyll recorded as the water is carried past it by the current (Platt, 1972). Or it may be towed at a fixed depth from a moving ship (Denman and Platt, 1975; Fasham and Pugh, 1976; Denman, 1976; Horwood, in press). Two samplers were towed at two fixed depths simultaneously by Denman (1976). Whether the data are collected in the time or space domain, it is not possible to identify either time or space as the single or unique independent variable. In spatial studies we cannot ignore the time-evolution of the chlorophyll field no more than we can ignore vertical displacements of the water column and advection in temporal studies.

For spectral analysis, the continuous records are reduced to a sequence (a *time series*) of digitized chlorophyll values equally spaced in either distance or time. The application of this method of statistical analysis to phytoplankton data is still quite recent (Platt *et al.*, 1970; Platt, 1972; Denman and Platt, 1975; Powell *et al.*, 1975; Fasham and Pugh, 1976; Denman, 1976; Lekan and Wilson, in press; Horwood, in press). A review of the potential of the method for ecological data in general has been given by Platt and Denman (1975).

The end product of a spectral analysis is called a power spectrum. It is a plot of the distribution of variance as a function of time-scale (or space-scale) calculated in such a way that the area under the spectrum equals the total variance in the original time series. The scales at which the component variances are calculated form a somewhat arbitrary set, depending on the length and sample spacing of the original series. To interconvert between time- and space-scales when discussing these spectra we invoke the Taylor hypothesis, also known as the frozen-field hypothesis (Hinze, 1959). An essential condition for this to hold is that the advection velocity is large compared to a typical turbulent velocity. We are not always justified in making this assumption (Denman and Platt, 1975).

## SPECTRAL THEORIES OF PHYTOPLANKTON PATCHINESS

Before discussing observational examples of power spectra for chlorophyll, it will be useful to consider some theoretical studies directed towards predicting the shape these spectra should have. (The historical development of the field was actually the reverse of this.) One of the motivations for a power spectral approach to the theory of patchiness was a historical preconception about the factors controlling the structure of real phytoplankton populations (Margalef, 1967; Platt, 1972). It was thought that since phytoplankton cells could be considered as passive contaminants of the physical flow field, their detailed distribution would be determined by turbulence. But how to demonstrate this?

A successful way to make progress in the theory of turbulence has been through the use of dimensional analysis (Corssin, 1956). It was applied by Kolmogorov (1941), who deduced that under certain drastic assumptions about the turbulent field (homogeneity and isotropy in all three dimensions), the logarithmic plot of the power spectrum of velocity fluctuations should be a straight line with a slope $-5/3$ over a range of length-scales called the inertial subrange.

Power spectra of *in vivo* chlorophyll (fluorescence) collected at a fixed station in the Gulf of St. Lawrence (Platt, 1972) had a slope of $-1.66 \pm 0.10$ for length scales from 10 m to 1000 m. Although the isotropy condition, on which Kolmogorov's hypothesis was predicted, could not be justified, one was tempted to make the connection that this value of the spectral slope was consistent with the idea of local phytoplankton abundance being controlled by turbulence. Precedents from physical oceanography included work by Ozmidov (1965) and Webster (1969) who had observed Kolmogorov-type velocity spectra in situations where the Kolmogorov conditions did not hold.

Later results for chlorophyll spectra gave essentially similar results (Denman and Platt, 1975; Powell *et al.*, 1975; Fasham and Pugh, 1976; Denman, 1976; Lekan and Wilson, in press; Horwood, in press), but the slope of the spectra was by no means always equal to -5/3. Slopes between -1 and -3 were observed. After the initial flush of enthusiasm about -5/3 and turbulence controlling patchiness, the enduring aspect of these results was that they showed that the variance of chlorophyll measurements depended in a systematic way on length-scale, and that this dependence was representable by a simple power function, regardless of the particular slope of that power function.

A theory of the chlorophyll spectrum was published by Denman and Platt (1976) that took into account the fact that phytoplankton cells were not conservative passive contaminants of the flow field: they had the ability to reproduce. They considered three main subranges characterized by the time-scale $\tau$ for the turbulent eddies of a given length-scale to transfer their kinetic energy (velocity variance) to eddies of half the size. The other important time-scale in the model is $r^{-1}$ where $r$ is the exponential growth-rate of the phytoplankton.

In the subrange $\tau \ll r^{-1}$, structures in the phytoplankton distribution on scales defined by the sizes of turbulent eddies satisfying this inequality are destroyed by mixing into a much larger volume of fluid as the kinetic energy is transferred to smaller scales. Reproduction has a negligible effect and the spectral shape for chlorophyll should not be different from that predicted for a conservative passive scalar, such as temperature.

In the subrange $\tau \gg r^{-1}$, features in the phytoplankton distribution are accentuated by reproduction before the eddies transfer their energy to smaller scales. Reproduction cannot now be neglected and the chlorophyll spectrum is expected to differ from that predicted for temperature.

The subrange $\tau \gtrsim r^{-1}$ is an intermediate regime between the two extreme subranges.

Applying dimensional analysis to the system described by these three subranges leads to the hypothetical spectrum shown in Figure 1. At low wavenumbers ($\tau >> r^{-1}$), the chlorophyll spectrum is predicted to have the form $E_\beta(k)$:

$$E_\beta(k) = C\chi_\beta r^{-1}k^{-1} \tag{1}$$

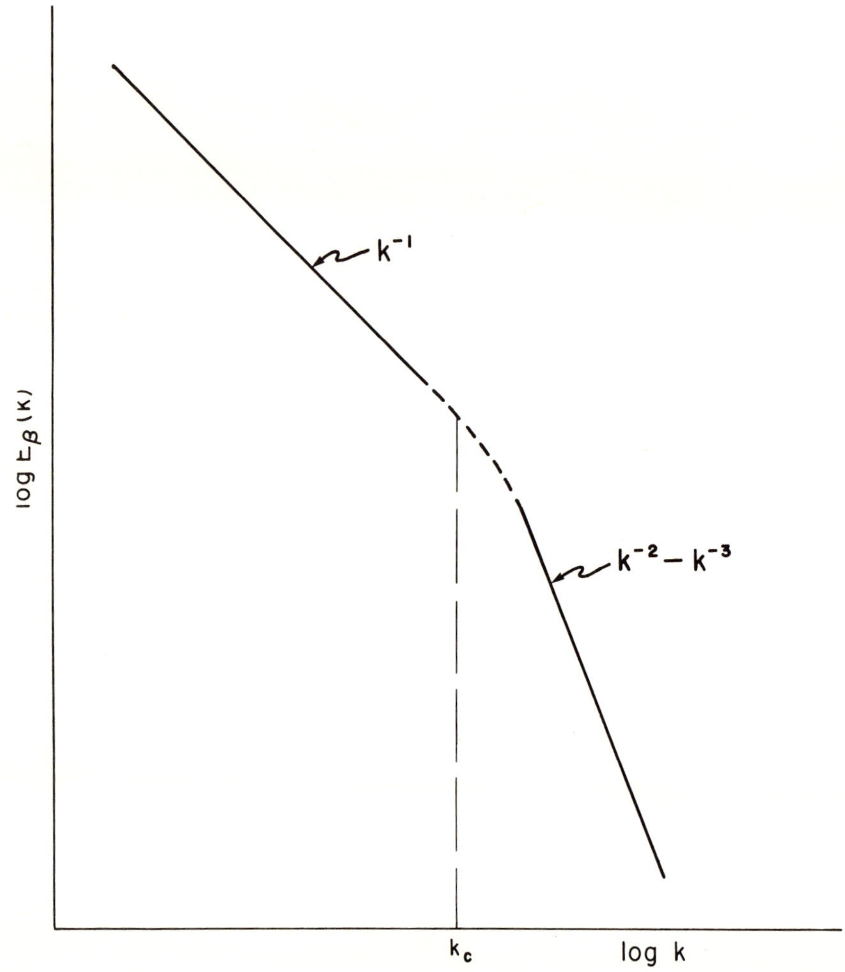


Figure 1. Theoretical chlorophyll power spectrum after Denman and Platt (1976).

Here $E$ denotes variance, $\beta$ is a suffix specifying chlorophyll, $C$ is a dimensionless constant, $\chi_\beta$ is a parameter specifying the rate of destruction of variance in $\beta$ at high wavenumbers (Corssin, 1951) and $k$ is the wavenumber. The corresponding spectrum for temperature is

$$E_\Theta(k) = A' \chi_\Theta \varepsilon^{-1/3} k^{-5/3} \qquad (2)$$

where $\chi_\Theta$ is analogous to $\chi_\beta$, $A'$ is a dimensionless coefficient, and $\varepsilon$ describes the viscous dissipation of velocity fluctuation such that it is analogous to both $\chi_\beta$ and $\chi_\Theta$. Comparing (1) and (2) we see that at low wavenumbers, $E_\beta$ should decrease less rapidly with increasing $k$ than should $E_\Theta$. The high-wavenumber boundary for this subrange is $k_c = (r^3/\varepsilon)^{1/2}$, analogous to a Kierstead and Slobodkin scale, and of magnitude about $1(\mathrm{km})^{-1}$. At lower wavenumbers, individual plankton species with different growth rates $r$ will share a common spectral slope $k^{-1}$. At higher wavenumbers ($\tau << r^{-1}$), the chlorophyll spectrum will share the same slope as the temperature spectrum, equal to $k^{-5/3}$ as in (2) if three-dimensional isotropic turbulence is the dominant physical process or $k^{-2}$ or $k^{-3}$ if two-dimensional turbulence or other physical processes are dominant.

The basic assumption of the work just described was that there existed a range of wavenumbers (i.e., for $\tau >> r^{-1}$) within which biological processes dominate over physical processes to the extent that the physical parameter could be dropped in making the dimensional solution to the spectral slope. A quite different approach to the chlorophyll spectrum, involving no explicit assumption about the relative importance of physical and biological processes, was developed by Denman *et al.* (1977).

Chlorophyll $\beta$ is assumed to be growing at exponential rate $r$ in a homogeneous, isotropic turbulent field. The calculation is restricted to the inertial-convective region, where molecular diffusion can be neglected.

$$\frac{\partial \beta}{\partial t} + u_i \frac{\partial \beta}{\partial x_i} = r\beta$$

Separating the chlorophyll field into mean and fluctuating components

$$\beta(t) = \langle\beta\rangle + \beta'(t) \qquad (3)$$

(and similarly for the velocity components $u_i$), leads to the following fundamental equation for the variance of the chlorophyll fluctuation:

$$\frac{\partial\langle\beta'^2\rangle}{\partial t} + \langle u_i\rangle \frac{\partial\langle\beta'^2\rangle}{\partial x_i} = -2\langle\Theta u_i'\beta_i'\rangle\frac{\partial\langle\beta\rangle}{\partial x_i} - \frac{\partial}{\partial x_i}\langle u_i'\beta'^2\rangle + 2r\langle\beta'^2\rangle. \tag{4}$$

After simplifying by application of the homogeneity condition, equation (4) yields a generalized estimate $J_\beta(k)$ for the flux of chlorophyll variance across any wavenumber $k$.

$$J_\beta(k) = \frac{E_\beta(k)}{\tau(k)} \times k \times e^{-2rt(k)} \tag{5}$$

where $\tau(k) = k^{-3/2}E_\nu^{-1/2}$ is the characteristic time taken to transfer variance from wavenumber $k$ to wavenumber $2k$, and $t(k)$ is the time taken for variance to migrate to wavenumber $k$ from some reference or source wavenumber $k_L$.

Making use of the condition that in the inertial convective region the flux $J_\beta(k)$ should be independent of $k$, equation (5) leads to the following solution for $E_\beta(k)$:

$$E_\beta(k) = E_\beta(k_L)\left(\frac{k}{k_L}\right)^{-5/3} exp\{3r\varepsilon^{-1/3}(k_L^{-2/3}-k^{-2/3})\}, \tag{6}$$

where

$$E_\beta(k_L) = B\varepsilon^{-1/3}k_L^{-5/3} \tag{7}$$

and $B$ is a constant. This function is sketched in Figure 2. Defining once more a characteristic wavenumber $k_c \equiv (3r\varepsilon^{-1/3})^{3/2}$ and choosing $k_L \ll k_c$ we see that in the range $k \gg k_c$, $E_\beta(k) \propto k^{-5/3}$, and the chlorophyll spectrum follows the velocity spectrum. In the range $k \ll k_c$, the spectrum cannot be represented by a simple power law. It is an increasing function of $k$ and passes through a maximum at some wavenumber $k_m$. It turns out that $k_m \sim 0.25k_c$. Only in the vicinity of $k_c$ does the slope of the chlorophyll spectrum take the value -1.

Clearly the important difference between this solution and the previous one (Denman and Platt, 1976) is the presence of a peak in the spectrum $E_\beta(k)$, implying that there is a preferred wavenumber region for chlorophyll variance. The mathematical difference between these two calculations is that in the earlier one, certain assumptions are made at the outset about the region $k \ll k_c$, whereas in the later one, a general solution is found for

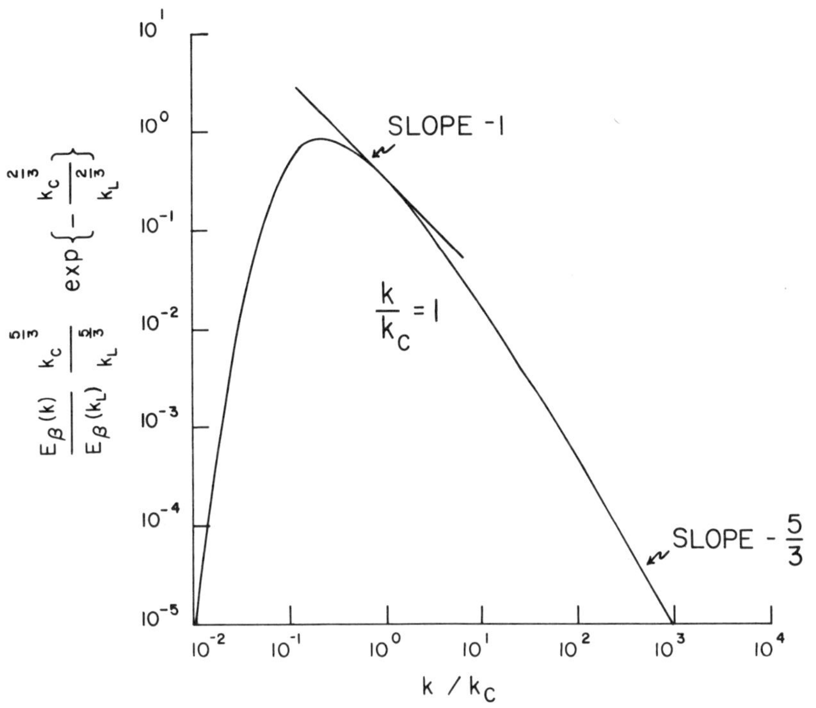


Figure 2. Theoretical chlorophyll power spectrum after Denman, Okubo and Platt (1977).

$E_\beta(k)$ with the low-wavenumber behavior found by taking the limit $k << k_c$.

## OBSERVATIONAL POWER SPECTRA OF CHLOROPHYLL CONCENTRATION

As an example of typical chlorophyll spectra observed in the sea, we consider those reproduced in Figure 3 taken from Denman (1976). They show a change in slope at a reciprocal wavenumber of about 1 km. At higher wavenumbers the spectra follow the temperature spectra fairly closely. At lower wavenumbers the chlorophyll spectrum had a smaller negative slope than the temperature spectra. In these respects the spectra correspond quite well to the shape predicted by the early theory of Denman and Platt (1976) described above.

A significant proportion of the spectra, however, did not correspond to the expected shape. These are separated out from the rest of Figure 3. In general they are the spectra for transects in which the total chlorophyll variance was small.

Horwood (in press) analyzed five sets of chlorophyll data collected at different times specifically to test the theory of Denman and Platt (1976). Of these five spectra, one showed a slope of -1 at low wavenumbers breaking at a reciprocal wavenumber ∿100 m to a slope of -2. None of the other four spectra could be said to reproduce the shape predicted by Denman and Platt. It *may* be significant that the one spectrum that did appear almost as a diagrammatic representation of the theoretical spectrum was that for which the total chlorophyll variance was the highest of all five spectra.

In another set of four spectra taken out to larger length-scales (reciprocal wavenumber ∿20 km), Horwood observed no break in slope whatsoever, a slope of about -2 being maintained throughout the range. Horwood now made the following provocative observation. As spring phytoplankton bloom progressed, the bulk of the standing stock was concentrated into an increasingly smaller area. At the height of the spring bloom in 1971, 50% of the standing stock was concentrated in 19% of the surface area. Presumably, these

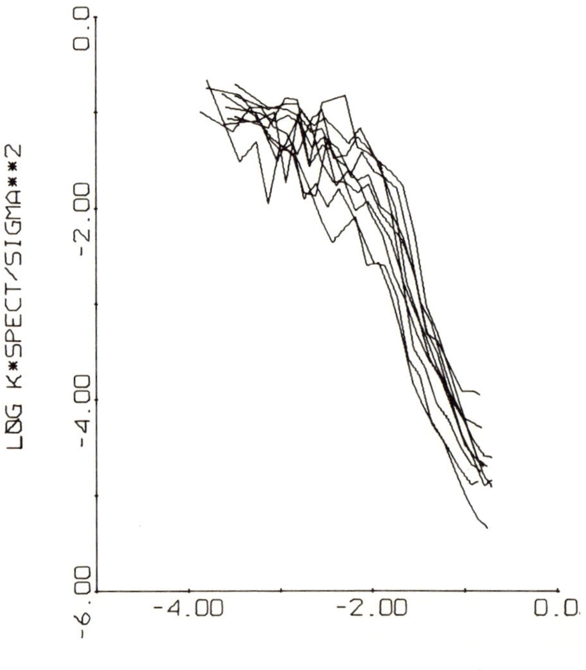


Figure 3. Typical chlorophyll power spectra after Denman (1976).

conditions are those giving rise to the greatest total variance in chlorophyll transects. But more importantly, such a progressive change in the way phytoplankton biomass is distributed in space must reflect some strong component of biological origin in the forces shaping the population structure of phytoplankton. Similar changes were recorded from observations made by Horwood in three different years. The mechanism cannot be entirely physical, unless we can rationalize a progressive change in the turbulence or internal wave field concommitant with the development of the spring bloom. Denman (1976) was able to separate out a group of chlorophyll spectra in which this total variance was considerably greater than could be expected from the observed temperature variance, on the assumption that all chlorophyll fluctuations were correlated with temperature fluctuations. He considered these to correspond to cases of high phytoplankton activity.

Horwood (in press) concluded that at least in the range of length-scales from 10 m to 1000 m, a knee-shaped spectrum is not characteristic of chlorophyll fluctuations. If there is any merit at all in the concept of a spectral break of slope near some characteristic wavenumber, the resolution of theory and observation has to be that the characteristic wavenumber frequently takes a much smaller magnitude than 1 $km^{-1}$; i.e., that the characteristic length-scale is no longer than can be observed by the length of transects commonly used so far in patchiness studies. (This, incidentally, is also the reason why the spectral theory of Denman *et al.* (1977) has not yet been tested by observation.) Apart from variations in $\varepsilon$, which can be considerable (Crawford, 1976), the range of variability in the phytoplankton growth parameters $r$ can be of decisive importance. Recall that the growth parameter has to refer to a *realized* rate of increase rather than a maximal or intrinsic rate of increase, and we should therefore include the effect of grazing by zooplankton. This was done by Platt and Denman (1975) who showed that when grazing rate was nearly in balance with division rate in the phytoplankton, the characteristic length-scale for patchiness would increase. Extreme changes in realized rate of increase from zero (or less in some cases) up to a maximum of $\sim$1 or several $day^{-1}$ for the unimpeded growth of the smaller species of algae could account for a wide variety of observed spectra, including those that showed no break at all in the range of length-scales so far seen. In other words, the values of $r$ that have been used up to now have been such that the values of the characteristic wavenumbers so deduced should be considered as upper bounds. Because grazing is the rule rather than the exception in the sea, we expect that the realized rate of increase will be lower than that commonly employed in the estimation of a characteristic scale; this scale, if it exists, will be correspondingly larger than has been thought to be the case.

Platt and Denman (1975) suggested that the characteristic scale would change in a systematic way with the rise and decline of the spring phytoplankton bloom, increasing rapidly towards the end of the bloom as nutrients become exhausted and zooplankton populations begin vigorous growth. The feedback between the biological system and the physical system cannot be ignored, however. Development of the phytoplankton bloom increases the optical attenuation coefficient which in turn leads to increased vertical stability through stronger heating (Denman and Miyake, 1973). Thus, unfolding of the seasonal biological cycle can, itself, lead to changes in the internal wave field, and therefore affect the relative shapes of the chlorophyll and temperature spectra.

## SUMMARY

The historical development is reviewed briefly of the study of spatial structure in phytoplankton population by means of spectral analysis of time series data on chlorophyll concentrations. Recently published theoretical predictions for the shape of the power spectrum of chlorophyll fluctuations in wavenumber space are described and compared with each other, and with observations by a number of authors in different sea areas. It is stressed that the growth rate of phytoplankton used to calculate characteristic length scales should be the realized rate of increase and not the maximum intrinsic rate of increase of the populations. This bears directly on the problem of comparing observed spectra with theoretical predictions.

## REFERENCES

Corrsin, S., 1951: On the spectrum of isotropic temperature fluctuations in an isotropic turbulence. J. Appl. Phys., 22, 469-473.

Corrsin, S., 1956: Dimensional analysis and similarity. In Fluid Models in Geophysics, R. R. Long (ed.), pp. 1-17, Washington, D.C.

Denman, K. L., 1976: Covariability of chlorophyll and temperature in the sea. Deep-Sea Res., 23, 539-550.

Denman, K. L. and M. Miyake, 1973: Upper layer modification at ocean station 'Papa': observations and simulation. J. Phys. Oceanogr., 3, 185-196.

Denman, K. L., A. Okubo and T. Platt, 1977: The chlorophyll fluctuation spectrum in the sea. Limnol. Oceanogr., 22, 1033-1038.

Denman, K. L. and T. Platt, 1975: Coherences in the horizontal distributions of phytoplankton and temperature in the upper ocean. Mem. Soc. Roy. Sci. Liege, 6e serie, 7, 19-30.

Denman, K. L. and T. Platt, 1976: The variance spectrum of phytoplankton in a turbulent ocean. J. Mar. Res., 34, 593-601.

Fasham, M. J. and P. R. Pugh, 1976: Observations on the horizontal coherence of chlorophyll a and temperature. Deep-Sea Res., 23, 527-538.

Herman, A. W. and K. L. Denman, 1977: Rapid underway profiling on chlorophyll with an in situ fluorometer mounted on a 'BATFISH' vehicle. Deep-Sea Res., 24, 385-397.

Hinze, J. O., 1959: Turbulence. McGraw-Hill, New York, 586 pp.

Horwood, J. H., 1978: Spatial distributions in pelagic fish. In press.

Kolmogorov, A. N., 1941: The local structure of turbulence in an incompressible viscous fluid for very large Reynolds number (in Russian). Dokl. Akad. Nauk, U.S.S.R., 30, 299-303.

Lekan, J. R. and R. E. Wilson, 1978: Spatial variability of phytoplankton biomass in the surface waters of Long Island. Estuarine Coastal Mar. Sci., in press.

Lorenzen, C. J., 1966: A method for the continuous measurement of in vivo chlorophyll concentration. Deep-Sea Res., 13, 223-227.

Margalef, R., 1967: The food web in the pelagic environment. Helgöland. wiss. Meeresunters, 15, 548-559.

Ozmidov, R. V., 1965: The scales of oceanic turbulence. Oceanology, 6, 325-328.

Platt, T., 1972: Local phytoplankton abundance and turbulence. Deep-Sea Res., 19, 183-187.

Platt, T. and K. L. Denman, 1978: Patchiness in phytoplankton distribution. In The Ecology of Phytoplankton, I. Morris (ed.), Chapter 14, in press.

Platt, T., L. M. Dickie and R. W. Trites, 1970: Spatial heterogeneity of phytoplankton in a near-shore environment. J. Fish. Res. Bd. Canada, 27, 1453-1473.

Powell, T. M., P. J. Richerson, T. M. Dillon, B. A. Agee, B. J. Dozier, D. A. Godden and L. O. Myrup, 1975: Spatial scales of current speed and phytoplankton biomass fluctuations in Lake Tahoe. Science, 189, 1088-1090.

Webster, F., 1969: Turbulence spectra in the ocean. Deep-Sea Res., 16, 357-368.

# COLLECTION AND ANALYSIS OF UNDERWAY DATA AND RELATED PHYSICAL MEASUREMENTS

Kenneth L. Denman and David L. Mackas

Institute of Ocean Sciences
P. O. Box 6000
Sidney, B.C., Canada V8L 4B2

## INTRODUCTION

Our ability to describe and understand the marine environment is limited both by the type and detail of the environmental data we can acquire, and by the power and resolution of the analysis methods we can apply to that data. Much of our recent understanding of the spatial variability in the planktonic community has been achieved from the analysis of series of essentially continuous measurements, taken either at a point in space or along a vertical or horizontal line. For a few parameters, well-established methods allow resolution of spatial structure over all length scales from centimeters to thousands of kilometers. Measurements of physical (and some chemical) scalar variables (e.g. temperature and salinity) provide the best examples. However, for biological parameters, many of the desired spatial "windows" (see Angel, 1977) are less adequate. Because these windows have in general been opened only recently by new methodologies, their technical limitations are poorly known. They also tend to be narrower (Figure 1) and yield data that are even more subject than the physical scalars to the confounding effects of variations that occur along the temporal and unsampled spatial axes. For this reason various authors (e.g. Denman, 1976) have suggested that one-dimensional sampling schemes are inadequate to describe the spatial structure fully. As discussed in the introductory chapter (Steele), horizontal structure often cannot be separated from vertical structure, either observationally or in terms of operative mechanisms, and there is a serious need for the collection of continuous biological, physical and chemical data in both dimensions simultaneously.

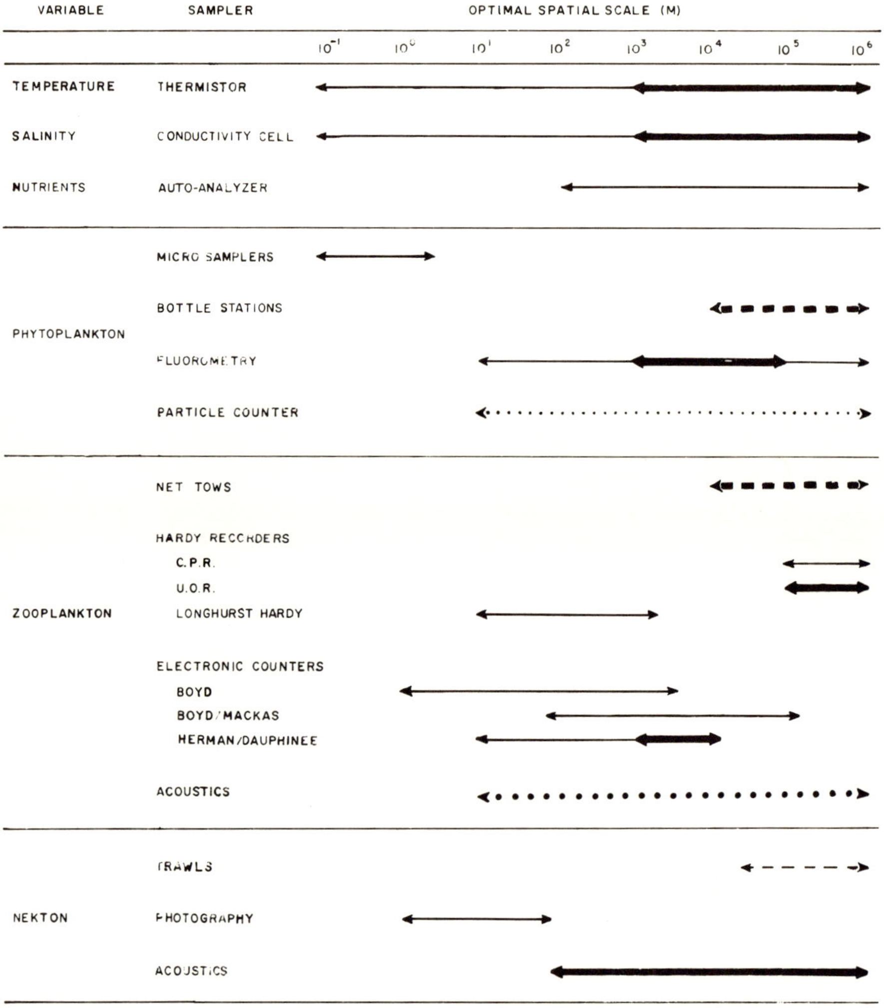


Figure 1. Sampling windows for the measurement of oceanic patchiness, showing the length scales (in the horizontal dimension) resolved by various samplers. Broken lines indicate poor resolution and/or aliasing due to widely-spaced samples. Dotted lines indicate very new or unproven methods. Heavy lines indicate two dimensional sampling. Although information at higher wavenumbers than indicated can often be extracted from undulator records, the data are essentially discontinuous and therefore less reliable.

In the remainder of this chapter we trace briefly the recent evolution of methods for sampling the horizontal and vertical variability of plankton concentrations. We then describe two approaches to the representation and analysis of the two-dimensional data provided by undulating samplers. Finally, we discuss the measurement of physical parameters that characterize the detailed dynamics of dispersion and vertical mixing.

## TOWED INSTRUMENT SYSTEMS

The continuous measurement of phytoplankton chlorophyll *a* by *in vivo* fluorescence has a short history, dating back only to Lorenzen (1966). The time series representation of such data (Platt and Denman, 1975) has progressed from a single series at a fixed point (Platt, 1972a) to cross spectra between series of chlorophyll and a physical parameter either at a fixed point (Fasham and Pugh, 1976) or along horizontal transects (Denman and Platt, 1975; Powell *et al.*, 1975; Denman, 1976). Horizontal mapping at a single depth (Platt, 1972b; Kelley, 1975) and time series of vertical profiles at a single station (Denman, 1977) have also been obtained. Denman and Platt (1975) and Denman (1976) were able to obtain simultaneous information on horizontal and vertical structure, but that was only at two depths.

Recently, however, two vehicles have been developed that will undulate along a sawtooth pattern in the vertical plane when towed from a moving ship. They have the potential of acquiring the simultaneous horizontal and vertical data necessary to study the structure and interactions in the marine ecosystem and its environment. One of these, the Undulating Oceanographic Recorder (UOR) can traverse a pattern with a cycle of 3 to 30 km (Bruce and Aiken, 1975). It is equipped with a continuous plankton recorder (Hardy, 1936) and a conductivity-temperature-depth (CTD) probe, and has been designed to be towed by "ships of opportunity." Aiken, Bruce and Lindley (1977) describe its use in a study of the hydrography and zooplankton distributions in waters north of Scotland. However, its cycle length of 3 km is too large to resolve many of the mesoscale phenomena in the ocean that are biologically relevant. The other undulating vehicle, the Batfish (Dessureault, 1976), has a cycle length as short as 0.3 km, making it more suitable for studying oceanographic features on these scales of 1 to 10 km.

Because these vehicles pass through the water at speeds of 3-5 m $s^{-1}$, the sensors they carry must have response times <1 s if they are to resolve structure on scales of meters. For conductivity, temperature and depth, this constraint can be satisfied without problem, but for chlorophyll *a* fluorescence, even an *in situ* fluorometer such as the Variosens (Impulsphysik) must be modified to give a sufficiently short response time ($\sim$0.5 s) while retaining an acceptably low noise level (Herman, 1975; Herman and Denman, 1977). A schematic of the Batfish vehicle equipped with a Variosens fluorometer and a CTD is shown in Figure 2. Herman and Denman (1977) describe the calibration and operation of this system and present examples of data obtained at sea.

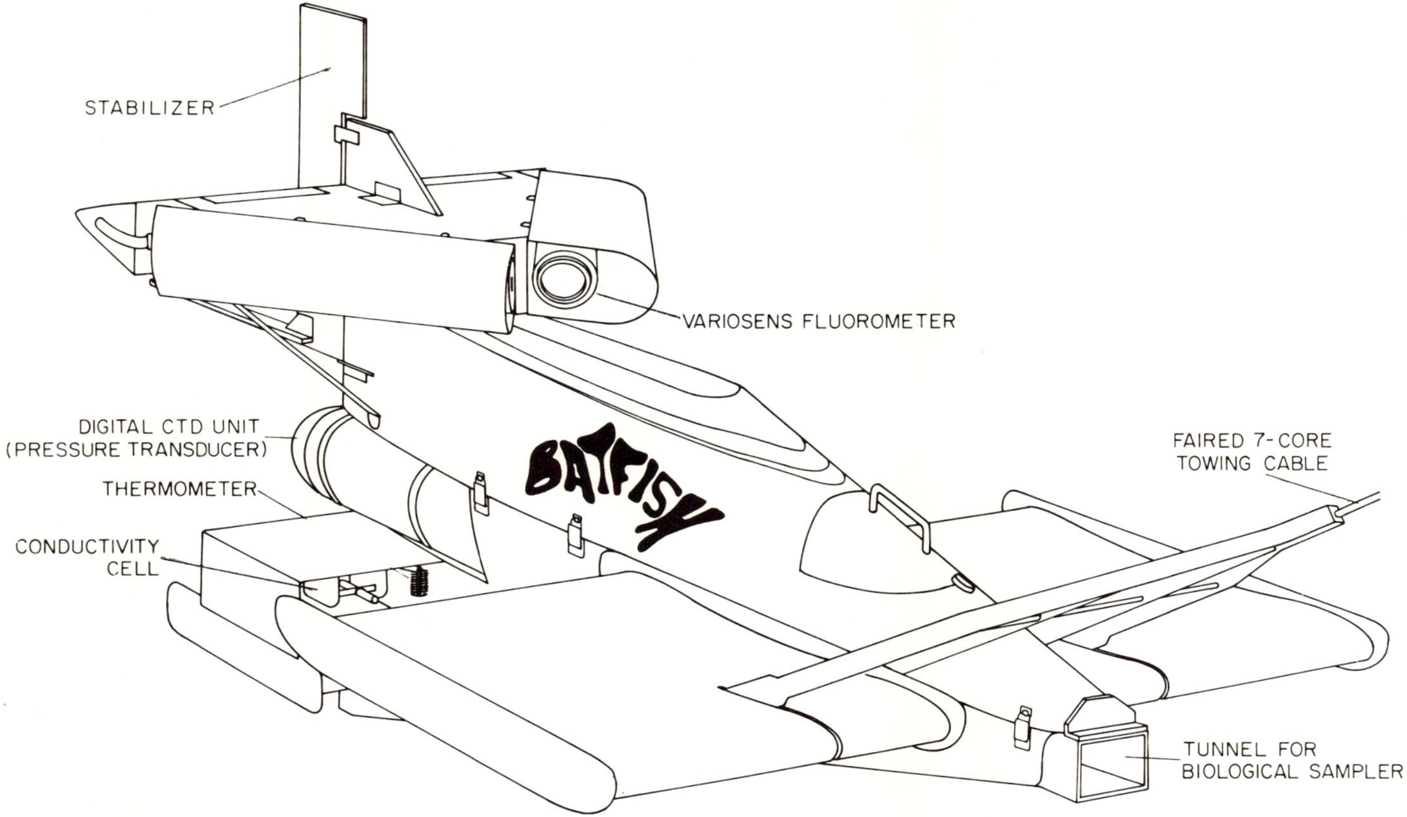


Figure 2. Batfish vehicle with mounted *in situ* Variosens fluorometer and CTD (conductivity, temperature and depth) unit. The CTD sensors are mounted outside the turbulent flow of the wings (from Herman and Denman, 1977).

Techniques for the continuous counting of zooplankton have not advanced in parallel with techniques for the continuous measurement of phytoplankton biomass, despite the earlier design of a prototype by Maddux and Kanwisher (1965). However, in the last several years an electronic zooplankton system has been developed (Boyd, 1973) that will count particles with diameters (of equivalent spherical volumes) from 0.5 to 2.5 mm. The device can discriminate particles separated in space by less than 1 m, but tow lengths are usually restricted to about 10 km because the concentrating net, necessary to obtain statistically meaningful numbers of counts, clogs if phytoplankton are at all abundant in the area.

Mackas (1977) has avoided this constraint by building a modification of the Boyd system that employs four counting cells in parallel, through which water is forced under pressure. The system is designed to be used with either a towed hose inlet or a through-hull sampling port, and includes a Turner-Designs Model 10 fluorometer and a thermistor. Thus, simultaneous measurements of zooplankton and phytoplankton abundance along a horizontal transect can be obtained. Figure 3 (from Mackas, 1977) shows the data for a transect (depth $\sim$3 m) over Georges Bank on the southern edge of the Gulf of Maine. The ship passed through a distinct temperature front separating Georges Bank water from Slope or Gulf Stream waters. The coherence and phase spectra for the run are presented in Figure 4. Zooplankton and chlorophyll *a* tended to covary in phase with each other but out of phase with temperature. The peak near 1 km$^{-1}$ in all three spectra may result in part from a superposition of shorter sinusoids necessary to reconstruct the frontal discontinuity in Figure 3. In these and in other data series, the zooplankton show more intense fine-scale variability than do the phytoplankton. This observation suggests that towed undulating samplers, such as a counter-equipped Batfish described by Herman (1977) and Dauphinee (1977), will require a very short cycle wavelength (100 m or less) if they are to resolve adequately the horizontal spatial structure of the zooplankton.

Electronic particle counting is less effective for larger animals (macrozooplankton and nekton) because of their lower abundance and their greater avoidance ability. For animals longer than $\sim$1 cm (and perhaps eventually for the smaller zooplankton as well), acoustic scattering is an attractive alternative for two reasons. Firstly, the volume sampled per unit distance along the transect is considerably larger. Secondly, and probably more important, the resulting data set is continuous in two dimensions. Estimates of the local abundance of recruited fish are now made routinely from acoustic data; a discussion of applications and analysis is given by Horwood and Cushing (this volume). For macrozooplankton and larval fish, the corresponding technology is still being developed. The smaller body size of the animals requires higher acoustic frequency which in turn results in higher attenuation losses and more

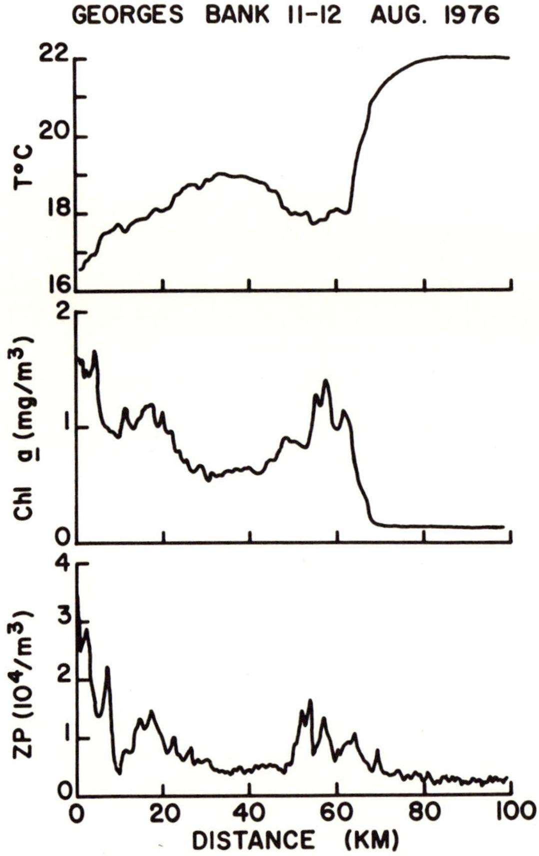


Figure 3. A transect at 3 m depth of temperature, chlorophyll *a* and zooplankton abundance taken from a through-hull intake. The direction was southwest roughly parallel to the outer edge of Georges Bank (from Mackas, 1977).

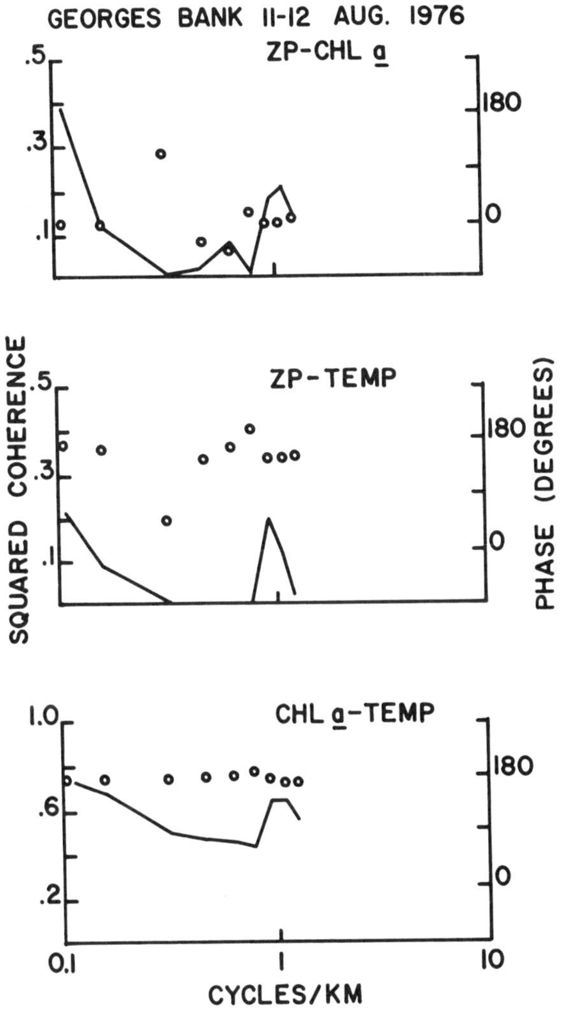


Figure 4. Spectral plots of squared coherence (lines) and phase (circles) for the series in Figure 3 (from Mackas, 1977).

interference from non-biological targets (particularly air bubbles). Thus, signal return is a poorly characterized function of size, abundance, orientation, morphology, and the sound frequency used; most current work (e.g. Sameoto, 1977; Pieper, 1977) is directed toward improving its calibration against biological "ground-truth" obtained from high resolution samplers.

## ANALYSIS OF DATA FROM TOWED VEHICLES

The spectral analysis of time or space series usually assumes equispaced data. However, the sawtooth sampling pattern of the Batfish presents a complex set of varying data intervals. For example, the aliasing problem that occurs when a process is not sampled often enough to resolve its fluctuations is not, in the case of the Batfish, amenable to analytic description. Consider, for example, a monochromatic sinusoidal internal wave, uniform in the vertical but with a horizontal propagation direction. If the Batfish frequency is, for example, 5/2 the wave frequency, then we would expect from sampling theory that the waves (corresponding to internal waves) should be adequately resolved. If the Batfish frequency is lower, say 5/4 the wave frequency, the waves are not well resolved. As indicated in Figure 5, the sawtooth sampling scheme then alters the phase and shape of the distortion as a function of depth. A large scale pattern evolves that bears no resemblance to the wave field and becomes more pronounced with decreasing Batfish frequency.

Typically, for a depth range of 50 m, the Batfish cycles over a wavelength of 0.5 km, causing the smallest resolvable horizontal structure to have a length scale of about 1 km. Although there are internal waves with wavelengths less than 1 km, the internal wave spectrum on these scales is usually very "red" (Garrett and Munk, 1975), implying that the longer wavelength (and therefore resolvable) waves have much larger amplitudes than the unresolved shorter waves. Thus, we expect that in most instances the dominant pattern imposed by internal waves will be of sufficiently large scale to be resolved by the Batfish.

Various analysis techniques have been applied to Batfish data. For example, Denman and Platt (1978) describe the reduction of these data to a format suitable for time series analysis. Briefly, the data map for each transect can be transposed from the Batfish sawtooth grid onto a rectangular array. An appropriate method is the objective mapping technique (Gandin, 1963; Liebelt, 1967; Bretherton, Davis and Fandry, 1976), first used to transform observations from a non-uniform grid of stations to a regular array for input into computer forecasting models. The algorithm produces minimum error variance linear estimates at the new grid points from the autocovariance properties of the original data set. The result of this procedure for Batfish data would be a rectangular data array $F(z, x)$ where $z$ $(=1,\ldots,N)$ indexes the vertical grid and $x$ $(=1,\ldots,T)$ indexes the horizontal spacing of the interpolated vertical profiles.

The goal of the subsequent analysis is to compress the data to the point where the significant pattern in the two-dimensional maps can be represented by a very few series in the horizontal.

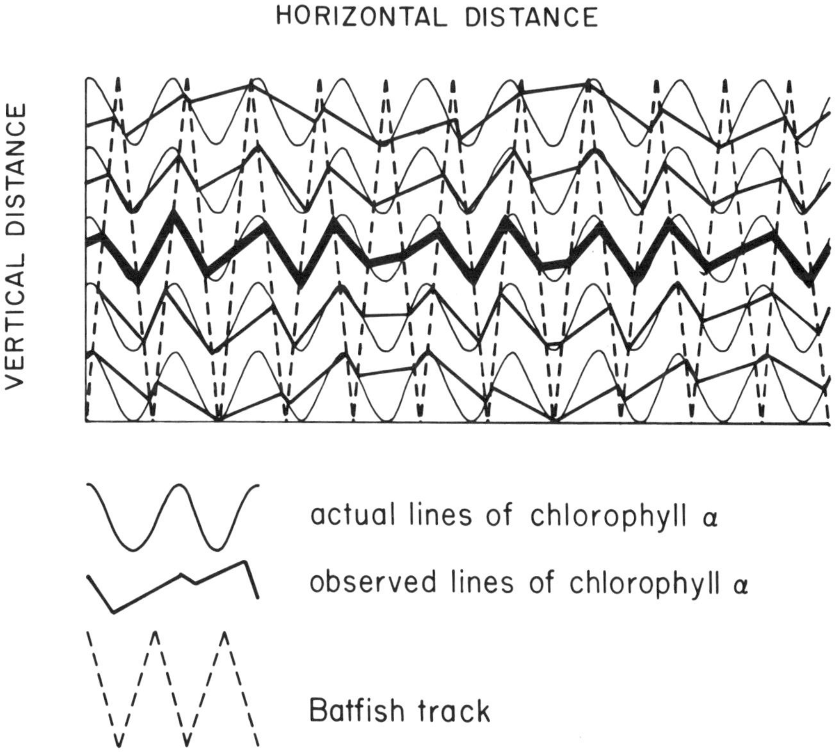


Figure 5. Simulated contour plot of a Batfish transect through a monochromatic internal wave field. Batfish cycle wavelength is 4/5 the internal wave length. The pattern repeats every $m$ Batfish cycles because the frequencies of the Batfish cycling to the internal wave are in a constant ratio $m/n$, where $m$ and $n$ are integers, a highly unlikely occurrence.

For this purpose, a suitable technique is the method of empirical orthogonal function analysis (Lorenz, 1956; Liebelt, 1967; Davis, 1976). The data field $F(z, x)$ is expressed in terms of functions or modes $f_n(z)$ and amplitudes $a_n(x)$ according to

$$F(z, x) = \sum_{n=1}^{N} a_n(x)\, f_n(z).$$

Both the $f_n(z)$ and the $a_n(x)$ satisfy suitable orthogonality conditions. If $z$ is indexed over a set of different parameters (species of organisms for instance), the technique corresponds exactly to the more familiar principal components analysis. If the functions $f_n(z)$ are chosen to be the eigenvectors of the covariance matrix formed from the correlation coefficients

$$C(z_1, z_2) = \frac{1}{T} \sum_{x=1}^{T} F(z_1, x)\, F(z_2, x)$$

where the eigenvectors and eigenvalues $\lambda_n$ are ordered such that $\lambda_n \geq \lambda_{n+1}$, then a fixed number $M \leq N$ of eigenfunctions provides the most efficient approximate representation of the data in the sense that no other expression with the same number of terms can have a lower mean square error. Since each eigenvalue $\lambda_n$ is proportional to the fraction of the overall variance

$$\sum_{z=1}^{N} \sum_{x=1}^{T} F(z,x)^2$$

accounted for by the corresponding eigenvector $f_n(z)$, we can expect that much of the pattern in the data field $F(z,y)$ can be depicted by an approximation

$$\sum_{n=1}^{M} a_n(x)\, f_n(z)$$

employing only the first $M$ empirical functions $f_n(z)$. If, say, 70% or more of the total variance can be represented by only $M \ll N$ terms, then the empirical orthogonal function technique has reduced significantly the number of data points needed to describe the multi-dimensional field. The amplitude $a_n(x)$ of the function $f_n(z)$ for the $x$-th vertical profile in the data field is just the scalar product of that function with the data field

$$\sum_{z=1}^{N} f_n(z)\, F(z,x).$$

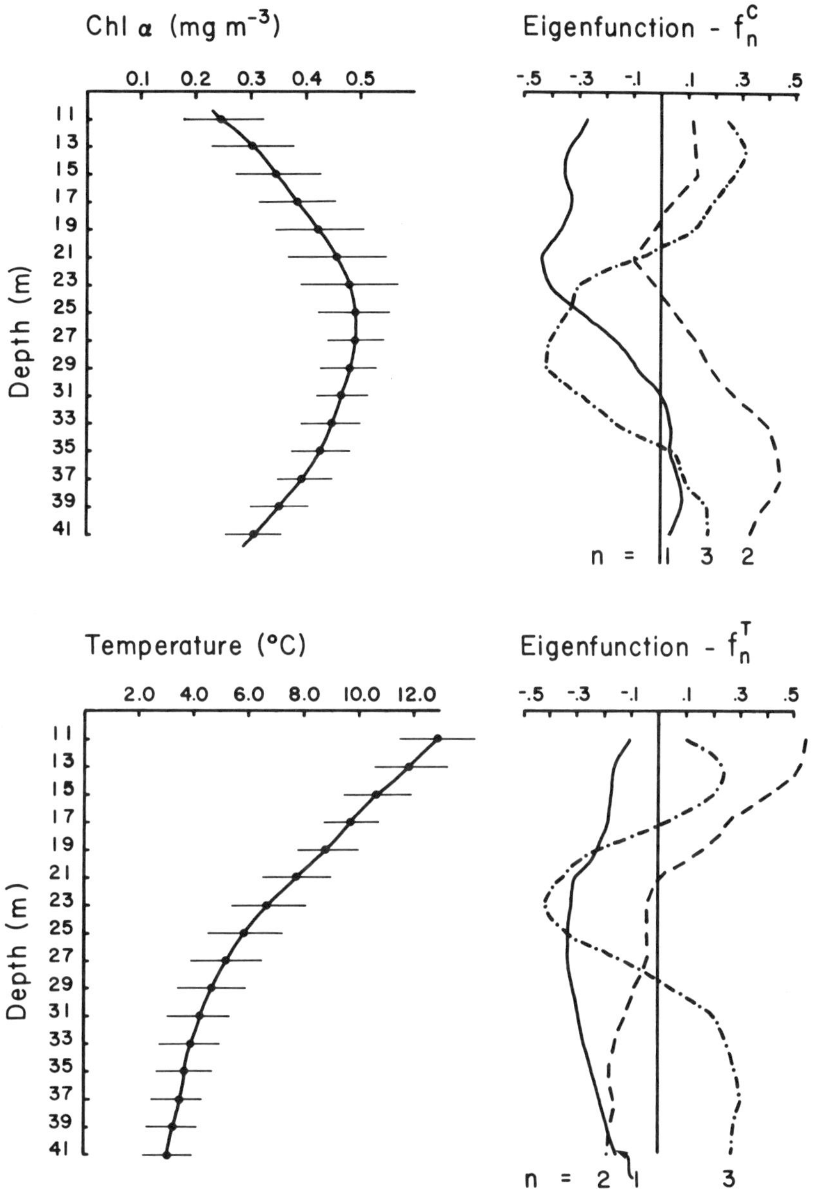


Figure 6. On the left, mean vertical profiles of chlorophyll *a* concentration and temperature for a 35 km Batfish transect on the continental shelf off Nova Scotia. The horizontal lines represent ±1 standard deviation (from 114 profiles). On the right, the first three empirical orthogonal functions $f_n(z)$ for each parameter (from Denman and Platt, 1978).

We can then perform various analyses on the $M$ space series $(a_1(x), a_2(x),\ldots,a_M(x); x=1,\ldots,T)$.

In Denman and Platt (1978) an example was presented of chlorophyll and temperature fields from a 35 km Batfish transect. The data had been averaged over 16 two-meter intervals resulting in 16 orthogonal functions $f_n(z)$. The mean profiles (which were subtracted from the data maps) and the first three eigenfunctions (which then represented deviations from the mean profiles) are shown in Figure 6. An obvious interpretation can be provided only for the first temperature eigenfunction $f_1^T$: such a uniformly positive or negative fluctuation over the depth range of each profile would be expected for vertical displacements by internal waves that extended over the whole of the water column sampled. The first three of these eigenfunctions accounted for 60%, 23% and 7% (90% total) of the variance in the temperature field and 55%, 14% and 11% (80% total) in the chlorophyll field. They were thus able to represent 80 to 90% of the variance or pattern in the original data set with a data set only 3/16 as large.

The objective technique described above is of value in studying the pattern of structure, but it would be of limited value when the Batfish cycling pattern is irregular. For example, Figure 7 (from Denman and Herman, submitted) shows depth-distance sections of chlorophyll $a$ and $\sigma_t$ (specific density) for a Batfish transect over the continental shelf in the eastern Gulf of Maine parallel to the coastline of western Nova Scotia. The bottom of the cycling pattern was varied from about 30 m at either end of the transect to almost 45 m in the middle to provide a safety margin as the ship passed over varying bottom topography. Transformation of this transect into a rectangular grid would either result in the loss of all the sections of data deeper than 30 m, or it would require extrapolation to 45 m such that the synthesized data below 30 m at either end of the transect would be essentially meaningless.

The contour maps of chlorophyll and $\sigma_t$ plotted in Figure 7 are dominated by very large vertical displacements (also in temperature and salinity) in the contour lines, indicative of strong internal wave activity. Overall patterns in the data, such as the relatively light, high chlorophyll surface water (shaded area) between 43°35' and 43°40'N, are masked by the internal wave effects. A straightforward transformation of the data that allows the larger scale patterns to emerge is that of plotting the chlorophyll, temperature and salinity data on sections of $\sigma_t$ versus horizontal distance (Figure 8). Not only are horizontal gradients in the vertical structure now visible, but our concern over inadequate resolution of the internal wave field is eliminated.

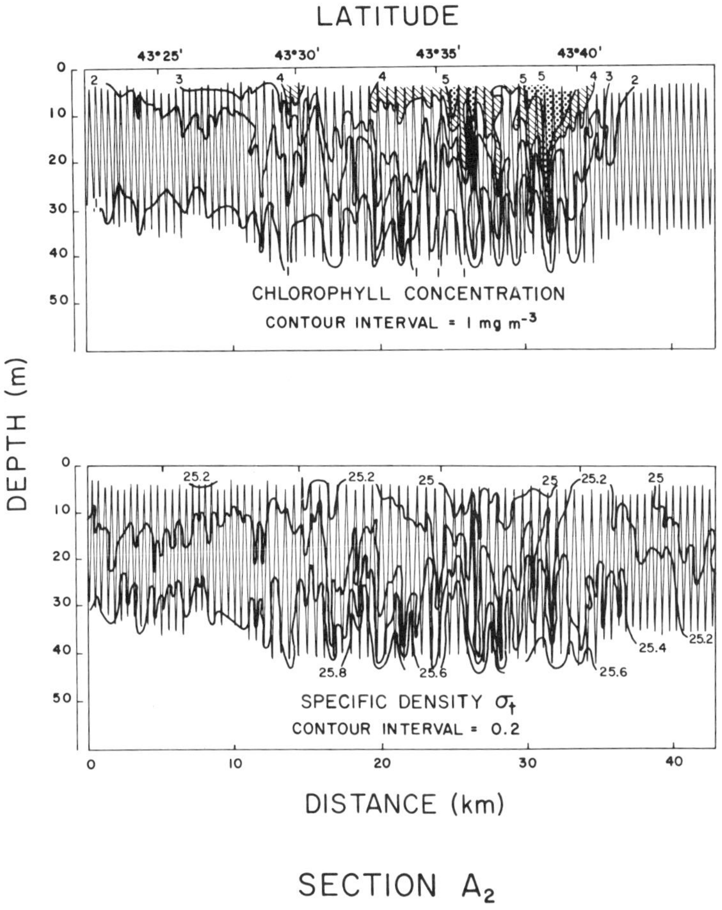


Figure 7. Contour plots of chlorophyll *a* concentration and sigma-t on graphs of depth against distance for a Batfish transect in the eastern Gulf of Maine parallel to the coastline of Nova Scotia. The sawtooth pattern indicates the Batfish track (from Denman and Herman, submitted).

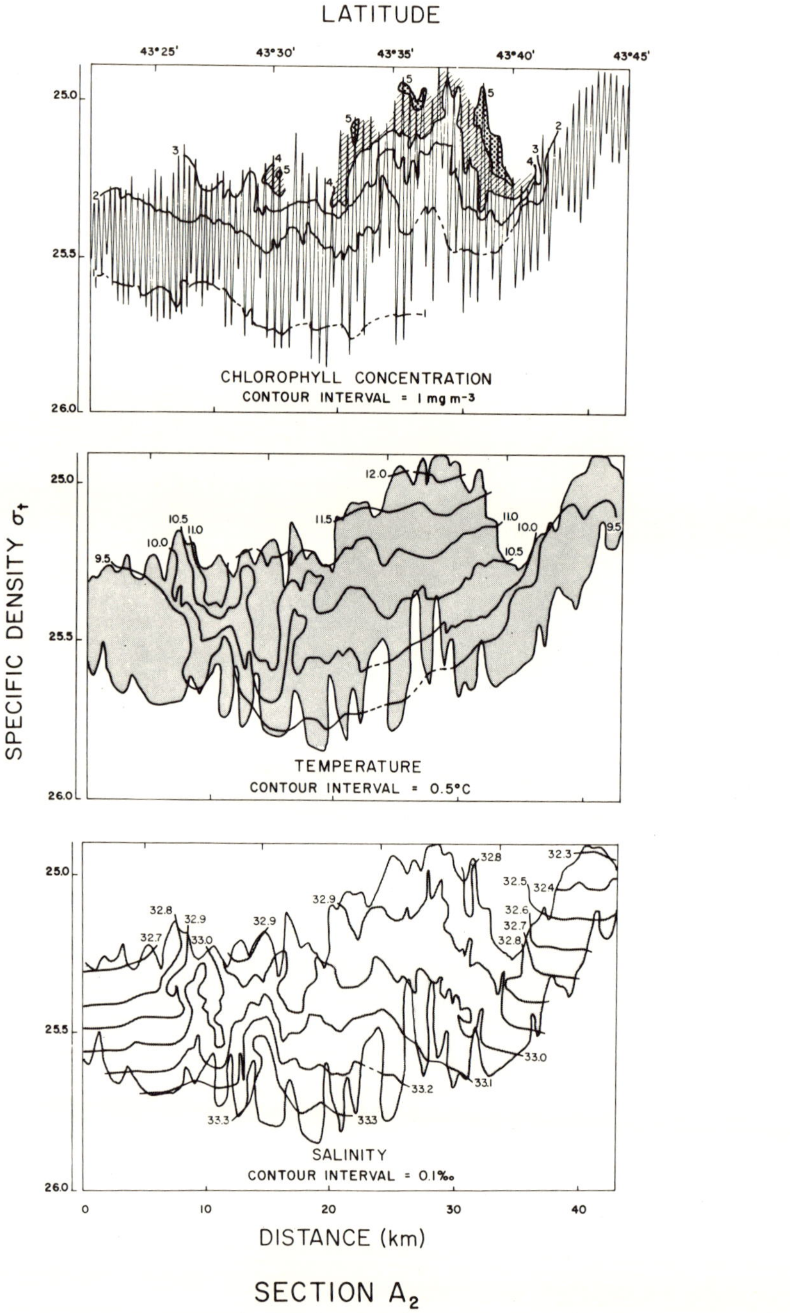


Figure 8. Contour plots of chlorophyll *a* concentration, temperature and salinity on graphs of sigma-t against distance for the transect of Figure 5 (from Denman and Herman, submitted).

Several other features in the data now become apparent. Firstly, the trace of the upper limit of the Batfish pattern represents the $\sigma_t$ values at 4 to 5 m depth. Thus the high surface chlorophyll concentrations (>4 mg $m^{-3}$) between latitudes 43°32' and 43°40'N in Figure 8 obviously occur in a lens of anomalously light, warm surface water. Secondly, the stratification at the center of the transect results from the vertical gradients of both salinity and temperature, but at the ends of the transect it is a function mainly of salinity. Thirdly, the large distortions in temperature and salinity in the first half of the transect are density-compensated and are not reflected at all in the chlorophyll structure, both suggestive of horizontal intrusions rather than vertical motion. Analysis by empirical orthogonal functions of maps of density versus distance is appealing, but inspection of Figure 8 shows that only for $\sigma_t \approx 25.3$ could one generate with confidence series that would be a significant fraction of the whole transect. The surface anomaly, which would be of interest to biologists, is delineated by waters of $\sigma_t < 25.3$; they are found over only a small portion of the 43 km transect such that a series on the $\sigma_t = 25.0$ surface, for example, would be too short to be useful.

## RELEVANT PHYSICAL MEASURES OF TURBULENT MIXING

The new capabilities for multidimensional mapping of the upper ocean allow us to see, for the first time with confidence, the spatial distribution of planktonic organisms and their physical and chemical environment. However, in many of the biologically more productive regions of the ocean, such as shallow seas, continental shelves, or coastal upwelling areas, mixing and advection are critical to the dynamics of the planktonic food chain (e.g. Pingree, this volume; Wroblewski, 1977; Walsh, 1977; Winter *et al.*, 1975). Although biologists have long been aware of their need for physical measurements of the advective and dispersive motions in the sea, they have often relied on rather crude parameters. To describe the vertical mixing, for example, they have used parameters detailed in Sverdrup, Johnson and Fleming (1942): a static stability criterion $E \approx 10^{-3}\ d\sigma_t/dz$, and a vertical eddy viscosity dependent on the wind speed that dates back to Ekman (1905). However, the advances in physical oceanography in the last decade have provided new information and new instruments that are of immediate use to biological oceanographers. In a recent NATO Advanced Study Institute on Modelling and Prediction of the Upper Layers of the Ocean (E. B. Kraus, editor), Woods (1977) reviewed the parameterization of the unresolved motions responsible for most of the turbulent mixing. These concepts are relevant to the study of biological pattern or structure, in the following section, we expand on several that offer the greatest insights into the physical control of biological structure.

In the conservation equation for any parameter $\Theta$ in the ocean, the net flux of $\Theta$ in the $x$ direction due to fluctuating motion is usually parameterized by the product of a turbulent transfer coefficient $K^{\Theta}$ and the gradient $\partial\Theta/\partial x$. If represents a velocity component of the water, then $K_m$ is called a viscosity, otherwise a diffusivity. Theories of plankton patchiness originating from Kierstead and Slobodkin (1953), and reviewed by Okubo and by Platt (this volume), use a horizontal eddy diffusivity $K_x$. Theories of vertical migration and horizontal-vertical interaction reviewed by Evans and by Kullenberg (this volume) use a vertical diffusivity or mass transfer coefficient $K$. Estimates of these parameters have been acquired largely from dye detection experiments; a summary of the results for $K_x$ in the upper ocean up to 1971 is given by Okubo (1971). The development of towed vehicles like the Batfish, equipped with *in situ* fluorometers, allows us to follow the evolution of dye patches in three dimensions (Herman, 1975), a potential that might in turn allow us to study further the interactions of $K_x$ and $K$.

Kullenberg (this volume) reviewed models of vertical mixing of the upper ocean by wind forcing. He relates $K$ to wind speed at 10 m, and to vertical gradients of current and density immediately below the sea surface. Although the near surface layer is extremely important biologically, the summer subsurface chlorophyll maximum usually occurs near the base of the mixed layer, in the seasonal thermocline. There, the turbulent mixing is less directly forced by the wind, but rather is dependent on the current shear and the static stability. They can be expressed together as a gradient Richardson number

$$Ri = -\frac{(g/\bar{\rho})\,\partial\rho/\partial z}{(\partial u/\partial z)^2}$$

where the numerator is a measure of the static stability and the denominator a measure of the rate of generation by turbulent kinetic energy from the vertical shear of the horizontal current. Kullenberg discusses the dependence of the vertical mass transfer coefficient $K$ on $Ri$ and the relationship between $K$ and $K_m$, the momentum transfer coefficient.

In addition, however, we wish to map the occurrence of turbulent and stable regions in a manner analogous to the maps of biomass, temperature and salinity discussed earlier. According to the theorem of Miles and Howard (1964), active turbulence will occur whenever $Ri \leq \frac{1}{4}$ long enough for instabilities to grow in the shear flow. Thus maps of $Ri$ in space and in time would be of great value in delineating turbulent regions. The problem in the ocean is to obtain simultaneous estimates of $\partial\rho/\partial z$ and $\partial u/\partial z$ over sufficiently small vertical separations to resolve the scales on which the mixing actually occurs.

Jones (1973) succeeded in measuring current shear with two current meters lowered on a taut wire 10 m apart. He took a CTD before and after each current profile, thereby allowing the calculation of a *Ri* profile with one point every 10 m. Halpern (1974) computed *Ri* values every 15 minutes for a month over 3 depth intervals from 4 current meters on a mooring. Richardson numbers varied over a range of $\sim 10^3$, spending periods as long as a day below the critical value (chosen by Halpern to be 1 rather than ¼). However, these *Ri* values probably have high uncertainties because of near surface effects. A more reliable technique is that of Johnson, Van Leer and Mooers (1976) who calculated *Ri* on 5 m spacing from simultaneous current and CTD measurements with an unattended continuous profiler (the Cyclesonde) that can cycle roughly once each hour for periods up to about 5 days. In Figure 9 we show one of their series as a depth-time contour plot of *Ri* with depth profiles of the time-averaged Richardson number and of the cumulative distribution of *Ri* values. The mean plot suggests greater overall stability than was actually present. They estimated a relative error in *Ri* values of 50-75%, but for their observed dynamic range >10 on the continental shelf, they felt that that error was acceptable.

The Richardson number describes the tendency for turbulence to occur, but a more dynamic indicator of turbulent activity is the rate of viscous dissipation of turbulent kinetic energy $\varepsilon = \langle \frac{1}{2}\nu(\partial u_i/\partial x_j + \partial u_j/\partial x_i)^2 \rangle$ (Tennekes and Lumley, 1972), where $\nu$ is the kinematic viscosity and $\langle \; \rangle$ represents appropriate spatial averaging. For a scalar parameter $\Theta$, the analogous quantity is $\chi_\Theta$, the rate of destruction of $\Theta^2$ variance or structure by turbulent mixing of the water mass. In the recent models of plankton patchiness by Denman and Platt (1976) and Denman, Okubo and Platt (1977) that use a spectral representation, it should be noted that the relevant parameter describing the turbulent processes is no longer $K$ but $\varepsilon$ or $\chi_\Theta$.

Measurement of $\varepsilon$ in the ocean has only been possible since the advent of instrumentation that can resolve the fine (1 to 50 m) and microstructure (1 m to 1 mm) scales. The first estimates of $\varepsilon$ from small scale velocity measurements in the ocean (by a spectral technique) were made by Grant, Stewart and Moilliet (1962). In a further study, Nasmyth (1970) showed that the intermittent nature of $\varepsilon$ could be used to estimate the percent of the ocean fully turbulent at various depths. The development of vertical profilers to measure the microscale gradient of horizontal velocity now allows an estimate of $\varepsilon$ directly from the definition (Osborn, 1974). These measurements have been refined to yield routine vertical profiles of $\varepsilon$ with a 2-m resolution, such as those shown in Figure 10 (from Crawford, 1976). In addition, Osborn and Cox (1972) developed a new method to calculate $K$, from a free-falling vertical

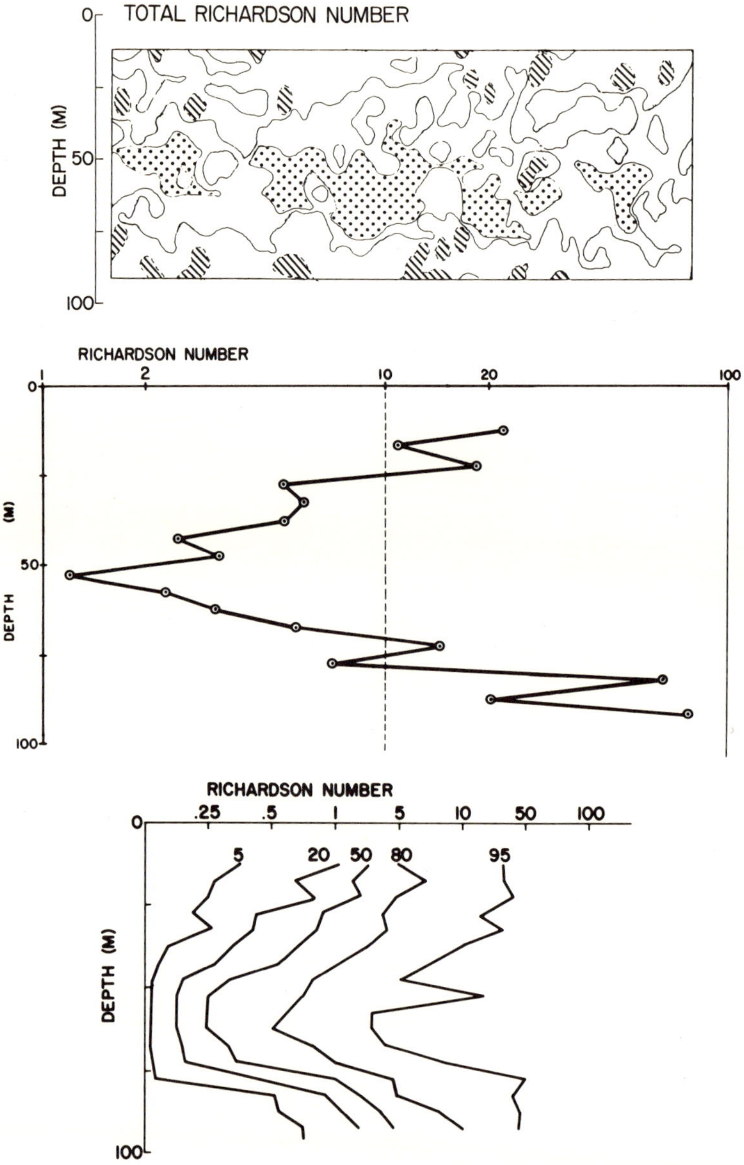


Figure 9. Gradient Richardson numbers calculated from a Cyclesonde moored in 100 m of water off the coast of Oregon. Top panel: time-depth contours (contour lines at $Ri$=0.25, 1.0 and 10.0; areas with $Ri \leqslant 0.25$ are stippled; areas with $Ri \geqslant 10.0$ are hatched). Middle: $Ri$ for the time-averaged horizontal velocity versus depth. Bottom: cumulative distribution of instantaneous $Ri$ versus depth. (Percentile isopleths are plotted; i.e., for line "x", $Ri \leqslant Ri$ ("X") for x% of the values at a particular depth.) With permission from Johnson, Van Leer and Mooers (1976).

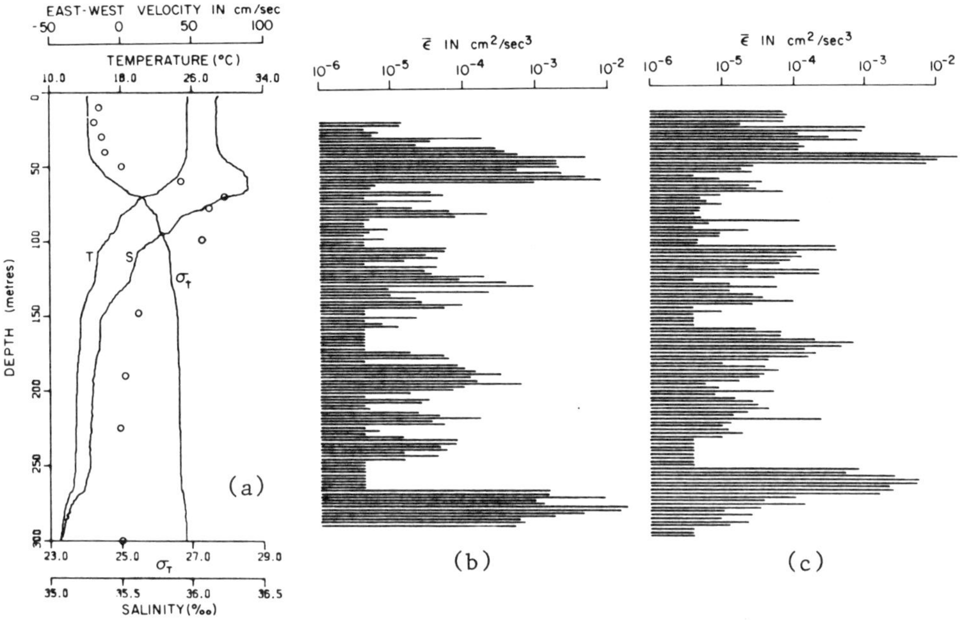


Figure 10. Vertical profiles of dissipation ε in the Atlantic Equatorial Undercurrent (b and c). Zonal current relative to 300 m, temperature, salinity and sigma-t for the same day (a). With permission from Crawford (1976).

temperature profiler. Their estimates and those of subsequent investigators (reviewed by Gregg, 1975) are all smaller than previous estimates from dye studies, suggesting, at the very least, that $K$, like $K_x$ (Okubo, 1971), is a scale-dependent parameter.

The microstructure profilers for velocity, temperature and salinity developed recently by Simpson (1972) and Osborn (1974) have the potential to measure $Ri$ on scales down to ∿30 cm. In fact, Simpson (1972) obtained such a profile from Loch Ness where salinity need not be measured. But Simpson (1975) was unsuccessful in salt water because of the difficulties in matching the velocity and CTD sensors. Certainly, reliable estimates of $Ri$ calculated from velocity, temperature and salinity, and averaged over 1 m scales, will be available in a year or two.

With the capability to map $Ri$ and ε on scales down to a meter in the vertical, we will be better able to relate biomass maps to dynamic physical processes. Two general results should evolve from this capability: firstly, a statistical small scale picture of how high biomass areas relate to areas of high and low intensities of turbulent mixing; and secondly, a geographical and seasonal "climatology" of the occurrence of physically active depths and regions, and the related biomass structure to be expected.

## CONCLUSIONS

We have seen in the last decade a dramatic development of instrumentation for continuous measurement of the physical and biological structure of the marine ecosystem. While there must be a continuing search for instruments that measure these parameters with increasing precision and spatial bandwidth, we also need to extend our measurement and analysis capabilities into new areas.

For data comprised of series in one dimension, conventional spectral analysis techniques have been extremely useful in describing approximately linear interactions between parameters. However, we need methods suitable for non-linear interactions between non-stationary variables. Also, the multidimensional data maps provided by undulating samplers, while providing us with many new subjective insights, do not yet yield gracefully to any simple and objective analytic treatment. We have outlined several currently available procedures for analysis of these data, but considerable effort must continue to be directed toward extracting the maximum information (and the minimum misinformation) from multidimensional data sets.

We will also want to extend the range of variables that are routinely measured. The new ability of physical oceanographers to make reliable measurements of the fine- and microscale structure of velocity, temperature and salinity has resulted in markedly improved estimates of turbulent mixing parameters. These measurements have not yet been fully utilized to explain the patterns of productivity in the aquatic ecosystem. The works of Pingree and of Holligan (this volume) on tidal fronts and plankton blooms exemplify the value of a synthesis of biological measurements, turbulent rate parameters, and predictive theory.

The biological parameters that can, at present, be measured continuously are all essentially estimates of standing stock. There are no continuous measures of biological rates analogous to the physical measures of currents, energy dissipation, and boundary fluxes. Per capita rates of growth, nutrient uptake, grazing and sinking may have scales of variability similar to those of biomass (Therriault and Platt, in press), but present measurement techniques usually permit only widely-spaced point estimates. Ideally, there must be future work on continuous biological rate measurements from towed bodies. Technological difficulties, however, make this a relatively more distant objective.

ACKNOWLEDGEMENTS

We wish especially to thank our coworkers, Carl Boyd and Alex Herman, for their contributions to our own research. Bill Crawford, Tom Osborn, Walter Johnson, John Van Leer and Chris Mooers all kindly allowed us to present their data; and John Steele and Grant Gardner made valuable editorial comments.

## SUMMARY

Recent advances in studying the variability in the marine environment have been aided by new developments in instrumentation and data analysis techniques. Sampling windows for biological oceanography, while they remain narrower than those used for physical oceanography, nevertheless have become sufficiently broad that sophisticated analysis procedures are required to extract the maximum information from the data. Towed undulating sampling vehicles perhaps offer us the greatest challenge, both in their ultimate potential and in their requirement for new techniques of data reduction. We discuss their sampling characteristics as well as the capabilities of sensors for fluorescence estimation of phytoplankton chlorophyll biomass and for electronic and acoustical estimation of zooplankton numbers and biomass. Two dimensional data maps provided by undulating samplers may be interpreted various ways: we give examples for data compression by objective fitting techniques and for elimination of confounding effects of internal waves by plotting on density surfaces. Finally, we review new techniques for estimating turbulent mixing and stability parameters critical to the creation and control of biological structure: eddy transfer coefficients, turbulent energy dissipation rates and gradient Richardson numbers.

## REFERENCES

Aiken, J., R. H. Bruce and J. A. Lindley, 1977: Ecological investigations with the undulating oceanographic recorder: the hydrography and plankton of the waters adjacent to the Orkney and Shetland Islands. Mar. Biol., 39, 77-91.

Angel, M. V., 1977: Windows into a sea of confusion: sampling limitations to the measurement of ecological parameters in oceanic mid-water environments. In Oceanic Sound Scattering Prediction, N. R. Anderson and B. J. Zahuranec (eds.), pp. 217-248, Plenum Press, New York.

Boyd, C. M., 1973: Small scale spatial patterns of marine zooplankton examined by an electronic in situ zooplankton detecting device. Netherlands Journal of Sea Research, 7, 103-111.

Bretherton, F. P., R. E. Davis and C. B. Fandry, 1976: A technique for objective analysis and design of oceanographic experiments applied to MODE-73. Deep-Sea Res., 23, 559-582.

Bruce, R. H. and J. Aiken, 1975: The undulating oceanography recorder - a new instrument system for sampling plankton and recording physical variables in the euphotic zone from a ship underway. Mar. Biol., 32, 85-97.

Crawford, W. R., 1976: Turbulent energy dissipation in the Atlantic Equatorial Undercurrent. Ph.D. Thesis, University of British Columbia, Vancouver, Canada.

Dauphinee, T. M., 1977: Zooplankton measurements using a conductance cell. MTS-IEEE Oceans '77 Conference Record, Vol. 2, 39B.

Davis, R. E., 1976: Predictability of sea surface temperature and sea level pressure anomalies over the North Pacific Ocean. J. Phys. Oceanogr., 6, 249-266.

Denman, K. L., 1976: Covariability of chlorophyll and temperature in the sea. Deep-Sea Res., 23, 539-550.

Denman, K. L., 1977: Short term variability in vertical chlorophyll structure. Limnol. Oceanogr., 22, 434-441.

Denman, K. L. and A. W. Herman, 1978: Space-time structure of a continental shelf ecosystem measured by a towed porpoising vehicle. Submitted.

Denman, K., A. Okubo and T. Platt, 1977: The chlorophyll fluctuation spectrum in the sea. Limnol. Oceanogr., 22, 1033-1038.

Denman, K. L. and T. Platt, 1975: Coherences in the horizontal distributions of phytoplankton and temperature in the upper ocean. Mem. Soc. r. Sci. Liege, 7, 19-30.

Denman, K. and T. Platt, 1976: The variance spectrum of phytoplankton in a turbulent ocean. J. Mar. Res., 34, 593-601.

Denman, K. and T. Platt, 1978: Time series analysis in marine ecosystems. In Time Series and Ecological Processes, H. L. Shugart, Jr. (ed.), pp. 227-244, SIAM, Philadelphia.

Dessureault, J. G., 1976: "Batfish": a depth controllable towed body for collecting oceanographic data. Ocean Eng., 3, 99-111.

Ekman, V. W., 1905: On the influence of the earth's rotation on ocean currents. Ark. Math. Astr. och Fys., 2, Stockholm.

Fasham, M. J. and P. R. Pugh, 1976: Observations on the horizontal coherence of chlorophyll a and temperature. Deep-Sea Res., 23, 527-538.

Gandin, L. S., 1963: Objective Analysis of Meteorological Fields. Gidromet. Jerusalem, Israel Program for Scientific Translations (1965).

Garrett, C. and W. Munk, 1975: Space-time scales of internal waves: a progress report. J. Geophys. Res., 80, 291-297.

Grant, H. L., R. W. Stewart and A. Moilliet, 1962: Turbulence spectra from a tidal channel. J. Fluid Mech., 12, 241-263.

Gregg, M. C., 1975: Oceanic fine and microstructure. Rev. Geophys. and Space Phys., 13, 586-591.

Halpern, D., 1974: Observations of the deepening of the wind-mixed layer in the northeast Pacific Ocean. J. Phys. Oceanogr., 4, 454-466.

Hardy, A. C., 1936: The continuous plankton recorder. Discovery Reports, 11, 457-509.

Herman, A. W., 1975: Chlorophyll and dye detection with the Variosens fluorometer. Rep. Series BI-R-75-2/Feb. 1975. Bedford Inst. of Oceanography, Dartmouth, Nova Scotia.

Herman, A. W., 1977: In situ chlorophyll and plankton measurements with the Batfish vehicle. MTS-IEEE Oceans '77 Conference Record, Vol. 2, 39D.

Herman, A. W. and K. L. Denman, 1977: Rapid underway profiling of chlorophyll with an in situ fluorometer mounted on a 'Batfish' vehicle. Deep-Sea Res., 24, 385-397.

Johnson, W. R., J. C. Van Leer and C.N.K. Mooers, 1976: A cyclesonde view of coastal upwelling. J. Phys. Oceanogr., 6, 556-574.

Jones, J. H., 1973: Vertical mixing in the equatorial undercurrent. J. Phys. Oceanogr., 3, 286-296.

Kelley, J. C., 1975: Time-varying distributions of biologically significant variables in the ocean. Deep-Sea Res., 22, 679-688.

Kierstead, H. and L. B. Slobodkin, 1953: The size of water masses containing plankton blooms. J. Mar. Res., 12, 141-147.

Liebelt, P. B., 1967: An Introduction to Optimal Estimation. Addison-Wesley Publ. Co., Reading, Massachusetts.

Lorenz, E. N., 1956: Empirical orthogonal functions and statistical weather prediction. Scientific Report No. 1, Statistical Forecasting Project, MIT, 48 pp.

Lorenzen, C. J., 1966: A method for the continuous measurement of in vivo chlorophyll concentration. Deep-Sea Res., 13, 223-227.

Mackas, D. L., 1977: Horizontal spatial variability and covariability of marine phytoplankton and zooplankton. Ph.D. Thesis, Dalhousie University, Halifax, Canada.

Maddux, W. S. and J. W. Kanwisher, 1965: An in situ particle counter. Limnol. Oceanogr., 10, R162-R168.

Miles, J. W. and L. N. Howard, 1964: Note on a heterogeneous shear flow. J. Fluid Mech., 20, 331-336.

Nasmyth, P. W., 1970: Ocean turbulence. Ph.D. Thesis, University of British Columbia, Vancouver, Canada.

Okubo, A., 1971: Oceanic diffusion diagrams. Deep-Sea Res., 18, 789-802.

Osborn, T. R., 1974: Vertical profiling of velocity microstructure. J. Phys. Oceanogr., 4, 109-115.

Osborn, T. R. and C. S. Cox, 1972: Oceanic fine structure. Geophys. Fluid Dyn., 3, 321-345.

Pieper, R. E., 1977: Some comparisons between oceanographic measurements and high-frequency scattering of underwater sound. In Oceanic Sound Scattering Prediction, N. R. Anderson and B. J. Zahuranec (eds.), pp. 667-678, Plenum Press, New York.

Platt, T., 1972a: Local phytoplankton abundance and turbulence. Deep-Sea Res., 19, 183-188.

Platt, T., 1972b: The feasibility of mapping the chlorophyll distribution in the Gulf of St. Lawrence. Fish. Res. Bd. Canada, Tech. Rep. 332.

Platt, T. and K. L. Denman, 1975: Spectral analysis in ecology. Ann. Rev. Ecol. Syst., 6, 189-210.

Powell, T. M., P. J. Richerson, T. M. Dillon, B. A. Agee, B. J. Dozier, D. A. Gcdden, and L. O. Myrup, 1975: Spatial scales of current speed and phytoplankton biomass fluctuations in Lake Tahoe. Science, 189, 1088-1090.

Sameoto, D. and S. Paulowich, 1977: Zooplankton detection with high frequency sonar. MTS-IEEE Oceans '77 Conference Record, Vol. 2, 39A.

Simpson, J. H., 1972: A free fall probe for the measurement of velocity microstructure. Deep-Sea Res., 19, 331-336.

Simpson, J. H., 1975: Observations of small scale vertical shear in the ocean. Deep-Sea Res., 22, 619-627.

Sverdrup, H. U., M. W. Johnson and R. H. Fleming, 1942: The Oceans: their physics chemistry and general biology. Prentice-Hall Inc., Englewood Cliffs, New Jersey.

Tennekes, H. and J. L. Lumley, 1972: A First Course in Turbulence. The MIT Press, Cambridge, Massachusetts.

Therriault, J. C. and T. Platt, 1978: Spatial heterogeneity of phytoplankton biomass and related factors in the near-surface waters of an exposed coastal embayment. Limnol. Oceanogr., in press.

Walsh, J. J., 1977: A biological sketchbook for an eastern boundary current. In The Sea, E. D. Goldberg et al. (eds.), Vol. 6, pp. 923-968, John Wiley, New York.

Winter, D. F., K. Banse and G. C. Anderson, 1975: The dynamics of phytoplankton blooms in Puget Sound, a fjord in the North Western United States. Mar. Biol., 29, 137-176.

Woods, J. D., 1977: Parameterization of unresolved motions. In Modelling and Prediction of the Upper Layers of the Ocean, E. B. Kraus (ed.), pp. 118-140, Pergamon Press, Oxford.

Wroblewski, J. S., 1977: A model of phytoplankton plume formation during variable Oregon upwelling. J. Mar. Res., 35, 357-394.

# THE APPLICATION OF PARTICLE COUNTING TO AN UNDERSTANDING OF THE SMALL-SCALE DISTRIBUTION OF PLANKTON

P. R. Pugh

Institute of Oceanographic Sciences
Wormley, Godalming
Surrey, U.K.

## INTRODUCTION

In the context of the modeling of the marine ecosystem, a knowledge of the spatial distribution and patchiness of the organisms is of great importance for, as Platt, Denman and Jassby (1977) considered, the spatial heterogeneity of plankton is an essential element for system stability and it provides the raw materials by which new system structures can emerge, through the nonlinearities in the state equations. The continuous measurement of chlorophyll *a* fluorescence has been used for the investigation of the spatial distribution of phytoplankton (e.g. Platt, 1972; Denman and Platt, 1975; Fasham and Pugh, 1976), and *in situ* particle counters for the distribution of the zooplankton size ranges (e.g. Boyd, 1973). However, Haury *et al.* (this volume) have pointed out that these crude measures of biomass give no information on the species composition which makes up this biomass, and that it is more important in marine ecology to understand how the individual species variabilities interact. In addition, there are other drawbacks to the use of chlorophyll *a* fluorescence as a measure of phytoplankton biomass, some of which are associated with the variability in the relationship between the fluorescence levels and the actual concentration of pigment present. Strickland (1968) noted large interspecific variations in this relationship, and Kiefer (1973b) found intraspecific changes according to the growth state of the cells. Light inhibition of cellular fluorescence also can produce up to tenfold variations in the fluorescence number; i.e., units of fluorescence per unit chlorophyll *a*, according to the intensity and variation in the incident irradiance (Kiefer, 1973a; Loftus and Seliger, 1975). Thus, without frequent calibration, often it is difficult to relate fluorescence levels directly to chlorophyll *a* concentrations, and thereby biomass, particularly in areas where large changes in the physical and biological conditions occur; e.g. upwelling and frontal regions (see Holligan, this volume).

Steele (this volume) pointed out the importance of considering the food chain relationships of marine organisms, as these are factors which may affect the spatial structure of the ecosystem. For this, a knowledge of the size distribution and taxonomy of the phytoplankton and zooplankton species present would be required. Further, Steele and Frost (1977) concluded that the size structure of the plankton community is as important as, and probably more significant than, the total biomass in determining the modes of energy transfer between the various trophic levels. In this context, it is the purpose of this paper to consider briefly some aspects of the use of particle counters for the study of the small-scale distribution and spatial variability of particle size spectra in the oceans, with particular reference to upwelling and frontal regions, while recognizing the drawbacks in the method *per se*, especially in its inability to provide data on the specific identification of these particles.

## THE USE OF PARTICLE COUNTERS IN MARINE BIOLOGY

Particle counters, especially the Coulter Counter, have been in use in the marine field for over fifteen years. One important application for these instruments has been in the study of the feeding patterns of herbivorous zooplankton. The earlier work on particle size selection by herbivores was summarized by Parsons and LeBrasseur (1970), who considered further the use of these size spectra for the investigation of the relative importance of different available foods to the secondary and tertiary producers. In the hope that future research would make it possible to extrapolate the results obtained from feeding experiments to natural assemblages of plankton, these authors concluded that information on, for instance, the patchiness of food organisms, and the concentration and size spectra of predator-prey associations *in situ* would be required. Some such work has been carried out in recent years but, as Frost (1977) points out, the interpretation of *in situ* particle size spectra may be complicated by certain factors; for example, it is not clearly understood how the different species and sizes of herbivores interact while feeding on natural assemblages of phytoplankton. Each individual herbivore can feed on a certain size range of particles, but when many species of grazers are present, it would be difficult to assess the impact of an individual species unless it numerically predominated.

Another major line of research, using particle counters, has been in the study of the distribution of seston in the world's oceans. As an example, Sheldon *et al.* (1972) studied the frequency distribution of particle sizes for both surface and deep waters in the Atlantic and Pacific Oceans, and they found that the shapes of the size spectra varied predictably both geographically and with depth. Their data led them to the hypothesis that there were roughly equal concentrations of material present in all the size ranges from bacteria to whales, such that the number of particles in any size range was a decreasing function of the particle's diameter. If this pattern of standing stock is to be maintained in the oceans, then it is necessary for the rate of particle production to have an inverse relationship with particle size, and this appears to be borne out from the available data. Kerr (1974) proposed a simple model, based on trophic processes, to explain these apparent regularities in the size composition of the pelagic ecosystem and, recently, Platt and Denman (1977) produced a more rigorous model to calculate the steady state distribution of biomass as a function of organism size.

Margalef and his coworkers (e.g. Margalef, 1974; Estrada and Vallespinòs, 1975) have made extensive studies of the particle size distributions in the upwelling region off northwest Africa, and their work is of especial importance as the particle counts were carried out in conjunction with taxonomic and other studies. Most of these and other data have been obtained from discrete water samples, and Margalef (1974) recognized that in order to study the small-scale distribution of seston it was desirable to obtain particle counts continuously, using an 'on-line' system. Unfortunately, he found that too much electrical noise was generated in his system when water was passed continuously through the sensors. However, other instruments for the continuous counting of particles in the zooplankton size range have been developed, e.g. an optico-electronic system for particles in the 1-15 mm range (Cooke *et al.*, 1970) and an *in situ* counter based on the electrical principle (Maddux and Kanwisher, 1965; Boyd and Johnson, 1969). Boyd (1973) described some of the results obtained using this latter instrument for the horizontal and vertical distribution of particles in the 531-2550 μm size range. Recently, a shipborne modification of this instrument has been used by Mackas (1976) during the Fladen Ground Experiment (FLEX). These data are apparently the first on which power spectra could be calculated and the small-scale distribution of the zooplankton-sized particles investigated, and the results of these analyses and for some more recent data are discussed by Denman (this volume).

## THE SMALL-SCALE DISTRIBUTION OF PARTICLES IN THE PHYTOPLANKTON SIZE RANGE

Earlier it was suggested that, although the measurement of chlorophyll *a* fluorescence is a useful tool for the investigation of the spatial distribution and patchiness of phytoplankton, such data should be treated with some caution because of the variable relationship between fluorescence levels and biomass, as calculated from the chlorophyll *a* concentration. A further problem in the measurement, by this method, of the biomass of material available to the herbivorous zooplankton lies in the fact that it does not account for the large numbers of colorless heterotrophic cells which also may be present. Some data (Holligan, personal communication) from a station in the middle of the English Channel may illustrate this point (Figure 1). Size spectra in the 2-150 μm diameter range were obtained from water samples collected from the top and bottom mixed layers and at the fluorescence peak in the thermocline. It is evidence that the concentration of particles of any size is low in the top mixed layer and that this is reflected by the low chlorophyll *a* concentration there. The slightly higher fluorescence levels in the bottom mixed layer are probably associated with the peak in the particle size spectrum for particles of ca. 56 μm diameter, which were identified from filtered samples as being mainly diatom cells. On the other hand, the size spectrum for the particles in the thermocline peak in fluorescence was completely dominated by a large concentration of ca. 9 μm diameter cells. During the vertical profiling of the fluorescence levels (Figure 1, lower) using a submersible pump system, discrete water samples were taken at frequent intervals and the concentrations of particles in these were measured. It was not feasible, in the time available, to obtain a complete size spectrum for each sample and so only the 8.0-10.1 μm diameter size range was investigated. It can be seen that the relationship between particle concentrations and chlorophyll *a* fluorescence, which was so evident at the fluorescence peak, breaks down at the shallower depths and that the subsidiary peaks in the particle concentration are not mirrored by changes in the fluorescence level. Microscopic observations of these water samples have shown that the increase in particle concentrations is not necessarily due to the presence of detrital material, which may accumulate in the pycnocline, but to the presence of a large number of colorless heterotophic cells, mainly flagellates and dinoflagellates. It is evidence that such organisms also must play an important role in the trophic relationships of the plankton community and that in order to study the entire spectrum of biomass which may be available to the herbivorous zooplankton a combination of both fluorescence and particle counting methods may be useful.

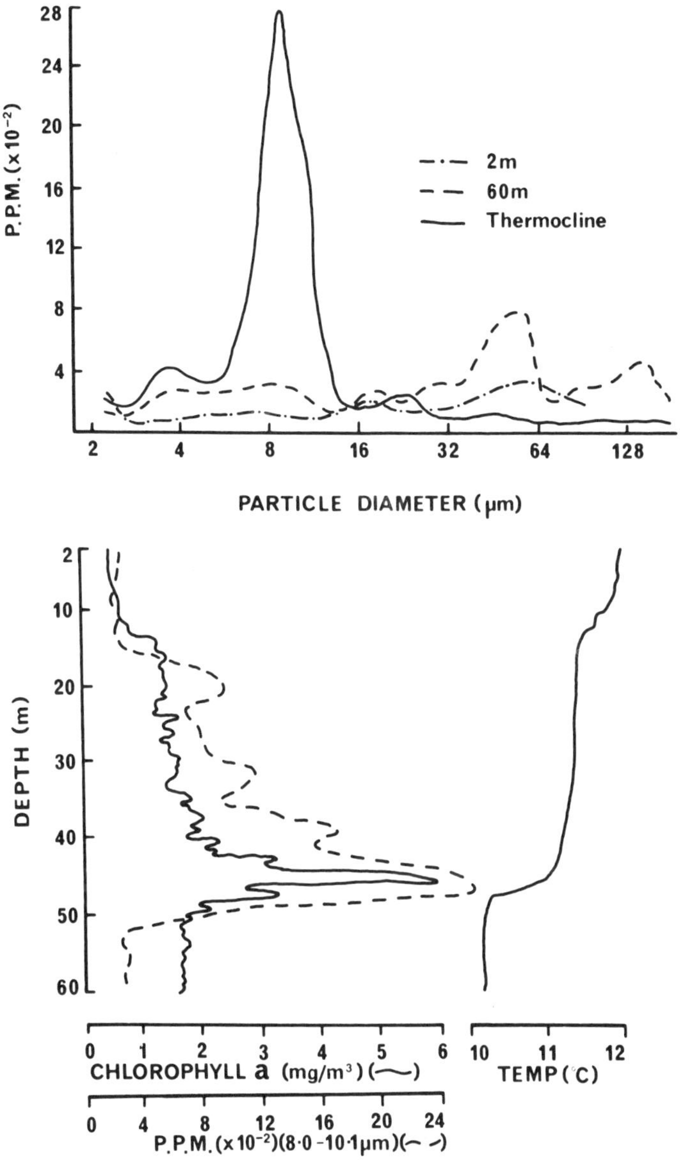


Figure 1. (Top) Particle size spectra (2-150 μm diameter range) for samples taken, at one station, in the top and bottom mixed layers and in the thermocline. (Bottom) Vertical profiles of chlorophyll *a*, temperature and particle counts (8.0-10.1 μm diameter) (from discrete samples) at the same station.

# THE CONTINUOUS OR SEMI-CONTINUOUS COUNTING OF PARTICLES IN THE PHYTOPLANKTON/MICROZOOPLANKTON SIZE RANGES

## Methodology

Some of our results for the horizontal distribution of particles in the 1-600 μm diameter size range, obtained during a 'Discovery' cruise to the upwelling region off northwest Africa, have been discussed by Fasham (this volume). These data were obtained on a semi-continuous basis using a 9-channel Hiac Model PC320 automatic particle size analyzer, which operates on the principle of light beam interruption by a particle. As this type of instrument appears to be new to the marine field and its full potential and precision in this context has only begun to be assessed, a brief account of the operational procedure is pertinent. A detailed account of the use of this and other instruments in the pharmaceutical application of contamination limit testing has, however, been given by Groves and Wana (in press). These authors concluded that, although there were differences in the results obtained, these could be accounted for, in the most part, by differences in the operational procedures and methods of instrument calibration. Although, simplistically, the attenuation of light due to the presence of a particle in a collimated light beam bears a relation to the projected area of that particle, the situation actually is more complicated and the degree of attenuation will also depend on the refractive index of the particle and the wavelength of light used. These factors could be important when, for instance, considering the colorless heterotrophic plankton cells, which may be undersized by the instrument. Because the size ranges used in this present study were broad, it is probable that some degree of sizing inaccuracy will not affect the results to any great extent, although this drawback should not be totally ignored.

There are a variety of sensors available with the instrument, such that, at present, there is a potential to count and size particles in the 1-9000 μm diameter size range, and the ratio of the diameters from the smallest to the largest particle which can be measured by an individual sensor is 1:60. Two sensors (1-60 μm and 10-600 μm equivalent spherical diameter size range) were incorporated into an 'on-line' system, the design of which is shown schematically in Figure 2. Sea water was taken either directly from the ship's pure sea water supply, for horizontal profiling, or from the outflow of a submersible pump (see Fasham and Pugh, 1976) during vertical profiling. Some prefiltering of the sea water was necessary to avoid sensor blockage, and the "plumbing" for this was arranged in two parallel systems so that the flow could be transferred from one to the other without significant interruption, if clogging of a filter occurred. Part of the flow was fed, via a bypass system, to a Turner Model 111 Fluorometer so

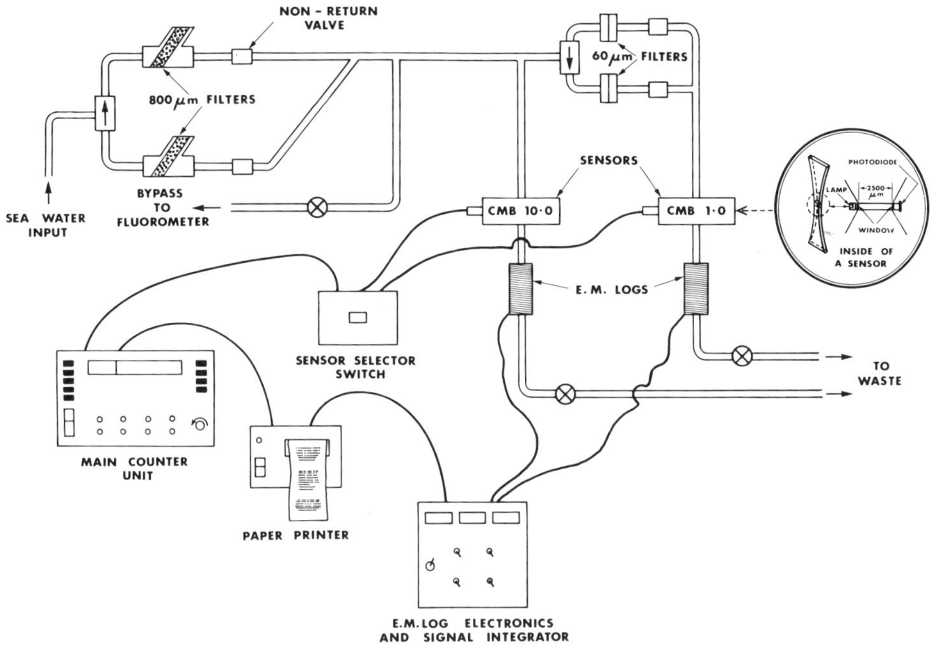


Figure 2. A schematic diagram of the 'on-line' and electronic system for particle counting with the Hiac Model PC320 automatic particle size analyzer.

that a continuous record of chlorophyll *a* fluorescence could be obtained in conjunction with the particle counts. Various other parameters, such as temperature, salinity and nutrient concentrations, were also recorded.

As particle counts in the various size ranges were accumulated over fixed time intervals, usually for 60 seconds every five minutes during horizontal profiling, rather than in a fixed volume, it was important to monitor the flow rate accurately. This was done using single-component electromagnetic logs (designed and built at the Institute of Oceanographic Sciences) and the flow was displayed continuously by digital voltmeters. It was found, during the calibration of the sensors, that there was no significant change in the size calibration curve over a range of flow rates from 150 to 600 ml min$^{-1}$ (for the 10-600 μm sensor) and so a signal integrator also has been constructed so that the total volume of water which passed through the sensor during the counting period can be logged on the printout together with the particle count data. A fuller account of the problems involved with the incorporation of the sensors into an 'on-line' system, and the precision of the instrument will be published elsewhere (Pugh, in preparation).

The channel threshold settings were set up approximately on the basis of the octave or $\log_2$ scale, or subdivisions thereof, as suggested by Sheldon and Parsons (1967). As Platt and Denman (1977) discuss, the use of this scale, where the equivalent spherical diameter at each threshold is double that at the previous one, is both practically expedient and mathematically convenient. The usual way of displaying particle count data is to consider them as a volume concentration (ppm) in each size range, but there are problems associated with the conversion of a surface area or equivalent diameter measurement, within a given range, to an average volume concentration. For the present we have assumed that, on average, there are equal concentrations of particles in logarithmically equal size ranges (Sheldon *et al.*, 1972) so that the volume concentration of particles is inversely proportional to the particle diameter. In this way, the mean volume and diameter for each channel setting can be calculated. In addition, all the count data were corrected for coincidence using the formula:

$$N = n + \frac{S}{2} N \cdot n$$

for a 1 ml sample (Allen and Marshall, 1972) where $N$ is the true count; $n$ the observed count and $S$ is the sensing volume (= $9 \times 10^{-4}$ $cm^3$ for the 10-600 μm sensor).

## The Horizontal Distribution of Particles

The raw data, the spectra of which were discussed by Fasham (this volume), for a transect between 20°50'N, 17°40'W and 22°30'N, 17°40'W, are shown in Figure 3 as particle concentration (ppm) in six logarithmically equal size ranges (i.e., 10-20 μm, 20-40 μm, etc.). Counts were made over 60 second periods every five minutes throughout the transect. Also plotted on the $y/z$ axes are the corresponding data for temperature, chlorophyll *a* fluorescence and nitrate concentration. Although it is difficult to display such large amounts of data, it is possible to discern directly some trends in the results. Large variations occur in the concentration of particles, particularly those in the 10-80 μm size ranges and, in general, there are corresponding variations in the chlorophyll *a* fluorescence levels. The relationships between these particle counts and the other parameters have been investigated by means of the cross spectrum. This consists of two parts, the coherence spectrum, which basically measures the correlation between two variables at a given wavenumber, and the phase spectrum, which measures the offset. The cross spectra and their standard errors were calculated as described in Fasham and Pugh (1976). It was found that the squared coherences between the fluorescence values and the particle counts in the 10-20 μm or 20-40 μm equivalent diameter size ranges showed significant non-zero values for wavelengths between 10 and 100 km or 5 and 100 km respectively, and that the two parameters tended to covary in phase. As would be expected, perhaps, the squared coherences for the larger particle

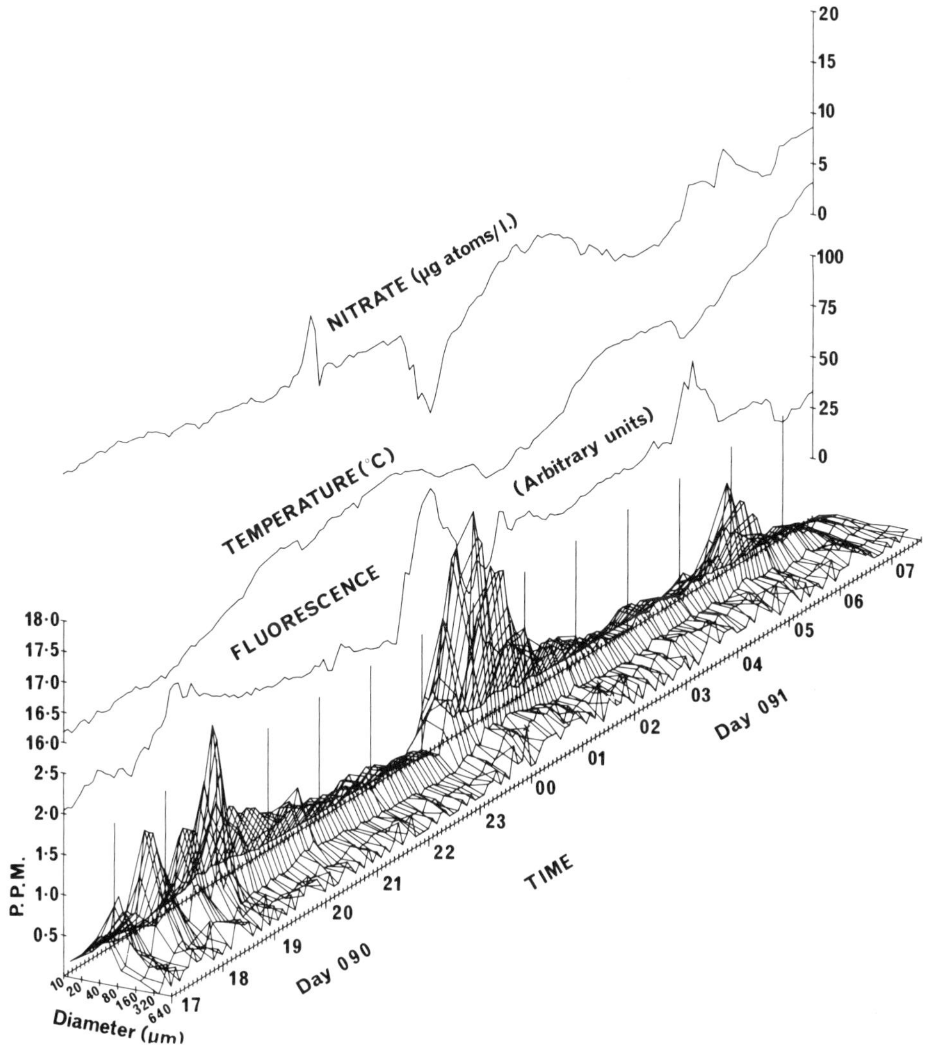


Figure 3. The changes in the particle size spectra, chlorophyll *a* fluorescence, temperature and nitrate concentration in the surface water between 20°50'N, 17°40'W and 22°30'N, 17°40'W.

size ranges and the fluorescence were not significantly different from zero at most space scales. Temperature, however, was only significantly correlated with fluorescence at a wavelength of ca. 15 km, at which point the covariance was $180^{\circ}$ out of phase. Similarly, the nitrate concentration and particle counts in the 10-40 μm size range tended to covary out of phase, and the squared coherences between these two parameters were significantly different from zero at wavelengths approximately between 10 and 100 km.

Although there was general agreement between the variations in the chlorophyll *a* fluorescence and the concentration of particles in the smaller size ranges during this transect, this does not necessarily mean that there is no additional information to be gained from a study of the particle size spectra. This can be illustrated by considering some particle count data collected during a cruise on R/V SARSIA in the English Channel, July 1977. The particle count data, as counts $ml^{-1}$, for the three smallest size ranges, namely 10-20, 20-40 and 40-60 μm equivalent diameter, are shown in Figure 4, together with the temperature and fluorescence profiles. The counts were again made over 60 second periods every five minutes during a 12-hour transect across the Channel between the Île d'Ouessant and Start Point. The hydrography of the area (Pingree *et al.*, 1975) is such that generally, for that time of year, the two ends of the transect are in well-mixed waters while the central half represents a transitional zone between well-mixed and stratified waters. On this occasion such a zonation was not clearly defined as a result, presumably, of the prevailing weather conditions. However, it can be seen that the fluorescence trace follows the general pattern of the particle counts but that it cannot be associated exclusively with any one size range. For instance, during the first part of the transect the fluorescence values are low, and any variations in them are generally reflected by changes in the numbers of 10-20 μm diameter particles. Between 03.30 and 06.30 the fluorescence levels are higher and there is a marked changeover in the abundance of particles in the three size ranges shown. The number of 10-20 μm particles decreases and the changes in the fluorescence values are reflected by variation in the numbers of 20-60 μm diameter particles. At 06.30 a front appears to have been crossed as there is a small but pronounced temperature decrease, with a corresponding rapid decline in the particle counts and fluorescence. These variations in the relative contributions of the different size ranges of particles could be associated with known changes in the phytoplankton population. Thus, although for this particular transect the hydrography of the area was rather complicated, it is clear that there is much small-scale spatial variability in the specific phytoplankton population, and that this variability is not necessarily reflected by changes in the chlorophyll *a* fluorescence readings.

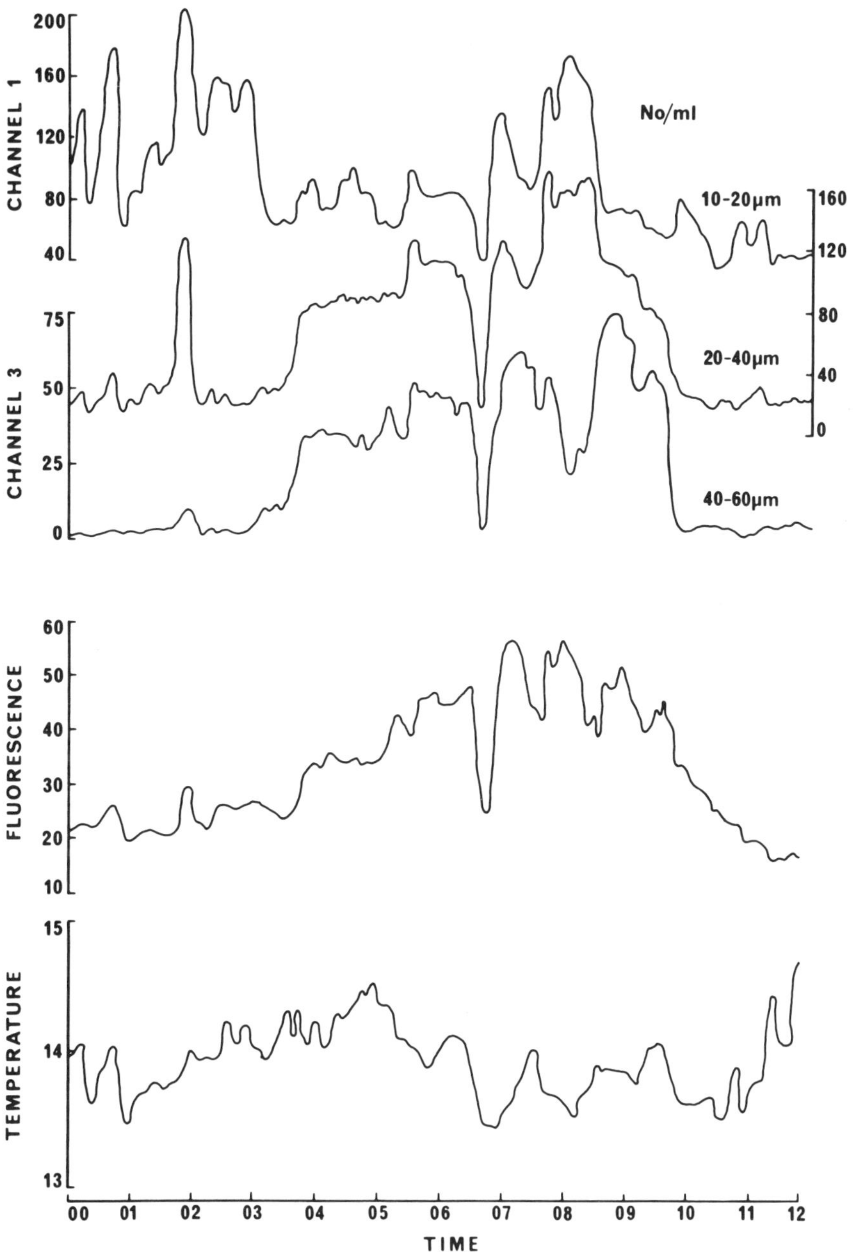


Figure 4. The particle counts, fluorescence and temperature values during a 12-hour transect of the English Channel between the Île d'Ouessant and Start Point.

## The Vertical Distribution of Particles

Such variability in the distribution of the particle size classes is also apparent when vertical profiles of chlorophyll *a* fluorescence and particle concentration are considered. We have carried out profiling of this kind at eighteen stations in the northwest African upwelling region, using the pumping system described in Fasham and Pugh (1976), and at four stations in the North Sea. An example from each locality is presented here. Figure 5 shows the results obtained at 'Discovery' station 9524 (21°05'N, 17°59'W), where vertical profiling was carried out over the 3 to 100 m depth range. The CTD, to which the inlet of the hosing was attached, was lowered and raised at ca. 0.1 m $sec^{-1}$, so that the particle counting period of 50 seconds every minute gives rise to an integrated particle count for a depth range of approximately 5 m. During this experiment the temperature structure of the water column changed quite markedly and this resulted in a change in the vertical distribution of the chlorophyll *a* fluorescence (Figure 5a). There were two fuorescence peaks in the thermocline during the "down" run, but the "up" run showed that

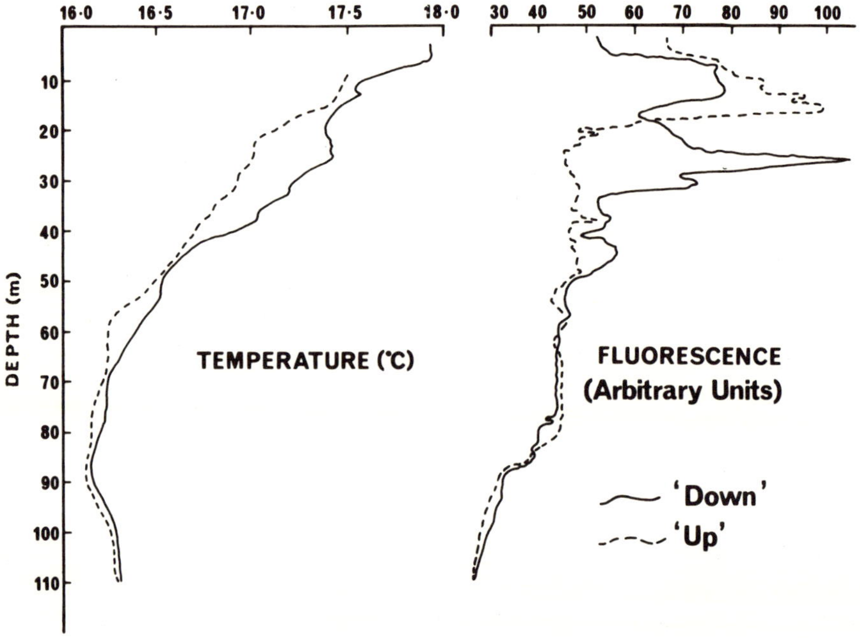


Figure 5a. Vertical profiling at 'Discovery' Station 9524. Temperature and fluorescence values during the "down" and "up" runs.

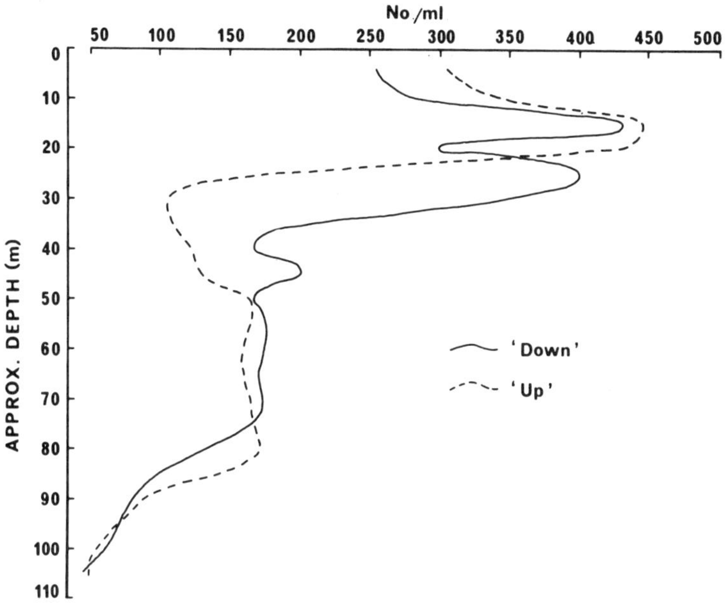


Figure 5b. Vertical profiling at 'Discovery' Station 9524. Total particle counts/ml (10-600 μm diameter).

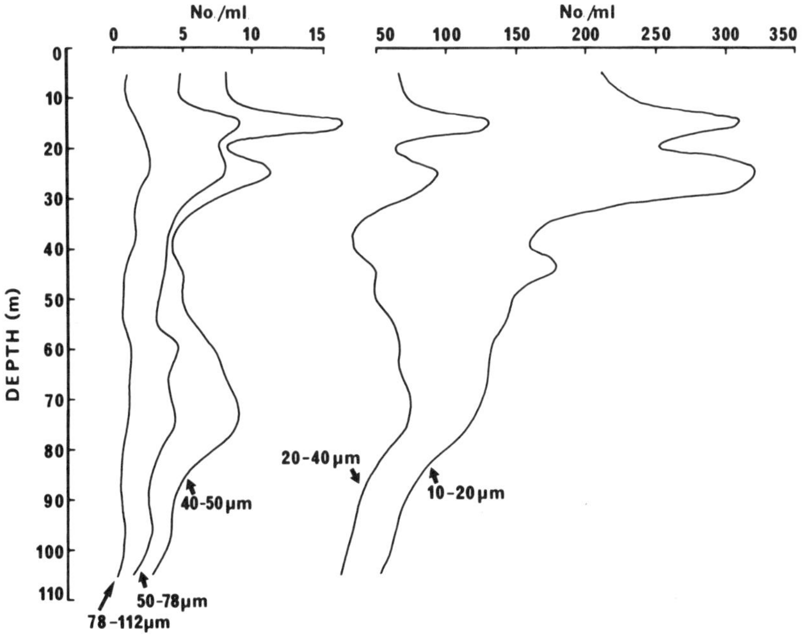


Figure 5c. Vertical profiling at 'Discovery' Station 9524. Individual particle counts/ml for 5 size ranges during the "down" run.

the deeper of these had either disappeared or become amalgamated with the shallower one. The changes in the depth structure of the chlorophyll *a* fluorescence were reflected by differences in the total particle count profiles, even though these are somewhat smoothed (Figure 5b). The counts in the individual particle size ranges (10-112 μm equivalent diameter), for the "down" run only (Figure 5c), show that the relative contributions of each size class to the phytoplankton biomass varies with depth and, as with the horizontal transects, it appears that there are considerable small-scale variations in the phytoplankton population. Although this is not necessarily surprising, as changes in the specific phytoplankton population with depth are well known, it is interesting to note that, within 10 m in the thermocline the two fluorescence peaks appear to be associated with different populations as their size composition of particles shows.

One of the vertical profiles carried out in the North Sea shows this point more strikingly. The particle counts were made over 15 second periods every half minute, which was equivalent to a depth resolution of 1 m. Although the chlorophyll *a* fluorescence changes were small (Figure 6), two peaks were present in the thermocline, while the size spectra of particles in each of these peaks were again very different. Whereas, in the shallow peak, there was a pronounced increase in the number of 10-80 μm equivalent diameter particles, no such increase was evident for the deeper peak. A sample of water taken at this deeper peak showed that most of the cells present were of a size less than 10 μm and

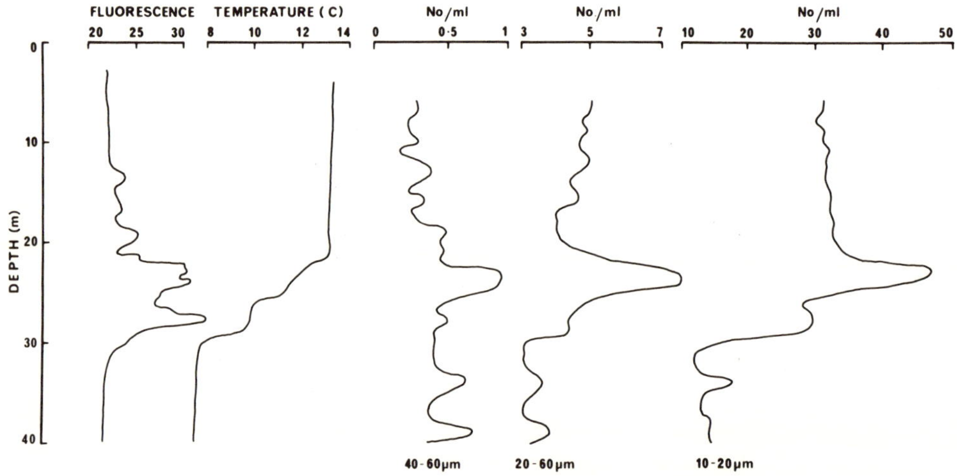


Figure 6. Vertical profiling at a station in the North Sea. Fluorescence, temperature and particle counts/ml in three size ranges.

so were not measured by the sensor. However, it can be inferred that within a depth range of 5 m in the thermocline, there were two distinct specific populations of phytoplankton present. Alternatively, Steele (1964) found that the carbon to chlorophyll ratio of the phytoplankton cells changed with depth and that, in the case of the deep chlorophyll maximum, generally there was an increase in the intracellular chlorophyll concentration relative to its carbon. Other observations have indicated that, at low light intensities, there was this increase in the chlorophyll content of cells, e.g. Eppley *et al.* (1973), from measurements of ATP and chlorophyll; Beers *et al.*, (1975), who used taxonomic data to assess directly the abundance of the cells; and Kiefer *et al.* (1976). In contrast to these studies, Venrick *et al.* (1973) found that there was a significant increase in the number of diatom cells, with high species diversity, in the chlorophyll maximum layer. Similarly, Hobsen and Lorenzen (1972) found no significant change in the carbon to chlorophyll ratio of phytoplankton with depth, and concluded that either their ratios can vary in the water column as a function of the algal taxa dominating the phytoplankton, or that the methods used for determining the dominant taxa were inadequate. Clearly, these apparent discrepancies in the results obtained deserve further attention, but these authors have mainly been considering the deep oceanic chlorophyll maximum layer and intuitively it is felt that, for the present data where the thermocline and fluorescence peaks are very shallow, there are probably two different phytoplankton populations present in these peaks, rather than there being a change in the biochemical composition of two, specificially similar phytoplankton populations within a depth range of 5 m in the thermocline.

## CONCLUSIONS

The few examples given in this paper illustrate that much additional information can be gained, for the investigation of small-scale distribution of the phytoplankton, from a study of the size spectra of the particles in the sea, in conjunction with fluorescence measurements. It is evident that chlorophyll *a* fluorescence is not always an accurate measure of phytoplankton biomass and does not necessarily reflect accurately the changes in the size distribution and thereby the specific population of the phytoplankton. Although emphasis has been laid here on the advantages of data on particle size spectra as means of providing information on the trophic relationships in marine ecosystems, where particle size may possibly be more important to a herbivore than the actual specific identification of the particle (although there is some evidence to the contrary), it is still obvious that, wherever possible and practical, these measurements should always be made in conjunction with taxonomic studies. So far the present work with the Hiac particle counter has been restricted to a study of the particles in the phytoplankton and microzooplankton size ranges, but the instrument does have the potential, with an additional sensor, to analyze particles of sizes up to 9 mm, and Dr. D. Tungate at MAFF, Lowestoft, has been experimenting with such a system. Of course, the work of Boyd and Mackas (as discussed by Denman, this volume) already has considerably enhanced our knowledge of the spatial distribution of particles in this zooplankton size range, and the statistical analyses of their results have produced some interesting spectral data. Fasham (this volume), nonetheless, has pointed out that the spectra derived from the statistical analysis of the results for semi-continuous processes, e.g. the counting of particles over discrete time periods, require careful interpretation as the different spectral slopes can be explained by a single underlying stochastic process. Attempts should thus be made to reduce the period over which particle counts are accumulated to as short a time as possible.

## SUMMARY

Although a study of the interactions of the species composition of the plankton is of major importance in the context of marine ecology, it may be possible to gain some information on the trophic interrelations of the plankton through studies of the size distribution spectra of particles. It has been shown that the food chain relationships of marine organisms may affect the spatial structure of the ecosystem, and that this size structure of the community is probably more significant than total biomass in determining modes of transfer of energy between trophic levels. Information on the small-scale distribution of particle size spectra in the oceans is thus of some relevance in this context and the information on these distributions, and the factors which may control them, are discussed. The introduction of continuous or semi-continuous methods for the determination of particle size spectra is considered as a means by which such information may be achieved, and comparisons are made with the determination of chlorophyll distributions to illustrate what possible additional information may be ascertained.

REFERENCES

Allen, T. and K. Marshall, 1972: The electrical sensing zone method of particle size measurement. University of Bradford, 105 pp.

Beers, J. R., F.M.H. Reid and G. L. Stewart, 1975: Microplankton of the North Pacific Central Gyre. Population structure and abundance, June 1972. Int. Revue ges. Hydrobiol., 60, 607-638.

Boyd, C. M., 1973: Small scale spatial patterns of marine zooplankton examined by an electronic in situ zooplankton detecting device. Neth. J. Sea Res., 7, 103-111.

Boyd, C. M. and G. W. Johnson, 1969: Studying zooplankton populations with an electronic zooplankton counting device and the LINC-8 computer. Trans. Applications of Sea Going Computers Symposium; Mar. Tech. Soc.,83-90.

Cooke, R. A., L.D.B. Terhune, J. S. Ford and W. H. Bell, 1970: An opto-electronic plankton sizer. Fish. Res. Bd. Canada, Tech. Rep. 172, 40 pp.

Denman, K. L. and T. Platt, 1975: Coherences in the horizontal distributions of phytoplankton and temperature in the upper ocean. Mem. Soc. R. Sci. Liege, 7, 19-30.

Eppley, R. W., E. H. Renger, E. L. Venrick and M. M. Mullin, 1973: A study of plankton dynamics and nutrient cycling in the central gyre of the North Pacific Ocean. Limnol. Oceanogr., 23, 1199-1208.

Estrada, M. and F. Vallespinós, 1975: Consideraciones estadísticas sobre algunos parámetros oceanográficos en la región de afloramiento del NW de Africa. Res. Exp. Cient. B/O Cornide, 4, 175-183.

Fasham, M.J.R. and P. R. Pugh, 1976: Observations on the horizontal coherence of chlorophyll a and temperature. Deep-Sea Res., 23, 527-538.

Frost, B. W., 1977: Feeding behaviour of Calanus pacificus in mixtures of food particles. Limnol. Oceanogr., 22, 472-491.

Groves, M. J. and D. Wana, 1977: A comparison of instrumental techniques for the determination of contamination in filtered aqueous solutions. Powder Technol., in press.

Hobson, L. A. and C. J. Lorenzen, 1962: Relationships of chlorophyll maxima to density structure in the Atlantic Ocean and Gulf of Mexico. Deep-Sea Res., 19, 297-306.

Kerr, S. R., 1974: Theory of size distribution in ecological communities. J. Fish. Res. Bd. Canada, 31, 1859-1862.

Kiefer, D. A., 1973a: Fluorescence properties of natural phytoplankton populations. Mar. Biol., 22, 263-269.

Kiefer, D. A., 1973b: Chlorophyll a fluorescence in marine centric diatoms: responses of chloroplasts to light and nutrient stress. Mar. Biol., 23, 39-46.

Kiefer, D. A., R. J. Olsen and O. Holm-Hansen, 1976: Another look at the nitrite and chlorophyll maxima in the central North Pacific. Deep-Sea Res., 23, 1199-1208.

Loftus, M. E. and H. H. Seliger, 1975: Some limitations of the in vivo fluorescence technique. Chesapeake Sci., 16, 79-92.

Mackas, D., 1976: Horizontal spatial heterogeneity of zooplankton on the Fladen Ground. I.C.E.S., C.M. 1976/L:20, Plankton Committee, 10 pp.

Maddux, W. S. and J. W. Kanwisher, 1965: An in situ particle counter. Limnol. Oceanogr., 10 (suppl.), 162-168.

Margalef, R., 1974: Distribution du seston dans la région d'affleurement du Nord-ouest de l'Afrique en Mars 1973. Tethys, 6, 77-88.

Parsons, T. R. and R. J. LeBrasseur, 1970: The availability of food to different trophic levels in the marine food chain. In Marine Food Chains, J. H. Steele (ed.), pp. 325-343, Oliver and Boyd, Edinburgh.

Pingree, R. D., P. R. Pugh, P. M. Holligan and G. R. Forster, 1975: Summer phytoplankton blooms and red tides along tidal fronts in the approaches to the English Channel. Nature, 258, 672-677.

Platt, T., 1972: Local phytoplankton abundance and turbulence. Deep-Sea Res., 19, 183-187.

Platt, T. and K. R. Denman, 1977: Organisation in the pelagic ecosystem. Helgolander wiss. Meeresunters, 30, 575-581.

Platt, T., K. R. Denman and A. D. Jassby, 1977: Modelling the productivity of phytoplankton. In The Sea, E. D. Goldberg et al. (eds.), Vol. 6, pp. 807-856, John Wiley and Sons.

Sheldon, R. W. and T. R. Parsons, 1967: A practical manual on the use of the Coulter Counter in marine research. Coulter Electronics Sales Co., Toronto, 66 pp.

Sheldon, R. W., A. Prakash and W. H. Sutcliffe, 1972: The size distribution of particles in the ocean. Limnol. Oceanogr., 17, 327-340.

Steele, J. H., 1964: A study of production in the Gulf of Mexico. J. Mar. Res., 22, 211-222.

Steele, J. H. and B. W. Frost, 1977: The structure of plankton communities. Phil. Trans. Roy. Soc. B., 280, 485-534.

Strickland, J.D.H., 1968: Continuous measurement of in vivo chlorophyll; a precautionary note. Deep-Sea Res., 15, 225-227.

Venrick, E. L., J. A. McGowan and A. W. Mantyla, 1973: Deep maxima of photosynthetic chlorophyll in the Pacific Ocean. Fish. Bull., 71, 41-52.

# THE APPLICATION OF SOME STOCHASTIC PROCESSES TO THE STUDY OF PLANKTON PATCHINESS

M.J.R. Fasham

Institute of Oceanographic Sciences
Brook Road
Wormley, Godalming
Surrey GU8 5UB

## INTRODUCTION

Since its first use by Platt (1972) spectral analysis has come to play a central role in the statistical analysis and understanding of plankton patchiness. For a number of reasons it has replaced earlier statistical models that were based on the distribution of counts/sample (Cassie, 1963). Firstly, the statistical distributions used to analyze the distribution of counts/sample (e.g. Neyman type A and Negative Binomial) assumed that the clusters or patches were small compared to the sample size (Skellam, 1958). When this is not true the estimated value of the distribution parameters and the goodness-of-fit will vary with quadrat size (Pielou, 1957). Secondly, the use of counts/sample data alone discards valuable information about the spatial covariance properties of the data being studied. Thirdly, it is fairly well accepted that the causes of patchiness are likely to be different at different space scales and this effect should be observable as a change in shape of the spectrum at different wavenumbers.

There are thus very good reasons for using spectral analysis but the application of the method so far has been fairly empirical. Spectra of chlorophyll fluorescence and possibly other variables such as temperature or nutrient concentration have been calculated and compared, with the intention of determining space scales over which the various variables are, or are not, related (Denman and Platt, 1975; Fasham and Pugh, 1976; Denman, 1976). Denman and Platt (1976) have attempted to provide a stronger theoretical basis for the topic by postulating a form for the wavenumber spectrum of chlorophyll fluorescence. This, in turn, has stimulated experimental work designed to test their hypothesis (Horwood, 1976, 1978).

There is obviously a need for further work on the theoretical spectra of various stochastic processes in order to provide a firm theoretical background for the evaluation of the empirical results. It is intended in this paper to provide a start to this process by considering a few particular stochastic processes and comparing their spectra with some of the data presently available. The first process is a stochastic modification of the Kierstead-Slobodkin (1953) and Skellam (1951) diffusion models, while the remaining models relate to the analysis of integer count data which raises some special problems.

## DATA ANALYSIS

Before considering the theoretical spectra of the chosen stochastic processes, the definition and estimation of the power spectrum for a continuous variable will be briefly reviewed. A more detailed discussion of this topic can be found in Jenkins and Watts (1968) and Bartlett (1966).

Let the covariance function of the continuous variable be $c(\tau)$, then the power spectrum $g(\nu)$ (hereafter called the spectrum) is defined by

$$g(\nu) = \int c(\tau)\ exp\ (-i\ \nu \cdot \tau)\ d\tau,\ -\infty<\nu<+\infty\ , \qquad (1)$$

where the integration is over the whole n-dimensional space and $\nu$ is related to the wavenumber by the equation $\nu = 2\pi k$. In most marine applications the observations are obtained along a one-dimensional transect and so attention will be focussed on this case although the calculation of spectra in higher dimensions is straightforward. The spectrum can be estimated from the sample periodogram (Bartlett, 1966, p. 305). If there are $n$ observations of the variable $Z_q$ then its Fourier transform is

$$J(\nu_p) = 1/\sqrt{n} \sum_{q=1}^{n} Z_q\ exp\ (-\ i\ \nu_p\ q), \qquad (2)$$

where $\nu_p = 2\pi p/n,\ p = 0,1,\ldots.n/2$. The periodogram is then defined as

$$I(\nu_p) = J(\nu_p)\ \ J^*(\nu_p). \qquad (3)$$

The factor $1/\sqrt{n}$ is chosen so that the mean value of $I(\nu_p)$ for a completely random series will equal its variance. It should be noted that with this choice of normalization the integrated spectrum will be equal to $n\sigma^2$ where $\sigma^2$ is the sample variance. It can be shown (Bartlett, 1966, p. 305) that $E\{I(\nu_p)\}$ tends to $g(\nu_p)$ as $n$ increases. It is well-known, of course, that the variance of $I(\nu_p)$ does not decrease as $n$ increases and so it is necessary to take the average of $m$ adjacent values of $I(\nu_p)$ to give a smoothed estimator $\bar{I}(\nu_p)$. The choice of $m$ leads to a conflict between estimation accuracy and spectral resolution. A constant value is often chosen, but if the spectrum is to be plotted on a log scale, this leads to a concentration of points at the high frequency or wavenumber end of the spectrum. This can be avoided by beginning with a low averaging interval and then gradually increasing its value for successive estimates. The sequence of intervals used in the spectra calculated in this paper were 3,3,4,5,6,7,8,10,12,15, 18,22,26,32,38....

If the $Z_q$ are normally distributed then the smoothed spectral estimate will be distributed as a $\chi^2$ with 2 $m$ degrees of freedom, which enables confidence intervals to be calculated. If the distribution of $Z_q$ is non-normal, the distribution of $\bar{J}(\nu)$ will be asymptotically normal by virtue of the central limit theorem and so the distribution of $\bar{I}(\nu)$ will still be approximately $\chi^2$.

If the original series of observations have a linear trend, this must be removed before calculating the spectra; otherwise, the spectral estimates at low wavenumbers will be biased. The two most commonly used methods of trend removal are least squares and first difference filtering. If the latter is used, it is essential to recolor the calculated spectrum as this filter will considerably attenuate the amplitudes at low wavenumbers which is often the range of interest.

## THE REACTION-DIFFUSION STOCHASTIC MODEL

Let the concentration of phytoplankton be given by $p(r,t)$ where $r$ is the space vector and $t$ is time. In most practical experiments it is only possible to estimate $p$ along one-dimensional profiles and so further argument will be restricted to the one-dimensional case, although it is easily generalized to higher dimensions (Fasham, 1978). If the phytoplankton is envisaged as growing exponentially in a turbulent environment, then $p$ will satisfy the equation

$$\frac{\partial p}{\partial t} = \alpha p + D \frac{\partial^2 p}{\partial x^2}, \qquad (4)$$

where $\alpha$ is the growth rate and $D$ the eddy diffusion coefficient. A two-dimensional version of this equation was first studied independently by Kierstead and Slobodkin (1953) and Skellam (1951). They showed that, if the plankton is growing in a cylindrical volume of water surrounded by water unsuitable for its development, there is a minimum radius below which the concentration of $p$ will tend to zero. This critical radius is given by

$$R_c = 2.4048 \sqrt{D/\alpha} \; . \qquad (5)$$

This model was used by Kierstead and Slobodkin to explain the phenomenon of 'red tides' but is rather unsatisfactory as a model for general plankton patchiness. The concept of 'unsuitable' water is of doubtful applicability in an oceanic situation (Platt, Denman and Jassby, 1977) and also the steady state solution is unstable (Skellam, 1951). Finally, it does not provide a mechanism whereby patchiness can be started in an initially homogeneous plankton distribution.

Some of these criticisms can be answered by adding a random forcing function $E(x,t)$ to equation (4) giving

$$\frac{\partial p}{\partial t} = \alpha p + \frac{\partial^2 p}{\partial x^2} + E(x,t) \qquad (6)$$

We will not, in this paper, attempt to identify the source of this randomness although storm mixing or turbulent upwelling of nutrient-rich water through the thermocline are possible candidates. Equation (6) can be solved by the method of Fourier transforms (Sneddon, 1957, p. 299). If the simplifying assumption is made that the function $E$ is uncorrelated in time, then the spectrum of $p$ as a function of $\nu$ and $t$ (assuming that $p$ is initially zero throughout all the space) is given by

$$f(\nu,t) = \phi(\nu) \left[1 - exp\{-2(D - \alpha)t\}\right]/2(D\nu^2 - \alpha), \qquad (7)$$

where $\phi(\nu)$ is the spatial spectrum of the process $E$. If we consider first the case when $\alpha$ is negative, which might apply during the winter or at the end of a spring bloom, then as $t \to \infty$ the spectrum tends to the expression

$$f(\nu) = \phi(\nu) \;/2(D\nu^2 + |\alpha|). \qquad (8)$$

This spectrum was first derived by Whittle (1962) in an investigation of the covariance of crop abundance. If it is further assumed that $E$ is not spatially coherent, then

$$f(\nu) = \bar{\phi}/2(D\nu^2 + |\alpha|), \qquad (9)$$

where $\bar{\phi}$ equals the variance of the uncorrelated random forcing function. The form of this spectrum for some typical values of $\alpha$ and $D$ is shown as curve A in Figure 1.

There are obvious qualitative similarities between the shape of the Whittle spectrum (A) and that postulated by Denman and Platt (1976) in which the turbulence is regarded as an energy cascade transferring variance from higher to lower wavenumbers. This is in contrast to the Whittle model (with a random noise function) in which the "turbulent" variability is spread equally over the whole wavenumber range. This difference between the two models is reflected in the low wavenumber part of the spectrum. The Denman-Platt spectrum has a $\nu^{-1}$ power law at the low wavenumber end of the spectrum where growth is assumed to dominate over turbulent diffusion. At high wavenumbers the Denman-Platt spectrum steepens to a $\nu^{-2}$ or $\nu^{-3}$ power law, depending on the assumed form of the turbulent motion. The high wavenumber boundary of the $\nu^{-1}$ region is given by

$$\nu_c^1 = 2\pi(\alpha^3/\varepsilon)^{\frac{1}{2}}, \tag{10}$$

where $\varepsilon$ is the rate of destruction, at higher wavenumbers, of variance in velocity. By contrast, the Whittle spectrum tends to a constant value at very low wavenumbers which has the useful mathematical advantage that the integral of the spectrum is finite. At high wavenumbers the Whittle spectrum tends to a $\nu^{-2}$ power law and the boundary between the low and high wavenumber regions can be defined by the quantity

$$\nu_c = (|\alpha|/D)^{\frac{1}{2}}, \tag{11}$$

at which point the tangent to the spectra has a slope proportional to $\nu^{-1}$.

Both the Whittle spectrum and the Denman-Platt spectrum assume equilibrium conditions whereas marine biologists are often interested in nonequilibrium situations such as the spring bloom. In this case, $\alpha$ will be positive and equation (7) shows that for $\nu > \nu_c$ the spectrum will converge to (9), while for $\nu \leqslant \nu_c$ the spectrum will tend to infinity as time increases (curve B, Figure 1). In reality, this will not happen as limiting factors such as herbivore grazing and nutrient depletion will become more important and (6) will cease to be a valid description of the dynamics. A possible way of modelling this situation is to replace the exponential by a logistic growth term giving

$$\frac{\partial p}{\partial t} = p(\alpha - \beta\, p) + \frac{\partial^2 p}{\partial x^2} + E(x,t). \tag{12}$$

This nonlinear model is not solvable by the Fourier technique and so resort must be had to simulation methods. Following Wroblewski, O'Brien and Platt (1975) we first rewrite (12) in a non-dimensional

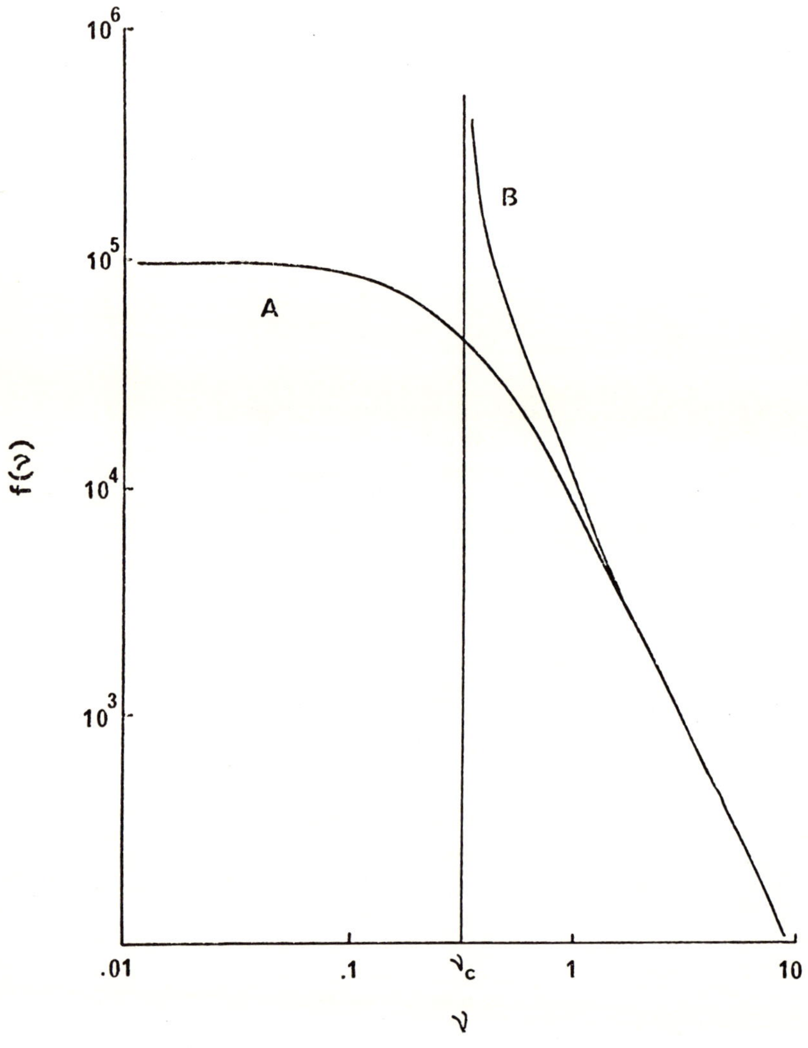


Figure 1. Theoretical chlorophyll wavenumber spectrum derived from equation (7) with $\phi(\nu)$ constant. The values of $D$ and $\alpha$ were $10^6 cm^2$/sec and $10^{-5}$ divisions/second respectively. Curve A is for negative $\alpha$ and B for positive $\alpha$.

form defining

$$\alpha t = \tau$$

$$x = y\sqrt{D/\alpha}$$

$$\gamma = \beta/\alpha$$

$$F = E/\alpha,$$

then (12) can be rewritten

$$\frac{\partial p}{\partial \tau} = p(1 - \gamma p) + \frac{\partial^2 p}{\partial y^2} + F(y,t). \qquad (13)$$

This equation was solved using the modified Crank-Nicholson method described in Adam and Dubois (1975). As the equations are intended to model the situation in a subspace of an open homogeneous ocean, the choice of boundary conditions presents some problems. As noted by Evans, Steele and Kullenberg (1977), choosing pre-set boundary values or reflecting boundaries results in the evolution of the model being eventually driven by the boundary conditions. We have, therefore, followed these workers in considering the one-dimensional strip as being tied together in a ring. This means that there is effectively no boundary and the model can evolve without exterior effects. The time change of the spatial distribution of $p$ will be demonstrated by calculating the spatial power spectrum of $p$ at different time intervals. The spatial range of the model was set as $-15 \leqslant y \leqslant +15$ and the $y$ increments and $\tau$ increments were set as 0.1 and 0.01 respectively. Unless otherwise stated, the initial conditions are $p(y,0) = 0$, $y = (-15, 15)$. The noise function $F$ was assumed to be uncorrelated in time and space and was simulated using a computer-generated, pseudo-random number sequence.

As a test of the program, an initial run was made with $\gamma=0$, $\alpha=-1$. The resulting spectra for $\tau = 0.5$ and 2.0 (Figure 2) show the form predicted by the Whittle spectrum (9) and this establishes some confidence in the numerical method. To simulate uninhibited growth, a run with $\alpha=1$, $\gamma=0$ was made and the results are shown in Figure 3. It can be seen that as time proceeds the low wavenumber spectral values increase dramatically (as predicted by equation (7)) giving rise to a $k^{-3}$ slope, while at the high wavenumber end the slope remains at $k^{-2}$.

In the next run (Figure 4), $\gamma$ was set to 0.3 thus simulating the effect of nutrient limitation or herbivore grazing at the later stages of a bloom. It can be seen that in the initial stages ($\tau = 2.0$) of the bloom the spectrum has a $k^{-2}$ slope throughout the wavenumber range with no trace of a change in slope. However, as limiting factors become important ($\tau = 8$), the low wavenumber

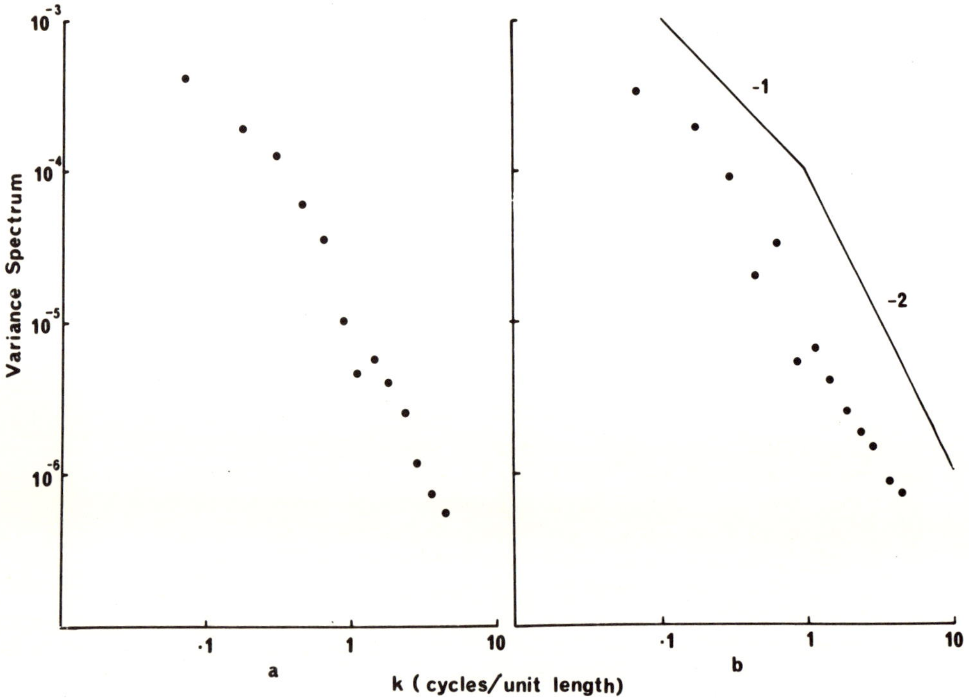


Figure 2. Simulated chlorophyll wavenumber spectra for the case $\gamma=0$, $\alpha=-1$ at times (a) $\tau=0.5$ and (b) $\tau=2.0$.

spectral values become attenuated leading eventually to the familiar Whittle spectrum.

In the final run, the simulation was run until $\tau=2$ with the parameters $\gamma=0$, $\alpha=-1$ and then the parameters were altered to $\gamma=0.3$, $\alpha=+1$ and run until $\tau=6$. This was intended to roughly simulate the transition from overwintering conditions, through the spring bloom and into the post-bloom situation. It can be again seen (Figure 5) that the spectrum changes from the Whittle form to a form having a constant slope and then back to the Whittle form.

These results suggest that during the exponential growth phase of the spring bloom the chlorophyll spectra will not show any change in slope from $k^{-1}$ to $k^{-2}$ , as suggested by the simple critical length theories and there is some observational evidence to support this contention. Fasham and Pugh (1976) made some observations of chlorophyll and temperature spectra in the Bay of Biscay during the spring bloom and also in the tropical eastern Atlantic during the summer. The chlorophyll spectra from the tropical area followed a -2 power low at high wavenumbers

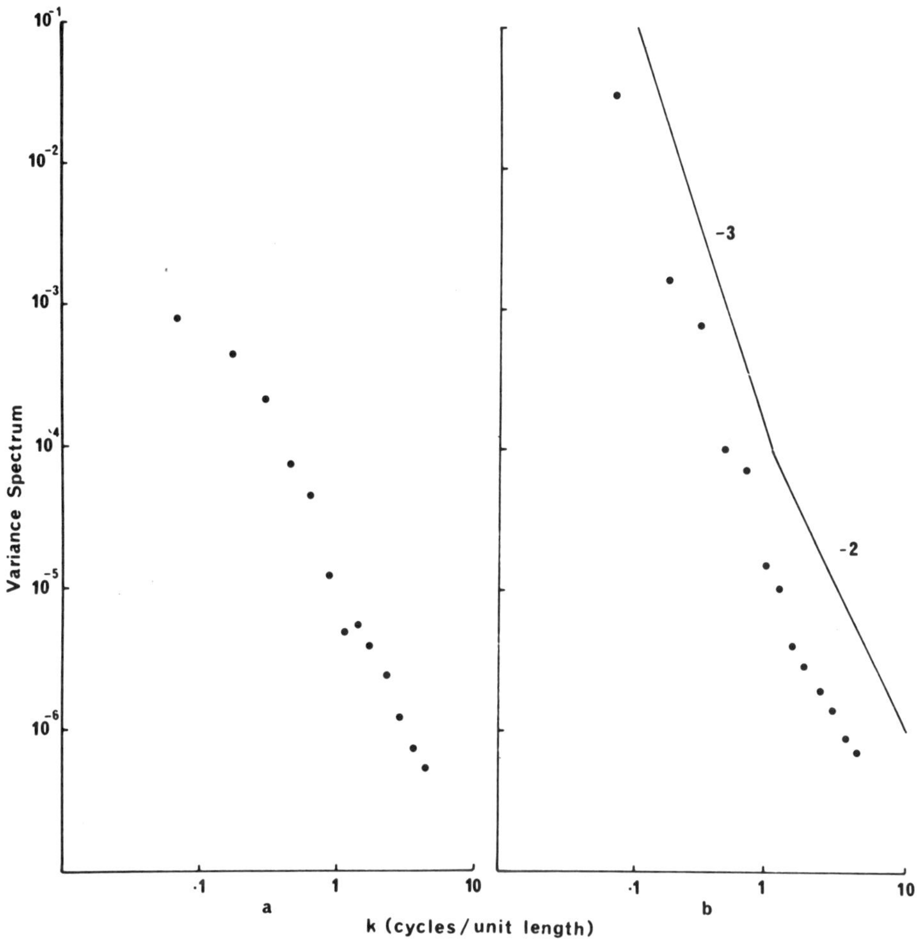


Figure 3. Simulated chlorophyll wavenumber spectra for the case $\gamma$=0, $\alpha$=1 at times (a) $\tau$=0.5 and (b) $\tau$=2.0.

flattening to a -1 power low at the lowest wavenumbers, which is in agreement with the predictions of the Denman-Platt and Whittle spectra. However, the chlorophyll spectra from the Bay of Biscay were generally of the same form as the temperature spectra with fairly constant slopes between -2 and -3. This difference between the two areas cannot be explained in terms of the simple equilibrium critical length theory, as the growth rate in the Bay of Biscay situation was presumably greater than that in the tropical area. The results of the simulations would suggest, however, that the absence of the knee-shaped spectra in the Bay of Biscay could be attributed to the non-equilibrium spring bloom conditions in contrast to the equilibrium conditions in the tropical area.

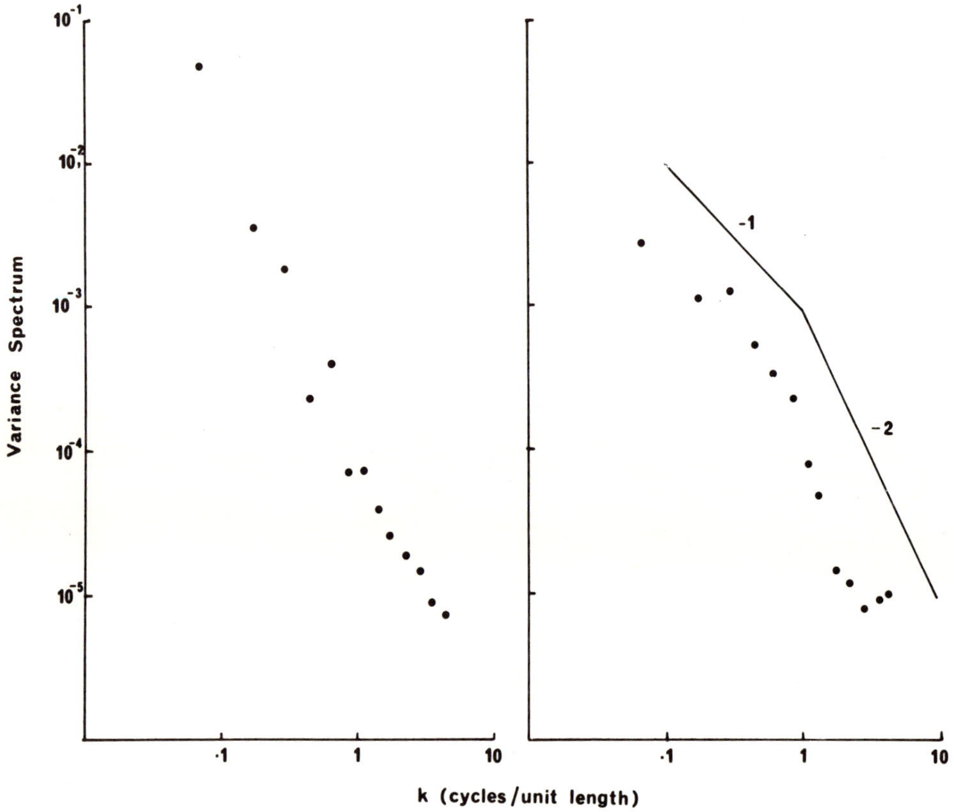


Figure 4. Simulated chlorophyll wavenumber spectra for the case $\gamma=0.3$, $\alpha=1$ at times (a) $\tau=2.0$ and (b) $\tau=8.0$. The initial conditions were zero chlorophyll concentration at all $y$ values.

Some other evidence has been provided by Horwood (1976, 1978) who made a number of transects at a fixed locality in the North Sea at regular intervals between February and April 1976. These experiments were designed to investigate the validity of the Denman-Platt spectrum and the typical 'knee-shaped' form of the spectrum was only observed on the first transect which was taken before the bloom developed. Thereafter, the chlorophyll spectra showed a number of varying forms but there was no sign of the 'knee-shaped' form. These results might also be explained in terms of the model described above. Unfortunately, no observations were made in the post-bloom situation and so confirmation of the reappearance of the knee-shaped spectrum could not be obtained.

If the conclusions derived from these simulations are valid, they suggest that chlorophyll spectra observed in non-equilibrium growth conditions are likely to show a similar slope to temperature spectra. Thus the fact that the spectra of these two variables have

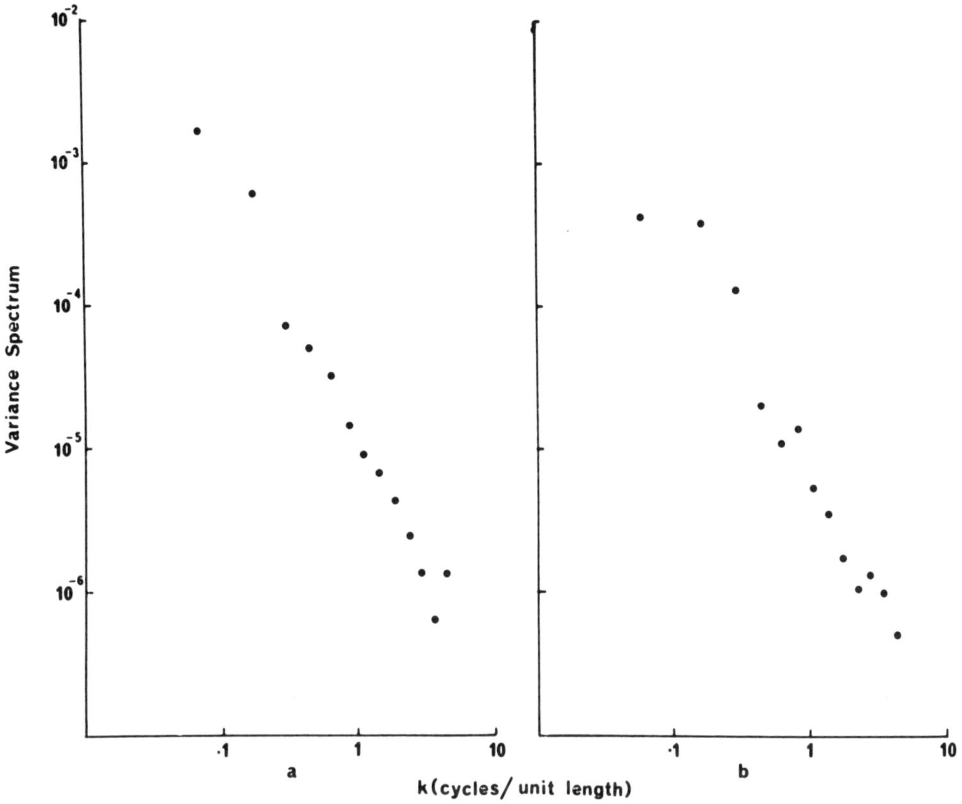


Figure 5. Simulated chlorophyll wavenumber spectra for the case $\gamma=0.3$, $\alpha=1$ at times (a) $\tau=2.0$ and (b) $\tau=3.0$. The initial conditions were the spatial pattern obtained at $\tau=2.0$ by the simulation with $\alpha=-1$, $\gamma=0$.

the same slope cannot be used to infer that the chlorophyll distribution is controlled physically rather than biologically.

## STOCHASTIC POINT PROCESSES

There is a fundamental difference between the spectral analysis of a quantity such as fluorescence or temperature and that of species or particle counts, in that the latter quantities can only take integral values. It will be seen that this distinction may have important effects on the shape of the spectra especially when the mean count level is low. The derivation of the spectrum of integral data requires a knowledge of the theory of stochastic point processes which will now be briefly summarized; further details on point processes are contained in Bartlett (1969, 1975) and Lewis (1970).

Let us initially assume that plankton is distributed in one dimension and does not change significantly with time during the period of observation. If the number of plankton occurring between some arbitrary origin and the point $x$ is $N(x)$, we can define the quantity

$$dN(x) = N(x + dx) - N(x). \tag{14}$$

Now, unlike a continuous variable, $dN(x)$ is a process that is zero everywhere except where an organism occurs, where it takes the value one. The function $dN(x)$ (termed a point process) is thus not continuous and its treatment presents mathematical complications that have probably inhibited the understanding and greater use of point processes by ecologists. If multiple events at the same value are assumed not to occur and if the probability that an individual will occur between $x$ and $x + dx$ is $\lambda dx$, then the expectation of $dN(x)$ is

$$E\ \{dN(x)\} = \lambda dx. \tag{15}$$

The quantity $\lambda$ is called the density of the point processes. In order to obtain any useful information on a point process, it is necessary to consider higher moments. By analogy with the continuous case we define a covariance density function (Bartlett, 1966, p. 83) $\mu(x, x + dx)$ as

$$\mu(x, x + dx) = \frac{E\ \{dN(x) \cdot dN(x + dx)\}}{(dx)^2} - \lambda^2 \tag{16}$$

If the process is stationary, then $\mu$ will be a function of $dx$ only. When we consider the value of $\mu$ for $dx = 0$, a complication arises. As the function $dN(x)$ can only take the values 0 or 1 and as multiple events at the same $x$ are excluded, then

$$E\ \{dN(x)^2\} = E\ \{dN(x)\}. \tag{17}$$

This degeneracy can be accounted for by using the Dirac delta function to define a covariance function for all $\tau = x - x'$ giving

$$\mu_c\ (\tau) = \lambda\delta(\tau) + \mu\ (\tau). \tag{18}$$

In order to obtain information on the process $dN(x)$ it is necessary to be able to sample the exact position of the planktonic organisms. With the possible exception of the Boyd *in situ* particle counter (Boyd, 1973), most plankton sampling devices only provide estimates of the integrated number $N(x)$ of plankton within a subvolume of the three-dimensional space. It is thus necessary to be able to compute the mean and covariance of the integrated count within a specified subvolume. A full treatment of this problem requires the use of three-dimensional point processes (Fasham, 1978). However, as in most cases, samples are obtained along a one-dimensional profile, the spectrum of $N(x)$ will be derived using a simplified one-dimensional theory. If it is assumed that the net travels a distance $l$ while the subsample $N(x)$ is obtained, then, bearing in mind the degeneracy referred to earlier, the mean and covariance of $N(x)$ is given by (Thompson, 1955; Bartlett, 1975),

$$E\ \{N(x)\} = \lambda l \tag{19}$$

$$Cov\ \{N(x),\ N(x+\tau)\} = \lambda\phi + \iint_{-l/2}^{+l/2} \mu(r_1+\tau-r_2)dr_1 dr_2, \tag{20}$$

where $\phi$ is a length measure of the overlap, if any, between the two samples. Obviously, the variance of $N(x)$ is a special case of (20) for which $\phi = l$. A Poisson or random point process is one for which $\mu = 0$ and therefore $var\ \{N(x)\} = \lambda l$ for this process as would be expected.

The power spectrum $g(\nu)$ of the point process $dN(x)$ can be obtained by taking the Fourier transform of $\mu_c(\tau)$ giving (Bartlett, 1975)

$$g(\nu) = \lambda + \int \mu(\tau)\ \ exp\ (-i\nu\tau)\ d\tau. \tag{21}$$

It can be seen that the delta function in the definition of $\mu_c(\tau)$ gives rise to the constant term $\lambda$ in the spectrum $g(\nu)$ and it is this property that distinguished the spectrum of a point process from that of a continuous variable. For a Poisson process the spectrum will be a constant $\lambda$.

As has been stated, in most cases only the integrated count $N(x)$ can be estimated and so it is necessary to derive the spectrum of this quantity as a function of $g(\nu)$. As an example, let us consider the spectrum $D(\nu)$ of counts from successive contiguous subsamples a distance $l$ apart as might be obtained by a Longhurst-Hardy net (Longhurst *et al.*, 1966). By taking the Fourier transform of (20) we obtain

$$D(\nu) = \lambda l + [4sin^2(\nu l/2)/\nu^2]\ h(\nu), \tag{22}$$

where

$$h(\nu) = \int \mu(\tau)\ \ exp\ (-i\nu\tau)\ d\tau. \tag{23}$$

We will now consider two special classes of point processes that have useful applications to the study of patchiness.

## CLUSTERING PROCESSES

In a clustering process organisms are envisaged as being grouped in clusters whose centers are randomly distributed in space with Poisson parameter $\lambda_0$. This process might provide a useful model of plankton swarming if the animals have sufficient mobility to overcome the effects of fluid turbulence. The number $n$ of organisms in a cluster is assumed to follow a distribution law $p(n)$ and the organisms are distributed in space about the cluster center $x_c$ according to the probability density $f(x - x_c)$. Bartlett (1975) has shown that if the characteristic function of $f(x)$ is $C(\nu)$ then the spectrum of the point process $dN(x)$ is

$$g(\nu) = \lambda\ (1 \quad + C(\nu)\ C^*(\nu)\ [m_o^{\ 2} + v_o - m_o]/m_o) \tag{24}$$

where $m_o$, $v_o$ are the mean and variance of the number of organisms per cluster and $\lambda = \lambda_o\ m_o$. As an example, if $f$ is a Gaussian distribution and $p$ is Poisson, then

$$g(\nu) = \lambda\ [1 + m_o\ exp\ (-\sigma^2\nu^2)]. \tag{25}$$

It should be noted that the dependence of the spectrum on $\nu$ is determined only by the form of $f(x)$ and so information on the shape of the clusters can be obtained independently of any assumption about the form of $p(n)$. The spectrum of the integrated count $N(x)$ is given by

$$D(\nu) = \lambda l + 4\lambda m_o\ [sin^2\ (\nu l/2)/\nu^2]\ exp\ (-\sigma^2\nu^2). \tag{26}$$

Depending on the relative magnitudes of $\sigma$ and $l$, the terms in square brackets in equation (26) can have a considerable effect on the spectrum. This is demonstrated by Figure 6, where spectra for different values of $\sigma$ have been calculated. It can be seen that for values of $\sigma < l$ the spectrum is virtually uniform over the wavenumber range from zero up to the Nyquist wavenumber. Thus, equation (26) states mathematically the otherwise obvious fact that if the patch size of the plankton is smaller than the sampling interval of the Longhurst-Hardy net (LHPR) or pump, then no information on the patch size can be obtained from the count data.

To see how common this sampling situation might be in practice, some zooplankton data obtained by the LHPR (Fasham, Angel and Roe, 1974) have been analyzed. Spectra were calculated for all the non-randomly distributed species (tested using the index of

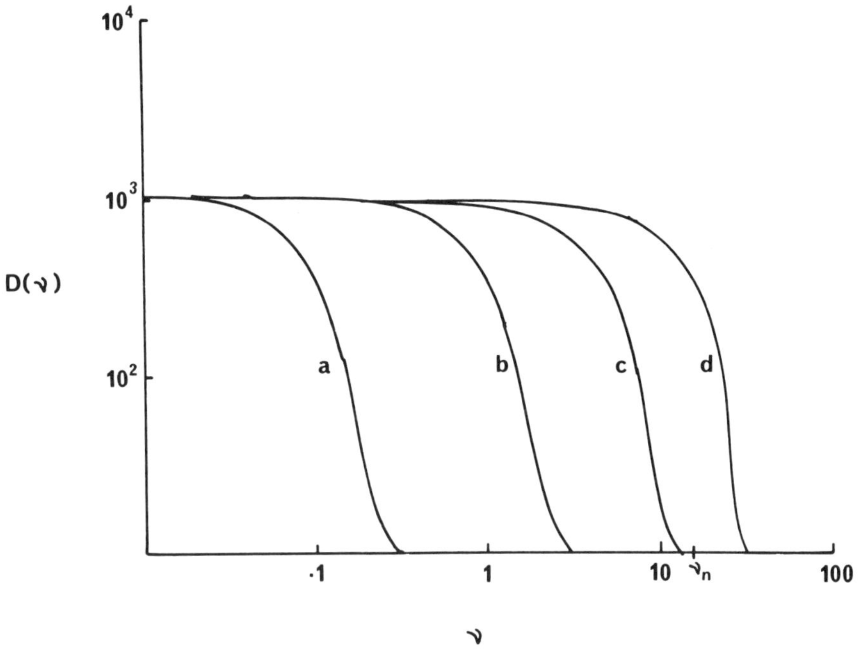


Figure 6. Theoretical spectra of integrated plankton counts calculated from equation (26). The parameter values used were $\lambda=5 \times 10^5$, $m_0=20$, and $l=0.2$ km. The values of $\sigma$ in km for the four curves were (a) $\sigma=10$, (b) $\sigma=1$, (c) $\sigma=0.2$ and (d) $\sigma=0.02$.

dispersion) from the longest tow having 76 subsamples. For these data a constant Fourier averaging value of 4 was used to calculate the smoothed spectral estimates. The wind-on period of the netting on the sampler was 60 seconds which, at a speed of 2 knots, meant that $l$ was equal to 60 m. It was found that very few of the spectra showed a clear decrease with wavenumber as is observed for chlorophyll spectra. The spectra were tested for uniformity using the Kolmogorov-Smirnov test (Jenkins and Watts, 1968, p. 234) and out of 19 species analyzed only 4 had non-uniform spectra suggesting that in most cases the patch size was less than 60 m.

The spectra of the four species having non-uniform spectra are plotted on a semi-log scale in Figure 7. These plots show a fairly linear trend which suggest that they might be fitted using $h(\nu)$ of the form

$$h(\nu) = k_1 \exp(-\sigma|\nu|) \tag{27}$$

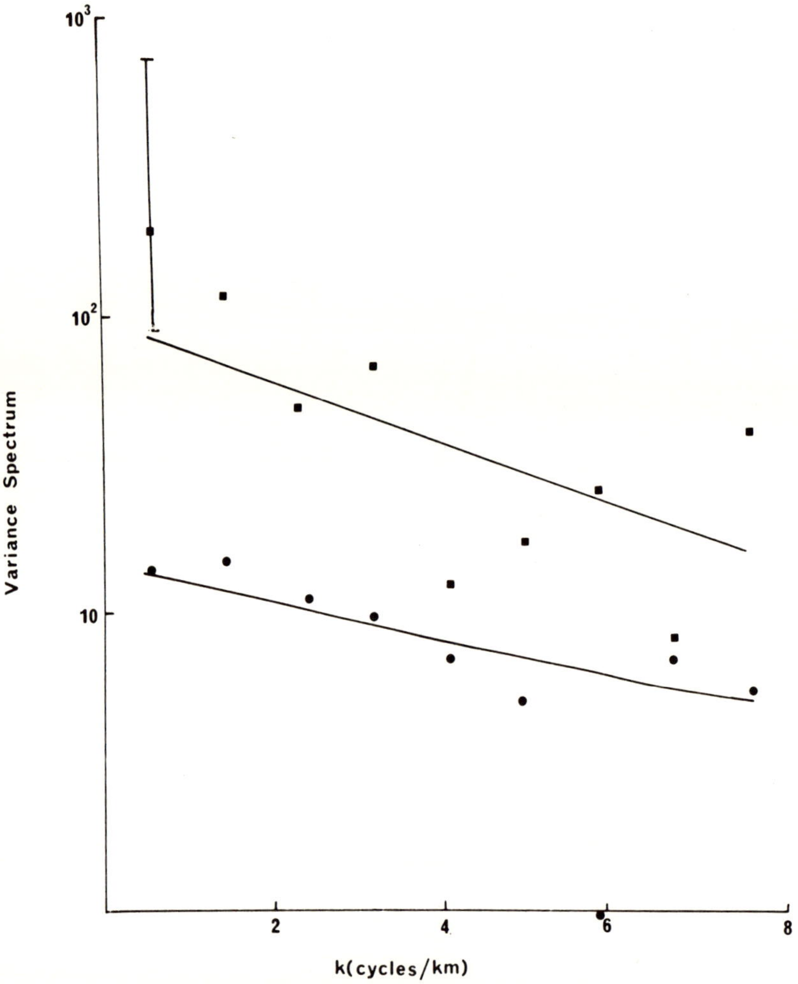


Figure 7a

Figure 7. Observed zooplankton spectra obtained from LHPR samples. The species and estimated values of σ are as follows; (a) Top spectrum, *Paraeuchaeta norvegica*, σ = 39.5 m; bottom spectrum, *Conchoecia elegans*, σ = 32.3 m, (b) *C. ametra*, σ = 33.5 m, (c) *C. lophura*, σ = 15.08. The 95% confidence limits are indicated for the lowest wavenumber and are the same for all wavenumbers.

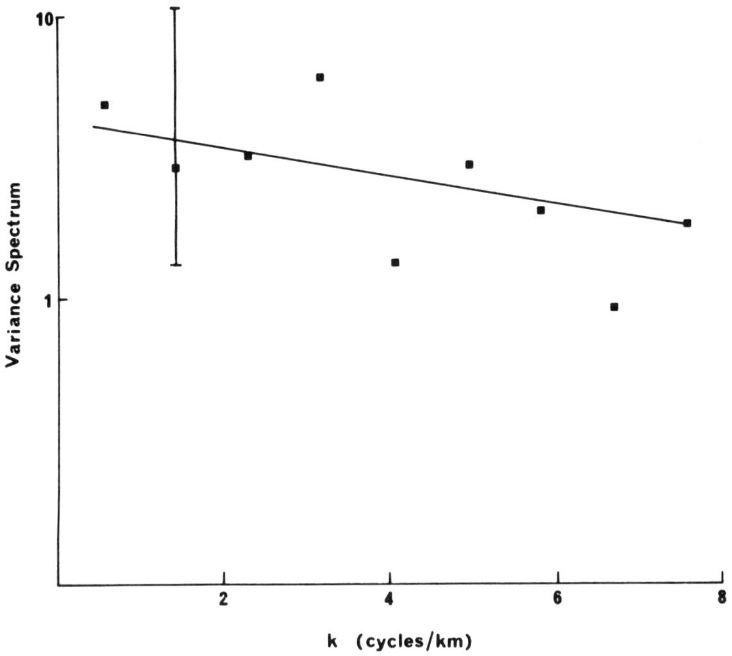


Figure 7b

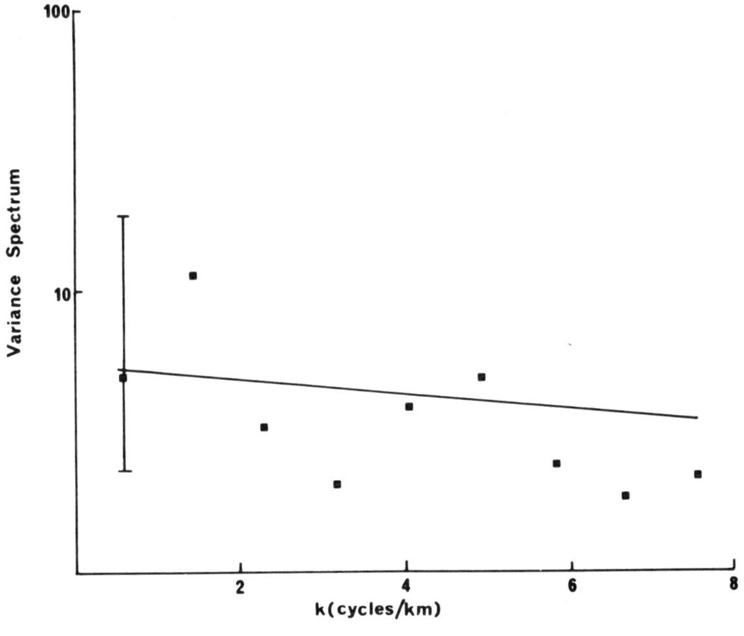


Figure 7c

where $k_1$, $\sigma$ are constants. The spectrum of the integrated count will thus be given by

$$D(\nu) = \lambda l + 4\ k_1\ [sin^2\ (\nu l/2)/\nu^2]\ \ exp\ (-\sigma|\nu|) \qquad (28)$$

It is possible then to estimate $\sigma$ from an observed spectrum by using the regression of the variable $Y(\nu) = log\ \{(D(\nu) - \lambda l)\ \nu^2/ sin^2\ (\nu l/2)\}$ against $\nu$. A numerical complication arises in that due to estimation errors some of the spectral values may be less than the mean ($\lambda l$) and so $Y$ cannot be calculated; these spectral estimates have to be excluded from the regression. Despite this the theoretical spectra give a reasonably good fit to the observed spectra with values of $\sigma$ ranging between 15 and 40 m.

Using equation (24) it is possible to relate $\sigma$ to the cluster size of the plankton. It is well known that the characteristic function of the Cauchy distribution with zero mean and parameter $\beta$ is $exp\ (-\beta|t|)$ (Johnson and Kotz, 1970) and, therefore, if the distribution $f(x)$ is assumed to be a Cauchy distribution with parameter $\sigma/2$, this will give rise to the spectrum given by equation (28). The distribution function for the Cauchy distribution shows that 85% of the population lies between $\pm 2\beta$. Thus for the four species discussed, 85% of their population will be contained in patches of dimensions between 30 and 80 m.

A similar analysis has been carried out on some unpublished zooplankton data obtained by pump sampling (Pugh, personal communication) and similar conclusions were obtained. The practical consequences of these results are that there is now considerable doubt whether the traditional sampling methods of LHPR and pump sampling are capable of resolving the patch structure of zooplankton. This may have to await the further development of *in situ* zooplankton counters.

It should be mentioned that in their original analysis of the LHPR data, Fasham, Angel and Roe (1974) used a clustering process that assumed that $f(x)$ was an isotropic Gaussian distribution. Spectral methods were not used and the parameters of this model were estimated from the variance of the difference between adjacent counts. Goodness-of-fit tests showed that the distribution was capable of explaining the observed distribution of counts/sample. Despite this fact, however, the spectral analyses described above show that the Gaussian model was incorrect. The Gaussian cluster process would predict a spectrum proportional to $exp\ (-\sigma^2\nu^2)$ (see equation (25)), whereas the spectra analyzed were proportional to $exp\ (-\sigma|\nu|)$. This demonstrates the superiority of spectral analysis as a means of elucidating the underlying model.

DOUBLY STOCHASTIC PROCESSES

If we consider a Poisson process in which the intensity $\lambda$ is not a constant but is itself a stochastic process $\Lambda(x)$, then the resulting non-random process is called a doubly stochastic process. The covariance function for this process is (Bartlett, 1963),

$$\mu_c(\tau) = \lambda\delta(\tau) + \mu(\tau), \tag{29}$$

where $\mu(\tau)$ is the covariance function of $\Lambda(x)$. The spectrum of $dN(x)$ can be obtained using equation (21). Before proceeding further we will first introduce some data which will later be analyzed using a doubly stochastic process. These data were obtained from a 9-channel HIAC particle counter connected on-line to the pumped sea water supply on RRS 'Discovery'. The particle size range for each channel is shown in Table 1 and the experimental details of the system are described by Pugh (in preparation). The data used here were obtained during a 15-hour transect off the coast of West Africa from 20°50'N, 17°40'W to 22°30'N, 17°40'W at an average speed of 13.2 km/hr. Particle counts were made every 5 minutes with a counting period of one minute. The resulting data were detrended using least squares and the spectra calculated in the usual way (Figure 8).

Table 1

Channel settings for HIAC particle counter, equivalent surface area diameters

| Channel | Diameter Range (μm) |
|---|---|
| 1 | 10-20 |
| 2 | 20-40 |
| 3 | 40-50 |
| 4 | 50-78 |
| 5 | 78-112 |
| 6 | 112-145 |
| 7 | 145-202 |
| 8 | 202-300 |
| 9 | 300-600 |

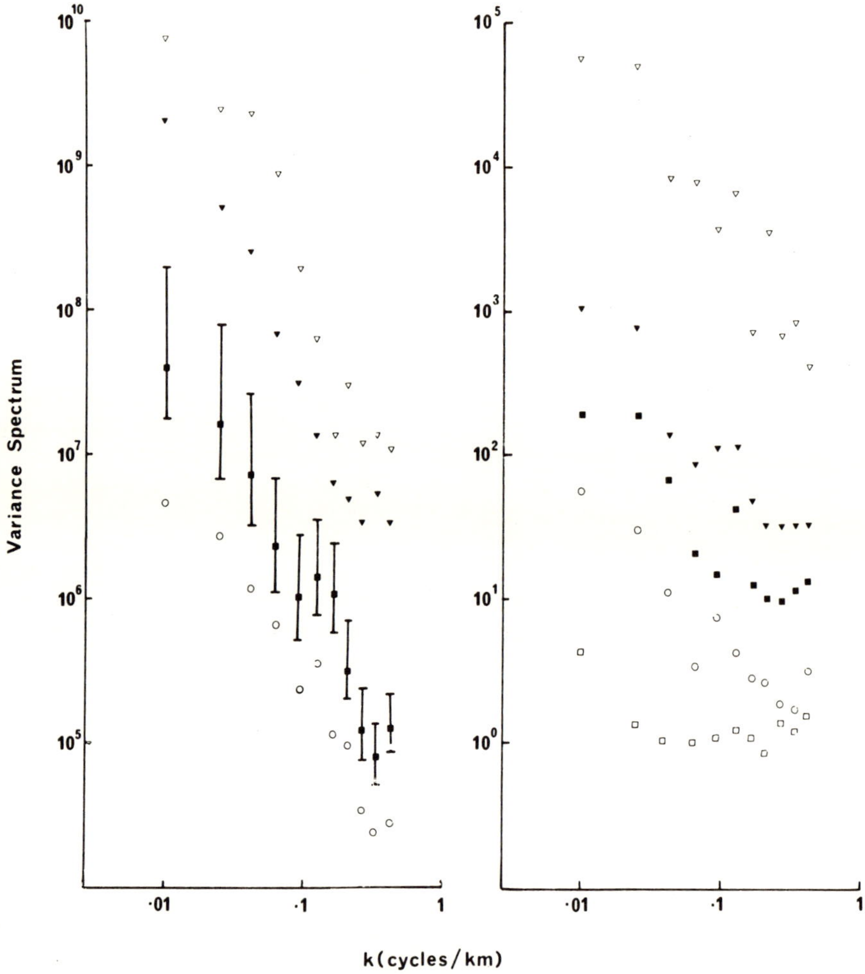


Figure 8. Observed spectra of counts from a HIAC particle counter. The spectra for the nine channels are in descending order of variance beginning with channel one at the top of the lefthand graph. 95% confidence limits are indicated for channel three.

A first glance at these spectra shows that they change their shape systematically going from channel 1 to 9. Thus the spectra of the first four channels possess, as might be expected, the familiar -2 power law form observed in other chlorophyll spectra. However, the spectra for the higher channels, which represent micro-zooplankton, show power laws closer to -1, while that for channel 9 is virtually random. This difference between chlorophyll and zooplankton spectra has also been observed by Mackas (1976) who used the Boyd particle counter to sample the zooplankton.

It is tempting, therefore, to suggest that the factors governing the distribution of micro-zooplankton are different from those governing phytoplankton. This may, in fact, be the case but, as will be shown below, considerable caution should be exercised before jumping to this conclusion.

Let us consider a one-dimensional, doubly stochastic process in which the intensity $\Lambda(x)$ is assumed to be produced by the reactive-diffusive process described by equation (6). By substituting the spectral form of equation (9) in equation (23) we obtain the following expression for the spectrum of the integrated count

$$D(\nu) = \bar{x} + 4\,[\sin^2(\nu l/2)/\nu^2]\,\bar{\phi}\,/2\,(D\nu^2 + |\alpha|) \qquad (30)$$

where $l$ is the distance run during each sample and $\bar{x}$ is the mean of the series of integrated counts. Let us further assume that $\bar{\phi}$, the variance of the forcing function $E(x,t)$, is equal to $k\bar{x}^2$ where $k$ is a constant. It should be emphasized that this assumption cannot at the moment be justified but it is introduced because of the interesting properties of the resulting spectrum. This spectrum is now given by

$$D(\nu) = \bar{x}\,[1 + 4\bar{x}k\sin^2(\nu l/2)/2\nu^2(D\nu^2 + |\alpha|)] \qquad (31)$$

This spectrum has been calculated, using the same values of $\alpha, D$ used for curve A in Figure 1, with $k = 10^{-3}$ and $\bar{x}$ taking values from 1 to 10,000. The resulting curves (Figure 9) show that a series of curves with different power laws can be obtained depending on the value of $\bar{x}$ even though the underlying spatial model is the same. This form of behavior is the direct result of dealing with discrete rather than continuous variables which gives rise to the constant $\bar{x}$ term in the resulting spectra. In practice, of course, even continuous variables are quasi-discrete in that they are subject to the minimum resolution of the measuring instruments.

There are obvious similarities between the curves of Figure 9 and the observed channel spectra shown in Figure 8 and it is tempting to suggest that the differences between the spectra might be explained by a doubly stochastic process as described above. However, there is one aspect of these spectra that cannot be explained by this process. There is evidence from the spectra of channels 5 to 8 of a flattening at the lowest wavenumbers that is not observed for the spectra of the lowest four channels. This cannot be explained by assuming a smaller value of $\alpha$ for organisms sampled by the higher channels as this would have the reverse effect on the spectra. Other possible explanations are that the increased mobility of the micro-zooplankton gives rise to a lower value of the effective eddy diffusivity or the effect may be caused by predation from schooling organisms.

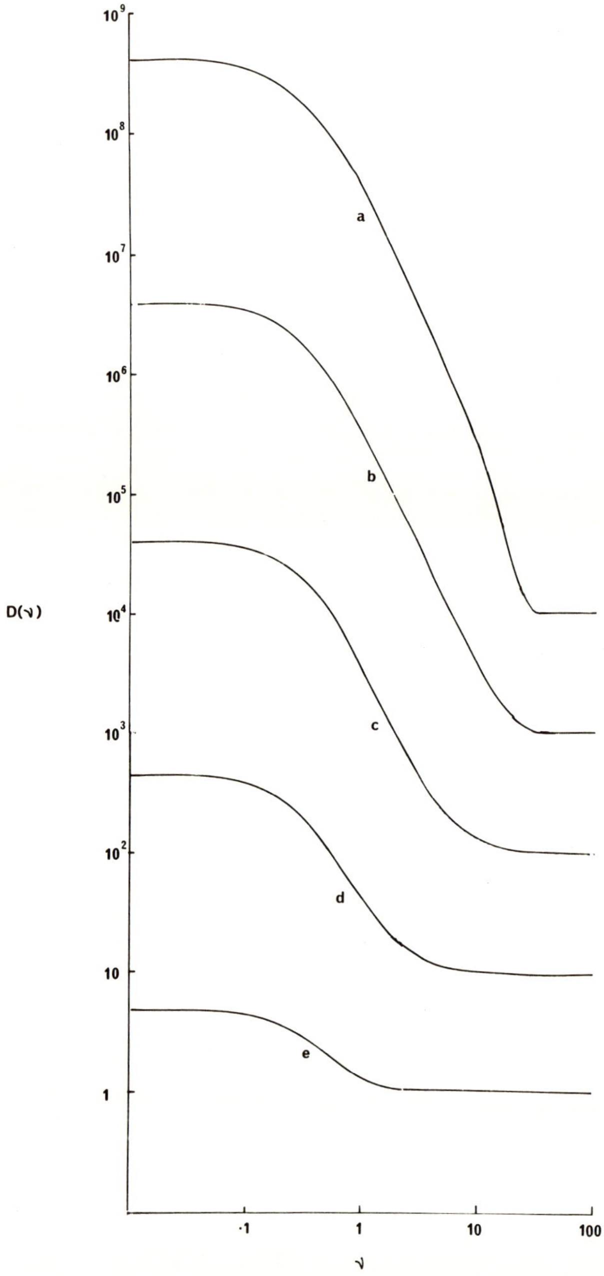


Figure 9. Theoretical count spectra derived from equation (31) with $k=10^{-3}$, $l=0.2$ km, $\alpha=10^{-5}$ divisions/sec and $D=10^6$ cm$^2$/sec. The values of $\bar{x}$ are (a) $\bar{x}=10^4$ (b) $\bar{x}=10^3$ (c) $\bar{x}=100$ (d) $\bar{x}=10$ (e) $\bar{x}=1$.

## CONCLUSIONS

I have attempted in this paper to introduce some stochastic processes that may not be too well known by marine biologists and to show how these can be used to interpret data. I also hope I have shown that, especially when using discrete count data, a firm understanding of the theory of the stochastic processes is essential if observed spectra are not to be misinterpreted. The processes discussed have by no means exhausted the possibilities and other possible candidates will no doubt arise in the future. Two such cases immediately suggest themselves and it is hoped to carry out analysis of these in the near future. The first concerns the effect of zooplankton swarming (Okubo, 1972) in conjunction with the reactive-diffusive process and the second concerns the extension of the reactive-diffusive process to cover two or more trophic levels. A start on the latter has already been made by Steele and Henderson (1977) and Evans, Steele and Kullenberg (1977).


## SUMMARY

Spectral analysis has now become a widely used method for investigating the spatial pattern of fluorescence and its relationship with other variables. However, apart from the work of Denman and Platt, there have been no attempts to derive theoretical spectra which can be tested against the data. This paper derives the theoretical spectra for three stochastic processes which may prove useful in plankton patchiness. The first is a stochastic modification of the Kierstead-Slobodkin plankton bloom equation, the second a clustering stochastic point process and the last a doubly stochastic point process. The study of these spectra show that considerable caution should be used in the interpretation of spectral slopes. For a given model this slope may change with time, in the case of a spring bloom and also, in the case of zooplankton counts, may be critically affected by the sampling methods.


## REFERENCES


Adams, Y. and D. M. Dubois, 1975: Space time simulations of a theoretical model of predator-prey equations. Unpublished manuscript.

Bartlett, M. S., 1963: The spectral analysis of point processes. J. Roy. Statist. Soc., 25B, 264-296.

Bartlett, M. S., 1966: An introduction to stochastic processes. Cambridge University Press, 362 pp.

Bartlett, M. S., 1975: The statistical analysis of spatial pattern. Chapman and Hall, London, 90 pp.

Boyd, C. M., 1973: Small scale spatial patterns of marine zooplankton examined by an electronic in situ zooplankton detecting device. Neths. J. Sea Res., 7, 103-111.

Cassie, R. M., 1963: Microdistribution of plankton. Oceanogr. Mar. Biol., 1, 223-252.

Denman, K. L., 1976: Covariability of chlorophyll and temperature in the sea. Deep-Sea Res., 23, 539-550.

Denman, K. L. and T. Platt, 1975: Coherences in the horizontal distributions of phytoplankton and temperature in the upper ocean. Mem. Soc. Roy. Sci. Liege, 7, 19-30.

Denman, K. L. and T. Platt, 1976: The variance spectrum of phytoplankton in a turbulent ocean. J. Mar. Res., 34, 593-601.

Evans, G. T., J. H. Steele and G.E.B. Kullenberg, 1977: A preliminary model of shear diffusion and plankton populations. Scot. Fish. Res. Rept., 9, 20 pp.

Fasham, M.J.R., 1978: The statistical and mathematical analysis of plankton patchiness. Oceanogr. Mar. Biol., 16, 43-79.

Fasham, M.J.R., M. V. Angel and H.S.J. Roe, 1974: An investigation of the spatial pattern of zooplankton using the Hardy-Longhurst plankton recorder. J. Exp. Mar. Biol. Ecol., 16, 93-112.

Fasham, M.J.R. and P. R. Pugh, 1976: Observations on the horizontal coherence of chlorophyll a and temperature. Deep-Sea Res., 23, 527-538.

Horwood, J. W., 1976: A critical investigation of small scale variability in phytoplankton. I.C.E.S. Contributions to Statutory Meeting, CM 1976/L:7, 6 pp.

Horwood, J. W., 1978: An analysis of the spatial heterogeneity of surface chlorophyll in one and two dimensions. J. Mar. Biol. Ass. U.K., in press.

Jenkins, G. M. and D. G. Watts, 1968: Spectral analysis and its applications. Holden-Day, San Francisco, 523 pp.

Johnson, N. L. and S. Kotz, 1970: Continuous univariate distributions - 1. Houghton Miflin Co., Boston, 300 pp.

Kierstead, H. and L. B. Slobodkin, 1953: The size of water masses containing plankton blooms. J. Mar. Res., 12, 141-147.

Lewis, P.A.W., 1970: Remarks on the theory, computation and application of the spectral analysis of a series of events. J. Sound Vib., 12, 353-376.

Longhurst, A. R., A. D. Reith, R. F. Bower and D.L.R. Seibert, 1966: A new system for the collection of multiple serial plankton samples. Deep-Sea Res., 13, 213-222.

Mackas, D., 1976: Horizontal spatial heterogeneity of zooplankton on the Fladen Ground. I.C.E.S. Contribution to Statutory Meeting, CM 1976/L:20, 5 pp.

Okubo, A., 1972: A note on small organism diffusion around an attractive centre: a mathematical model. J. Oceanogr. Soc. Jap., 28, 1-7.

Pielou, E. C., 1957: The effect of quadrat size on the estimation of Neyman's and Thomas' distributions. J. Ecol., 45, 31-47.

Platt, T., 1972: Local phytoplankton abundance and turbulence. Deep-Sea Res., 19, 183-187.

Platt, T., K. L. Denman and A. D. Jassby, 1977: Modelling the productivity of phytoplankton. In The Sea, E. D. Goldberg, I. N. McCave, J.J. O'Brien and J. H. Steele (eds.), pp. 807-856, Wiley-Interscience, New York.

Skellam, J. G., 1951: Random dispersal in theoretical populations. Biometrika, 38, 196-218.

Skellam, J. G., 1958: On the derivation and applicability of Neyman's type A distribution. Biometrika, 45, 32-36.

Sneddon, I. N., 1957: Elements of partial differential equations. McGraw-Hill, New York, 327 pp.

Steele, J. H. and E. W. Henderson, 1977: Plankton patches in the northern North Sea. In Fisheries Mathematics, J. H. Steele (ed.), pp. 1-19, Academic Press.

Thompson, H. R., 1955: Spatial point processes with applications to ecology. Biometrika, 43, 102-115.

Whittle, P., 1962: Topographic correlation, power-law covariance functions and diffusion. Biometrika, 49, 305-314.

Wroblewski, J. S., J. J. O'Brien and T. Platt, 1975: On the physical and biological scales of phytoplankton in the ocean. Mem. Soc. Roy. Sci. Liege, 7, 43-57.

# BIOLOGICAL EFFECTS OF VERTICAL-HORIZONTAL INTERACTIONS

Geoffrey T. Evans

Marine Laboratory
Victoria Road
Aberdeen AB9 8DB, Scotland

## INTRODUCTION

This paper discusses some effects of vertical structures and processes on horizontal plankton patterns. It describes briefly how vertical patterns arise, and how horizontal variation in the timing of these patterns leads to horizontal structure. My main concern is with vertical current shear, and the ways in which it can influence the development of patterns, in combination with two other vertical processes: zooplankton migration and turbulent mixing. I present results of a mathematical model of shear effects. The model can produce patchy plankton distributions in a uniform environment.

## VERTICAL PATTERNS AND THE FORMATION OF PATCHES

Properties of the sea usually change more rapidly in the vertical direction than in the horizontal. Figure 1 shows similar variability in chlorophyll from a 50 m vertical profile and a 50 km horizontal transect. Vertical patterns are less surprising than horizontal ones, considering the asymmetry of the driving forces. The inputs which drive a water column, both physically (kinetic energy, heat, fresh water) and biologically (light, heat) arrive mainly at the surface. Even if these inputs, and the turbulence which carries them to deeper water, were constant, most heating and primary production would occur near the surface, producing a layer of high temperature and chlorophyll. But the inputs are not constant: they have regular daily and yearly rhythms, and others less predictable but just as important. All these rhythms combine to produce a pattern in time of surface inputs, which is carried down

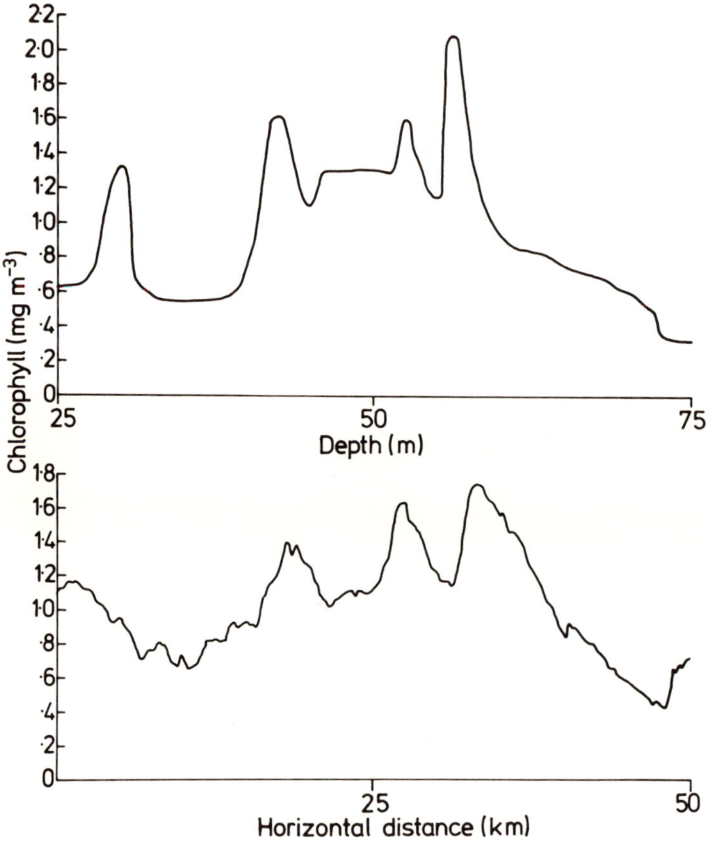


Figure 1. Top: vertical chlorophyll profile between 25 and 75 m depth, redrawn from Strickland (1968). Bottom: horizontal profile of chlorophyll at 3 m depth, measured on Friedrich Heincke during FLEX '76.

to make a changing pattern in space of temperature, chlorophyll and the like. Nor is this all: the turbulence which carries the other properties down is itself a property subject to the same variations.

In winter the sea is well mixed vertically because of strong winds and low heating. Any given phytoplankton cell will spend most of its time in deep water which is too dark for photosynthesis. In the spring, heating increases and winds abate until a layer of reduced mixing (a thermocline) forms near the surface. Plant cells above this layer are now trapped in the region of high illumination, and grow rapidly, leading to the spring phytoplankton bloom (Gran and Braarud, 1935; Riley, 1942). At the time of stabilization, the sea is especially sensitive to small variations in density or turbulent energy which can determine whether a thermocline forms.

Such variations could come from incomplete mixing of old water masses, or from variations in weather like patchy cloud or gusts of wind (Tully and Giovando, 1963). As soon as a thermocline does form, input heat is mostly confined to the surface water, reducing its density and increasing the stability of the thermocline. The phytoplankton bloom contributes to this by reducing the penetration of light and heat radiation (Denman and Miyake, 1973). Therefore, the thermocline can soon persist in the face of winds which would have prevented its formation. In other words, in early spring the sea is poised to make patterns: the time and depth of formation of the spring thermocline, and therefore the onset and speed of the spring outburst, can vary greatly in space as the result of small variations in the initial conditions.

This variation in the start of the production cycle will produce patterns of phytoplankton, and later perhaps zooplankton, which will persist even after the whole sea has stabilized. A head start of a few days could produce a noticeable pattern lasting for weeks.

## SHEAR AND VERTICAL MIGRATION

Horizontal currents in the sea are generally different at different depths: the result of competing wind and bottom friction modified by topography. This vertical current shear, interacting with other processes, has important effects on plankton. Hardy was the first to identify one: the interaction between vertical shear and vertical zooplankton migration (Hardy and Gunther, 1935; Hardy, 1953, 1967). If the surface layer is moving relative to deeper water, then in the course of its daily migration an animal will leave the surface layer, remain still as the surface moves over it, and ascend again well upstream, much further than it could have swum. Hardy suggested this as a major adaptive advantage of migration: it provides the zooplankton with a means of sampling widely separated areas of the sea with little cost in energy.

If animals want to sample different areas, it stands to reason that some areas will be less attractive. "It is suggested that the greater the phytoplankton concentration the deeper, *or possibly for a longer time*, will the zooplankton organisms penetrate into the layer below, and since the surface currents differ in speed and sometimes in direction also from those below, the organisms will be carried gradually away from the zone of high phytoplankton concentration." (Hardy and Gunther, 1953; their emphasis). Hardy later discarded the idea that dense phytoplankton patches might emit zooplankton repellent, but the same argument applies to any property of the water which could modify the animals' migration pattern: the interaction with vertical shear can produce zooplankton patches.

Vertical migration of zooplankton has been well reviewed by Longhurst (1976). We shall mention here only the features that seem directly relevant, remembering that the subject is extremely complex and not open to a single interpretation. Animals usually spend the night in the surface layers and the day in deeper water. They often appear to follow a certain level of light intensity as it moves up and down. The attraction of the surface layers, to a herbivore, is probably the presence of food, herbivores appear to congregate in layers of high chlorophyll (Anderson *et al.*, 1972; Mullin and Brooks, 1972). The attraction of deep water may be darkness and concealment from predators (hence the rise at night), or cold water and a more efficient metabolism (McLaren, 1963). If the only attraction of the upper layer is food, then a full animal might descend sooner as suggested by Hardy (above, 1967) and Conover (1968). This seems to have been observed by Pearre (1973) and Mackas and Bohrer (1976). Nauplii and young copepodites tend to migrate less than adult copepods, spending more time at the surface (Russell,1927; Nicholls, 1933). This may be because the energy required for migration is relatively greater, or because it takes them longer to become satiated.

## SHEAR AND VERTICAL MIXING

If, instead of the reasonably concerted motion of vertical migration, we consider the random motions of vertical turbulence, the interaction with vertical shear produces horizontal spreading of patches. Kullenberg (this volume) has reviewed work on the subject; here is a summary of the theoretical results.

Under the action of vertical shear dispersion, a patch which is introduced at a point will, after a long time, have a Gaussian distribution whose variance increases proportional to time, just as if the patch were being spread by real, Fickian horizontal diffusion. The effective horizontal diffusion coefficient is half the rate of change of variance with time. Even for small time when the Fickian analogy does not hold, the variance (or mean square particle separation) of the patch provides a good measure of its size; a measure easy to obtain both in experiments and in theoretical models.

Okubo (1967) developed a general model for the variance of a patch in a two-dimensional sea with a combination of steady and oscillating shears:

$$\frac{dv}{dz} = a_o + a \cos \omega t.$$

He showed that:

1. Real horizontal diffusion, steady shear and oscillating shear can be considered separately and their effects added to get the total variance. There is a characteristic time, $T$, for shear effects: the time since release of the patch (not a constant) for steady shear, and the period of oscillation for oscillating shear.

2. If $T$ is much less than the time $t_1$ required for complete vertical mixing, then the variance is proportional to $D^2kt$, where $k$ is the vertical mixing rate and $D$ is the maximum relative horizontal displacement of two layers one unit apart vertically: $a_0t$ or $a/\omega$.

3. If $T >> t_1$, the variance is proportional to $A^2t/k$, where $A$ is the maximum shear $a_0$ or $a$.

Kullenberg (1974) has extended the case $T << t_1$ to include two horizontal dimensions and two oscillating velocities. The results are basically the same: effects are additive, and the two horizontal dimensions can be considered separately. In both these models shear and mixing were constant vertically. In reality, both will vary, generally inversely. The two-layer model studied by Thacker (1976) and Evans (1977) can be considered the opposite extreme: two thick layers with no shear and infinite mixing separated by a very thin layer with high shear and finite mixing. For steady shear the variance can be calculated for all times: the small and large time approximations are qualitatively the same as Okubo's, and the numerical factors for large time are very similar.

Dissolved nutrients, and to a lesser extent phytoplankton, are simply carried along by water movements, and so will be subject to this form of horizontal dispersion. Zooplankton, on the other hand, can probably maintain their desired level in the water despite vertical turbulence; their vertical motion depends on different factors. It therefore follows that the shear-induced horizontal mixing of nutrients and phytoplankton will be different from that (if any) of zooplankton. This may be important for patch formation: when the mixing rates of two reacting components differ, diffusion can sometimes amplify patterns instead of smoothing them out (Turing, 1951; Segel and Jackson, 1972; Okubo, 1974 and this volume; Levin, this volume). Okubo's papers refer specifically to plankton patches, and suggest shear diffusion as a possible reason for different diffusion coefficients.

Phytoplankton, too, are not simply carried along with the water. Most diatoms sink at rates between 10 cm and 10 m per day, depending mainly on size. Eppley *et al.* (1967) found that non-growing cells sink about four times as fast as growing ones:

typical rates for a cell of 20 μ diameter were 2 and 0.5 m per day, respectively. See Smayda (1970) for a review. Flagellates can also swim up or down at speeds up to 2 m $h^{-1}$ (Eppley *et al.*, 1968). All this means that shear-induced spreading of phytoplankton is likely to differ from that of nutrients.

## A MODEL OF VERTICAL SHEAR AND PLANKTON PATTERNS

### Description

I have constructed a numerical model to illustrate some of the features discussed here. Its main purpose is to show off shear effects, not to reproduce exactly the behavior of some part of the sea. In John Steele's happy phrase, we want to produce not a picture but a caricature of reality, emphasizing certain points.

Other people have introduced shear as an important component of biological models. Kamykowski (1974, 1976) has a model in which vertically migrating phytoplankton interact with shears produced by internal waves. Riley (1976) and Dooley (1977) look at zooplankton transport as the result of vertical migration interacting with tidal shear.

This model simulates the flow of nitrogen through three trophic levels: dissolved nutrient (nitrate and/or ammonia), one species of phytoplankton, and one of herbivorous zooplankton (divided into three size classes: nauplius, copepodite and adult) (Figure 2a). For example, one might think of *Calanus* feeding on *Rhizoselenia* in the North Sea, although no effort has been made to model particular genera accurately. Most of the interaction terms are simple and common to many models. Primary production per unit plant biomass is a Michaelis-type function of nutrient density, less a constant respiration: $aN/(b+N) - c$. Plant cells also sink, at a velocity proportional to $b/(b+N)$. Herbivore grazing has a form similar to plant production, but with a threshold phytoplankton concentration below which no grazing occurs: the grazing rate per animal is $d(P-P_O)/(e+P-P_O)$. The food assimilated by, say, nauplii is not used to increase nauplius biomass, but rather to turn a few nauplii into copepodites, while the food assimilated by adults is used to produce new nauplii. In the results described here, a constant fraction of the herbivore numbers is removed by predation, although the model allows predation to be a more complicated function of herbivore numbers.

These interactions take place in a two-dimensional vertical slice through the water. The water is divided vertically into three biologically active layers in relative horizontal motion, and a deep reservoir layer (and horizontally into a large number of boxes for numerical purposes). The reservoir serves as a source of nutrients and as a place for animals to go during the day. There is mixing between vertically adjacent boxes, but not between horizontally adjacent ones. The ends of each layer are joined to form a ring: this allows the populations to develop without interference from external boundaries (Figure 2b).

I use the following model for vertical zooplankton migration. At noon all the animals are in the reservoir layer. At some set time they all rise and must decide which of the three active layers to enter. Each layer has a certain attraction, which is a non-decreasing function of phytoplankton density, and the animals

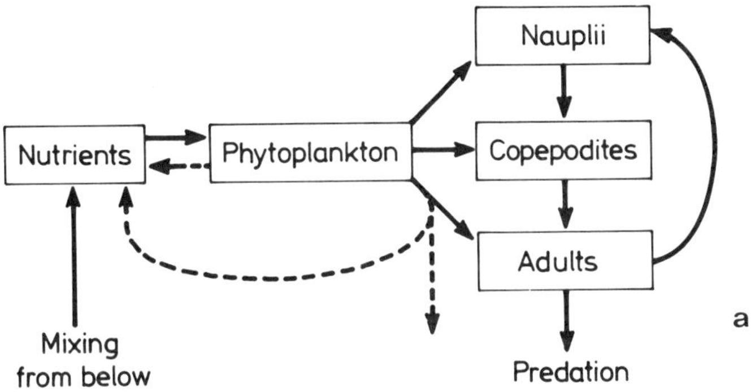


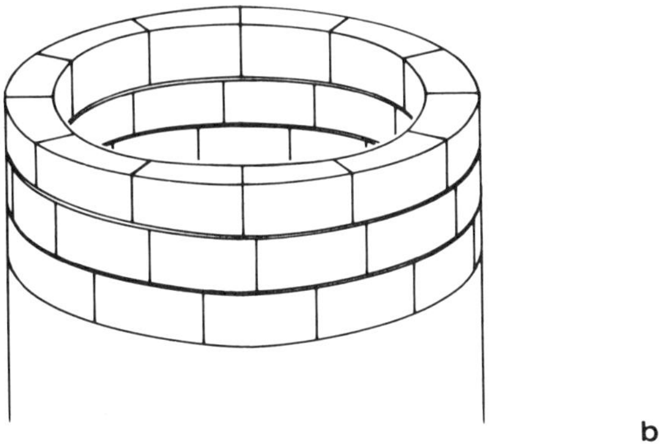


Figure 2. A model of vertical shear and plankton populations: (top) biological interactions; (bottom) physical framework. The soluble and insoluble losses during grazing, shown as dashed lines for adults, occur also for the smaller classes.

divide so that their concentrations in the three layers are in the same proportion as the attractions of the layers. Each animal stays in its selected layer until it descends into the reservoir again, either at a set time or when it has eaten a set amount of food, whichever comes first.

The model is a slight modification of the one described in detail by Evans *et al.* (1977). The following sections describe some of the results.

## Dye Runs

Nutrients and phytoplankton can be made to behave like passive contaminants, to test the model against theories of shear dispersion. The model agrees well with the two-layer steady shear theory of Evans (1977). I have also tested the effect of phytoplankton sinking on shear dispersion: the difference in horizontal variance is the same as would be produced by an increased vertical mixing rate. (Subsequent theoretical work confirms this.)

## Column Runs

The model can be run as a single vertical column, with no shear, to generate the main features of "smooth world" theories (Riley, 1963; Steele, 1974b). Starting from pre-outburst conditions, it generates successive peaks of phytoplankton and herbivores as expected. Chlorophyll values are generally higher in mid-water than at the surface, as predicted by Steele and Yentsch (1960). Total primary production over 200 days seems reasonable. No formal sensitivity analysis has been done, but our intuitive impression is that small (factor of two) changes in randomly selected parameters have little effect on the behavior with two exceptions: the rate of vertical mixing in the thermocline and the grazing threshold.

The most interesting feature of the column runs is the existence of solutions which oscillate in time. These can be induced by increasing the thermocline mixing rate or decreasing the grazing threshold (within limits). Figure 3 shows the results of four runs, each starting from pre-outburst conditions, for different thermocline mixing rates. The oscillatory behavior increases with the mixing rate; the division between oscillation and stable equilibrium is near 1.6 $cm^2\ s^{-1}$. The period of oscillation, 50 days, is roughly the same as the generation time for *Calanus* and also the same as in the model of Steele (1974b) which treats zooplankton size structure in a different way. However, more recent work shows that the same 50-day period arises in our model even when zooplankton size structure is effectively absent.

Although herbivore migration generally responds to phytoplankton concentrations, it is possible to choose parameters so that the animals migrate purely mechanically, independent of food. This need not destroy the oscillation although the mixing rate needed to sustain it is higher. In fact, the oscillation does not even depend on the vertical structure: it can occur in simple chemostat models of nutrients, phytoplankton and herbivores (Evans, unpublished).

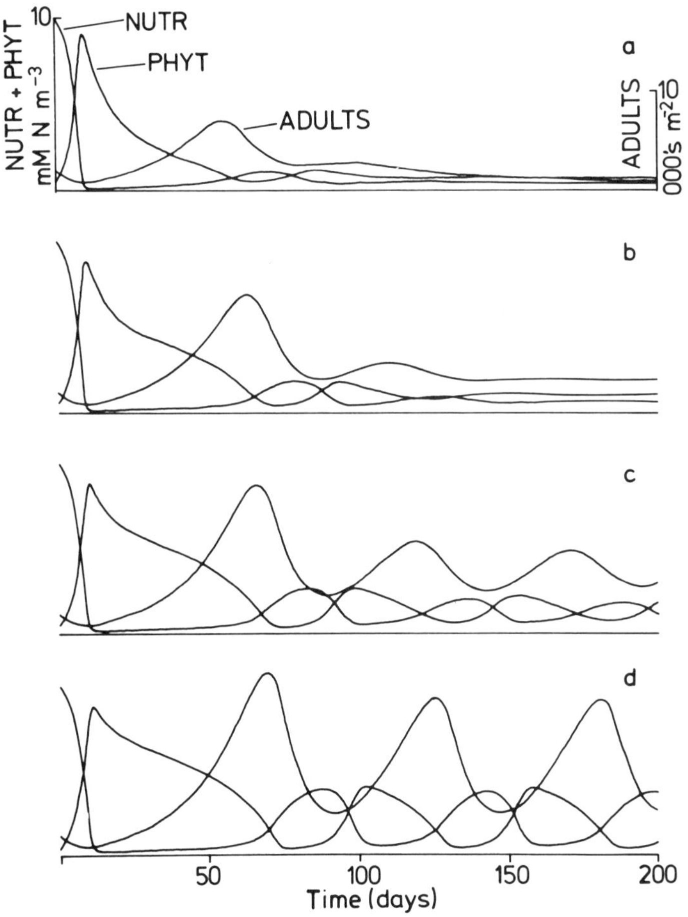


Figure 3. Output from column runs: surface nutrients, surface phytoplankton and adult herbivores, for four different vertical mixing rates: (a) 0.28 $cm^2$ $s^{-1}$; (b) 0.83; (c) 1.61; (d) 2.78.

## Full Model: Shear and Biology

Running the full model has shown that oscillating columns can combine with vertical shear to produce spatial waves. Figures 4-6 illustrate three runs (call them A, B and C) in which the biology of Figure 3d (the limit cycle) was used in a 200 km ring in which the top layer rotates once in 100 days and the middle one once in 100 (giving a shear of about $10^{-3}$ $s^{-1}$). The runs differ only in initial conditions: A starts with large amplitude patches of various lengths, each column representing a point on the limit cycle; B starts with almost flat initial conditions near a point on the limit cycle; C starts with a single sine wave in nutrients and phytoplankton. These figures show that spatial patterns can persist and grow, that short waves seem to be damped (compare days 25 and 175 of Run A) while long waves break up into shorter ones. They also suggest (to the eye of faith) that the statistical properties of the model may tend to a steady state.

Figure 4 also shows an apparent flaw in the model: the consistent strong negative correlation between nutrients and phytoplankton. Such a correlation is sometimes observed near the start of a bloom (Steele, this volume) but not throughout the year. This behavior in the model may be the result of homogeneous physical conditions: the rate of mixing and the deep water nutrient supply are both constant in space and time.

One way of getting a quick look at the behavior of a run is to look each day at the maximum and minimum concentrations of adults around the ring, and then plot these as a function of time, as in Figure 7. Such plots show that Run A behaves irregularly, and that the waves in Run B grow in amplitude. The oscillations of Run B have a period slightly shorter than the natural limit cycle. Run A has been followed for 2000 days: after about 500 days the max-min adult graph suddenly switches to a pattern more like Run B, appearing almost but not quite periodic, with period 207 days (four large oscillations). Although there is no hope of proving the system is 'chaotic,' it would not be surprising considering the results which Ruelle and Takens (1971) and May (1975) have proved for much simpler systems.

We can also see from Run A that the amplitude of the cycles in the full model is often greater than the amplitude of the limit cycle (the initial conditions for Run A): shear can amplify oscillations. This raises the exciting possibility that it may be able to create them in one of the ways described earlier. A set of parameters which in a column model would have a stable equilibrium may oscillate in space and time when inserted into a full model with shear. To investigate this, I looked at the parameter set of Figure 3c, which was on the borderline between equilibrium and oscillation. I used this set in a full 200 km model, with flat

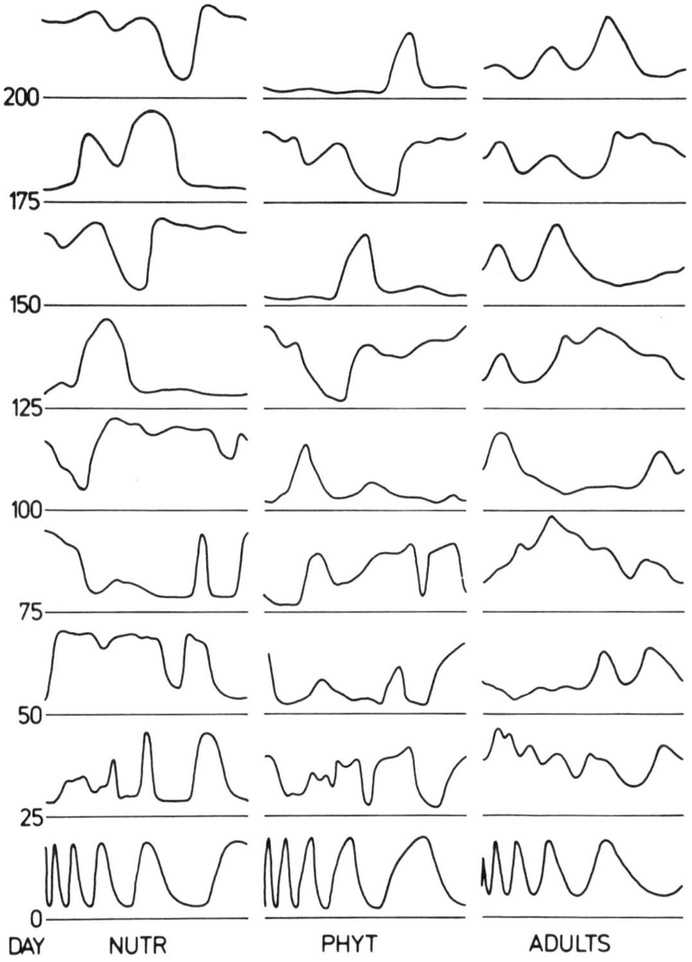


Figure 4. Output from a full (200 km ring) run of the model: Run A. The bottom set of graphs shows the initial distributions of surface nutrients, surface phytoplankton and adult herbivores, each plotted against distance around the ring. The maximum of nutrients and phytoplankton is about 4 mM N $m^{-3}$, of adults about 15000 under a square meter. Graphs further up the page show these same distributions about 25, 50, ....., 200 days.

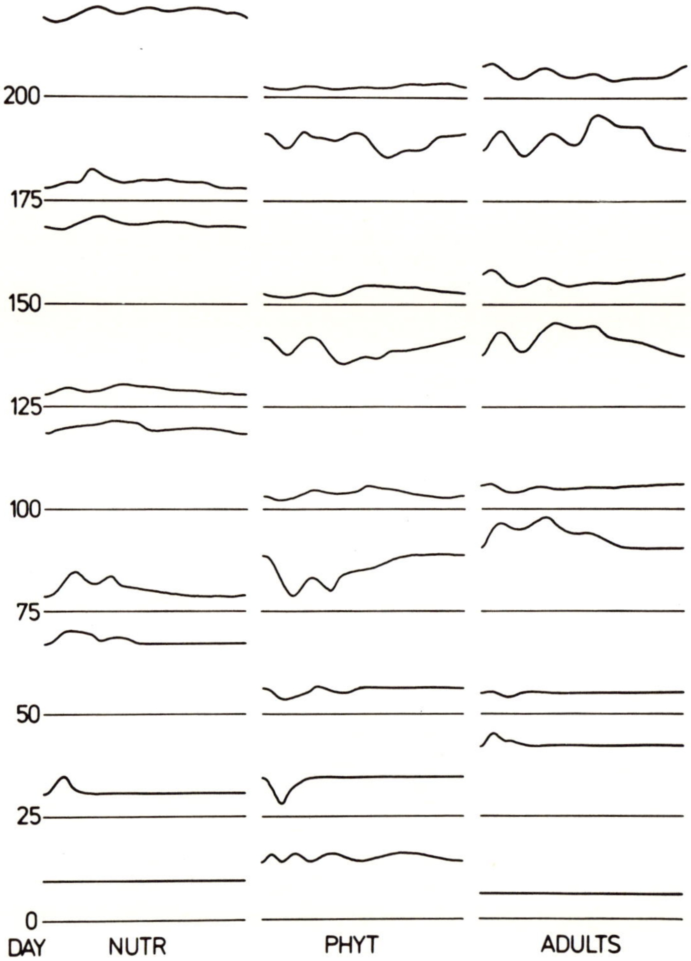


Figure 5. Output from Run B. See legend of Figure 4.

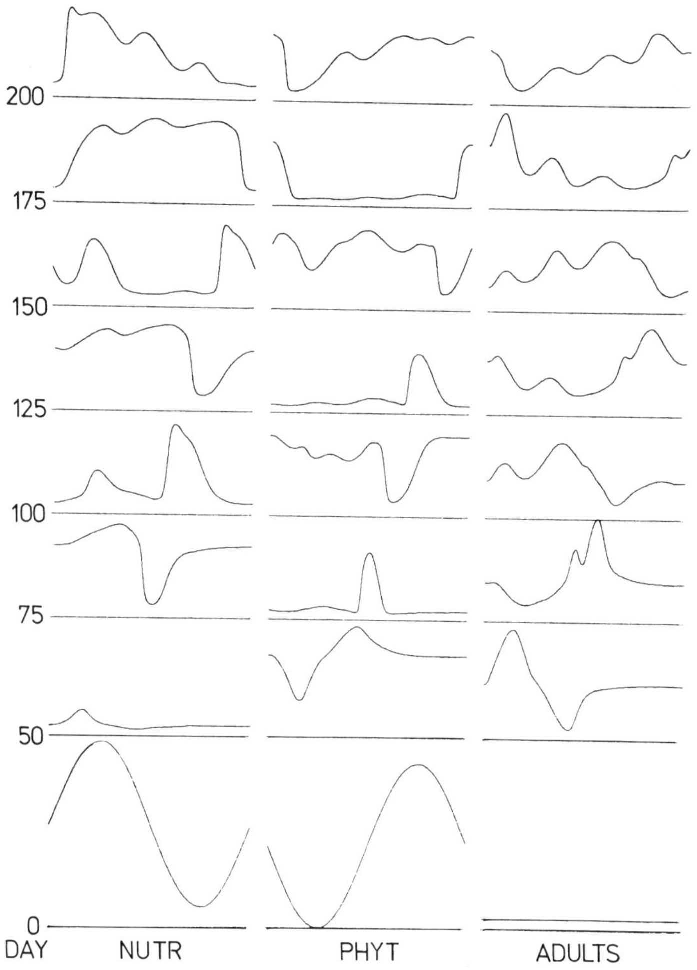


Figure 6. Output from Run C. See legend of Figure 4 (day 25 is omitted for reasons of space).

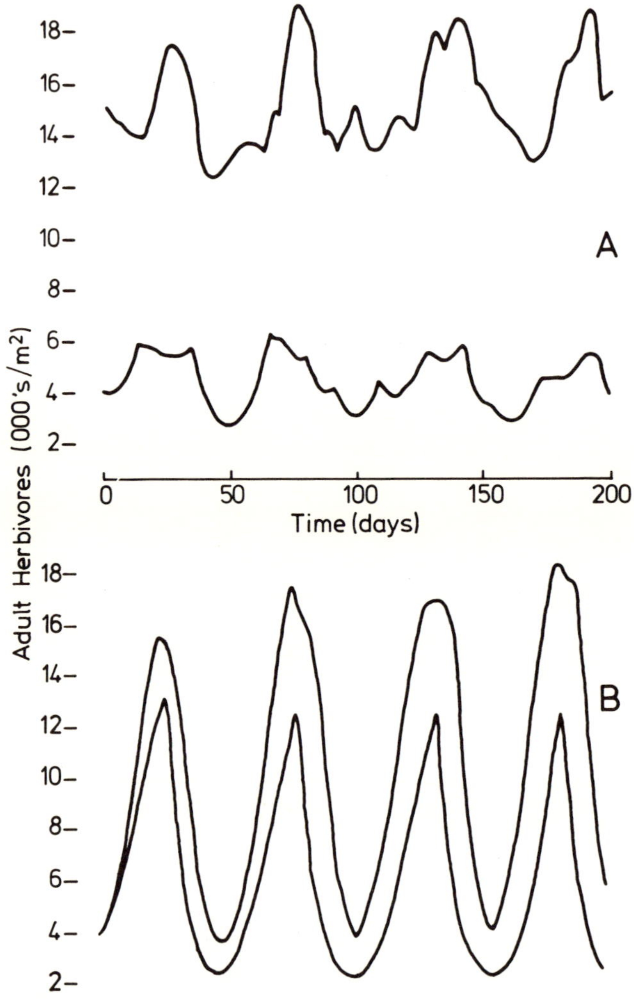


Figure 7. Maximum and minimum adult concentrations each day plotted against time, for runs A and B.

initial conditions near equilibrium apart from a small ripple in the phytoplankton. Figure 8 shows plots of maximum and minimum number of adults for two runs. In the first there was no shear; each column ran by itself. The initial variation in the phytoplankton is transmitted to the zooplankton, then damped out. The second run had the same shear as Run A and the variation increases.

The shear-induced patchiness might have been due to the effect of food on vertical migration, but I can show that it happens even when there is no satiation. This suggests that it is an example

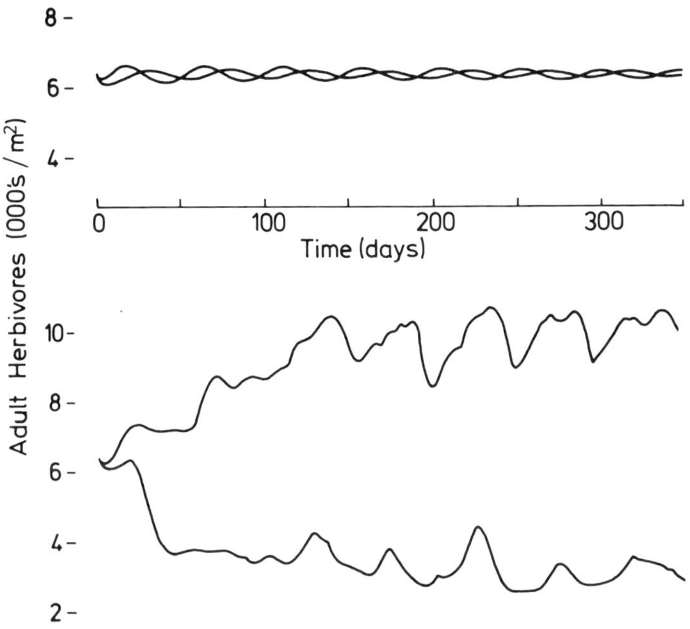


Figure 8. Maximum and minimum adults for two runs near the borderline of oscillating columns (see Figure 3c); top without shear, bottom with shear.

of diffusion-induced patchiness, although we cannot rule out the possibility of other, non-diffusive effects of shear.

One of the reasons for studying patches is the belief that they affect average levels. Figure 9 shows a comparison between Run C and a smoothed version (a column run whose initial conditions were the average of Run C's). Compared with the smoothed run, the average phytoplankton in Run C has oscillations which are about 35% smaller in amplitude, 5 days shorter, and with a time average 10% less.

## Spectral Analysis

Many questions about the behavior of the model, concerning statistical steady states for example, or the decay of short waves, can best be asked of its spectra (Platt and Denman, 1975). The variables are constrained to be periodic in space and are, therefore, precisely represented by their Fourier coefficients. Also, it may be reasonable to regard separate runs as realizations of a single stochastic process since the initial conditions, which can be chosen at random, have a long lasting effect on the results.

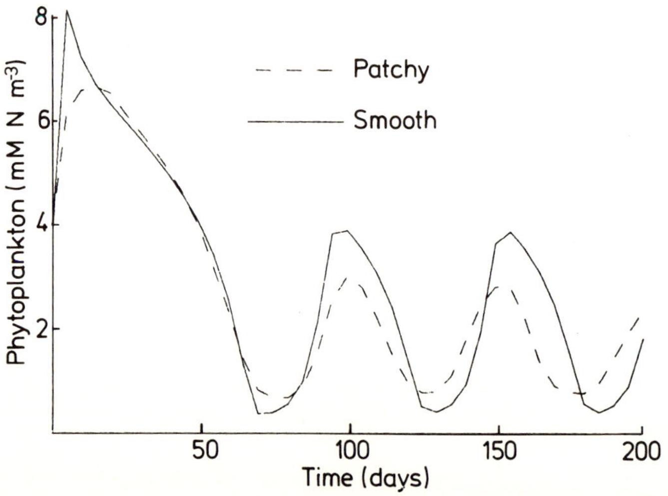


Figure 9. Spatially averaged phytoplankton from Run C (dashed line) and from the corresponding column run (solid line).

Spectra provide a useful means of comparison with other people's observations (Fasham and Pugh, 1976; Denman, 1976) or theories (Platt, this volume; Fasham, this volume). We must recognize, though, that they can be misleading in nonlinear interactions (Steele, 1974a). In Figure 4, for example, it often appears that nutrient and phytoplankton patches have a characteristic shape, with sharp edges; as Mackas (1977) points out, such a shape would contain short wavelength terms which have nothing to do with small patches.

Variance spectra for Run A show that the variation at scales of 15 to 40 km was much lower at day 175 than at day 25 (Figure 10). I have also looked at smoothed spectra (smoothed in increasing groups as suggested by Fasham in this volume) plotted on a logarithmic scale (Figure 11), for which Platt and Fasham predict slopes in the region of -1 to -3. These have much steeper slopes, more like -4 to -7: this may be because this model is set in a physically smooth world, while Platt and Fasham deliberately introduce short wavelength noise.

## DISCUSSION AND SPECULATION

Vertical shear, diffusion and migration are all common in the sea, and the model shows that they can have an important combined effect on horizontal patchiness: to understand horizontal

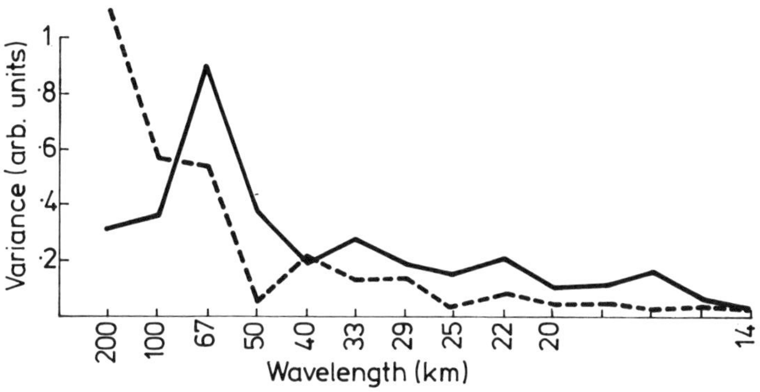


Figure 10. Raw spectrum of phytoplankton for days 25 (solid line) and 175 (dashed line) of Run A. Variance in arbitrary units.

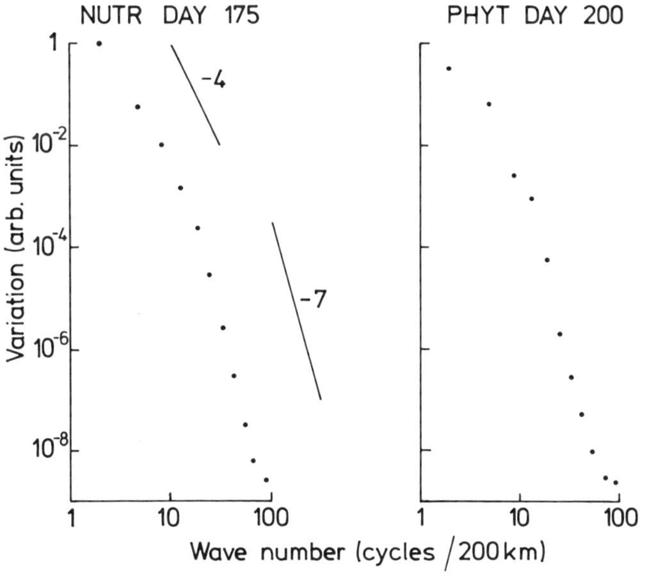


Figure 11. Smoothed spectra of nutrients from day 175 and phytoplankton from day 200 of Run A. The log wave number scale is expanded relative to the log variance.

structure we must keep vertical processes in mind. The model was constructed deliberately taking a very narrow view; in this section I broaden the view somewhat and speculate about interactions which have not been considered. One of the problems of any ecological model (indeed, of any work in theoretical ecology) is the need for 'closure terms,' which represent the effect on the model of parts of the ecosystem we have chosen to overlook. There are two explicit closure terms in the model: predation on the herbivores, and the deep water supply of nutrients.

In all the model runs so far, predation has been represented by removing a constant fraction of herbivore numbers each day. The model is written to allow various kinds of density dependent mortality which Segel and Jackson (1972), Landry (1976) and Steele (1976) have shown to have important effects on model behavior, but this aspect has not been tested yet.

At the other end of the food chain, we assume that the density of nutrients in deep water is constant. This assumption is reasonable in deep oceans, less so in the North Sea. Assumptions like constant nutrient supply and constant maximum photosynthetic rate limit the usefulness of the model as a long term simulation, but not as an exploratory tool.

A more important, related assumption is that the vertical mixing rate is constant in space and time. This rate varies a lot in the sea, and I have already mentioned that variations can cause patches in early spring. Later in the year, small scale variations are almost certain to influence the spectrum, and possibly the large scale visible features as well. Work is in progress on a model in which the vertical mixing rate can vary in space and time, in response to changing wind and heat and variations in water density. In the meantime, the present model shows how plankton patches can arise even in a smooth environment (although the thermocline mixing rates needed to achieve this may be unreasonably high).

The restrictions of the physical setting are probably not important. We only treat one horizontal dimension, but models from Skellam (1951) onwards have found little difference between one dimension and radial symmetry. The boundary conditions avoid interference from 'outside' water, and although each run is constrained to have a definite fundamental wavelength, this length can be changed from run to run.

We found that patchiness in the model had little effect on mean concentrations. However, the model does not consider competition within a trophic level, where we might expect the effects to be greater. This would be an interesting line to pursue.

The most dubious piece of biology in the model is the vertical zooplankton migration. I have mentioned that migration has several possible benefits, and so an optimum migration strategy is bound to be complicated. Migration is also difficult to study experimentally, and no clear pattern has emerged. Because of all this, there is no real chance of building the 'right' migration model into the full model. The next best thing is probably to try a lot of different models, to see how much difference they make to overall behavior. For example, we might suppose that the animals react to the phytoplankton concentration while they are swimming up, so that if food is plentiful in the lower active layers, few animals will bother to swim to the top; or we might introduce some spread in the timing of migration.

The outputs have concentrated on synoptic pictures of a 200 km ring. A ship might take 20 hours to survey such a ring -- W. C. Clark has suggested that it might be illuminating to examine such quasi-synoptic records in the model.

The division of herbivores into size classes is not really consistent, since all three classes are feeding on the same phytoplankton. Partly because of this, and partly because of unavoidable numerical problems, the size structure quickly settles into a nearly steady state. It may turn out that the interesting behavior of the model does not depend on size structure.

## SUMMARY

Given the importance of vertical shear in the sea, it is important to look at the vertical movements of plankton. This talk examined some of the possible patch-forming effects of these motions. One such effect (suggested by Hardy) is that vertically migrating zooplankton can move large horizontal distances and perhaps form aggregations, using shear. Another is that the different vertical motions of nutrients, phytoplankton and herbivores will lead to different shear-induced horizontal mixing rates, and hence possibly to the diffusion-induced patches suggested by Okubo and others.

A mathematical model was described of vertical shear interacting with a nutrient-phytoplankton-herbivore ecosystem. The model can produce plankton patches even in a physically smooth world.

## REFERENCES

Anderson, G. C., B. W. Frost and W. K. Peterson, 1972: On the vertical distribution of zooplankton in relation to chlorophyll concentrations. In Biological Oceanography of the Northern North Pacific Ocean, A. Y. Takenouti (ed.), pp. 341-345, Idemitsu Shoten, Tokyo.

Conover, R. J., 1968: Zooplankton - life in a nutrionally dilute environment. Am. Zoologist, 8, 107-118.

Denman, K. L., 1976: Covariability of chlorophyll and temperature in the sea. Deep-Sea Res., 23, 539-550.

Denman, K. L. and M. Miyake, 1973: Upper layer modification at ocean station Papa: observations and simulation. J. Phys. Oceanogr., 3, 185-196.

Dooley, H. D., 1977: The long term transport of diurnally migrating plankton. I.C.E.S. C.M. 1977: C24.

Eppley, R. W., R. W. Holmes and J. D. Strickland, 1967: Sinking rates of marine phytoplankton measured with a fluorometer. J. Exp. Mar. Biol. Ecol., 1, 191-208.

Eppley, R. W., O. Holm-Hansen and J. D. Strickland, 1968: Some observations on the vertical migration of dinoflagellates. J. Phycol., 4, 333-340.

Evans, G. T., 1977: A two layer shear diffusion model. Deep-Sea Res., 24, 931-936.

Evans, G. T., J. H. Steele and G.E.B. Kullenberg, 1977: A preliminary model of shear diffusion and plankton populations. Scot. Fish. Res. Rep. 9, 20 pp.

Fasham, M.J.R. and P. R. Pugh, 1976: Observations on the horizontal coherence of chlorophyll a and temperature. Deep-Sea Res., 23, 527-538.

Gran, H. H. and T. Braarud, 1935: A quantitative study of the phytoplankton in the Bay of Fundy and the Gulf of Maine. J. Biol. Bd. Canada, 1, 279-467.

Hardy, A. C., 1953: Some problems of pelagic life. In Essays in Marine Biology, the Richard Elmhirst Memorial Lectures, pp. 101-121, Oliver and Boyd, Edinburgh.

Hardy, A. C., 1967: Great Waters. Collins, London, 542 pp.

Hardy, A. C. and E. R. Gunther, 1935: The plankton of the South Georgia whaling grounds and adjacent water, 1926-1927. Discovery Report 11, 466 pp.

Kamykowski, D., 1974: Possible interactions between phytoplankton and semidiurnal internal tides. J. Mar. Res., 32, 67-89.

Kamykowski, D., 1976: Possible interactions between plankton and semidiurnal internal tides. II. Deep thermoclines and trophic effects. J. Mar. Res., 34, 499-509.

Kullenberg, G.E.B., 1974: An experimental and theoretical investigation of the turbulent diffusion in the upper layers of the sea. University of Copenhagen, Inst. of Physical Oceanography Rept. No. 25, 212 pp.

Landry, M. S., 1976: The structure of marine ecosystems: an alternative. Mar. Biol., 35, 1-7.

Longhurst, A. R., 1976: Vertical migration. In The Ecology of the Seas, D. H. Cushing and J. J. Walsh (eds.), pp. 116-137, Blackwell, Oxford.

Mackas, D., 1977: Ph.D. Thesis, Dalhousie University, Institute of Oceanography.

Mackas, D. and R. Bohrer, 1976: Fluorescence analysis of zooplankton gut contents and an investigation of diel feeding patterns. J. Exp. Mar. Biol. Ecol., 25, 77-85.

May, R. M., 1975: Biological populations obeying difference equations: stable points, stable cycles and chaos. J. Theor. Biol., 51, 511-524.

McLaren, I. A., 1963: Effects of temperature on growth of zooplankton and the adaptive value of vertical migration. J. Fish. Res. Bd. Canada, 20, 685-727.

Mullin, M. M. and E. R. Brooks, 1972: The vertical distribution of juvenile Calanus and phytoplankton within the upper 50 m of water off La Jolla, California. In Biological Oceanography of the Northern North Pacific Ocean, A. Y. Takenouti (ed.), pp. 347-354, Idemitsu Shoten, Tokyo.

Nicholls, A. G., 1933: On the biology of Calanus finmarchicus. III. Vertical distribution and diurnal migration in the Clyde Sea area. J. Mar. Biol. Ass. U.K., 19, 139-164.

Okubo, A., 1967: The effect of shear in an oscillatory current on horizontal diffusion from an instantaneous source. Int. J. Oceanol. and Limnol., 1, 194-204.

Okubo, A., 1974: Diffusion-induced instability in model ecosystems. Chesapeake Bay Institute Tech. Rept. 86, 17 pp.

Pearre, S., 1973: Vertical migration and feeding in Sagitta elegans. Ecology, 54, 300-314.

Platt, T. and K. L. Denman, 1975: Spectral analysis in ecology. Ann. Rev. Ecol. Syst., 6, 189-210.

Riley, G. A., 1942: The relationship of vertical turbulence and spring diatom flowering. J. Mar. Res., 6, 67-87.

Riley, G. A., 1963: Theory of food-chain relations in the ocean. In The Sea, M. N. Hill (ed.), Vol. 2, pp. 438-463, Wiley, New York.

Riley, G. A., 1976: A model of plankton patchiness. Limnol. Oceanogr., 21, 873-879.

Ruelle, D. and F. Takens, 1971: On the nature of turbulence. Comm. Math. Phys., 20, 167-192.

Russell, F. S., 1927: The vertical distribution of plankton in the sea. Biol. Rev., 2, 213-262.

Segel, L. A. and J. L. Jackson, 1972: Dissipative structure: an explanation and an ecological example. J. Theor. Biol., 37, 545-559.

Skellam, J. G., 1951: Random disperal in theoretical populations. Biometrika, 38, 196-218.

Smayda, T. J., 1970: The suspension and sinking of phytoplankton in the sea. In Oceanogr. Mar. Biol. Ann. Rev., H. Barnes (ed.), pp. 353-414, George Allen & Urwin, London.

Steele, J. H., 1974a: Spatial heterogeneity and population stability. Nature, 248, 83.

Steele, J. H., 1974b: The Structure of Marine Ecosystems. Harvard University Press, Cambridge, 128 pp.

Steele, J. H., 1976: The role of predation in ecosystem models. Mar. Biol., 35, 9-11.

Steele, J. H. and C. S. Yentsch, 1960: The vertical distribution of chlorophyll. J. Mar. Biol. Ass. U.K., 39, 217-226.

Strickland, J. D., 1968: A comparison of profiles of nutrient and chlorophyll concentrations taken from discrete depths and by continuous recording. Limnol. Oceanogr., 13, 388-391.

Thacker, W. C., 1976: A solvable model of shear dispersion. J. Phys. Oceanogr., 6, 66-75.

Tully, J. P. and L. F. Giovando, 1963: Seasonal temperature structures in the eastern subarctic Pacific Ocean. In Marine Distributions, M. J. Dunbar (ed.), pp. 10-36, University of Toronto Press, Toronto, Canada.

Turing, A. M., 1952: The chemical basis of morphogenesis. Phil. Trans. Roy. Soc. B., 237, 37-72.

# MIXING AND STABILIZATION OF PHYTOPLANKTON DISTRIBUTIONS ON THE NORTHWEST EUROPEAN CONTINENTAL SHELF

R. D. Pingree

Institute of Oceanographic Sciences
Wormley, Godalming
Surrey, U.K.

## INTRODUCTION

This chapter examines the effects of mixing and stabilization on phytoplankton distributions on the northwest European shelf (Figure 1). The European continental shelf is an area of generally large tide with associated currents frequently exceeding 1 m $sec^{-1}$. It has been estimated that approximately 10% of the total global dissipation of the $M_2$ tide in shallow seas occurs on the northwest European shelf. It is interesting to note that this large tidal energy dissipation is not evenly distributed over the area under consideration but varies geographically by nearly four orders of magnitude. Since the effects of wind mixing, on the other hand, are more uniform, in an initial simplification it is convenient to consider only the mixing due to tidal stirring on the spatial patterns of phytoplankton.

Mixing and stabilization determine the availability of nutrients and light energy necessary for phytoplankton growth. The availability of nutrients and light can largely be found from two parameters, the stratification parameter and the water depth, scaled appropriately by the extinction coefficient or Secchi Disc depth. This chapter considers in more detail how these two parameters can be used to predict and classify the development of spring, summer and autumn phytoplankton distributions commonly observed on the northwest European shelf. Unfortunately, few hydrographic surveys have been conducted with the spatial and temporal resolution necessary to allow unambiguous interpretations in terms of the stratification parameter and scaled water depth and, consequently, examples supporting this approach have been drawn mainly from the Celtic Sea and English Channel where the

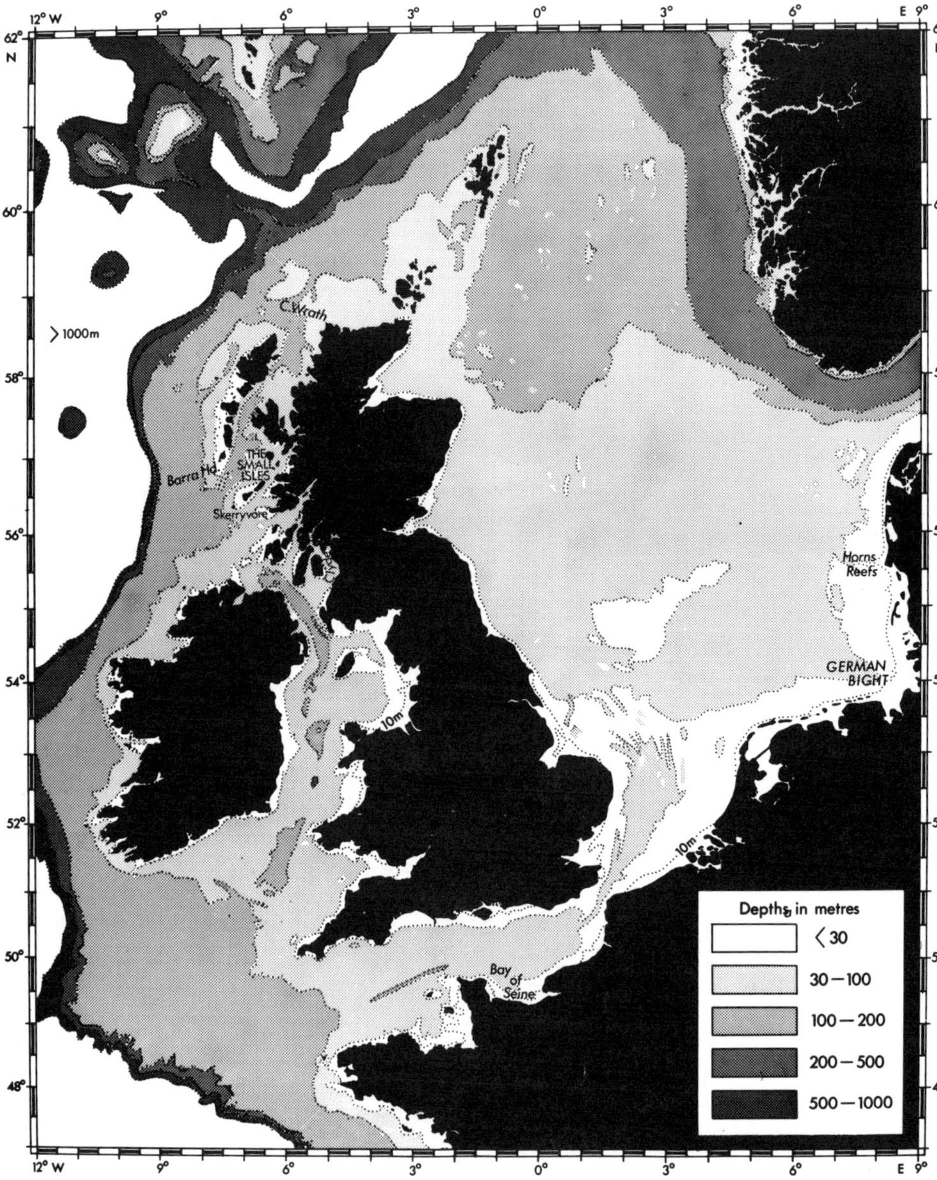


Figure 1. Bottom topography of the northwest European continental shelf and place names.

scientific cruise programs were planned with these concepts in mind. Since this is a first attempt at introducing the stratification parameter as a biological index, it is necessarily inadequate in many details and it is hoped that others may be stimulated to take the appropriate measurements in other areas, thus either confirming the generality of this approach or confining the conclusions to the Celtic Sea and English Channel.

## THE STRATIFICATION PARAMETER $S$ AND SCALED WATER DEPTH $H$

The stratification parameter represents a convenient measure of the tendency for the water column to stabilize under the influence of a surface heat flux, $Q_o$, in the presence of mixing due to tidal stirring. It may be defined by considering the ratio $R$ of the production of potential energy ($PE$) in maintaining well-mixed conditions to the rate of tidal energy dissipation ($TED$) in a water column of unit cross sectional area. Thus

$$R = \frac{PE}{TED} \qquad (1)$$

or

$$R = \int_o^h \frac{g\alpha Q(Z)}{Cp}\, dz \Big/ \rho C_D\ \overline{|u|^3} \qquad (2)$$

or

$$R = \frac{g\alpha Q_o}{2C_p\rho}\left[\frac{h}{C_D\ \overline{|u|^3}}\right] \qquad (3)$$

where $\alpha$ is the volume coefficient of expansion,
$C_p$ is the specific heat at constant pressure,
$g$ is the acceleration due to gravity,
$\rho$ is the density of sea water,
$C_D$ is the drag coefficient, $\sim 0.0025$,
$h$ is the water depth measured from the bottom to the free surface,
$Q(Z)$ is the heat flux at a height $Z$ above the bottom, thus
$$Q(Z) = \frac{z}{h} Q_o$$
$|u|$ is the magnitude of the instantaneous tidal stream velocity, $u$,
and —, the over bar, denotes the average value over one tidal cycle.

Simpson and Hunter (1974) have argued that for a constant surface heat flux, $Q_o$, then critical values of the stratification parameter, $S$, where

$$S = log_{10}\left[\frac{h}{C_D\ |u|^3}\right] \quad (c.g.s.\ units) \qquad (4)$$

will determine the position of summer boundary fronts separating well-mixed waters from well-stratified waters. Fearnhead (1975) developed these simple arguments to include some effects of wind mixing. For developing the arguments presented here it will be sufficient to derive quantitatively contours of the Simpson-Hunter stratification parameter from numerical techniques and then qualitatively to consider any departures from the basic theme caused by wind mixing.

Although initially a critical value of the stratification parameter is derived to determine the positions of the boundary regions separating well-mixed regimes from well-stratified regimes, this exclusive use of the stratification parameter is relaxed later. Thus regions characterized by high values of the stratification parameter may be associated with well-stratified conditions in the summer months, whereas low values represent areas where the water column remains well mixed throughout the year. The stratification scale covers, in effect, the range of conditions from where weather dominates mixing and stabilization, to conditions dominated by tidal stirring. If we now specify the scaled depth, $H$, as

$$H = water\ depth/Secchi\ Disc\ Depth\ (D),$$

it will be shown that these two parameters are sufficient to characterize phytoplankton development on the northwest European shelf since they largely control the availability of nutrients and light energy. The reader concerned only with the practical use of this approach rather than the formal basis from which it is derived may wish to proceed to the section on practical justification of the use of a critical value of the stratification parameter in frontal prediction.

## NUMERICAL MODELS

Although values for the stratification parameter can be derived from suitable hydrographic data on tidal streaming and water depth (Hughes, 1976), it is more convenient to derive the values from hydrodynamic numerical models of the principal semi-diurnal lunar tide ($M_2$) in the region under consideration (Pingree and Griffiths, 1978). This has an advantage in accuracy and resolution since the distribution of measured tidal streams is usually rather limited. Since much use is subsequently made of the results of the numerical model, it is worthwhile to summarize some salient features of the numerical technique.

To predict the positions of the summer frontal boundaries between well-stratified and well-mixed waters from numerical models we require only the tidal elevations from a few chosen positions bordering the open sea boundaries of the modelled region. These have been obtained from tide gauges at the edge of the continental shelf and are used as input to the numerical model, thereby specifying the elevation on the open sea boundaries at all times (Flather, 1976). In addition, it is necessary to apply the condition that the normal velocity at a coastal boundary is zero. The rules governing tidal motion are contained in the momentum equation describing the conservation of horizontal momentum and the continuity equation for conservation of mass. Due to the presence of frictional dissipation after a few tidal cycles, the numerical model will correctly simulate the $M_2$ tide irrespective of the initial interior values used to start the numerical model, provided certain stability requirements have been met.

Since the region of interest covers 15° of latitude (47°N-62°N) and 21° of longitude (12°W-9°E), variations of the Coriolis parameter with latitude and the curvature of the earth's surface are taken into account by using spherical coordinates. A grid of approximately 5 nautical miles (5' of latitude and 10' of longitude) was considered sufficient to give adequate spatial resolution (Figure 2). The model contained about $22 \times 10^3$ grid points of which about half were land, giving about $12 \times 10^3$ positions in the sea where tidal characteristics could be evaluated every minute of the tidal cycle. From the tidal stream values the tidal energy dissipation rates could be calculated.

## THE EQUATION DESCRIBING THE BALANCE OF TIDAL ENERGY

A simplified equation describing the energy balance for tidal motion can be written as

$$\frac{\partial(K.E.)}{\partial t} + \nabla.(K.E.)\,u + \frac{\partial}{\partial t}\,(T.P.E.) = -\nabla\rho.g(h+E)uE - \rho C_D|u|^3 \qquad (5)$$

where $u$ is the tidal stream velocity, $E$ is the tidal elevation and $(h+E)$ is the total depth of the water column. The first term on the lefthand side of equation (5) represents the local change of tidal kinetic energy with respect to time, the second term

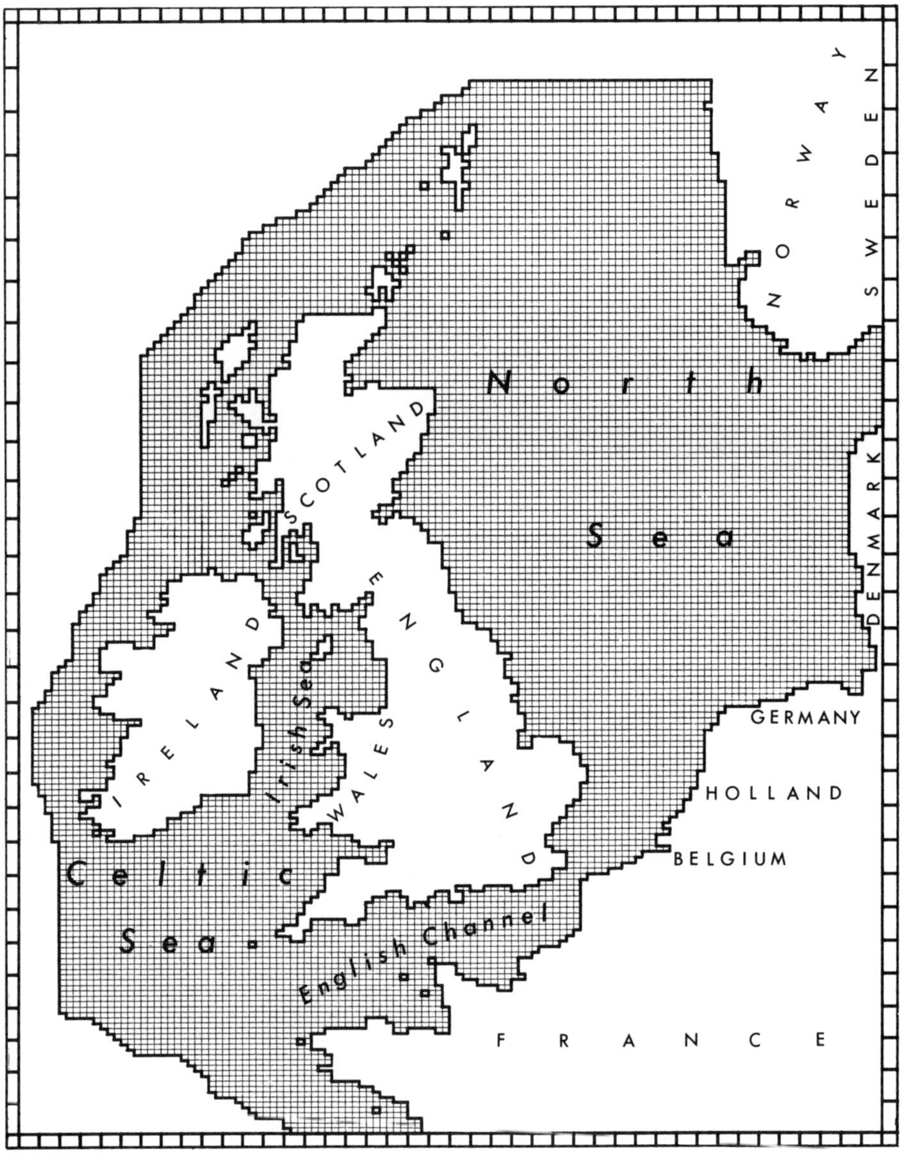


Figure 2. Finite difference grid use in the numerical model and place names.

represents changes in tidal energy due to advection and the third term considers changes in the potential energy of the tidal wave resulting from changes in sea surface elevation. The last term on the righthand side of equation (5), $-\rho C_D|u|^3$, represents the rate of dissipation of tidal energy per $cm^2$ due to work done against the frictional drag $-\rho C_D\, u^2$. In order to balance this continual loss of tidal energy on the continental shelf, energy enters the modelled region through the open sea boundaries as described by the integration over space and time of the divergence of energy flux, the first term on the righthand side of equation (5) (the second term on the lefthand side making a negligible contribution). Direct effects of the tide-generating force have been neglected (Garrett, 1975).

## THE NUMERICAL EVALUATION OF THE STRATIFICATION PARAMETER

If we now average over time, equation (5) reduces to

$$0 = \overline{-\nabla.\rho g(h+E)\; uE} \;\; -\rho C_D\overline{|u|^3} \qquad (6)$$

which simply states that the generation of tidal energy is balanced by the rate of tidal energy dissipation. Contours of the parameter

$$\left[\frac{h}{C_D\overline{|u|^3}}\right]$$

can now be conveniently derived from the numerical model by dividing the mass of the water column, $\rho h$, (Figure 1) by the mean tidal energy dissipation rate (see equations 1-4). It is clear that the inverse of this parameter has dimensions $L^2/T^3$ and can be identified as the tidal energy dissipation rate per unit mass ($cm^2$ $sec^{-3}$), an important parameter in the theory of turbulence.

## A PRACTICAL JUSTIFICATION OF THE USE OF A CRITICAL VALUE OF THE STRATIFICATION PARAMETER IN FRONTAL PREDICTION

A critical value of the Simpson-Hunter stratification parameter is

$$S = log_{10}\left[\frac{h}{C_D\overline{|u|^3}}\right] = 1.5 \qquad (7)$$

This value might be considered to represent the boundary regions between well-mixed and well-stratified water (Figure 3). The transitional zone between $S = 1$ and $S = 2$ is indicated by the relatively heavier shading surrounding the $S = 1.5$ contour.

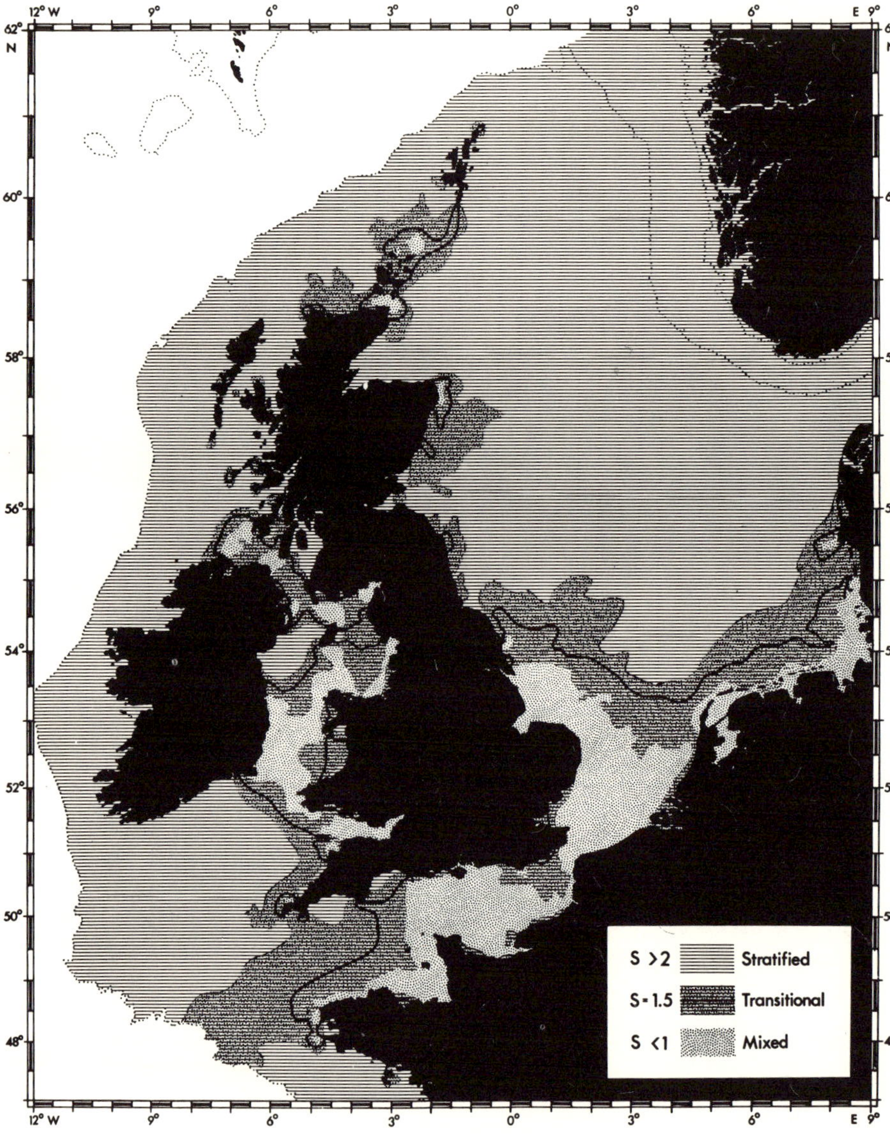


Figure 3. Predicted contours for $S$ = 1, 1.5 and 2 from the numerical model. The continuous line represents the predicted frontal boundary for which $S$ = 1.5. All these frontal regions can be identified from hydrographic studies (Figure 4) and satellite images (Figure 5) and represent important physical, biological and chemical boundaries.

The validity of this procedure for delineating transitional zones between well-mixed and stratified waters has been adequately confirmed by experience gained from measurements at sea. Figure 4 shows some sea surface temperature fronts measured on a single cruise of R/V SARSIA in July 1977. The frontal positions can readily be identified with the critical contour $S$ = 1.5 and these fronts are listed by name in Table I. In addition to measurements at sea, infrared satellite images of sea surface temperature (Figure 5) provide further confirmatory evidence of the importance of the tidal stratification parameter as an index of mixing on the northwest European continental shelf. Further frontal structures can be identified stretching across the southern North Sea, along the German Bight and near Horns Reefs off the coast of Denmark. A multitude of smaller scale fronts can also be observed, for example, in the region west of Scotland; near Skerryvore, off the Small Isles, at the entrance to the Clyde, in the Minch, off Barra Head and off Cape Wrath, just to name a few.

Frontal regions which are not due to tidal mixing can be observed off the Norwegian coast and these are associated with inflow into the Skagerrak and outflow of fresher Baltic water along the Norwegian coast. The stabilizing influence of this less dense Baltic water (Steele, 1957) will have a marked effect on the timing of phytoplankton blooms in this region and for simplicity such advective effects have not been considered in the next sections.

## USE OF THE *SH* DIAGRAM DURING SPRING OUTBURST

It is proposed to look in detail at the seasonal development of phytoplankton distributions in the Celtic Sea and English Channel and formally introduce the stratification, scaled depth, or *SH* diagram so that any conclusions and results may be generalized to other regions with similar *SH* characteristics.

The distribution of the tidal energy dissipation rate is illustrated in Figure 6. Dividing the depth of each grid point by the tidal energy dissipation rate (and multiplying by $\rho$, the sea water density) gives the required contours for the stratification parameter, $S$ (Figure 7).

For illustrative purposes, consider the region in the Celtic Sea where $S$ = 2.75 (i.e., say 51°N, 8°W). In this region the water depth, $h$, is about 100 m and the Secchi Disc depth, $D$, is about 17 m, so the scaled bottom depth, $H_B$, is given as

$$H_B = h/_D = \frac{100}{17} \sim 6 \qquad (8)$$

The point $S$ = 2.75 and $H$ = 6 is plotted on the *SH* diagram in Figure 8.

Table 1

| FRONT | REGION | $H_B$ |
|---|---|---|
| Ushant | Western entrance to English Channel French Coast | 8 |
| Lizard, Lands End | Western entrance to the English Channel English Coast | |
| Start Point | Western English Channel | 6 |
| Dungeness | Eastern English Channel | 7 |
| Flamborough Head | Central North Sea | 4 |
| Southern North Sea | North Sea | |
| Helgoland/Elbe River | Southern North Sea German Bight | |
| Horns Reefs | Central North Sea Danish Coast | |
| Farne Islands | Central North Sea English Coast | 5 |
| Aberdeen | Northern North Sea Scottish Coast | 5 |
| East Orkney, Shetland | Northern North Sea | 8 |
| West Orkney, Shetland | Western Shelf | 7 |
| Islay | Malin Head (Ireland) to Islay (Scotland) | 4 |
| Isle of Man | Irish Sea | |
| Celtic Sea | Northern Celtic Sea Waterford (Ireland) to Hartland Point (England) | |

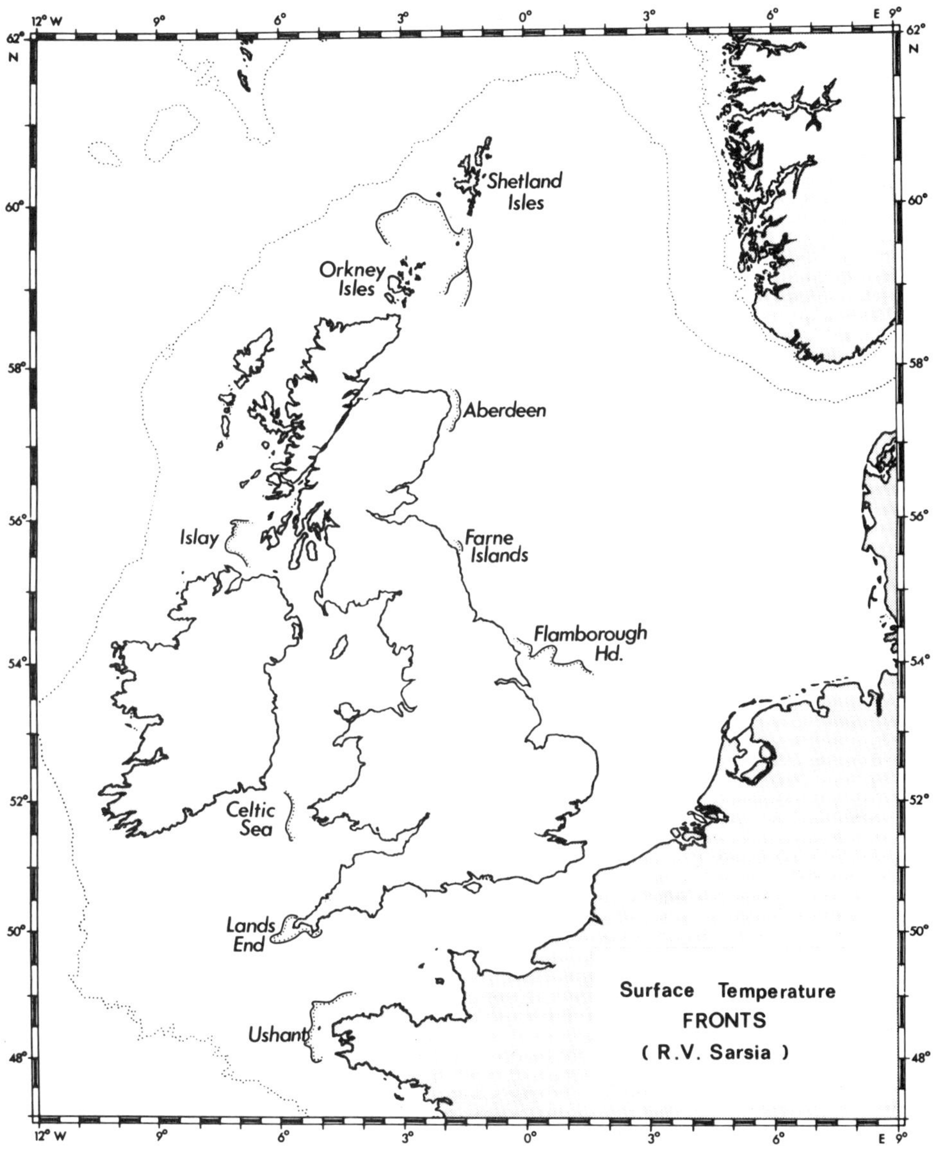


Figure 4. Frontal positions determined by R/V SARSIA, July 1977. The position of the Ushant front off the French coast is drawn from a more extensive survey in this region undertaken in July 1975. The continuous side of the line represents the warmer, more stratified side of the frontal boundary.

Figure 5. Infrared satellite images. Darker tone depicts warmer water and lighter tones represent colder regions. The frontal regions can be readily identified and compared with measurements taken at sea (Figure 4) and predicted positions derived from the numerical model (Figure 3).

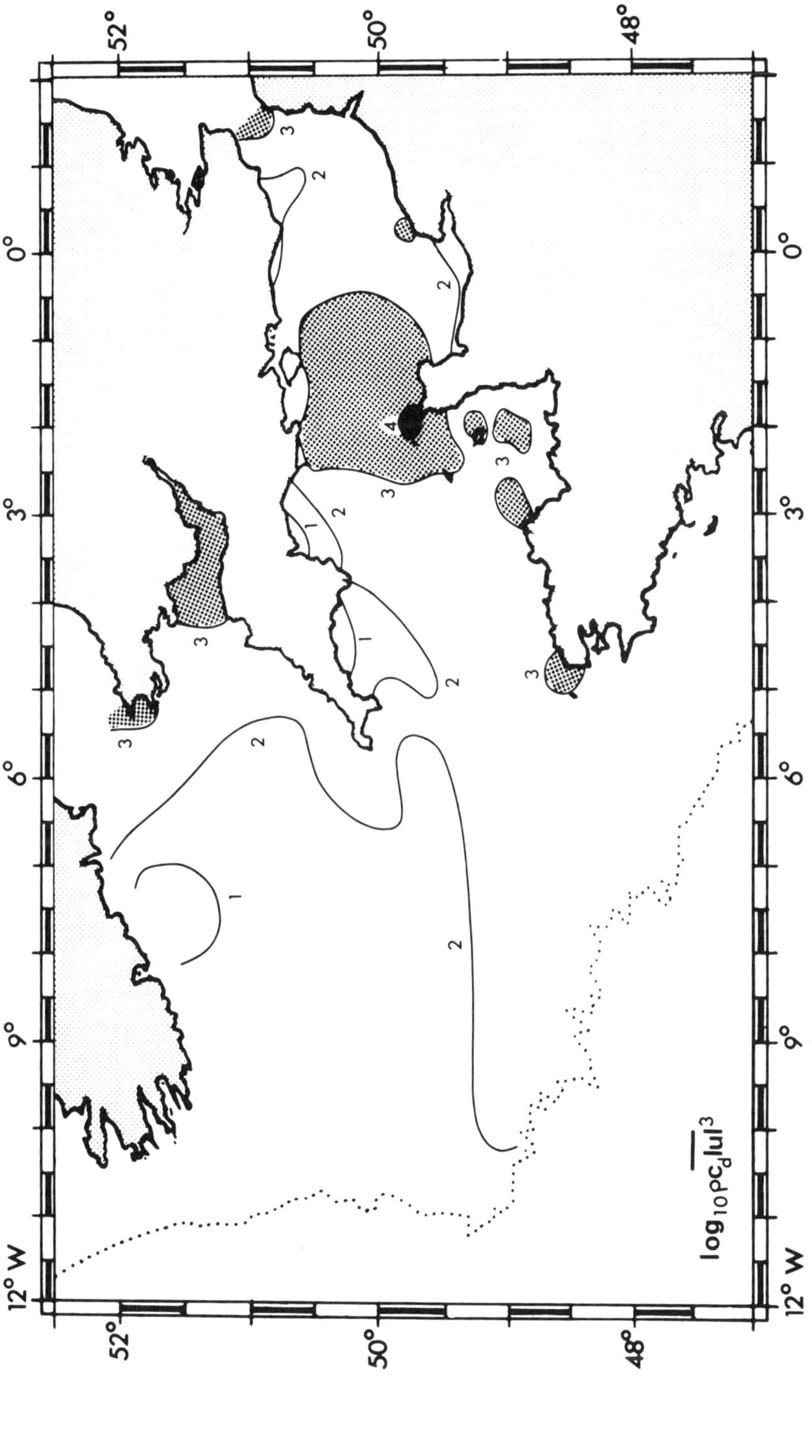


Figure 6. The distribution of tidal energy dissipation rate ergs $cm^{-2}$ $sec^{-1}$ in powers of ten. Notice that the values range through about three orders of magnitude and, therefore, this distribution is likely to be relatively more important than the water depth in determining the shape of the stratification contours (Figure 7).

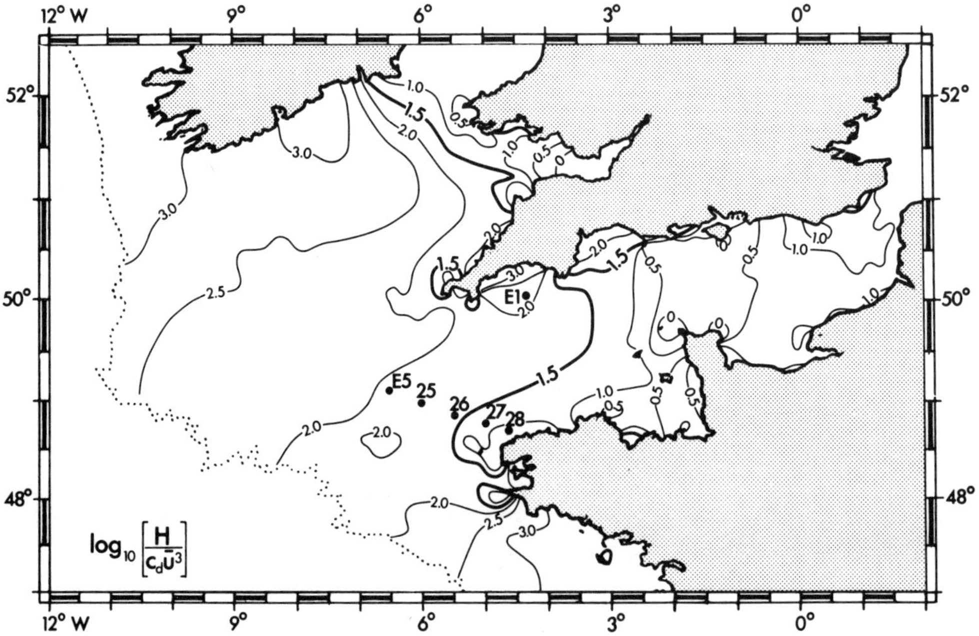


Figure 7. Stratification contours of $S = log_{10} \left[\frac{h}{C_D |u|^3}\right]$ in the Celtic Sea and English Channel. Low values are associated with well-mixed conditions and high values with well-developed summer thermoclines. The transitional region, characterized by summer fronts, can be considered to occur in the neighborhood of the $S$ = 1.5 contour (heavy line).

Sverdrup (1953) has defined the "critical depth" as the depth above which the total photosynthesis is equivalent to the total respiration for a uniform vertical distribution of phytoplankton. Critical depth values have been calculated on the basis that net production occurs when the mean light level exceeds 0.65 K cal cm$^{-2}$ month$^{-1}$ (Riley, 1957).

Indicated on Figure 8 is the scaled critical depth, $H_{CD}$, where

$$H_{CD} = critical\ depth/D \qquad (9)$$

The scaled critical depth values are approximately indicated by a set of monthly values and may be readjusted for different latitudes. (The total solar radiation decreases by a factor of about 2/3 from south to north across the northwest European shelf for the months of interest between March and September. This effect could be

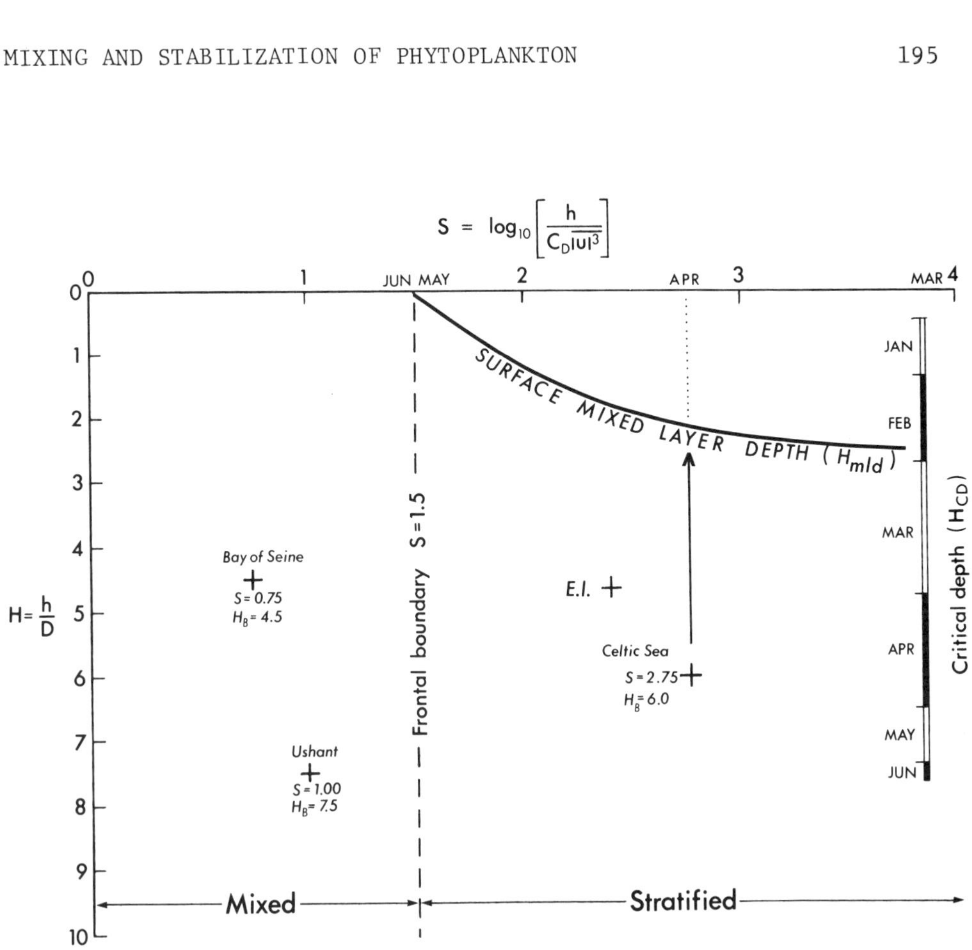


Figure 8. *SH* diagram illustrating spring phytoplankton development.

included in a more universal *SH* diagram by scaling the *H* scale with latitude; see section on general consideration concerning phytoplankton distributions, discussed later in this chapter.)

Now, if the water column were to remain well mixed, net production would occur during April. However, *S* is greater than 1.5 and so the water column is going to stratify during the seasonal heating cycle (Figure 9). When the water column stabilizes, the relevant depth becomes the surface wind mixed layer depth above the developing thermocline which in this region was observed to be ∿35 m at this time of year. Also indicated in Figure 8 is the scaled mixed layer depth, $H_{mld}$, where

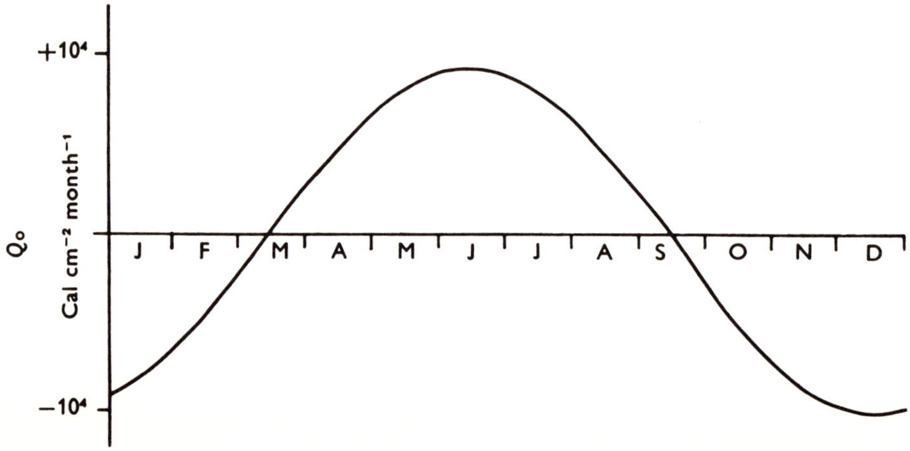


Figure 9. Annual surface heat flux, $Q_o$, in cal cm$^{-2}$ month$^{-1}$.

$$H_{mld} = mixed\ layer\ depth/D \tag{10}$$
$$= \frac{35}{17} \sim 2$$

(Notice this line varies with $S$ since it is expected that the thermocline will outcrop at the sea surface at the value $S = 1.5$ during the summer months when frontal characteristics are experienced (see Figure 3). Further adjustments will also be necessary seasonally.) Thus, when the water column stabilizes, our original point characterized by $S = 2.75$, $H = 6$ adjusts to $S = 2.75$, $H = 2$ where light levels as low as those experienced in February would be sufficient for active growth in the surface mixed layer.

The onset of stabilization can be estimated from the stratification parameter, $S$, and monthly values representing the onset of stabilization have been indicated in Figure 8 along the $S$ axis. For large values of $S$, for example $S > 3$, some stabilization will occur as soon as the seasonal heating cycle becomes positive, i.e., in March (Figure 9); for $S \sim 1.5$, as we have already seen, this will be delayed until June when the surface buoyancy flux is at a maximum, and for $S < 1.5$ no seasonal thermocline will develop.

Thus the Celtic Sea characterized by $S = 2.75$, $H_B = 6.0$ will stratify in April, and because little production will have already taken place, nutrients will still be readily available. Stabilization of the water column will thus provide greatly increased illumination for plant cells held in the surface layer, thereby triggering the spring bloom.

In a similar manner at position E.1. (50°02'N, 4°22'W), net production can be expected in March with a further increase in the production rate in late April when the water column stabilizes (Figure 8).

The consequence of these general considerations on the spatial scales of phytoplankton can be observed in the surface distribution of chlorophyll *a* obtained in April 1975 (Figure 10) (Pingree *et al.*, 1976). In the western entrance to the English Channel, where STD profiles showed the temperature of the water column to be homogeneous to 0.01°C, the surface chlorophyll *a* values were uniformly low ($\sim$1 mg chl *a* $m^{-3}$). Some small increase from the winter minimum values may be attributed to the steadily increasing illumination during the spring when the critical depth becomes greater than the water depth (see Figure 8). However, in the northern Celtic Sea where the values of $S$ were higher and the water column had started to stabilize, the chlorophyll *a* values reached typical spring bloom values $\sim$8 mg chl *a* $m^{-3}$. A surface to bottom temperature difference of only $\sim$0.5°C was sufficient to bring about this bloom condition (Figure 11). A corresponding reduction of the inorganic nutrients, nitrate, phosphate and reactive silicate, was also observed. In this region, under bloom conditions, 98% of the nitrate had been removed from the surface waters (Figure 12) and surface reductions of inorganic phosphate and reactive silicate were $\sim$70% and 85% respectively.

The spring outburst may be expected to flush across the region following the development of the thermocline (Figure 13) (Pingree, 1975), according to the values of the stratification parameter $S$ (Figure 7) and the surface heat flux (Figure 9) as described by equation (3), using a critical value $R \sim 0.01$.

The relative importance between thermocline development and increasing illumination on phytoplankton production during the spring will depend on the values of $S$ and $H_B$ used to characterize the region of interest. Clearly, for $S > 1.5$ and large values of $H_B$ the spring outburst will be determined by stabilization of the water column. However, for $S > 1.5$ and small values of $H_B$ the effects of stratification will be less marked, since production will be possible in well-mixed conditions before stabilization occurs.

For $S < 1.5$ no permanent seasonal thermocline will develop and stabilization will not play an important role in production. The region near Ushant has values of $S = 1$ and with $h = 98$ m and $D = 13$, $H_B = 7.5$ as indicated in the $SH$ diagram (Figure 8).

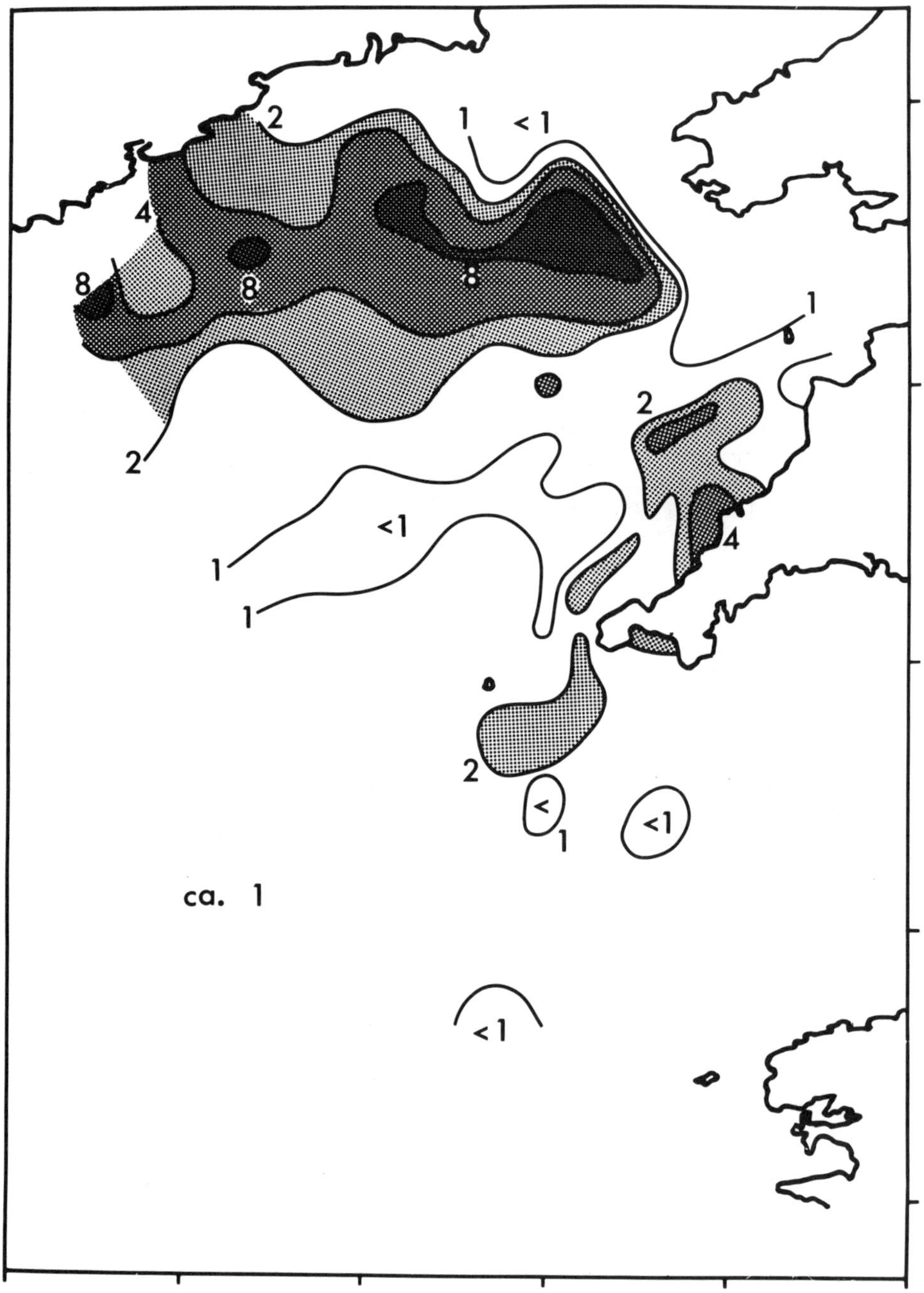


Figure 10. Surface distribution of chlorophyll *a* (mg $m^{-3}$) 9-21 April 1975.

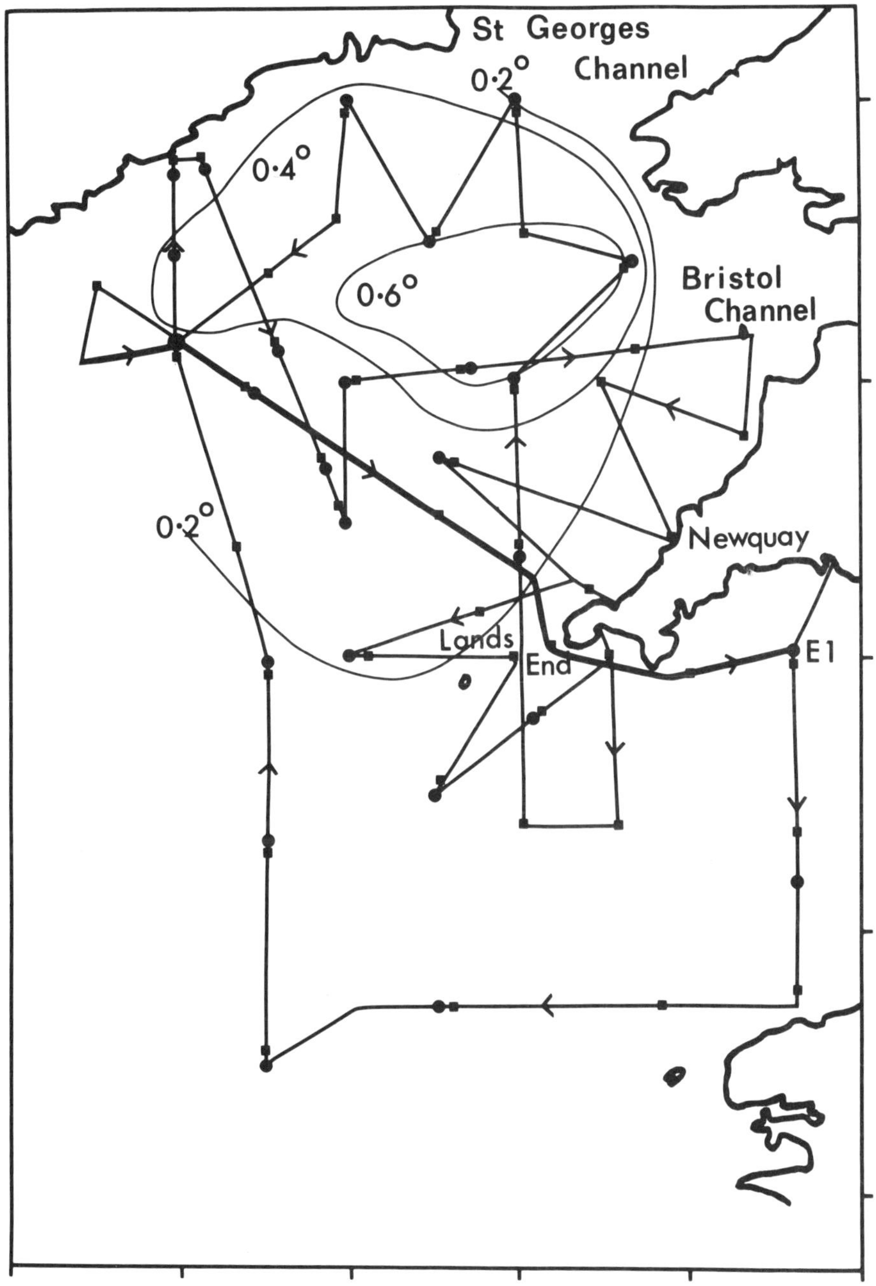


Figure 11. Track of R/V SARSIA, 9-23 April 1975, and surface to bottom temperature differences (°C) observed over the period 9-21 April 1975.

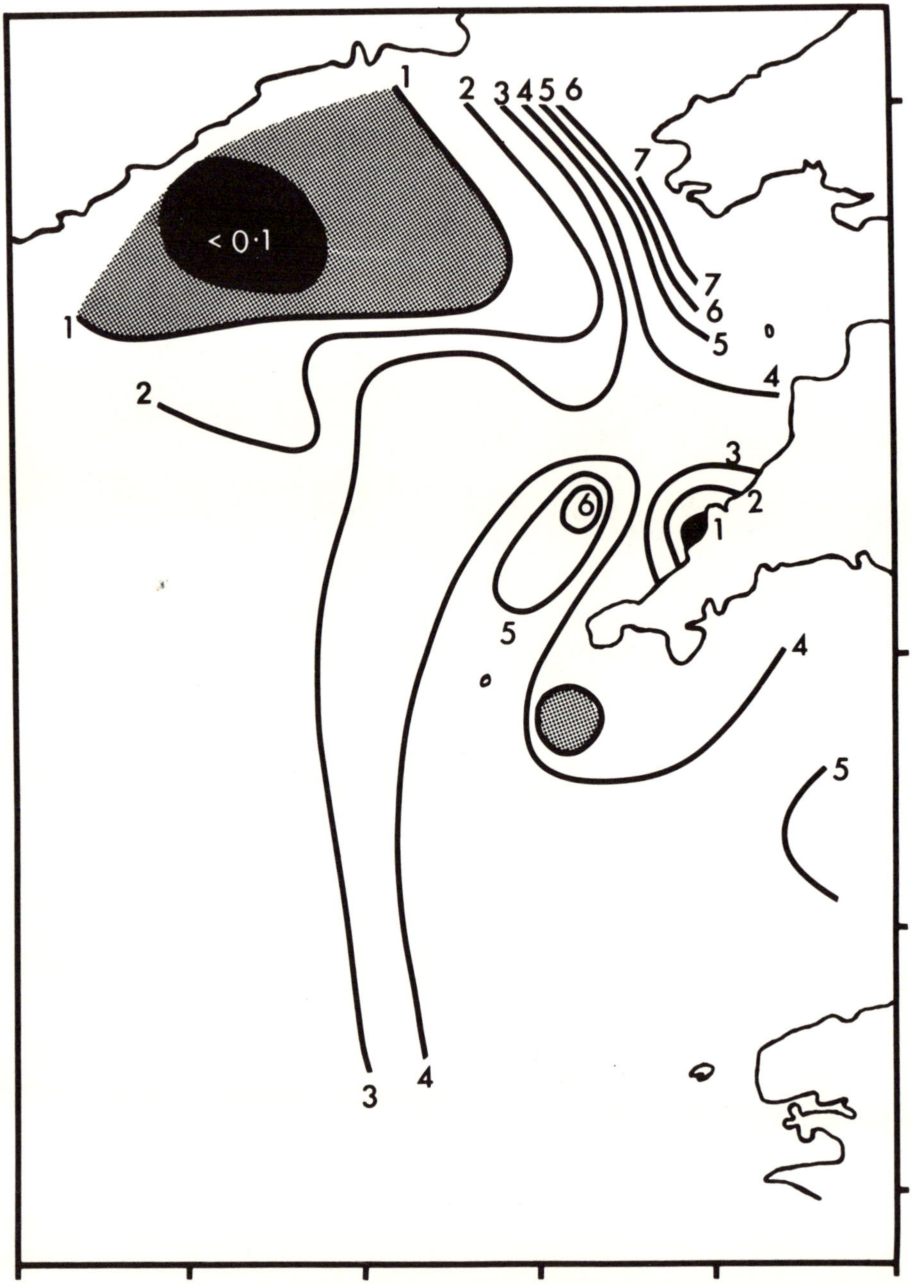


Figure 12. Surface distribution of nitrate (μg at $l^{-1}$N), 9-21 April 1975.

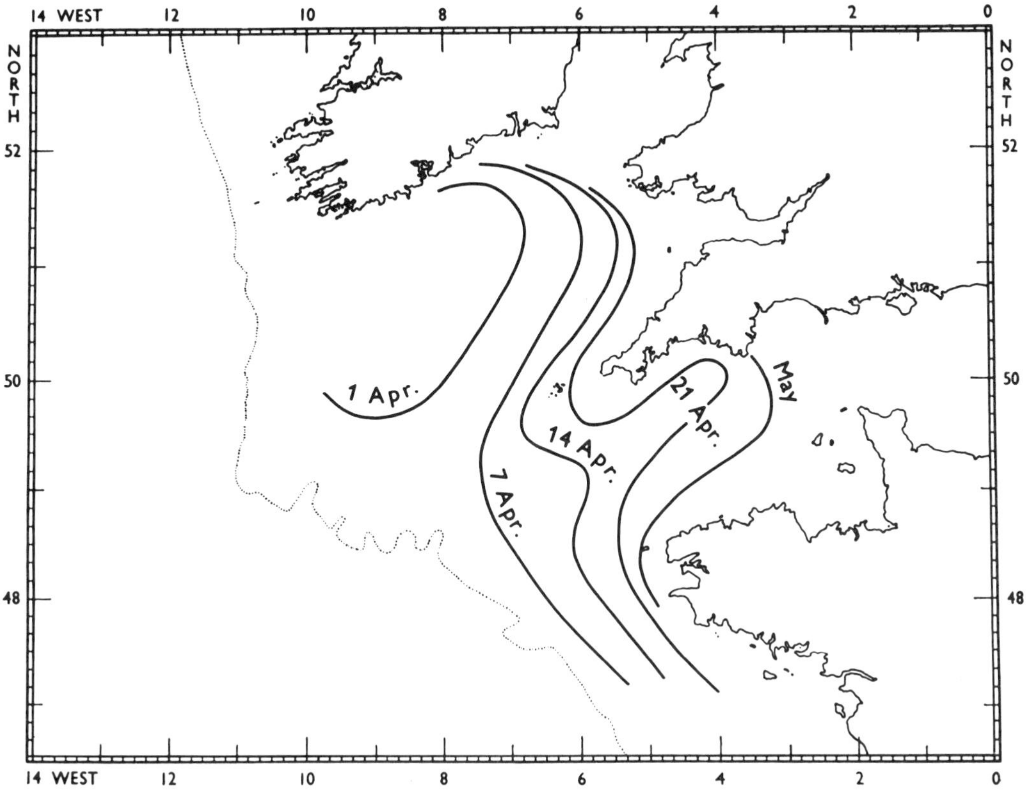


Figure 13. Development of the thermocline by weekly contours.

Production in this region will probably be delayed and no marked phytoplankton outburst is anticipated (see later, Figure 18). In the Bay of Seine, $S$ = 0.75 and with $h$ = 30 m and $D$ = 7, $H_B$ = 4.5 and this point is also illustrated on the $SH$ diagram. The spring bloom in this area may be expected during March as was indeed observed (Figures 14 and 15) (Holligan *et al.*, 1977).

Thus we see how the $SH$ diagram may be used to characterize phytoplankton production and how such diagrams may be used to predict the extent of spring phytoplankton development from a knowledge of $S$ and $H$ only. In effect, the stratification parameter defines a scale where for large $S$ the effects of weather are important in mixing and stabilization, for intermediate values ($1.5 < S < 3.5$) weather and tide interact, and for low values of $S$

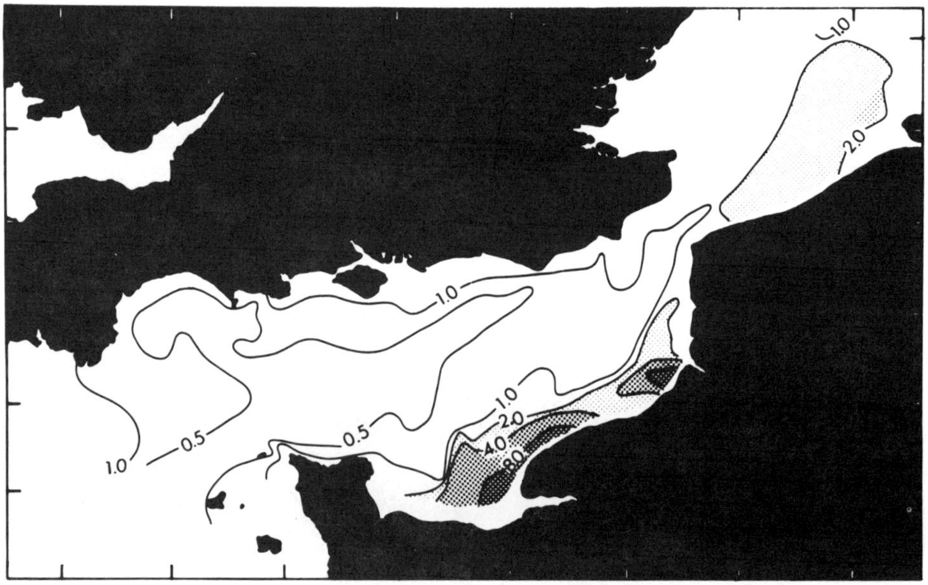


Figure 14. Surface distribution of chlorophyll $a$ (mg $m^{-3}$) in the eastern English Channel, 14-23 March 1976.

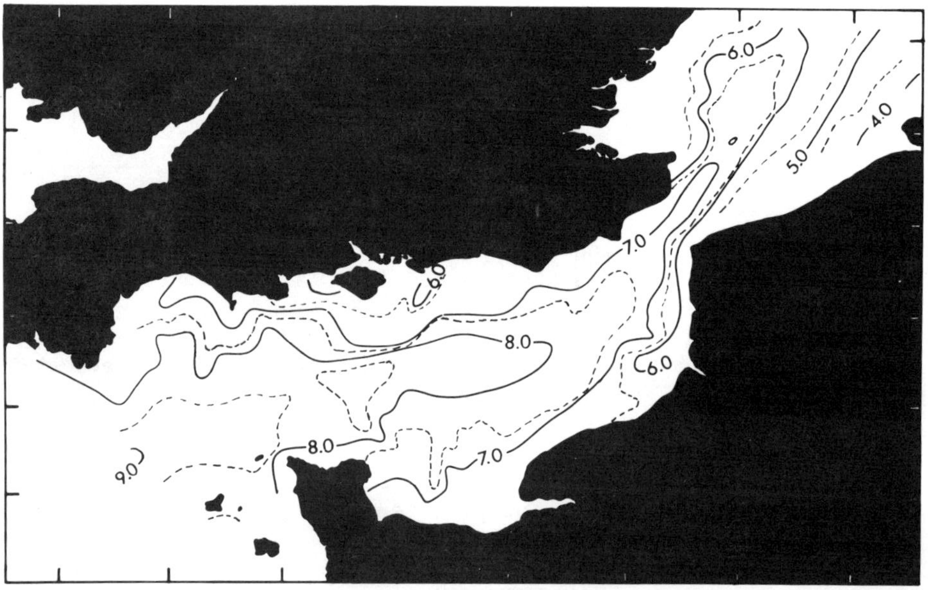


Figure 15. Surface distribution of temperature (°C) in the eastern English Channel, 14-23 March 1976.

($S < 1.5$) the effects of tides predominate. The scale $H$ determines whether or not plant cells can take advantage of the light levels associated with the mixing regime.

## SUMMER PHYTOPLANKTON DISTRIBUTIONS AND THE *SH* DIAGRAM

In a similar manner it is instructive to consider summer phytoplankton distributions in terms of $S$ and $H$.

After the spring outburst the levels of chlorophyll decrease in the stratified surface waters due to grazing and sinking. For high values of $S$ and $H_B$, subsequent growth of phytoplankton above the thermocline will be determined by the rate at which nutrients are regenerated *in situ*. Some production will be possible within the thermocline particularly near the base where the combination of nutrients mixed upwards by bottom mixing and a favorable light regime, maintained by stability, may create an environment suitable for periodic phytoplankton growth. The relevant parameters for studying thermocline production on the *SH* diagram are the scaled thermocline depth and the compensation depth (Figure 16). The compensation depth is the depth where the rate of photosynthesis just balances the rate of respiration. Evolution of patches in subsurface phytoplankton populations are dealt with in the chapter by Holligan.

For $S < 2$ nutrient transfer through the thermocline will play an increasingly important role in phytoplankton production. Indeed, the region between $S = 1$ and $S = 2$ (Figure 16 and Figure 3) may be considered as the productive region in the summer months. As we have seen, this region contains the frontal regions (Table I) and it is convenient to start our study of summer production by considering the phytoplankton distributions associated with the Ushant frontal system (Figure 17).

Repeated chlorophyll *a* traces through this frontal system have defined a region on the stratified side of the frontal boundary where dense phytoplankton blooms may persist throughout the summer months (Figure 18). On the unstratified side of the frontal region strong tidal stirring ensures complete vertical mixing and the plant cells in the water experience a much lower mean light level than those held in the surface layers on the stratified side of the frontal boundary. Whereas in the well-mixed region, levels of inorganic nutrients are relatively high, in the stable regions, surface nutrient values have remained low since being depleted in the spring bloom (Figure 10). Thus the frontal zone is a region where the combination of high nutrients and a good light regime, provided by a shallow thermocline, create conditions suitable for rapid phytoplankton growth (Figure 19) (Pingree *et al.*, 1975).

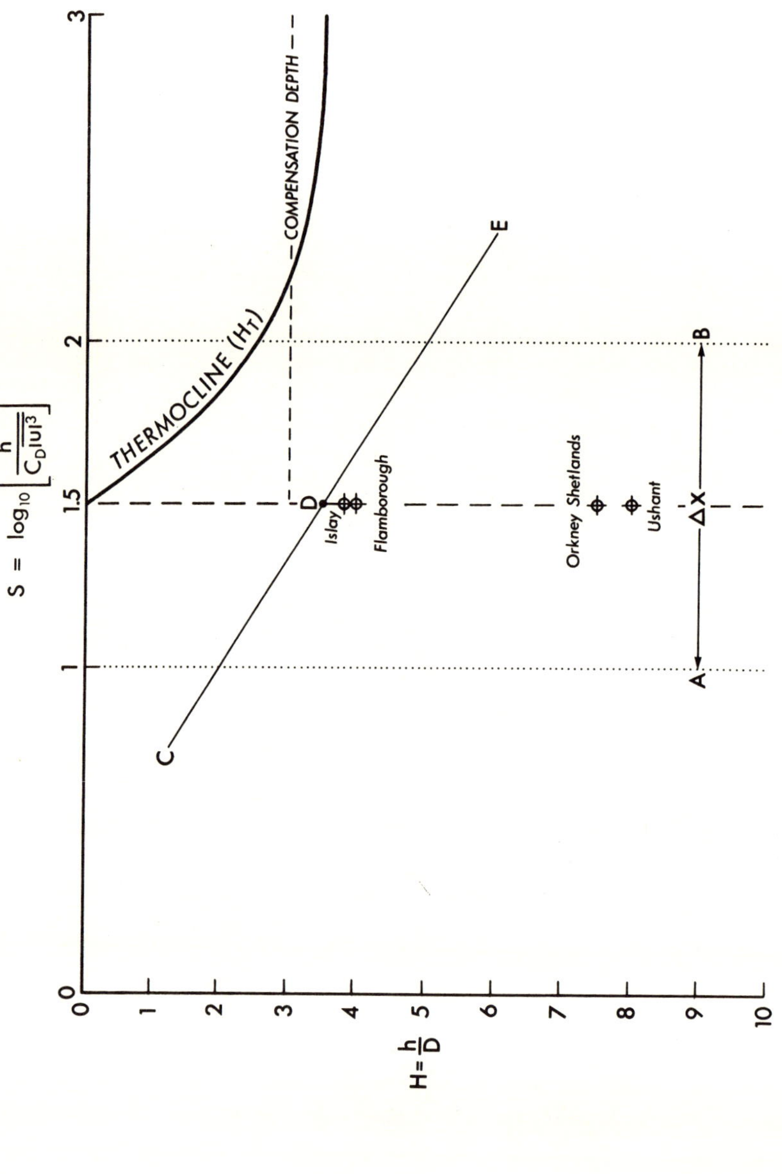


Figure 16. *SH* Diagram illustrating summer phytoplankton development.

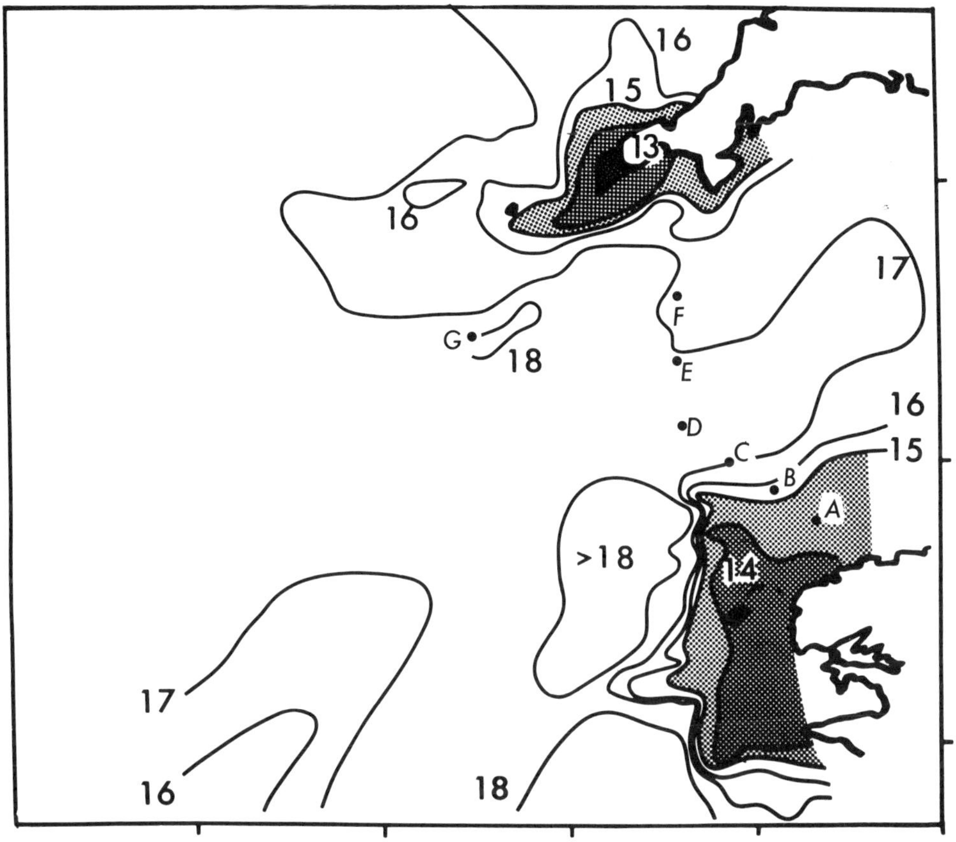


Figure 17. Surface temperature ($^{o}$C) showing frontal structure near Lands End and Ushant (25-31 July 1975).

During spring tide conditions it is expected that the frontal region will move slightly towards the more stratified areas ($S > 1.5$). During neap tide conditions marginal restabilization of the waters to the well-mixed side of the frontal region where nutrient renewal has taken place will provide conditions suitable for phytoplankton growth. Indeed, we may consider the contours of $S$ as defining regions in the summer where the water column remains stable for a longer and longer period of the tidal cycle. Now, since the tidal energy dissipation varies by a factor of 8 from neap tides to spring tides, it would seem reasonable to assume that contours of $S$ ranging from 1 to 2 will be involved to some extent in the mixing stabilization cycle controlling phytoplankton growth.

The Ushant front is plotted on the $SH$ diagram (Figure 16) and similarities can be anticipated between this region and the fronts

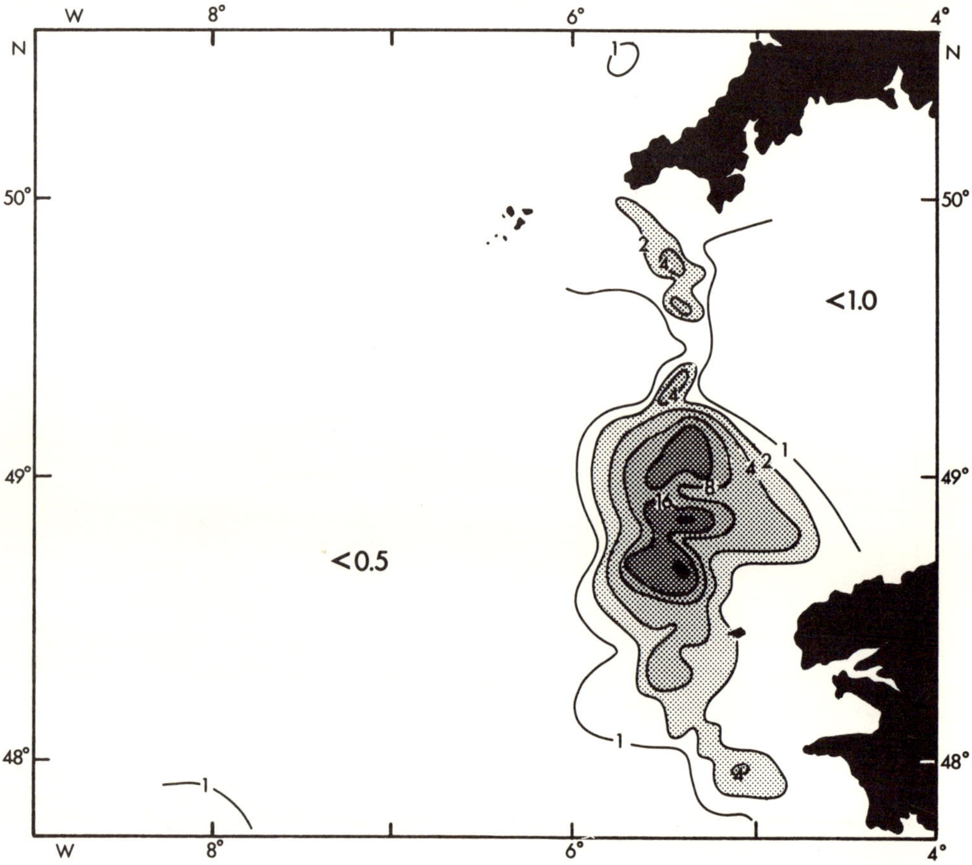


Figure 18. Surface chlorophyll *a* distribution (mg $m^{-3}$) corresponding to the temperature structure shown in Figure 17. While the gross structure is associated with the frontal region, it is suggested that the smaller scale patches along the frontal boundary are associated with frontal eddies (see text).

in the neighborhood of the Orkney Isles and Shetland Isles. The most productive fronts will fall on the line $S = 1.5$ towards higher values of $H$. Shallow water fronts (small $H_B$) may be relatively unproductive and show little chlorophyll *a* responses since nutrient levels will already be minimal. In addition to the value of $H$, summer fronts must also be characterized by the length scale, $\Delta X$ (see Figure 16, line AB), which represent the geographical distance for unit change in $S$, from $S = 1$ to $S = 2$. For the Ushant front, $\Delta X$ is of order 100 kilometers and this represents both a measure of the productivity per unit distance along the frontal boundary and gross length scale for summer phytoplankton distributions. For the Islay front, $\Delta X$ is $\sim$20 kilometers and consequently

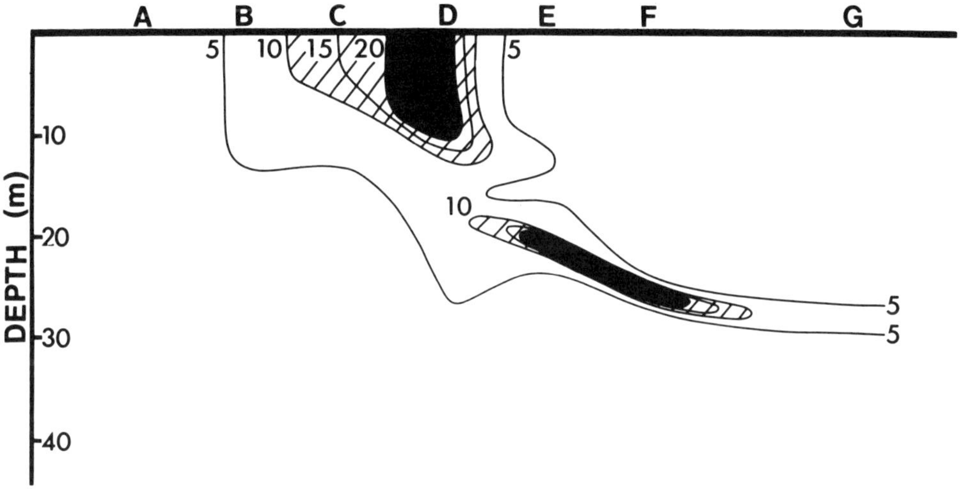


Figure 19a. Chlorophyll *a* section through the frontal region. The units of chlorophyll *a* are mg $m^{-3}$.

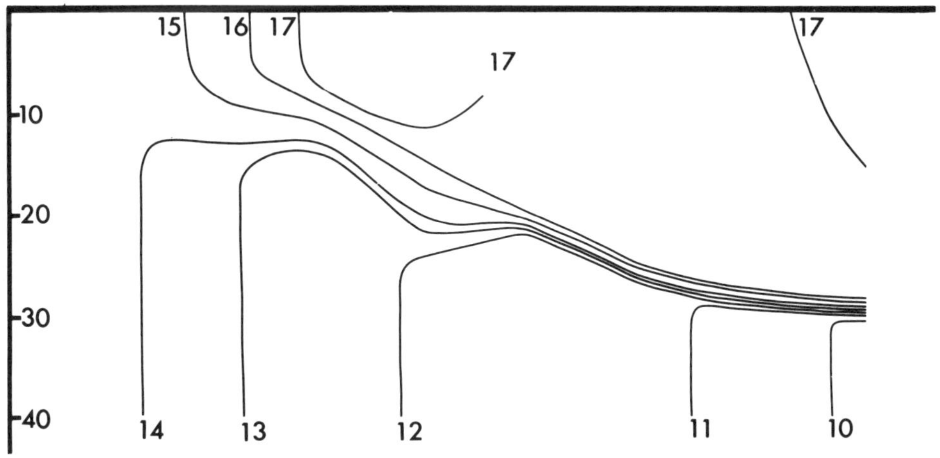


Figure 19b. The corresponding temperature section ($^{o}C$). Station positions A, B, C, D, E, F, G are as shown in Figure 17.

both the frontal position and the associated chlorophyll response may be expected to be more closely confined. This is to be contrasted with the situation off Flamborough Head where $H_B$ is also about 4 but $\Delta X$ ∿130 kilometers. For large $\Delta X$ the frontal boundary will also be more sensitive to the mixing and stabilizing effects of weather.

One further point to consider is the dynamics associated with some frontal systems. Upwelling may sometimes provide a source of nutrients for shallow, well-mixed regions which would otherwise be unproductive. This is sketched diagramatically (Figure 16) by the line *CDE*. *E* represents nutrient-rich water below the thermocline, well below the compensation depth, and *D* represents the frontal region separating well-mixed from well-stratified waters. The region *C* defines the shallow, well-mixed area where intermittent upwelling may be taking place. Although upwelling may be attributed to the usual effects of weather, sustained upwelling due to tidal streaming is also possible. For example, whenever the tide flows around a headland, promontory, bank or island, the flow in the vicinity of the headland increases rapidly. If the tides well offshore are weak, then a critical value of $S$ = 1.5 may be located in the neighborhood of the promontory. Since curvature of flow around a headland, together with increased tidal streaming, results in a lowering of mean sea level locally, nutrient-rich bottom water will tend to move toward the low pressure center (Garrett and Loucks, 1976). Thus a continuous supply of nutrients is available to replace nutrient uptake during photosynthesis in the shallow water region, *C*. Shallow headlands and promontories with deeper offshore stratified regions are likely to be very productive and may be associated with relatively high chlorophyll *a* levels.

All the frontal systems considered are sufficiently large that the gross motions will be along rather than across the frontal regions. Smaller scale baroclinic instabilities associated with the frontal system may occasionally result in a charactristic series of mainly cyclonic eddies along all the transitional regions. As many as six adjacent irregular eddies (10-30 km size, which may persist for a few days (2-3 days)) have been observed in infrared satellite images of the Ushant frontal system. In addition to tidal mixing and upwelling, cross frontal mixing and nutrient renewal associated with these cyclonic eddies must represent an important nutrient source for the stratified waters alongside the frontal boundaries. It is significant that the chlorophyll *a* patch size along the Ushant frontal boundary is also about 20 km (see Figure 18) and perhaps this scale is associated with the cyclonic eddies.

## AUTUMN PRODUCTION

During the autumn, mixing across the thermocline increases as a result of convective heat loss at the sea surface and increasing wind stress. Despite the greater availability of nutrients, plant populations in the thermocline are found to be unable to maintain themselves against vertical mixing and dispersal by turbulence. However, relatively high concentrations of chlorophyll *a* may still be observed in the surface waters where the frontal boundary marks the region between stratified and well-mixed waters in the autumn (Figure 20). Autumn blooms comparable with spring and summer outbursts are unlikely since, during the autumn period, the surface mixed layer becomes increasingly deeper and this coupled with falling surface light results in lower mean light levels experienced by the plant cells.

Autumn blooms may also be studied on the *SH* diagram (Figure 21) and clearly these will be confined to the region where $S > 1.5$. The region separating well-mixed water from stratified waters retreats towards higher values of $S$ as the winter is approached and the thermocline retreats westward across the Celtic Sea (Figure 22) (Pingree, 1975). In addition, the scaled mixed layer depth becomes deeper. The critical depth scale shows that while autumn blooms may be anticipated in October, by November falling light levels will not permit rapid phytoplankton growth despite increases in surface nutrients due to increased mixing. However, it must be

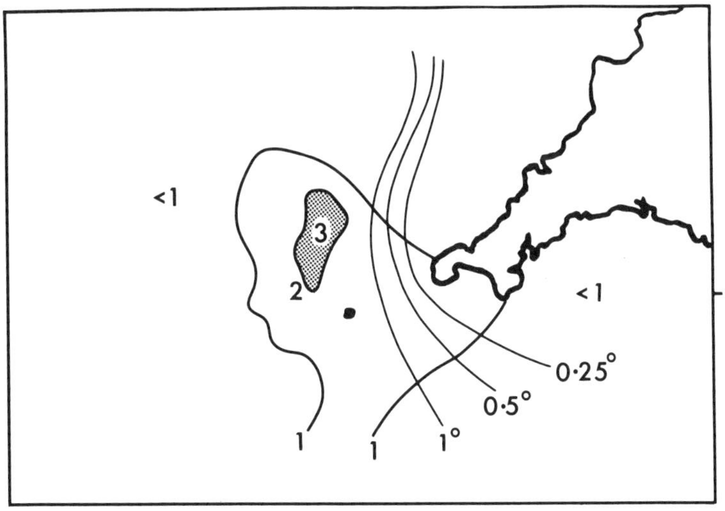


Figure 20. Surface to bottom temperature differences (°C) and the surface distribution of chlorophyll *a* (mg $m^{-3}$) around Lands End, October 1975.

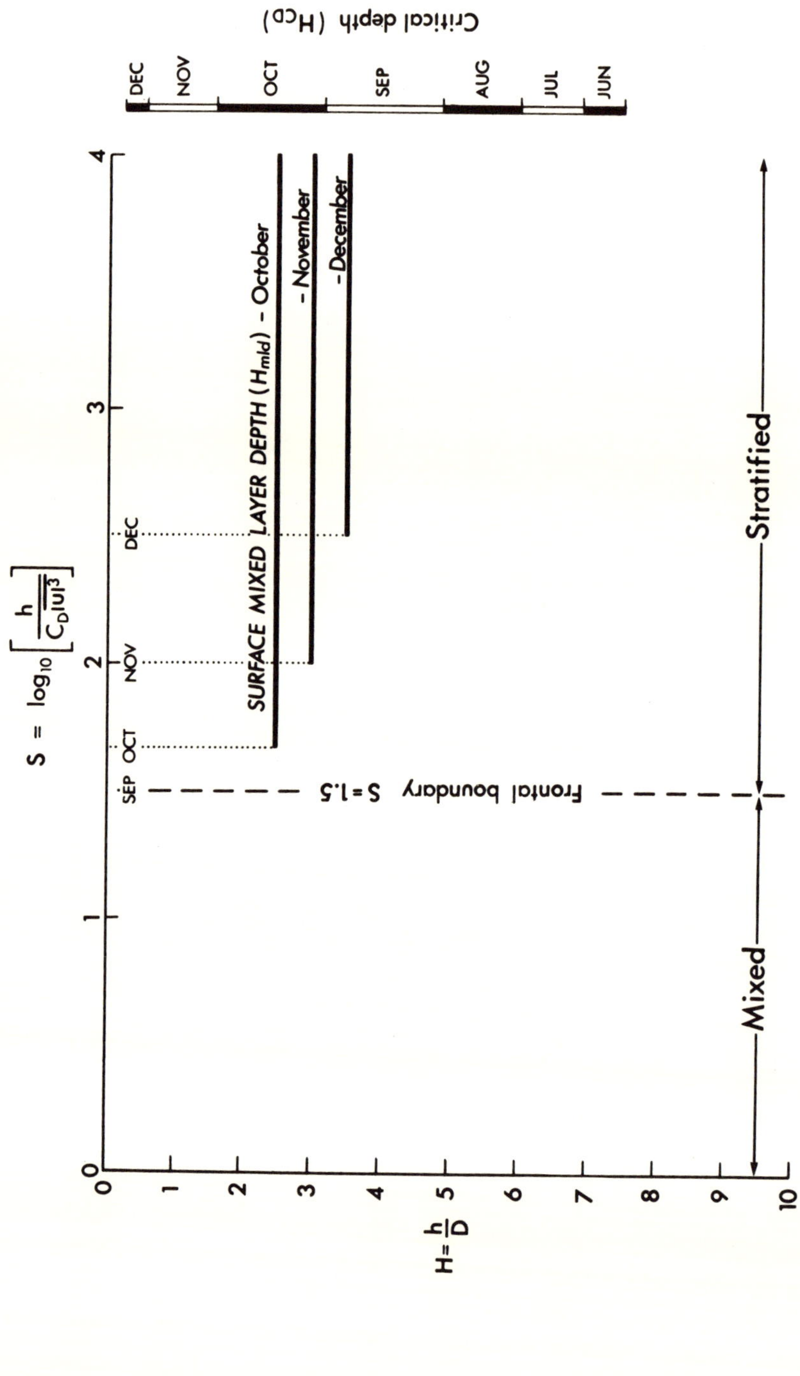


Figure 21. *SH* diagram illustrating autumn production.

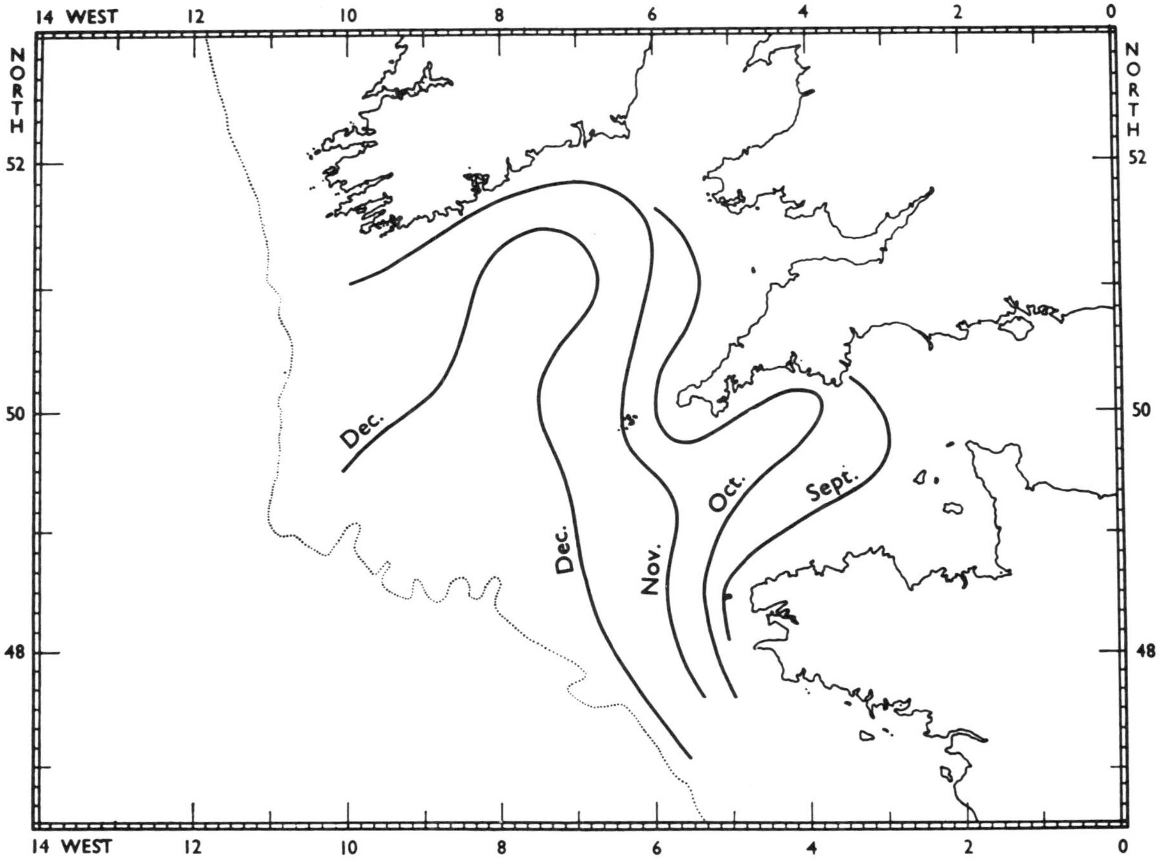


Figure 22. Erosion of the thermocline across the English Channel and Celtic Sea. Note marginal stratification still exists in December in the Celtic Sea where $S \sim 2.5$.

remembered that the mean values of incident light and the mixed layer depth may not be generally applicable when considering autumn phytoplankton distributions. For example, a period of sustained calm sunny weather during November may result in increases in chlorophyll *a* levels even if the mean light level and mean mixed layer depth appear adverse. The same arguments concerning mean values also apply to the conditions at the start of the productive cycle in the spring (Figure 8) and so marked variance due to weather may be anticipated (Steele, 1958).

## GENERAL CONSIDERATION CONCERNING PHYTOPLANKTON DISTRIBUTIONS

It has been shown that the gross scales for phytoplankton distributions may be determined from the stratification parameter and the scaled water depth. These two properties control the availability of light and nutrients necessary for phytoplankton growth. Where stratification is important ($S > 1.5$) the gross patterns of phytoplankton distributions will be determined by geographical variations in the stratification parameter. The manner in which phytoplankton patches move and develop across contours of $S$ will also depend on the time scale for cell division and grazing rates in addition to nutrient and light levels. A fast cell division rate will be particularly important in the summer allowing a production response for brief periods of stabilization thereby allowing plant cells to take advantage of conditions where the water column stabilizes for only a few days in the spring to neap cycle of tidal mixing. A set of contours of $S$ (with values of $S \leqslant 1.5$) will define regions where some measure of surface stability exists for given proportions of the full fortnightly cycle.

While this approach has been mainly concerned with gross scales, the *SH* diagram can also be used as a characteristic diagram for identifying regions exhibiting smaller scale structure. Records in the spring contain structures on many different scales and it is apparent that we are essentially dealing with a broad spectral band (Platt, 1972). Characteristically, records representative of spring conditions, where stabilization of the water column plays an important role in the spring outburst, show considerable fine structure. Temperature is generally inversely related to nutrient levels but positively correlated with chlorophyll *a* (Figure 23a). This reflects the basic theme of this chapter; after mixing has provided nutrients near the surface, then stabilization is subsequently necessary to allow growth; and this cycle of events may be repeated many times. The high chlorophyll, low nutrient structures define patches of water with a longer history of surface stability than the high nutrient regions. Older structures still may show low chlorophyll *a* values due to grazing and sinking. Thus positive, negative and uncorrelated chlorophyll *a* and temperature or nutrient relationships are

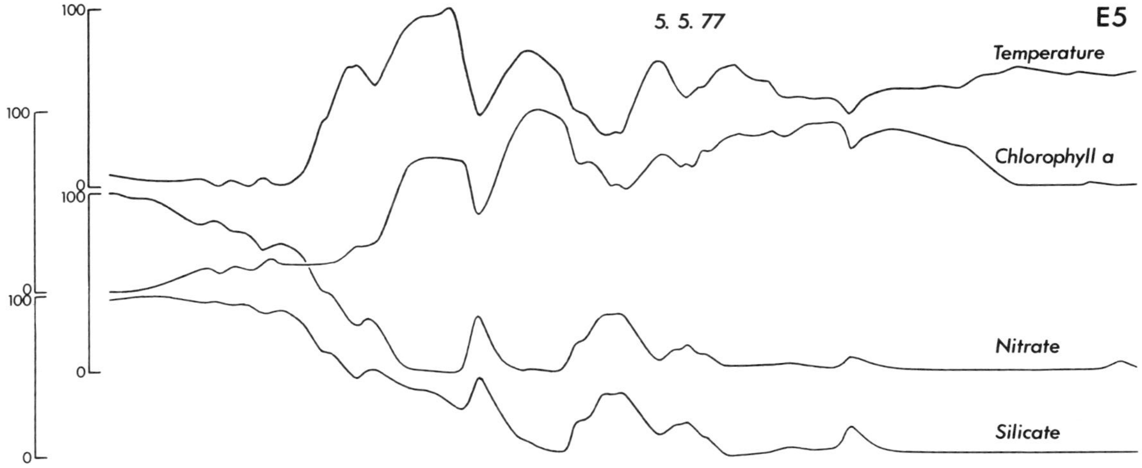


Figure 23a. Surface records of temperature, chlorophyll *a*, nitrate and silicate (relative scales) obtained in the neighborhood of E5 in the Celtic Sea (5.5.77). The records were obtained steaming approximately along contours of $S$ from $48^{\circ}23'$N, $6^{\circ}10'$W ($S \sim 1.8$) to E5 ($S \sim 2$) a distance of about 45 nautical miles. During this period of early thermocline development, nutrients are inversely related to chlorophyll *a*, which is positively correlated with temperature (see text).

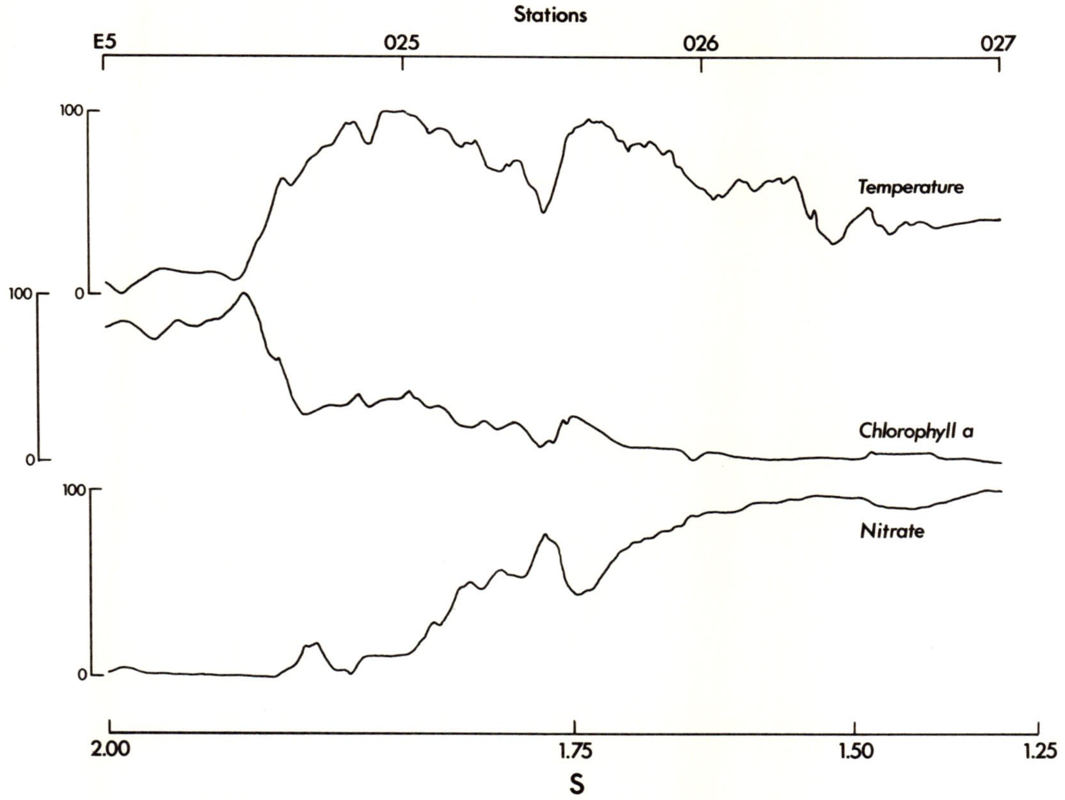


Figure 23b. Surface records of temperature, chlorophyll $a$ and nitrate (relative scales) obtained in May 1977. The track is shown in Figure 7 and the corresponding $S$ values are indicated along the records ($H_B \sim 8$).

possible, particularly when steaming across contours of *S* (Figure 23b). However, a region of high wavenumber variance will characterize the region of active growth. This region will move across the *SH* diagram from high *S* values to lower values as the season progresses. In the summer, therefore, "spring conditions" exist where $S = 1.5$ and at the base of the thermocline. Conditions are much more uniform where $S < 1$, and tidal mixing is intense, and also in the region characterized by $S > 2$ where, as a result of nutrient depletion, grazing and diffusion, chlorophyll values are also more uniform and low (Figure 24). Within the productive region, $1 < S < 2$, 10 km scale structures of chlorophyll *a* may be associated with the cyclonic eddies that tend to form along the frontal regions.

While the *SH* diagram has been used to study distribution patterns for chlorophyll *a* it seems likely that the same approach may be useful for studying distribution patterns for individual species and the seasonal succession. It may also be informative to follow the distributions of herbivorous zooplankton or young fish in relation to phytoplankton distributions on the *SH* diagram. The sequence of heat input at the sea surface, wind and tidal mixing (controlling thermocline development and the availability of nutrients and light energy), phytoplankton development and subsequent grazing by herbivorous zooplankton, may have many further links through to fish. Historically, the importance of advection has been emphasized in describing the distributions of larger zooplankton which may be used as indicators of water movement (Russell, 1935). As energy flows from the primary producers along the food chain, advection, diffusion and migration must play important roles in the redistribution of the higher trophic levels. Even so, the spatial pattern of the ultimate predator in the North Sea, the European fishing fleets (Figure 25) (Ramster *et al.*, 1973), reflects to some extent these characteristic *S* and *H* contours which suggest a similar approach for zooplankton may still be instructive. Particularly conspicuous is the large, relatively unfished region consisting of the central and northern North Sea which approximates to the well-stratified deep regions where $S > 2$. Also equally noteworthy is the region of most intense activity which generally corresponds to the shallow areas where $S < 2$.

Although fishing activity may also increase at the edge of the shelf, other mixing processes will be important here. In the summer, at the edge of the northwest European shelf, particularly along the continental slopes bordering the Celtic Sea, there exists a region about 100 km broad where intermittent upwelling and mixing creates a zone of marginally reduced surface temperature. It seems likely that the mixing-stabilizing processes associated with the shelf break region will also be reflected in increased phytoplankton production. Fournier (1977) has shown stabilization between different water mass boundaries (along the shelf break front

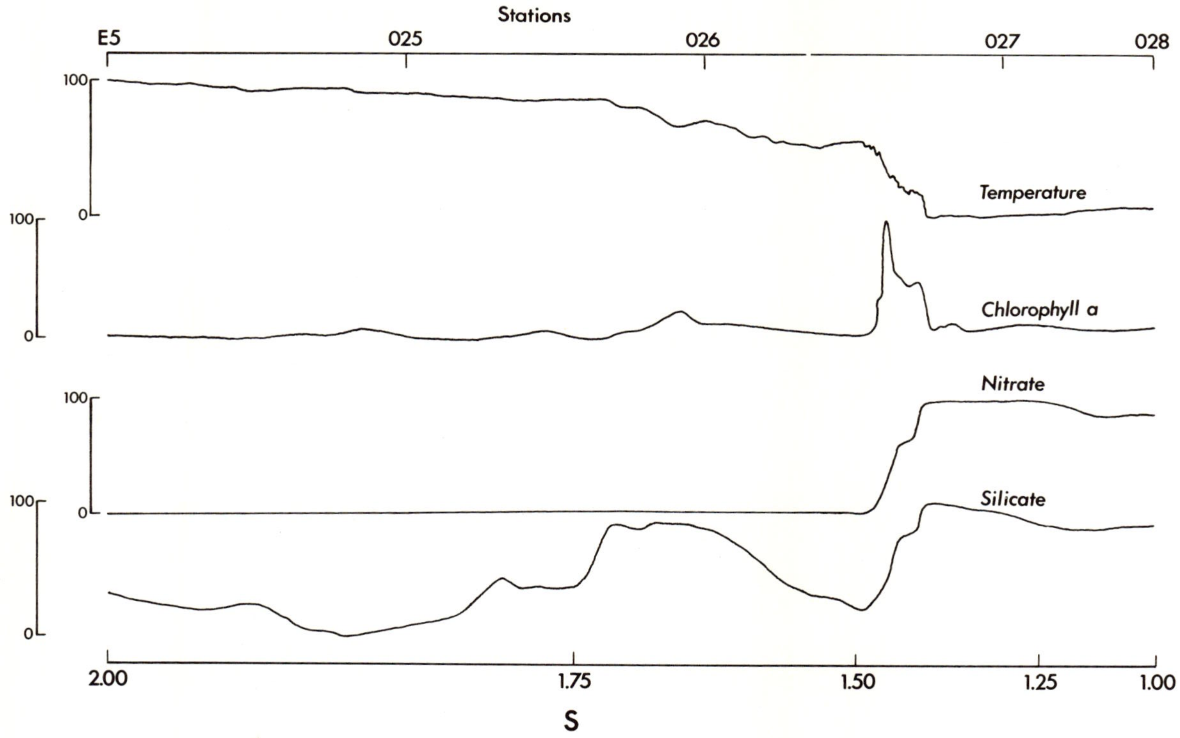


Figure 24. Surface records of temperature, chlorophyll $a$, nitrate and silicate (relative scales) obtained in August 1977 for the same runs shown in Figure 23b. Notice the sea surface temperature front near $S = 1.5$ and associated chlorophyll $a$ response. This region is the productive summer region and is characterized by more variance at higher wave number. Both in the well-mixed waters $S < 1.5$ and in the more stratified regions $S > 1.75$, conditions are more uniform (see text).

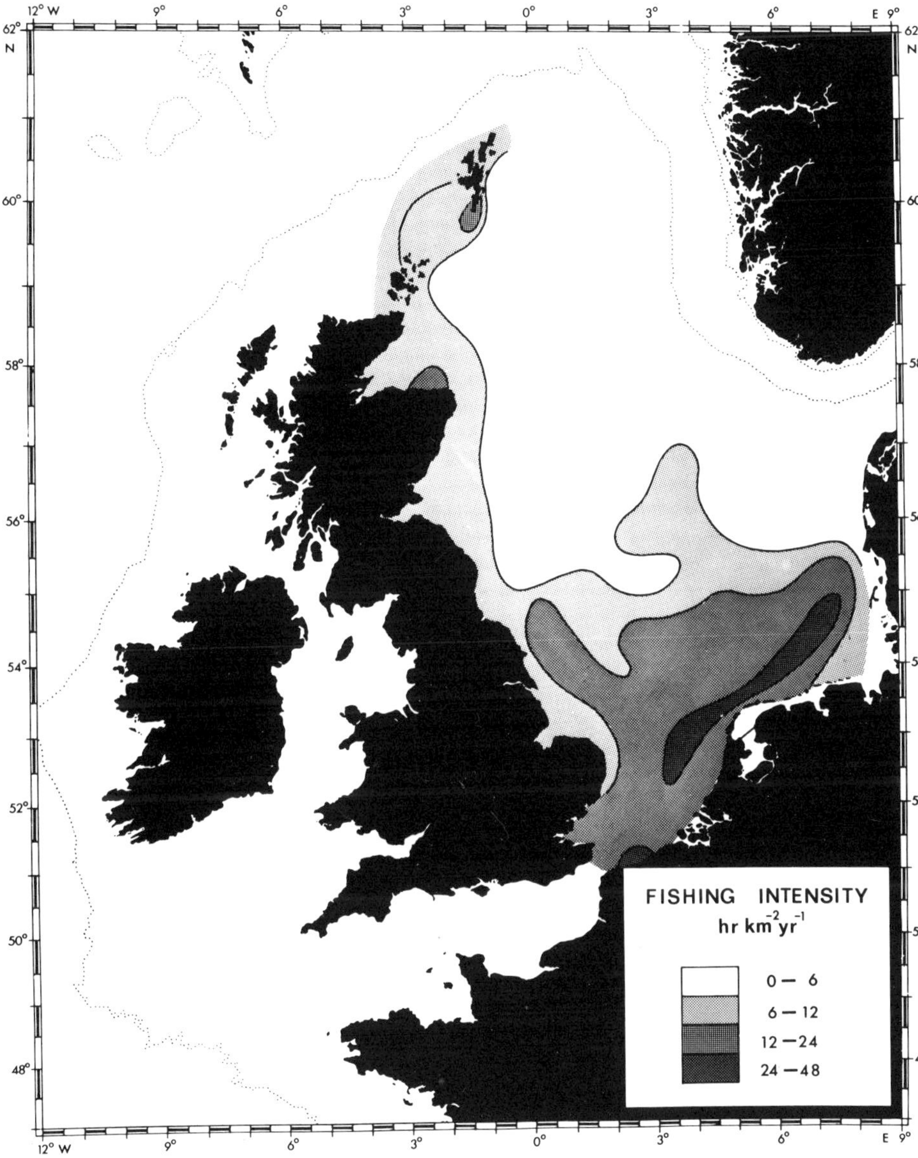


Figure 25. Schematic representation of the fishing intensity in the North Sea for 1971. Contours give a rough indication of how many hours each square kilometer of sea is fished in a year.

off Nova Scotia) may provide conditions suitable for phytoplankton growth even in late winter. While it is clear that the processes operating at the shelf break are different from those which have been considered, it is suggested that the same broad principles of mixing and nutrient renewal, followed by stabilization and growth, will also be applicable along the slope regions bordering the northwest European shelf and that this region is worthy of further study.

Finally, we might briefly consider whether there are any suitable modifications that can be made to the *SH* diagram to make it more universally applicable. Earlier it was noted that the solar radiation decreases by about a factor of $\sim 2/3$ from south to north across the northwest European continental shelf between March and September. This effect could be incorporated in the $H$ scale by changing to a new scale $H^+$ which includes the latitude factor. Similarly, it might seem desirable to make the $S$ scale non-dimensional with a new scale, $S^+$,

$$S^+ = \log_{10}\left[\frac{hB_o}{C_D\,\overline{|u|}^3}\right] \tag{11}$$

where $B_o$ is the maximum surface buoyancy flux, which will also change with latitude. In addition, it may be possible, ultimately, to account quantitatively for seasonal wind mixing in the redistribution of the buoyancy flux below the surface. Thus all fronts will be defined by the same value of $S^+$ irrespective of latitude. The new $S^+H^+$ diagram would be more universally valid for making comparisons between phytoplankton distributions in different latitudes and under varying tidal regimes. At the present early stage in the development of the approach presented here, it seems preferable to continue with the more clumsy *SH* diagram to illustrate how the *SH* values may be used to characterize phytoplankton distributions on a somewhat coarse average basis.

## SUMMARY

Mixing and stabilization of surface waters largely determine the availability of nutrients and light energy necessary for phytoplankton growth. The availability of nutrients and light can be found from two parameters, the stratification parameter, $S$, and the water depth, $H$, scaled appropriately by the extinction coefficient. The stratification parameter is a convenient measure for the tendency of a water column to stabilize under the influence of a surface heat flux in the presence of mixing due to wind or tidal stirring. It may be alternately considered as the inverse of the rate of energy dissipation per unit mass which is readily derived from hydrodynamic numerical models.

Spatial and temporal distributions of phytoplankton can be characterized by their stratification/scale depth properties and these values can be plotted on the characteristic *SH* diagram. The development of spring, summer and autumn distributions of phytoplankton in the English Channel and Celtic Sea are considered in these terms and it is suggested that the results may be generalized to other regions with similar *SH* characteristics.

## ACKNOWLEDGEMENTS

The satellite photographs were supplied by kind permission of Mr. P. E. Bayliss and Mr. R.J.H. Brush of the Department of Electrical Engineering and Electronics of the University of Dundee.

## REFERENCES

Fearnhead, P. G., 1975: On the formation of fronts by tidal mixing around the British Isles. Deep-Sea Res., 22, 311-321.

Flather, R. A., 1976: A tidal model of the north-west European continental shelf. Mem. Soc. R. Sci. Liege, 9, 141-164.

Fournier, R., 1977: Biological aspects of the Nova Scotian shelf-break fronts. Proceedings of MSRC Workshop, May 25-27, 1977, Bowman and Esaias (eds.).

Garrett, C.J.R., 1975: Tides in gulf. Deep-Sea Res., 22, 23-35.

Garrett, C.J.R. and R. H. Loucks, 1976: Upwelling along the Yarmouth Shore of Nova Scotia. Fish. Res. Bd. Canada, 33, 116-117.

Holligan, P. M., Pingree, R. D., Pugh, P. R. and G. T. Mardell, 1977: The hydrography and plankton of the eastern English Channel in March 1976. Annales Biologique, 33, 69-71.

Hughes, D. G., 1976: A simple method for predicting the occurrence of seasonal stratification and fronts in the North Sea and around the British Isles. I.C.E.S. C.M. 1976/C:1 Hydrography Committee.

I.C.E.S. Statistical News Letters. No. 55 Herring and Mackeral. No. 57 Demersal Species. Nominal catch and Fishing effort 1971, J. Moller (ed.), Christensen, Copenhagen, 1972.

Pingree, R. D., 1975: The advance and retreat of the thermocline on the continental shelf. J. Mar. Biol. Ass. U.K., 55, 965-974.

Pingree, R. D. and D. K. Griffiths, 1978: Tidal fronts on the Shelf Seas around the British Isles. J. Geophys. Res. (Oceans and Atmospheres). Chapman Conference Special Issue.

Pingree, R. D., P. M. Holligan, G. T. Mardell and R. N. Head, 1976: The influence of physical stability on spring, summer and autumn phytoplankton blooms in the Celtic Sea. J. Mar. Biol. Ass. U.K., 56, 845-873.

Pingree, R. D., P. R. Pugh, P. M. Holligan and G. R. Forster, 1975: Summer phytoplankton blooms and red tides along tidal fronts in the approaches to the English Channel. Nature, 258, 672-677.

Platt, T., 1972: Local phytoplankton abundance and turbulence. Deep-Sea Res., 19, 183-187.

Ramster, J. W., K. J. Meddler and J. A. Bedwell, 1972: Fishing intensity in the North Sea and the planning of moored current meter exercises. I.C.E.S. C.M. 1973/C:16 Hydrography Committee.

Riley, G. A., 1957: Phytoplankton of the north central Sargasso Sea, 1950-52. Limnol. Oceanogr., 2, 252-270.

Russell, F. S., 1935: On the value of certain plankton animals as indicators of water movement in the English Channel and North Sea. J. Mar. Biol. Ass. U.K., 20, 309-332.

Simpson, J. H. and J. R. Hunter, 1974: Fronts in the Irish Sea. Nature, 1250, 404-406.

Steele, J. H., 1957: The role of lateral eddy diffusion in the northern North Sea. J. Cons. Int. Exp. Mer., 22, 152-162.

Steele, J. H., 1958: Plant production in the North Sea. Marine Research, Scottish Home Department, 7, 36 pp.

Sverdrup, H. V., 1953: On conditions for the vernal blooming of phytoplankton. Journal du Conseil, 18, 287-295.

# PATCHINESS IN SUBSURFACE PHYTOPLANKTON POPULATIONS ON THE NORTHWEST EUROPEAN CONTINENTAL SHELF

P. M. Holligan

Marine Biological Association
Citadel Hill
Plymouth PL1 2PB, U.K.

## INTRODUCTION

In the western English Channel levels of chlorophyll *a* in surface water are relatively low ($\leqslant 0.5$ mg $m^{-3}$) in the summer after the decline of the spring diatom outburst. However, a chlorophyll-rich layer becomes established within the seasonal thermocline, generally during May or June, and persists for two to four months (Holligan and Harbour, 1977). Similar subsurface phytoplankton populations appear to be a characteristic feature of stratified waters around the British Isles at this time of year (e.g. Dooley, 1976), particularly in regions adjacent to tidal fronts (see Pingree, this volume, Figure 3).

On account of the high degree of vertical stability within the thermocline (Pingree, Pugh, Holligan and Forster, 1975), changes in phytoplankton biomass will tend to reflect the interaction of biological rather than physical processes with the basic growth-determining parameters, light and nutrient levels. Also, temporal and spatial variations in the growth rate and species composition of the phytoplankton in this layer are likely to be important factors in determining the fate of herbivore populations that build up at the time of the spring diatoms outburst. These general considerations form the basis for interpreting observations on patchiness in the subsurface chlorophyll maximum that are presented in this paper.

## THE SEASONAL DISTRIBUTION OF PHYTOPLANKTON

The area of stratified water in the western English Channel in the summer corresponds to the distribution of warm surface water and is clearly delineated in infrared photographs (Figure 1). Observations on chlorophyll *a* levels at stations E1 and E5 between 1975 and 1977 indicate that changes in phytoplankton abundance are primarily related to the vertical mixing on the availability of light and nutrients for plant growth (Pingree, Holligan, Mardell and Head, 1976). Data from E1 are summarized in Figure 2 to show the location in space and time of the spring, summer (subsurface) and autumn phytoplankton populations in relation to the mean vertical stability of the water column.

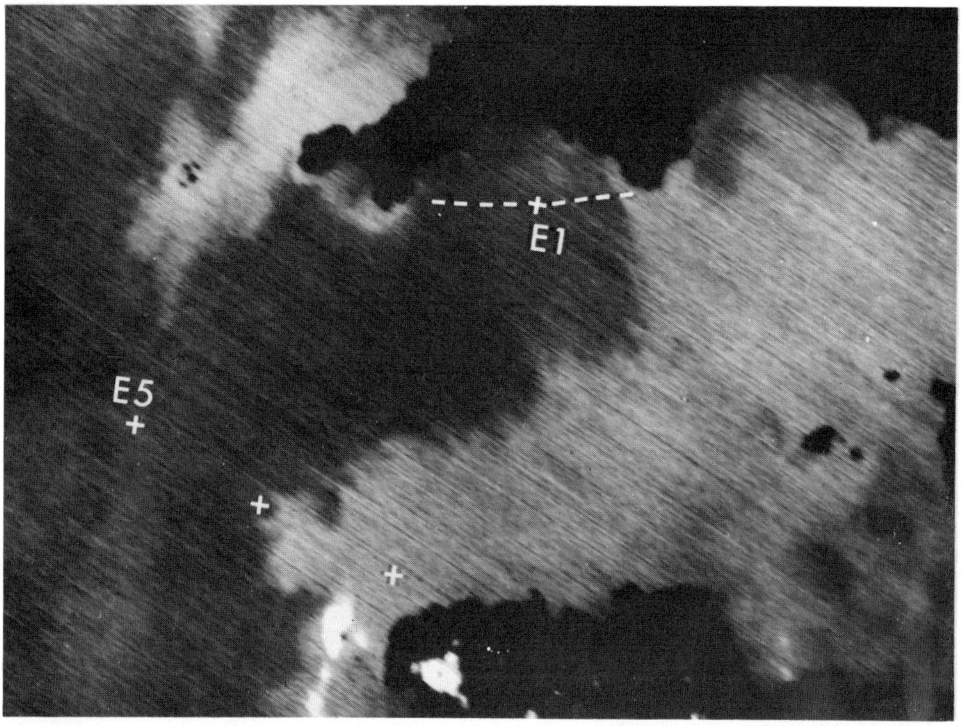


Figure 1. Infrared image taken from the NOAA 5 Satellite at 1008 h GMT on 4 July 1977. The positions of stations E1 and E5, and the section between Start Point and the Lizard (Figure 4) are shown. + marks sampling stations referred to in Table 2 (by permission of P.E. Baylis and R.J.H. Brush, University of Dundee).

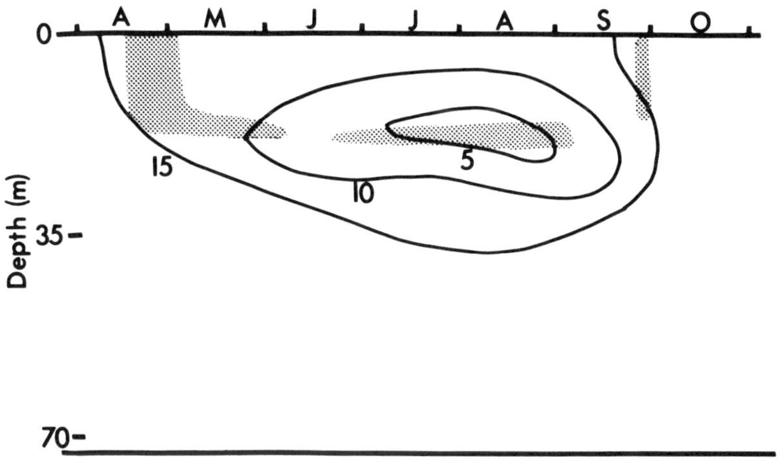


Figure 2. Mean vertical stability between April and October at station E1 in terms of the Brunt Väisälä period, $2\pi/N$ in minutes (from Pingree, Maddock and Butler, 1977). The stippled areas show the distribution of chlorophyll *a* concentrations >2 mg $m^{-3}$.

The three 'patches' of chlorophyll shown in Figure 2 can also be defined in terms of phytoplankton species and succession (Holligan and Harbour, 1977). In the spring, chain-forming diatoms are dominant. As the surface water becomes depleted of inorganic nutrients and the region of high chlorophyll restricted to the thermocline, the average size of the diatom cells decreases due to a reduction in cell size at each division as well as to the disappearance of large species. The colonial flagellate, *Phaeocystis pouchetii*, is frequently abundant at this time. In the summer the subsurface phytoplankton populations are largely composed of dinoflagellates, although under certain conditions coccolithophorids or other small flagellates may also be important. These form a thin layer in the thermocline overlying nutrient-rich bottom water. Finally, in the autumn, as the thermocline is eroded and inorganic nutrients mixed to the surface, there is a rapid decrease in the abundance of dinoflagellates and a reappearance of diatoms. This type of succession, which in temperate regions is determined by seasonal changes in solar radiation, is also seen in upwelling areas where diatoms are largely replaced by dinoflagellates between upwelling events (see Wangersky, 1977).

During the spring, fine scale spatial patchiness in the distribution of chlorophyll *a*, caused by the effects of intermittent stabilization and turbulence on the growth rates and sinking of diatoms, is well documented (see Steele, this volume). However, relatively little is known about coarse or fine scale variability in the distribution of plant populations in the thermocline, even though the subsurface chlorophyll maximum appears to be a characteristic feature of stratified shelf waters (Eppley, Reid and Strickland, 1970; Anderson, 1972; Denman, this volume). Off the coast of California, Lasker (1975) has shown that storms and upwelling can cause the rapid dispersion of dinoflagellate blooms. Observations in the western English Channel indicate that other factors may also lead to marked fluctuations in the biomass and species composition of phytoplankton within the thermocline. However, before these are described, consideration must be given to the problems of sampling subsurface layers and to environmental conditions that permit the establishment and maintenance of the subsurface chlorophyll maximum.

## SAMPLING METHODS

A submersible centrifugal pump is used to deliver water to the ship's deck at 100 $\ell$ min$^{-1}$ for the measurement of chlorophyll *a* fluorescence, the continuous analysis of inorganic nutrients, and the collection of plankton samples. The hose inlet is attached to an STD sensor which is raised or lowered at 2 m min$^{-1}$ to optimize the compatability of *in situ* temperature and salinity records with the output of the fluorometer and nutrient analyzer. However, vertical profiles of water properties become seriously distorted if, by chance, the STD sensor and hose inlet follow an internal wave motion within the thermocline (Figure 3). For this reason, profiles are generally repeated at least twice to check reproducibility. A faster vertical movement (e.g. 10 m min$^{-1}$ - see Figure 3) is unsuitable because of excessive smoothing of the fluorescence trace and also because insufficient water is obtained for reliable counts of the less numerous zooplankton species. These difficulties could be partly overcome by the use of larger pumps.

During each profile water is collected continuously from the fluorometer outflow to provide integral samples from the thermocline and top mixed layer for chlorophyll *a* determinations and phytoplankton cell counts. For more precise measurements of nutrient levels and maximum chlorophyll *a* and cell concentrations within the thermocline, the hose inlet is held at a constant depth so that gradual changes with the passage of internal waves can be correlated with one another (see Pingree, Holligan and Head, 1976).

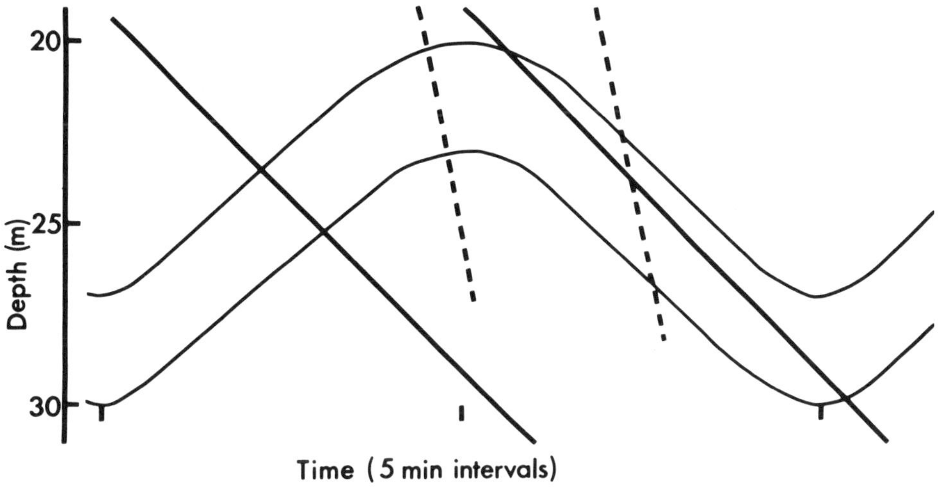


Figure 3. Sampling patterns across a chlorophyll-rich layer, 3 m thick, with a wave form of 5 m amplitude and 10 min period. This corresponds to a typical internal wave motion at station E5 in late summer. The tracks correspond to vertical speeds of 10 m $min^{-1}$ (----) and 2 m $min^{-1}$ (——) for the movement of the hose inlet.

## ESTABLISHMENT OF THE SUBSURFACE CHLOROPHYLL MAXIMUM

### The Distribution of Chlorophyll and Phytoplankton Species

The sequence of events induced by increasing vertical stability in the water column is best illustrated on a spatial rather than temporal scale. A section from Start Point towards the Lizard (see Figure 1) shows changes in chlorophyll *a* levels that are typical of the transition from mixed to stratified conditions during the spring and early summer (Figure 4). At station 17 a mixed diatom population was present, and high chlorophyll *a* values (>7 mg $m^{-3}$) below 15 m indicate that some sinking had occurred. Diatoms were also dominant at station 16, but here the maximum concentrations of chlorophyll *a* (∿12 mg $m^{-3}$) were recorded in the thermocline where *Rhizosolenia delicatula* and *R. stolterfothii* were the dominant organisms. The importance of sinking and *in situ* growth to the formation of the chlorophyll peak are not known.

Between stations 16 and 5, the diatoms largely disappeared in the thermocline and were replaced by *Phaeocystis* (station 11/E1) and then by the dinoflagellates, *Ceratium fusus* and *Dinophysis acuminata* (station 5). These changes, which are summarized in Table 1, are generally consistent with observations made at station E1 during 1976 (Holligan and Harbour, 1977), although a considerable difference in the timing of the succession is apparent. In 1976, by late June, the surface water was warmer than the bottom water by about 5°C and it is probable that the dinoflagellate population had become established two or even three weeks earlier than in 1977. Another difference between the two years is that at the end of May, 1976, before the dinoflagellates became abundant, there was no peak in chlorophyll fluorescence at the thermocline (cf. station 8, Figure 4). This suggests that the subsurface maximum

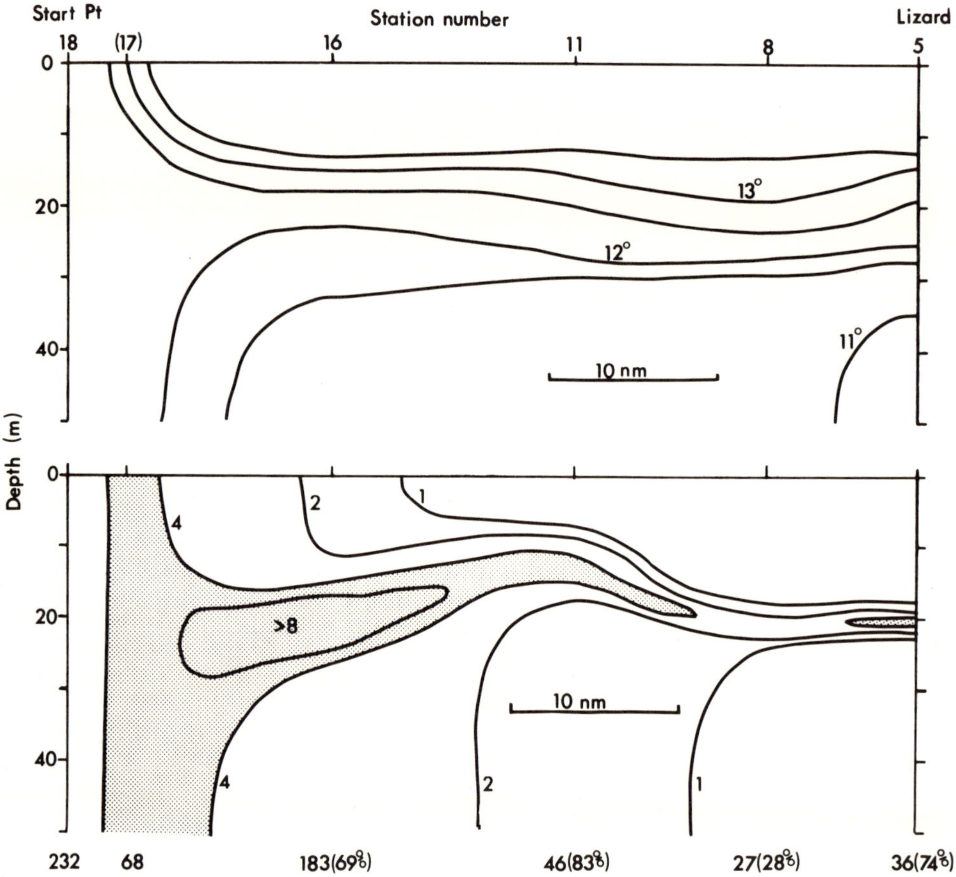


Figure 4. Temperature (°C) and chlorophyll *a* (mg $m^{-3}$) sections between Start Point and the Lizard (Figure 1), 28-29 June 1977. Data from station 17, which was not on a direct line between stations 18 and 16, is considered representative of the weekly stratified region. Values at the bottom of the diagram give total chlorophyll *a* (mg $m^{-2}$) in the top mixed layer and thermocline, with the proportion in the latter as a percentage.

Table 1

The effect of thermocline development on subsurface phytoplankton populations between Start Point and the Lizard, 28-29 June 1977.

| Station No. (see Figure 4) | | 16 | 11/E1 | 8 | 5 |
|---|---|---|---|---|---|
| Surface to bottom temperature difference ($^{o}$C) | | 2.2 | 2.3 | 2.7 | 3.0 |
| Chlorophyll *a* (mg m$^{-2}$) | | 126 | 38 | 8 | 27 |
| Flagellates | ($10^3$ cells m$^{-2}$) | 7920 | 28000 | 4400 | 5180 |
| Diatoms (A) | " | 3332 | 1261 | 111 | 47 |
| Dinoflagellates (B) | " | 43 | 43 | 73 | 222 |
| Ratio B/A | " | 0.01 | 0.03 | 0.66 | 4.72 |

tends to disappear under conditions of intermediate vertical stability as indicated in Figure 2. The rate at which diatoms are replaced by dinoflagellates on a given year is unlikely to be a steady function of time, but highly dependent on weather and tidal conditions. The difference in vertical stability between stations 16 and 5 is small, and it is probable that dinoflagellates later became more abundant as surface warming continued during the summer.

In terms of the distribution of temperature and chlorophyll *a*, the section shown in Figure 4 represents a frontal region at an early developmental stage. Although data on the phytoplankton for this section in the late summer are not yet available, results from Ushant area (see Figure 1) show that diatoms continue to dominate the phytoplankton in inshore mixed waters but in the front itself are replaced by dinoflagellates (Table 2).

## Growth of Phytoplankton Within the Thermocline

In the western English Channel the depth of the thermocline approximates to the 10% light level (i.e., somewhat deeper than the Secchi disc depth). As the subsurface chlorophyll maximum is formed, the gradient in nitrate concentration across the thermocline becomes determined by the distribution of plant cells rather than by mixing between the nutrient-rich bottom water and the nutrient-poor surface water (Figure 5). The same is probably also true for inorganic phosphate but, by contrast, as the diatoms disappear, the silicate gradient is increasingly related to the temperature structure and appears to become a function of physical rather than biological parameters. Levels of dissolved organic nitrogen (including ammonia) and phosphorus in the chlorophyll-rich layer are variable but often higher by as much as 50% than above or below (E. I. Butler, unpublished data).

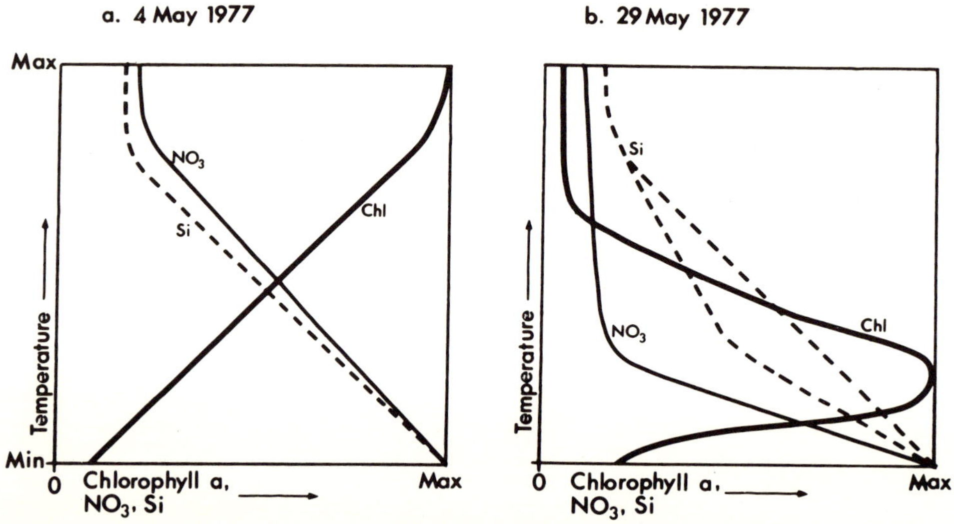


Figure 5. The distribution of chlorophyll *a*, nitrate and silicate across the thermocline at station E5 on 4 May and 29 May 1977. Each curve is based on more than 30 values that were obtained with the hose inlet held at constant depths in the temperature gradient. On 17 August 1977, the distributions of chlorophyll *a* and nitrate were very similar to those on 29 May, and that of silicate is shown by the upper curve. The surface to bottom temperature differences on these three dates were, respectively, 0.6, 2.4 and 7.5°C.

Although the environment of the thermocline in terms of light and nutrient levels is favorable for phytoplankton growth, it is not immediately apparent how the dense concentrations of plant cells are sustained. The motility of dinoflagellates reduces loss by sinking that affect diatom populations. It is also probable that microbial recycling processes within or close to the chlorophyll-rich layer provide a significant source of inorganic nutrients in addition to upward transport from bottom water (Pingree and Pennycuick, 1975). The abundance of dinoflagellates is indicative of a regenerative rather than a renewal system of nutrient supply (Wangersky, 1977).

## TEMPORAL PATCHINESS

Although observations on phytoplankton patchiness within the seasonal thermocline are, as yet, too few to allow any general quantitative analysis of population dynamics, they illustrate the scales of variability that may be found in shelf waters and also raise important ecological implications.

Table 2

Differences in the composition of phytoplankton populations across the Ushant front.

| Position | 48°45'N 04°40'W | 49°51'N 05°20'W | 49°06'N 06°32'W (E5) |
|---|---|---|---|
| Date | 27/7/76 | 28/7/76 | 2/8/76 |
| Sample Depth | 2 m | 2 m | 25 m |
| Chlorophyll *a* (mg $m^{-3}$) | 1.2 | 46 | 58 |
| Diatoms (cells $ml^{-1}$) | 27.9(27) | 0.1(1) | 0.1(3) |
| Dinoflagellates " | 2.1(15) | 2760(9) | 0.1(3) |
| Coccolithophorid " | 0 | 0 | 8000(1) |
| Flagellates " | 410 | no count | 760 |

Numbers of diatom and dinoflagellate species recorded from each sample are given in parentheses.

Table 3

Changes in phytoplankton populations in the thermocline at stations E1 and E5, July-August 1976.

| | Date | Maximum chlorophyll *a* concentration (mg $m^{-3}$) | Dominant phytoplankton species |
|---|---|---|---|
| E1 | 14 July | 2.5 | *Dinophysis acuta, Gyrodinium aureolum, Heterocapsa triquetra* |
| | 20 July | 5 | *D. acuminata, D. acuta, Ceratium lineatum* |
| | 4 August | 26 | *G. aureolum* |
| | 12 August | 17 | *G. aureolum* |
| E5 | 21 July | 5 | Flagellates |
| | 2 August | ∿80 | Coccolithophorid |
| | 11 August | ∿50 | *G. aureolum*, coccolithophorid |

In the summer of 1976, marked temporal changes in the phytoplankton populations at stations E5 and E1 were observed (Table 3). Increases of an order of magnitude or more in plant biomass can only be attributed to factors that affect directly the growth rate of the cells. An *in situ* $^{14}CO_2$-fixation experiment at station E5 on 11 August showed that light saturation of photosynthesis was not achieved at the depth of the thermocline even under optimal conditions (Figure 6). Thus, provided that nutrients are not limiting (there is no evidence to suggest that they might be), rates of

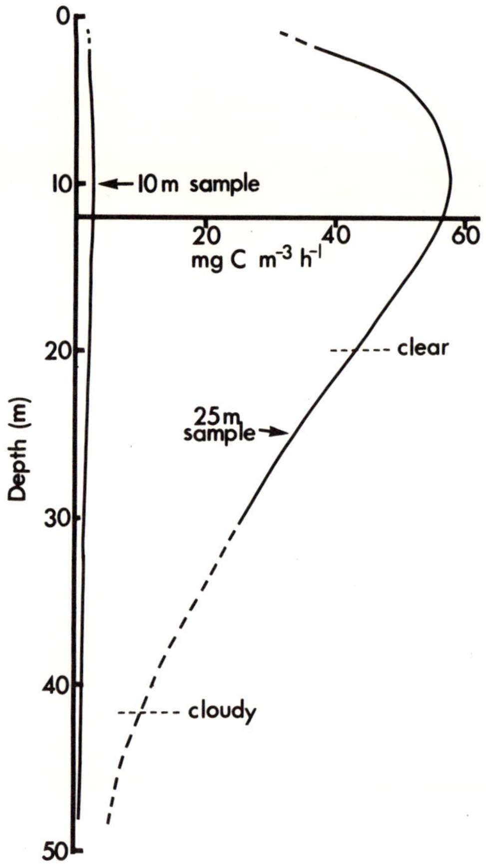


Figure 6. Curves for rates of carbon fixation against depth at station E5, 12.15 - 15.15 h GMT, 11 August 1976 (for original data see Holligan and Harbour, 1977). Water samples were collected at 10 m and 25 m as representative of the top mixed layer and thermocline phytoplankton populations. The probable range of values in the thermocline under extreme conditions of light intensity at mid-day is shown.

carbon fixation in the thermocline can be considered proportional to daily solar radiation values which vary by a factor of 5 or more between overcast and clear days. On this basis, subsurface chlorophyll concentrations >20 mg $m^{-3}$ in August appear to have been the result of a prolonged period of sunny weather; mean daily sunshine totals at stations in southwest England ranged from ∿6 h $day^{-1}$ in mid-July to >11 h $day^{-1}$ during the first part of August (data from the daily weather reports of the British Meteorological Office).

Diurnal changes in chlorophyll fluorescence in the thermocline have also been detected (Table 4) which suggest that as much as half the standing crop of phytoplankton can be consumed by herbivores during the night. Chlorophyll *a* calibration values and cell counts from water samples taken over this period showed comparable trends. Furthermore, rates of photosynthetic carbon fixation (Figure 6) when compared with estimates of plant carbon derived from cell volumes (Eppley, Reid and Strickland, 1970) were of the right order for a doubling of the plant population during each 24-hour period. Two herbivores (*Limacina* and *Oithona*) were aggregated in the chlorophyll-rich water during the afternoon and first part of the night. A similar set of observations was made at station E1 in June 1976, when the larger dinoflagellates *Ceratium fusus* and *Dinophysis acuminata* were the dominant phytoplankton in the thermocline. On this occasion the decrease in chlorophyll fluorescence at night was not so marked, and *Calanus* and the small copepods, but not *Oithona*, aggregated in or close to the chlorophyll-rich layer. It appears, therefore, that changes in the vertical distribution of a particular herbivore species are related to the type of plant food available.

The degree of vertical turbulence due to tides, as well as wind (Lasker, 1975), may also cause significant changes in the distribution of plant cells in the water column. Towards neap tides the vertical temperature gradient tends to become more relaxed and any inclusion of nutrient-rich water at the base of the thermocline will enhance phytoplankton growth. By contrast, increased bottom mixing at spring tides causes a sharpening of the temperature gradient, with dispersion of plant cells from the chlorophyll-rich layer into the bottom water (Table 5).

## SPATIAL PATCHINESS

The spatial patchiness of chlorophyll *a* and phytoplankton species within the thermocline can be described only in rather broad terms because of limitations imposed by pump sampling techniques and by short term changes in plant biomass due to grazing and turbulence. The light regime, as determined by the depth of the thermocline, the turbidity of the surface mixed layer and surface radiation, appears to be the major factor determining the

Table 4

Changes in chlorophyll *a* fluorescence and in zooplankton numbers in the thermocline at station E5, 10-11 August 1976.

| Time (BST) | 1730 | 2045 | 2400 | 0300 | 0605 | 0900 | 1230 | 1530 |
|---|---|---|---|---|---|---|---|---|
| Chlorophyll *a* fluorescence | | | | | | | | |
| Depth of subsurface maximum (m) | 35 | 28 | 25 | 27 | 35 | 25 | 25 | 25 |
| Integrated peak value (chart units[1]) | 62 | 48 | 45 | 25 | 18 | 54 | 70 | 60 |
| Maximum peak value (chart units[1]) | >10 | 6.3 | 6.2 | 4.1 | 5.2 | 6.0 | 8.1 | >10 |
| Zooplankton (relative abundance*) | | | | | | | | |
| *Limacina retroversa* (86) | 1.5 | 5.4 | 4.6 | 0.9 | 0.8 | 0.4 | 2.7 | - |
| *Oithona* spp. (658) | 5.8 | 5.4 | 5.2 | 1.7 | 1.1 | 3.6 | 1.6 | - |
| *Calanus* spp. (84) | 1.0 | 0.1 | 0.5 | 0.3 | 1.2 | 0.2 | 0.5 | - |
| Small copepods (965) | 1.1 | 0.5 | 0.2 | 0.3 | 0.1 | 0.5 | 0.8 | - |
| *Centropages* spp. (326) | <0.1 | <0.1 | <0.1 | <0.2 | <0.1 | <0.1 | <0.1 | - |

[1]Relative values are used as the fluorometer was not calibrated for every profile. Maximum observed chlorophyll *a* concentrations were ∿50 mg $m^{-3}$ (see Table 3).

*Ratio of no. $m^{-3}$ in chlorophyll-rich layer to average no. $m^{-3}$ between 2 m and 50 m. The latter value is given for each species in parentheses. Sample volumes, filtered on 170μ aperture nets from the pump outflow, were 0.5 and 2.5 $m^3$ respectively.

Table 5

Vertical distribution of photosynthetic dinoflagellates (excluding *Prorocentrum balticum*) at station E1 (water depth 73 m).

| | 22 June 1976 (neap tide) | | 14 July 1976 (spring tide) | |
|---|---|---|---|---|
| | Average number of cells $ml^{-1}$ | % total | Average number of cells $ml^{-1}$ | % total |
| Top mixed layer | 1.1 | 4 | 1.7 | 10 |
| Thermocline* | 57.0 | 95 | 36.0 | 47 |
| Bottom mixed layer | 0.1 | 1 | 1.7 | 43 |

*The depths of the subsurface chlorophyll maxima were 14-21 m in June and 13-16 m in July.

growth rate of the phytoplankton. Field observations support this hypothesis -- at any given time chlorophyll *a* levels are higher and the larger dinoflagellates more abundant in inshore areas where the thermocline is shallow, whereas in deeper thermoclines offshore, small dinoflagellates and flagellates are usually dominant (Holligan and Harbour, 1977). Also, in the fine summers of 1975 and 1976, chlorophyll levels were generally higher and dinoflagellates more abundant than in 1977, when more normal weather conditions prevailed and maximum chlorophyll *a* concentrations during July and August rarely exceeded 10 mg $m^{-3}$ (unpublished data).

Available information on the spatial distribution of individual species is more precise. Data from the continuous plankton recorder for the North Sea show that large dinoflagellates, such as *Ceratium fusus*, are most abundant during the summer months in areas with a relatively shallow thermocline (Tomczak and Goedecke, 1964) (Figure 7). Similarly, in July 1975, *Gyrodinium aureolum* was dominant both at the surface in frontal regions and further offshore in the thermocline across much of the western English Channel (Figure 8) and in a series of samples collected in July 1977 from the subsurface chlorophyll maximum along a line of stations in the northern North Sea, extending more than 100 km to the southeast of the Orkney Islands, the phytoplankton was composed almost exclusively of *Dinophysis norvegica, Ceratium lineatum* and *C. furca*. So for a hydrographically discrete region, such as the Celtic Sea or the North Sea, the distributional limits of dense populations of these common dinoflagellates appear to be determined inshore by fronts marking the boundary between mixed and stratified water masses and offshore by low light levels at thermocline depths greater than about 25 m. Dispersion due to horizontal shear and small-scale mixing events within the thermocline will lead to the rapid spread of individual species when growth conditions are favorable.

## DISCUSSION

Calculations based on chlorophyll *a* data and carbon assimilation values at differing light intensities (e.g. Figure 6) give rates of primary production in the thermocline between 0.05 and 2.5 gC $m^{-2}$ $day^{-1}$. These account for 30-80% of the total for the euphotic zone in the summer. On an annual basis, the subsurface chlorophyll maximum may yield as much particular organic carbon as the spring diatom outburst, if the light regime is favorable, and is probably important in determining gross differences in production between one year and the next. The high biomass of zooplankton that persists in temperate shelf waters through the summer months (Harvey, Cooper, Lebour and Russell, 1935; Cushing, 1959) must be largely dependent on plant production in the thermocline. Furthermore, as Lasker (this volume) has found, the type and concentration of plant cells in the thermocline can be critical for the survival

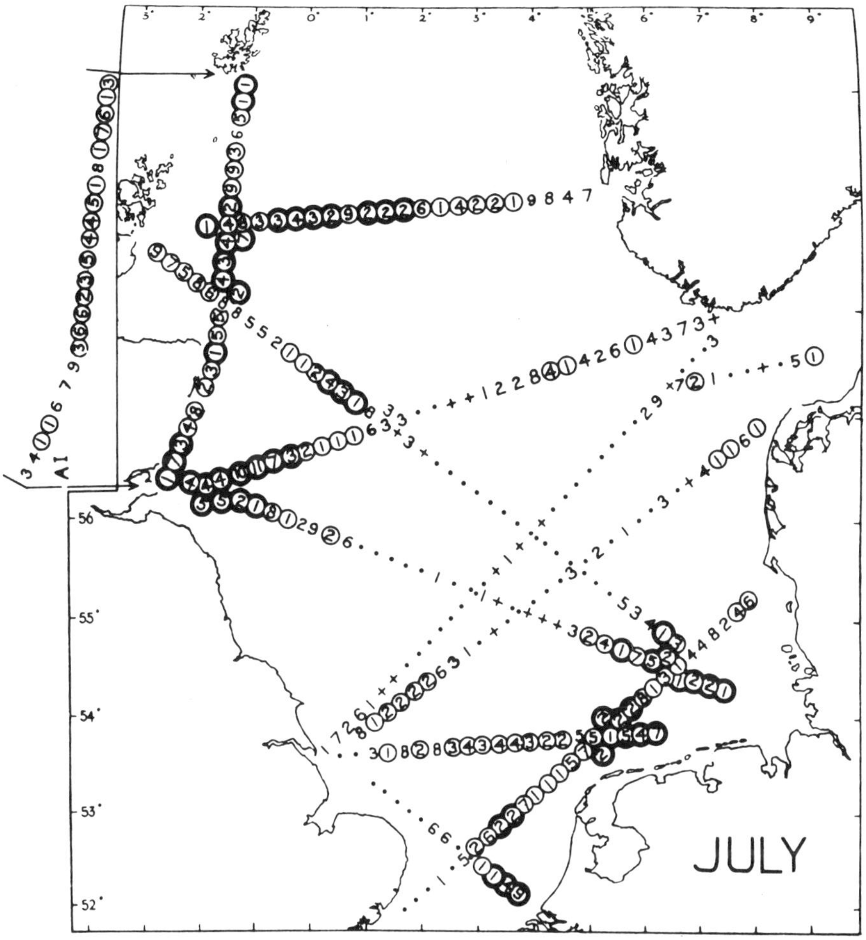


Figure 7. The distribution of *Ceratium fusus* in the North Sea in July 1938, as shown by continuous plankton recorder samples (from Lucas, 1942).

of certain types of zooplankton that do not feed effectively on diatoms.

In relation to food chain studies, temporal rather than spatial patchiness in subsurface phytoplankton populations appears to be of greater significance, although obviously in regions with strong horizontal gradients in water properties (i.e., close to fronts) spatial changes may be equally important. Changes in both the biomass and species composition of phytoplankton, at time scales ranging from a day to a month or more, will reflect the balance between light-dependent growth rates and mortality due to

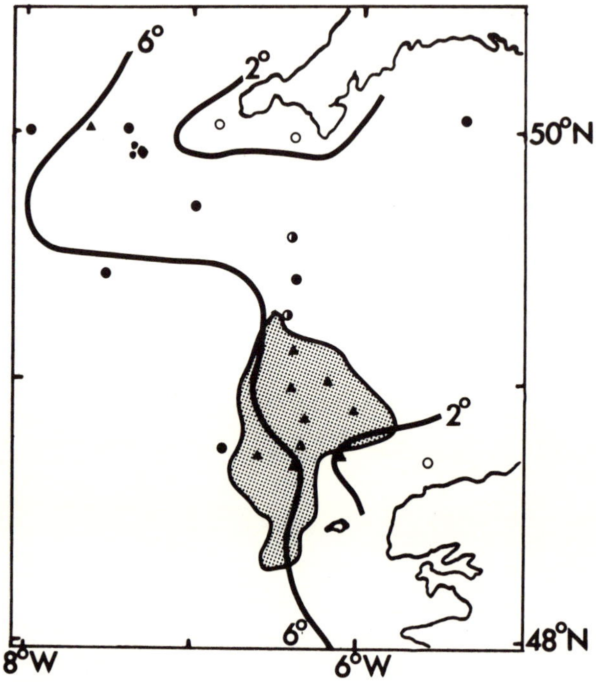


Figure 8. The distribution of *Gyrodinium aureolum* in the western English Channel in July 1975. Contours for surface to bottom temperature differences of 2° and 6°C are given. The stippled area shows the region where surface chlorophyll *a* values exceeded 4 mg $m^{-3}$. ● dominant in the thermocline; ▲ dominant in the top mixed layer; ○ not recorded; ◑,▲ indicate that other species of phytoplankton were also abundant.

grazing intensity and selectivity. Nutrients are considered relatively unimportant, partly because there is a constant source of nitrate and inorganic phosphate in the bottom water but also because regeneration within the thermocline itself is likely to be rapid. Certain physical conditions could disrupt the general pattern of plankton production, particularly early in the summer when, for example, cool weather will tend to delay the increase in dinoflagellates.

Models of plankton production are generally not extended to include plant production within the seasonal thermocline. However, further studies of plankton communities in the thermocline may prove valuable in assessing the validity of models such as those proposed by Steele and Frost (1977) which emphasize the size relationships between prey and predator. The reason for this is that the thermocline represents a physically stable environment in which biological factors such as grazing are probably dominant in determining community structure.

## SUMMARY

Over the continental shelf around the British Isles a chlorophyll-rich layer persists for two to four months within the seasonal thermocline. Under conditions of weak stratification diatoms are the dominant phytoplankton, and a well-defined subsurface chlorophyll maximum develops as the cells sink out of the nutrient-depleted surface water. With increasing vertical stability during the early summer the diatoms are replaced by dinoflagellates. However, despite a relatively stable physical environment at the level of the thermocline, temporal fluctuations in the concentration of chlorophyll *a* and in the abundance of particular species are considerable. These probably reflect changes in the balance between light-dependent growth rates of the plant cells and depletion by grazing. Spatial patchiness in chlorophyll levels is difficult to define due to sampling difficulties and to diurnal variation in grazing pressure, but the same dinoflagellate species may be dominant over distances of 100 km or more.

## REFERENCES

Anderson, G. C., 1972: Aspects of marine phytoplankton studies near the Clumbia River, with special reference to a subsurface chlorophyll maximum. In The Columbia River Estuary and Adjacent Ocean Waters, A. T. Pruter and D. T. Alverson (eds.), pp. 219-240, University of Washington Press.

Cushing, D. H., 1959: The seasonal variation in oceanic production as a problem in population dynamics. J. Cons. perm. Int. Explor. Mer., 24, 455-464.

Dooley, H. D., 1976: Phytoplankton standing crop adjacent to ocean boundaries in the northern North Sea. I.C.E.S. C.M. 1976/L:28, 8 pp.

Eppley, R. W., F.M.H. Reid and J.D.H. Strickland, 1970: Estimates of phytoplankton crop size, growth rate, and primary production. In The ecology of plankton off La Jolla, California, in the period April through September, 1967. Bull. Scripps Inst. Oceanogr., 17, 1-103.

Harvey, H. W., L. C. Cooper, M. V. Lebour and F. S. Russell, 1935: Plankton production and its control. J. Mar. Biol. Ass. U.K., 20, 407-442.

Holligan, P. M. and D. S. Harbour, 1977: The vertical distribution and succession of phytoplankton in the western English Channel in 1975 and 1976. J. Mar. Biol. Ass. U.K., 57, 1075-1093.

Lasker, R., 1975: Field criteria for survival of anchovy larvae: the relation between inshore chlorophyll maximum layers and successful first feeding. Fish. Bull. Nat. Ocean. Atmos. Admin. U.S., 73, 453-462.

Lucas, C. E., 1942: Continuous plankton records: phytoplankton in the North Sea 1938-39. Part II. Dinoflagellates, Phaeocystis, etc. Hull Bull. Mar. Ecol., 2, 47-70.

Pingree, R. D., P. M. Holligan and R. N. Head, 1977: Survival of dinoflagellate blooms in the western English Channel. Nature, 265, 266-269.

Pingree, R. D., P. M. Holligan, G. T. Mardell and R. N. Head, 1976: The influence of physical stability on spring, summer and autumn phytoplankton blooms in the Celtic Sea. J. Mar. Biol. Ass. U.K., 56, 845-873.

Pingree, R. D., L. Maddock and E. I. Butler, 1977: The influence of biological activity and physical stability in determining the chemical distributions of inorganic phosphate, silicate and nitrate. J. Mar. Biol. Ass. U.K., 57, 1065-1073.

Pingree, R. D., P. R. Pugh, P. M. Holligan and G. R. Forster, 1975: Summer phytoplankton blooms and red tides along tidal fronts in the approaches to the English Channel. Nature, 258, 672-677.

Steele, J. H. and B. W. Frost, 1977: The structure of plankton communities. Phil. Trans. Roy. Soc. Lond., 280, 485-534.

Tomczak, G. and E. Goedecke, 1964: Die thermische Schichtung der Nordsee auf Grund des mittleren Jahresganges der Temperatur in ½°- und 1° - Feldern. Deutschen Hydrographischen Zeitschrift, Erg.-Heft Reihe B(4°), Nr. 8, 182 pp.

Wangersky, P. J., 1977: The role of particulate matter in the productivity of surface waters. Helgolander wiss. Meeresunters., 30, 546-564.

# SPATIAL HETEROGENEITY IN CLOSED BASINS

Peter J. Richerson, Thomas M. Powell,
Mark R. Leigh-Abbott and John A. Coil

Division of Environmental Studies
University of California
Davis, California 95616

## INTRODUCTION

Everything varies in space and time. This truism appears to be as accurate for aquatic habitats as it is for terrestrial habitats. Whether one measures physical parameters, the concentrations of chemical species, or numbers of organisms, all are spatially and temporally heterogeneous.

Since this spatial (and temporal) variability is such an integral part of lakes, estuaries, and the ocean, two important questions confront aquatic scientists attempting to understand the ecology of organisms at any trophic level. First, what is the ecological significance of the observed spatial heterogeneity? Second, assuming that some significance can be attached to this pervasive phenomenon, what is the best method to both measure and describe spatial heterogeneity so that its importance can be assessed in each body of water?

Much important aquatic and terrestrial ecology has investigated the significance of spatial heterogeneity. Beyond the obvious effect that patchiness has in sampling design (the problems caused by the patchy distribution of organisms prompted the initial studies of the phenomenon), no consensus exists on the causes or the ecological effects of patches in aquatic systems. Indeed, the interaction of physical factors and chemical species upon organisms at various space and time scales is complex. The complications within even simple models of plankton growth that incorporate spatial variation (e.g. Wroblewski, 1977) allow for an enormous number of responses. Thus any simple, deterministic explanation of patchiness (e.g. that Langmuir circulations alone cause patches) is suspect (Steele, this volume). Moreover, spatial inhomogeneities at one trophic level can influence levels both above and below. The papers in this volume show one thing clearly; spatial and temporal heterogeneity is a product of many physical processes interacting with many biological processes. More experimental and theoretical work is needed to uncover basic facts about organisms, nutrients, and physical phenomena interacting in highly stochastic, but far from completely random, environments.

Two basic techniques have been used to describe spatial heterogeneity in aquatic systems. The earliest pictures of the non-random distribution of plankton depicted actual patches where plankton were growing, while outside, populations were low (Ricker, 1937). These early treatments gave way to summary statistics, such as variance/mean ratios or Fisher's $k$ (Cassie, 1959; Richerson *et al.*, 1970), in an attempt to describe the non-randomness that the distribution of organisms exhibited. With the availability of new fluorometric techniques that collect large amounts of data nearly simultaneously, the application of time series analysis became feasible. The presentation of spectra, autocorrelations, cross spectral and coherence spectral estimates is now common.

In an effort to find generalizations that might apply to the question of variability in aquatic systems, we have gathered physical, chemical, and biological information from a variety of lakes over the last five years. By looking at lakes of various trophic type (e.g. ultraoligotrophic, where zooplankton grazing is negligible, and eutrophic, where it is not), the effects of higher trophic levels might be isolated. Lakes of different size (and thus different physical characteristics) might show the effects of some specific physical process (e.g. turbulent mixing) in causing, maintaining or destroying heterogeneity. We had hoped that some assessment of the role that dissolved nutrients play in the distribution of phytoplankton might be elucidated with detailed transects as well as with widely-spaced stations, separated by scales at which nutrient variability is expected to be significant. We have expended substantial effort to collect data on the spatial distribution of individual phytoplankton species, and we have collected some zooplankton species information. Few generalizations are suggested by our data. Lakes, with distinct problems of their own, are as complicated as the mid-ocean, the continental shelves, and estuaries. It is unlikely that the simpler environments of lakes will provide the answers that are lacking for the marine realm. Though our data are interesting in their own right and for comparison to the marine data, our work thus far has not answered either of the two major questions that we proposed earlier.

This report summarizes the basic aspects of the physical and biological regimes in lakes that we believe to be most relevant to the fundamental ecological questions. These results are compared with those obtained by other investigators in other closed basins, coastal oceans, and the open ocean, and some directions for future work are suggested. In general, the results from this wide variety of habitats are remarkably similar, particularly in the dominance of spatial variability by large-scale patterns in space and the complex relationship between physical, chemical and biological processes.

## PHYSICAL AND BIOLOGICAL BACKGROUND

### The Physical Regime of Lakes

Several excellent reviews (Boyce, 1974; Mortimer, 1974; Csanady, 1975) of the physical characteristics of lakes relevant to biological considerations have recently appeared. It is not appropriate to repeat these works here. The interested reader can consult them, and the original literature cited there, for further details.

The most obvious and important physical characteristic of the water mass in lakes is stable stratification, a feature lakes share with the mixed layer-thermocline-abyssal structure of the ocean. All lakes stratify to some extent but to varying degrees and with different timing. A detailed classification of lakes according to their stratification type is often used (Hutchinson, 1957). Recent and continuing studies have also shown a detailed vertical temperature microstructure in lakes superimposed upon the gross stratification as in the ocean (e.g. Simpson and Woods, 1970; Thorpe *et al.*, 1977; Caldwell *et al.*, 1978). An example from Lake Tahoe, California-Nevada, USA, is shown in Figure 1. It is unclear what effect this small scale vertical microstructure has upon the vertical distribution of planktonic organisms that are largely at the mercy of water movements.

Lakes also have features superficially similar to the large horizontal structure of the mixed layer in the sea. Gyres exist, though they probably do not persist as the mid-oceanic gyres. Moreover, gyres in lakes, unlike those in the ocean, are caused by the variable topography of the basin sides and bottom, though the effect of the earth's rotation cannot be neglected. These are often referred to as topographic gyres (Csanady, 1975). There is

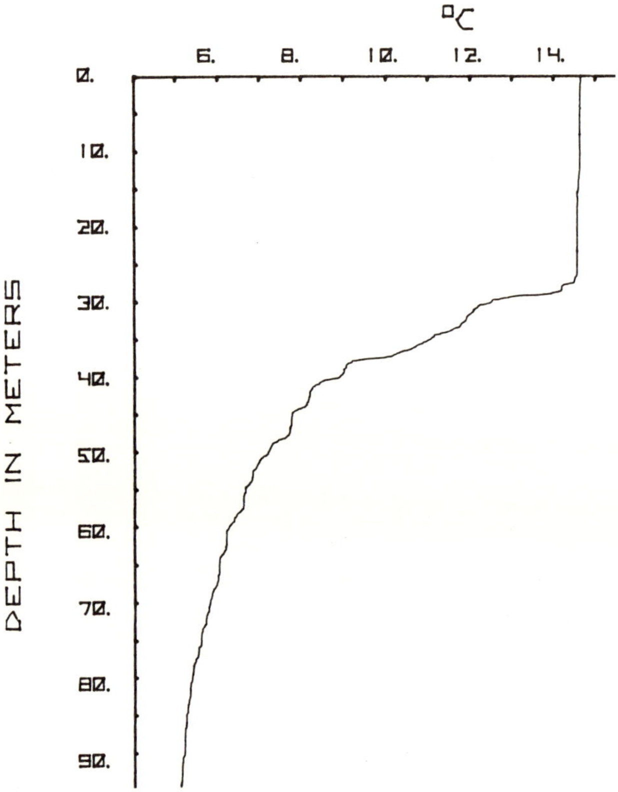


Figure 1. Vertical temperature microstructure in the thermocline of Lake Tahoe, 29 September 1977.

also a distinct division between the coastal zone and the middle region of a lake. (In large lakes it seems useful to define the coastal zone as lying within one Rossby radius of deformation of the actual coastline.) In particular, each responds in a distinct way to external forces, primarily the wind. This division resembles the differences between the coastal and mid-ocean; however, the effects of the closed basin geometry in lakes are so pervasive that little correspondence to oceanic responses is anticipated or seen.

When a strong wind stress excites the shallow coastal region of a lake, the integrated mass flow of water (i.e., the net flow of water when integrated from the surface to the bottom) is in the direction of the wind stress. When the wind stress subsides, low frequency phenomena, particularly Kelvin waves and coastal jets, are believed to dominate the flows in the coastal zone. The important point is that the time scale of response of the mixed layer of the coastal zone is of the order of several days.

In contrast, the midlake region responds more rapidly. During a strong wind event, the mixed layer responds as a classical Ekman layer. However, as the wind subsides, the dominant responses are resonant oscillations determined by the closed basin geometry; i.e., surface seiches in the barotropic mode, and Poincaré waves (resonant internal wave modes near the thermocline) in the baroclinic modes (Mortimer, 1974; Csanady, 1975). The vertical water movements in the internal Poincaré wave modes are much larger than in the surface seiches and are of considerably higher frequency than the responses of the coastal zone (typically the periods are near the inertial period ≃15-20 hours in large, temperate latitude lakes).

In summary, long-term current meter records (Blanton, 1974) show that kinetic energy in the coastal region is highest in the frequency band where the periods of oscillation center around 100 hours, while kinetic energy in the mid-lake region is dominated by contributions near the inertial period. The detailed geometry of the boundaries and bottom is also critical. It enters into coastal zone responses directly and mid-lake responses indirectly (because the dominant response is an oscillation in a closed basin). The earth's rotation cannot be neglected in large lakes because of its effects through topographic gyres and Poincaré waves and, indirectly, through the Rossby radius of deformation.

## The Biological Regime of Lakes

To describe the biology of lakes in a few short paragraphs is not possible. Most of the published literature in limnology is biological. However, there are some notable effects that are important to studies of spatial heterogeneity in closed basins.

The importance of inflowing streams cannot be overestimated. The nutrients carried in by these flows at various points on a lake's shore cause regions of increased phytoplankton numbers. Thus the streams act as agents for variability on a large, lakewide scale, producing variation reminiscent of the current-system related processes in the ocean. The measurements of Goldman *et al.* (1972) clearly show this effect in Lake Tahoe. It will be seen that this variability at the largest spatial scale is one of the dominant (if not the dominant) characteristics of spatial variation in lakes. Accordingly, large scale sources of variance in chlorophyll *a* or nutrients, due to stream inflows, are extremely important.

The effects of the sides and bottom in lakes, and all closed basins, are important. Since a littoral zone may occupy a substantial fraction of any lake's area, the organisms found there (e.g. periphyton) can be expected to play a substantial role in the nutrient and energy budgets of lakes. Benthic organisms can act as sources of variance that make themselves felt throughout the lake through various physical transport mechanisms.

The geographical isolation of one lake from another leads to an assortment of species assemblages that vary greatly from lake to lake. Various growth and death processes are important in creating, maintaining and destroying spatial heterogeneity. Thus a changed species composition within the phyto- and zooplankton communities would affect these processes through interspecific competition and grazing, leading to different spatial structure in different lakes. Data from Lake Tahoe show that, even within one lake, different phytoplankton species show a variety of spatial patterns, ranging from a random distribution to a highly organized patch structure. In Castle Lake (California), zooplankton are more heterogeneous than phytoplankton.

In summary, measurements show the sources of spatial variance in lakes occur at large, lakewide scales, and that various species will present significantly different pictures of spatial heterogeneity.

## Coupled Physical and Biological Effects

Given these scale characteristics of lakes, we need to examine how the interaction of biological and physical processes might produce patterns that differ from those in the ocean. Most of these processes in the sea (excluding those that depend solely upon salinity concentrations) occur also to a greater or lesser extent in freshwater lakes, despite the substantial contrasts between the two environments. Any comparison or contrast between spatial variation in the ocean and in lakes demand some knowledge of the scales of similar processes acting in both water bodies. For example, the actual dimensions of the littoral zone in the ocean are larger than those found in all lakes and estuaries. However, the percentage of a lake's total area that is in the littoral zone is much larger than the corresponding percentage in the ocean. Accordingly, the influence of littoral zone organisms should be more important at the largest spatial scales in lakes.

Physical and biological processes in all but the largest lakes occur on much smaller scales than in the ocean. The analogs of many of the relatively well-organized, large-scale features shown in the Stommel diagram in Haury, McGowan and Wiebe (this volume), are absent. The effect of the smaller size of the lakes upon spatial variation is unclear. One might think that lakes should be more homogeneous than the ocean (i.e., mixing can readily homogenize a small body of water). However, inflowing streams can be intense sources of variance in lakes at scales where there are no sources in the ocean. Mixing processes also scale with size, so that mixing is diminished in a lake relative to the sea. Accordingly, the critical length scale $(K/r)^{\frac{1}{2}}$, where $K$ is an eddy diffusivity and $r$ a growth rate, first proposed for marine organisms by Kierstead and Slobodkin (1953), should be smaller in lakes than in the ocean. Our data support this conclusion. Hence, many of the ecologically important, spatially varying phenomena seen in the ocean probably also appear in lakes. Due to the variability in $K$ and $r$ under different circumstances, comparisons between habitats depend on the estimation of the critical scales of the spatial patterns.

Concerning temporal scales, we have described the different frequencies at which kinetic energy is concentrated in the coastal and mid-lake regions. The effects these differing frequencies have upon the plankton are not clear. Moreover, effects dominated by an energy input in one frequency band may interact in a nonlinear and complicated fashion with those dominated by another frequency band as lake waters are diffused and advected. Since dominant scales of physical variance (1 day) are not too different from phytoplankton growth rates (doubling times 1-10 days), a whole range of complex and transient phenomena may exist.

One interesting contrast between oceanic and lake studies may become apparent when the differences between the coastal zone and mid-lake regions are compared to differences between the coastal ocean and marine pelagic regions. Walsh (1976), citing a personal communication from P. Niiler, has noted that more high frequency energy is present in the coastal regions than in the mid-ocean. This is in direct contrast to the lake measurements. A synoptic comparison of the time and space scales of heterogeneity between the coastal and mid-ocean (lake) regions could shed some light on this question. We will present some results on this in lakes.

## MEASUREMENTS IN LAKES AND ESTUARIES

### Methods

The investigation of heterogeneity in lakes and estuaries has made use of a wide variety of research designs. The simplest method is visual observation and measurement of dense concentrations of algae or zooplankton. This approach has been made more sophisticated with the advent of remote sensing, applied to lakes by Horne and Wrigley (1975), for example. The patterns of phytoplankton in eutrophic lakes are often seen in photographs taken by remote sensing and prepared for popular distribution.

Another classical technique involves collecting samples directly from the lake or estuary for statistical analysis. The simplest point sample designs are a series of samples taken at random from a body of water or from a short transect, and analyzed for departures from a Poisson distribution. The analyses of the resulting data test the departure of the distribution of final counts of plankton from the Poisson, and judge variances significantly larger than the mean to represent patchiness in the water body (e.g. Harris and Smith, 1977; McAlice, 1970; Bernhard and Rampi, 1965; Cassie, 1959; Richerson *et al.*, 1970). Some of these studies have not distinguished variance that might have resulted from the sampling procedures themselves from variance contributed by the actual distribution of organisms in the body of water. When claims are made that phytoplankton are patchy on small scales (1 cm to 10 m) the sampling and subsampling error in studies of this kind may be large relative to *in situ* variance, and a careful analysis of variance is required to detect non-random distributions. An example of this effect is given by Haury (1973) for the Longhurst-Hardy plankton sampler. (See Venrick, 1971, and Sandusky and Horne, 1977, for approaches to this problem.) In our own technique of preparing samples for enumeration, the variance introduced by the procedure itself is enough to make most species appear non-random at the 5% level, judged by a $\chi^2$ test, even when the samples are replicates from a well-stirred 100 liter vessel. As we shall see, the existence of heterogeneity of phytoplankton on small scales (<10 m) is unexpected under most circumstances and is seldom found by more sophisticated techniques. For example, Cassie (1959) did not detect autocorrelation in most of his data, despite departures of his data from Poisson. The departures from randomness were probably due to sampling artifacts, not small scale patchiness. Ecologically important or statistically detectable patches are likely to exist at small scales only in small basins with strong external sources of variance (Harris and Smith, 1977).

A sophisticated version of the point-sample approach, with replicated samples taken at randomly chosen stations, uses analysis of variance to estimate the magnitude of between-station differences. Platt and his co-workers have carried out a series of such experiments in St. Margaret's Bay, Nova Scotia (Platt *et al.*, 1970; Platt and Filion, 1973; Therriault, 1977). These studies are notable for their careful choice of sampling scale and wide variety of measurements collected with the same research design (including chlorophyll *a* productivity/biomass, ATP biomass, particulate carbon, and nitrogen species). We will report preliminary results for a series of similar stations occupied in Lake Tahoe.

In some cases, sufficient density of more or less equi-spaced samples were taken in lakes to obtain synoptic surveys of patterns (Horne, Sandusky and Carmiggelt, 1978; Goldman *et al.*, 1972; Platt, 1972a). If the variance at scales smaller than the sample spacing does not alias the data, synoptics will provide a reasonable picture of large scale patterns.

Another research design employed in closed basins has been the use of continuous transects along which *in vivo* chlorophyll *a* and other variables are measured (Platt, 1972b; Denman and Platt, 1975; Powell *et al.*, 1975; Richerson *et al.*, 1975). This design is described in detail by Steele in the introductory chapter of this book. A related design depends upon a stationary vessel to sample continuously as currents advect water past. (These two kinds of experiments can be related by Taylor's frozen turbulence hypothesis.) For example, Denman (1977) has used a series of closely spaced vertical profiles to study the way that internal waves affect fixed depth records. Other studies have used similar strategies to examine the two-dimensional structure of heterogeneity (Pingree *et al.*, 1975).

The connection between the patch pictures and the statistical descriptions is not clear. The two methodological approaches contrast deterministic (patches) and stochastic (spectral) conceptions of aquatic environments, neither of which can be neglected in any comprehensive description of ecological phenomena. Spectral analysis cannot be used on data collected along strings or in grids of stations because too few data points exist (e.g. Therriault, 1977; Sandusky and Horne, 1977). However, the variability in aquatic processes is evident in such data sets, and it is not clear how to interpret the small scale information that the spectral estimates produce in light of the large scale information produced by an analysis of variance of data from a number of stations. Obviously, some connection must exist since variance spectra and analyses of variance are two ways of looking at the same quantity. The comparison of these methods needs to be formalized.

## Results

*In vivo* chlorophyll *a*, as an indicator of biomass, has been used extensively to study spatial heterogeneity. A consistent result is a negative slope of the chlorophyll variance spectrum as a function of wavenumber. (See Platt and Denman, 1976, for a discussion of the use of spectral analysis in ecology.) Figure 2 shows a typical plot of chlorophyll *a* and temperature variance from a horizontal tow in Lake Berryessa, California. These plots indicate that most of the patchiness in chlorophyll *a* exists at large scales and that as one proceeds to smaller scales, patchiness

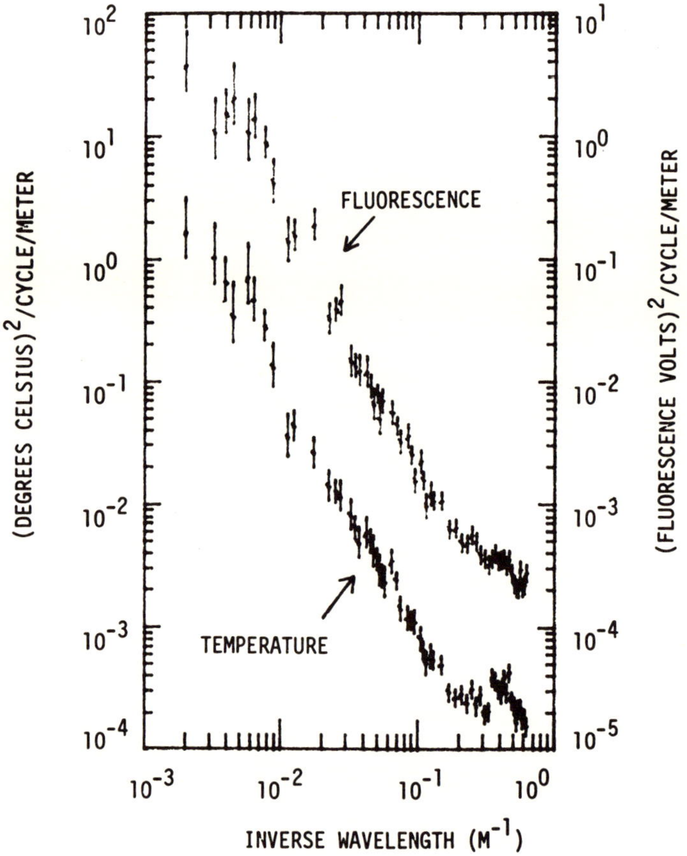


Figure 2. Variance spectra for fluorescence and temperature from a horizontal transect at 4 m depth in Lake Berryessa, 11 August 1976. Error bars are 80% confidence limits.

gets less intense. As can be seen from the temperature spectrum in Figure 2, there is a general resemblance to the spectrum of chlorophyll fluctuations. Much of the empirical and theoretical work surrounding the use of the spectral analysis technique has centered on the measurement and interpretation of this similarity. The main theoretical approach (Kierstead and Slobodkin, 1953; Wroblewski *et al.*, 1975; Okubo, 1976) has been to consider the patch size that can maintain itself by growth, in the face of turbulent diffusion. The characteristic length scale derived from the theory ranges from 100 m to 1 km, for reasonable values of growth rates and diffusivities. Thus patches larger than this size are likely to result from differential phytoplankton growth or death and probably persist for appreciable lengths of time. Smaller patches must be generated by processes acting more rapidly than differential growth and are unlikely to persist for any length of time. Verification of this theory depends upon interpreting the physical fluctuations, like temperature, as markers of turbulent mixing, while the chlorophyll fluctuations are interpreted as indicators of the scale of fluctuations in biological rates and measures of patch size.

The evidence collected from closed basins supports the theory. At small scales, the spectra of both chlorophyll and physical factors have slopes near -5/3 (on a log-log plot) as expected from Kolmogorov turbulence theory, indicating that chlorophyll is nearly a passive contaminant at these scales. The continued increase of chlorophyll variance beyond the critical length is evidence that large scale patches persist. Figure 3 (Powell *et al.*, 1975) illustrates the difference between spectra of velocity fluctuations (turbulence) and chlorophyll at larger scales. Apparently the theory is correct, and differential growth and death dominate the pattern of chlorophyll variation above the predicted critical length.

As yet, no experiments have been conducted to estimate directly the coherence (spectral correlation) between velocity and chlorophyll fluctuations. However, temperature fluctuations are easy to measure on horizontal transects and extensive use has been made of this measurement as an indicator of mixing processes. The similarity of the two spectra in Figure 2 indicates the limitations of using temperature fluctuations as markers of turbulent mixing at large scales. Temperature is not conservative as it can be changed by differential heating of the lake waters, and by inputs of colder or warmer water. Thus, its large scale distribution is not entirely dependent on turbulent mixing. It is not surprising that the temperature and chlorophyll variance spectra are similar at small scales where turbulent mixing dominates both variables, while at large scales they are still similar since there are sources of variance for both variables. However, experiments in closed basins have not always indicated high coherence between

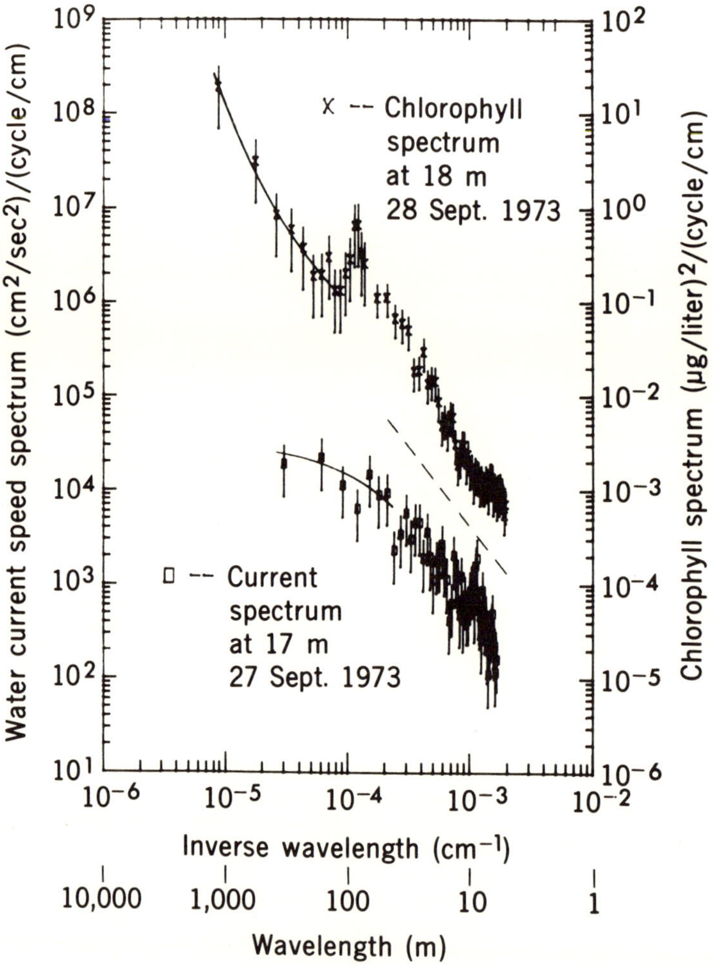


Figure 3. Plots of chlorophyll *a* and current variance spectra in Lake Tahoe. Error bars are 80% confidence intervals. Copyright 1975 by the American Association for the Advancement of Science.

temperature and chlorophyll, particularly in lakes. Our measurements in Lake Tahoe have seldom shown significant coherence at any scale, although occasionally, low frequency coherences are significant. In estuaries, Denman and Platt (1975) found high coherences between chlorophyll and temperatures at intermediate scales and concluded that from 50 m to a few km, both chlorophyll and temperature behave as passive contaminants. By this argument, biological processes dominate over diffusive transport to form patches only on scales above several km in the Gulf of the St. Lawrence. This interpretation is partly confounded because temperature variance can be introduced into estuarine systems at scales far above the peak of wind-generated turbulence fluctuations. Where temperature gradients are large, temperature may be a water mass indicator. If these water masses are also associated with conditions affecting phytoplankton growth, such as nutrient supplies and light regimes, phytoplankton biomass may also be a water mass indicator. Thus we would expect temperature differences to be closely associated with biomass differences. Pingree *et al.* (1975) have given an example of the process that will lead to coherence in a thermocline outcropping situation. If temperature differences are not associated with water masses of different phytoplankton growth potential, then temperature fluctuations may not resemble phytoplankton fluctuations. In lakes the size of Lake Tahoe, sources of temperature variance are smaller, so temperature is less often a good water mass marker and coherence is less frequently observed. The three-dimensional character of water bodies can also complicate the interpretation of horizontal transects. Strong temperature and biomass vertical gradients, combined with internal waves, may influence the record of chlorophyll fluctuations at a fixed depth (Denman, 1977).

We (Richerson *et al.*, 1975) have extended the transect sampling scheme to include enumeration of phytoplankton in lakes. The object of these observations is to obtain information on how the distribution of organisms is related to chlorophyll content, which is more easily measured, but biologically less informative. The phytoplankton count data from Lake Tahoe has a coefficient of variation about an order of magnitude higher than *in vivo* chlorophyll. The count sample density along transects is between one and two orders of magnitude less than chlorophyll, despite considerable effort to collect a large number of samples on each transect. Thus resolution is drastically reduced. Nevertheless, the same general patterns are apparent in the enumeration data. When a population does differ from a random distribution, the large scale variance dominates and, in favorable cases, negative slopes are apparent in the variance spectra (Richerson *et al.*, 1975). Figures 4, 5 and 6 show the raw data, autocorrelation function and variance spectrum for three phytoplankton populations in Lake Tahoe.

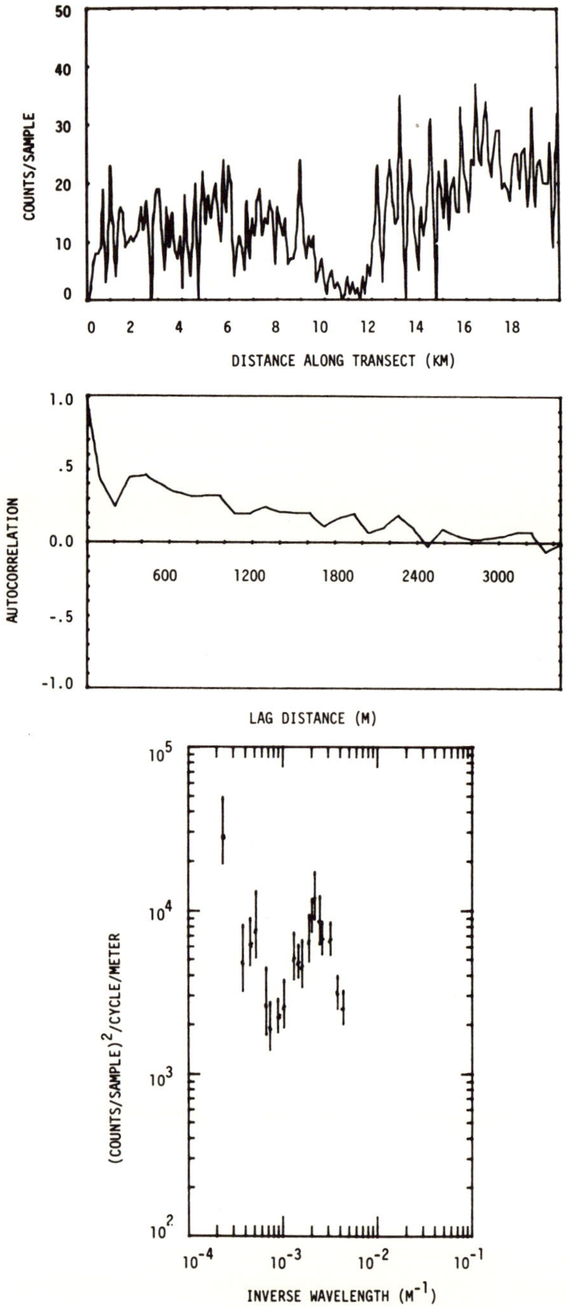


Figure 4. Top, raw data; middle, autocorrelation function; bottom, variance spectrum for a spatially organized population, *Asterionella formosa*, Lake Tahoe, 18 September 1974, 60 m depth.

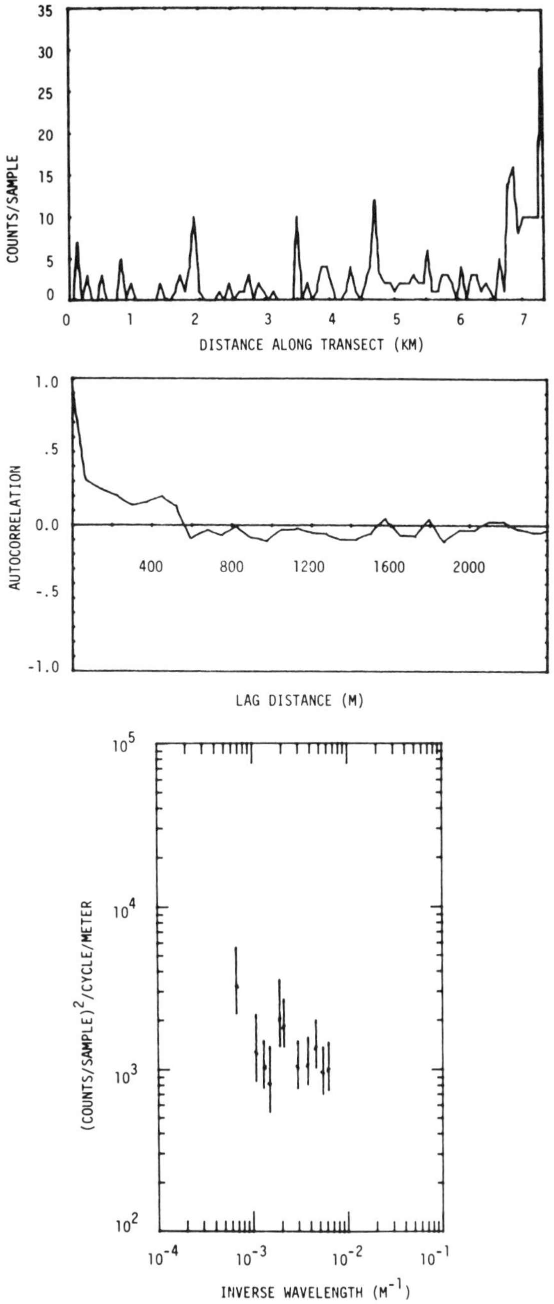


Figure 5. Top, raw data; middle, autocorrelation function; bottom, variance spectrum for a marginally organized population, *Fragilaria pinnata*, Lake Tahoe, 12 September 1974, 6 m depth.

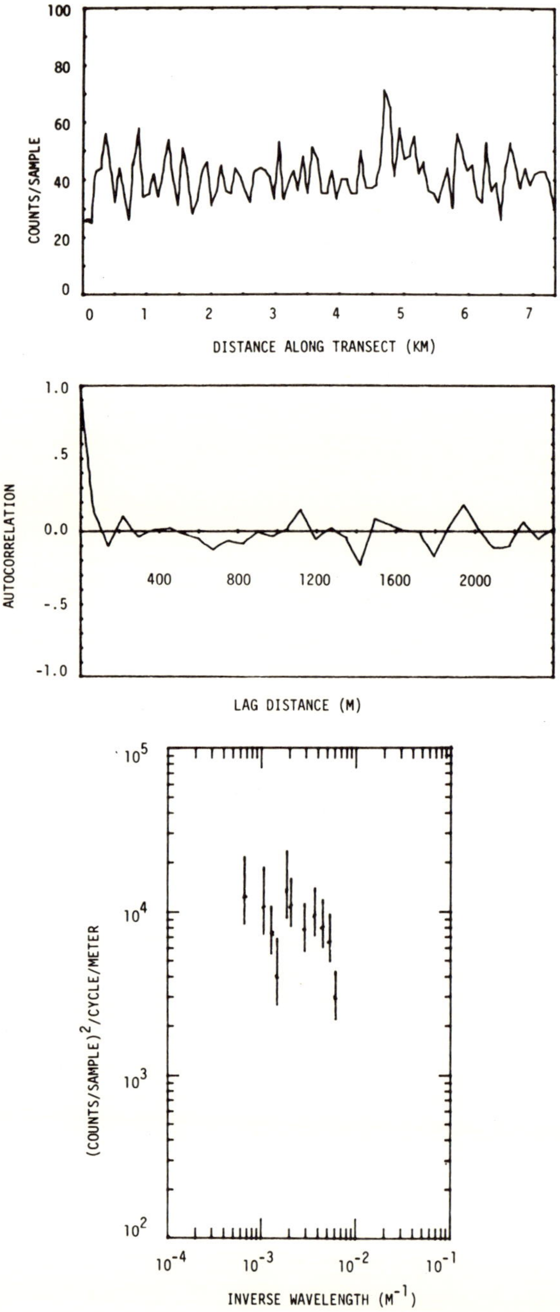


Figure 6. Top, raw data; middle, autocorrelation function; bottom, variance spectrum for an unorganized population, *Fragilaria crotonensis*, Lake Tahoe, 12 September 1974, 6 m depth.

From the raw data, it is apparent that spatial organization, when present, is predominantly large scale. Since the largest scales that can be examined by spectral analysis are only 1/4 to 1/10 of the transect length, it is apparent that most of the variance in the *Asterionella formosa* population plotted in Figure 4 cannot be shown in the spectral plot. Despite this, most populations that are well organized spatially show variance decreasing smoothly from large scales to small scales.

Because of the concentration of variance at large scales, and the small sample size and poor precision of the count data, we have relied more on autocorrelation patterns for interpretation. Examples of species that are less organized are shown in Figures 5 and 6. Of the eighty populations with high mean counts, from nine transects in Lake Tahoe (a large, oligotrophic lake) and one transect in Castle Lake (a small, mesotrophic lake), sixteen were well organized, resembling *A. formosa* in Figure 4 and twenty showed a pattern resembling *Fragilaria pinnata* in Figure 5. Coherence lengths (length scale where the autocorrelation function first crosses zero, interpretable as characteristic patch size) ranged from a few hundred to a few thousand meters in Lake Tahoe. In Castle Lake, where the maximum linear dimension of the lake is under 1 km, the coherence length of the most organized population was 60 m. The count data also permit examination of the relationship between species distributions. We have calculated the product-moment correlation coefficient between populations to estimate the internal organization of the plankton community. Table 1 gives an example of such calculations, although it is atypical in the large proportion of coefficients greater than .3. In general, significant correlations between the abundances of species are few and show a tendency to be positive. This tendency is probably a result of higher biomass in inshore regions where nutrient inputs are concentrated but may result from the correlation of enumeration errors.

Measurements other than chlorophyll, physical factors, nutrients and phytoplankton populations, have seldom been made on a continuous basis, particularly when several variables are measured simultaneously. We have data on zooplankton from Castle Lake where higher zooplankton abundance makes continuous collection with a hose-pump system feasible. Zooplankton populations have more spatial structure than typical phytoplankton populations (compare Figure 7 which is typical for zooplankton with Figure 4 that is well organized for phytoplankton). Thus the possibility arises that zooplankton may cause spatial variance in phytoplankton distributions by grazing (Steele, 1974; Steele, this volume; Wroblewski, 1977).

Nitrate data from Lake Tahoe, though not as highly organized as zooplankton, indicate both large and small scale heterogeneity. Data collected in replicate at several sets of stations

Table 1. Example of a 2 x 2 table of the product-moment correlation coefficient between abundances of species at given points on a transect. Underlined values are significant at the .05 level.

| | C.s. | F.c. | F.p. | C.p. | C.2 | C.3 | C.1 | f.2 | f.3 | l.c. |
|---|---|---|---|---|---|---|---|---|---|---|
| *Cyclotella stelligera* | 1.000 | <u>0.382</u> | 0.088 | <u>0.206</u> | <u>0.372</u> | <u>0.241</u> | -0.080 | 0.117 | 0.171 | -0.03 |
| *Fragilaria crotonensis* | | 1.000 | -<u>0.219</u> | 0.042 | <u>0.234</u> | 0.037 | -0.070 | -0.073 | -0.023 | -0.137 |
| *Fragilaria pinnata* | | | 1.000 | 0.088 | 0.077 | 0.128 | -0.060 | -0.044 | -0.102 | -0.044 |
| *Cocconeis placentula* | | | | 1.000 | -0.007 | -<u>0.206</u> | 0.007 | 0.058 | -0.107 | 0.126 |
| Coccoid 2 | | | | | 1.000 | 0.174 | <u>0.249</u> | <u>0.249</u> | <u>0.447</u> | 0.132 |
| Coccoid 3 | | | | | | 1.000 | 0.051 | 0.044 | <u>0.266</u> | -0.101 |
| Coccoid 1 | | | | | | | 1.000 | 0.134 | <u>0.556</u> | <u>0.520</u> |
| flagellate 2 | | | | | | | | 1.000 | <u>0.221</u> | 0.101 |
| flagellate 3 | | | | | | | | | 1.000 | <u>0.364</u> |
| large coccoid | | | | | | | | | | 1.000 |

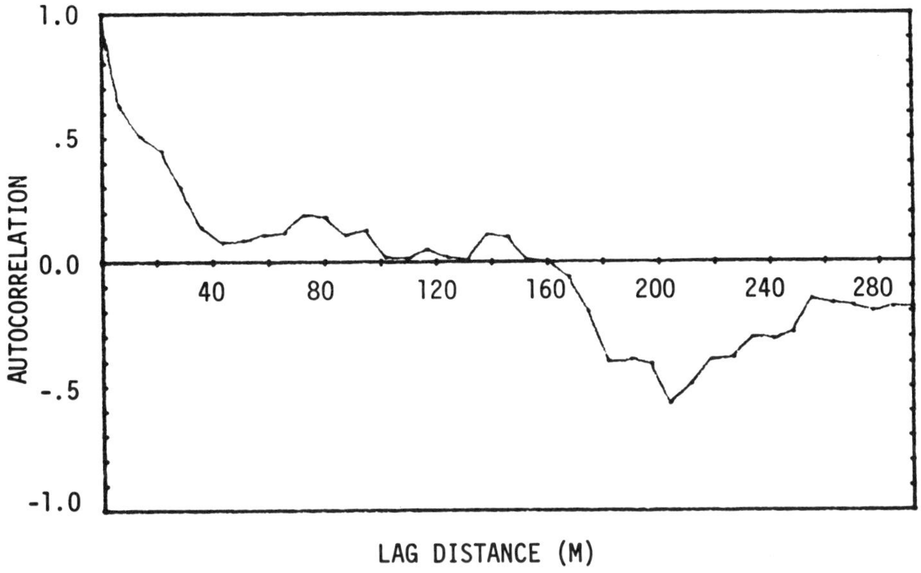


Figure 7. Autocorrelation function of *Diaptomus novamexicanus* in Castle Lake, 6 October 1974, 8 m depth.

randomly located along transects were reduced using a random-effects model analysis of variance. The resulting between- and within-station variance components indicate, in some cases, large-scale patchiness, dependent upon location, depth and date of collection. Table 2 is a summary of variance components we have derived from a series of such station collections, done in the summer of 1977. The data indicate both a difference between mid-lake and nearshore percent variation components at the same time (July) and a shift to a higher between-station variance components in both locations through time (August). These results suggest that the processes governing nitrate distribution in the coastal and limnetic zones occur on different scales in July and are more similar in August. This conclusion is consistent with a coastal zone discussion to be treated later. Within the same time period that the stations were sampled, continuous transects were carried out in the epilimnion during which nutrient samples were collected from the hose-pump system after the water had passed through the fluorometer. These samples (512 in mid-lake transects and 256 in nearshore transects) were analyzed and reduced by spectral analysis. An example of a mid-lake transect done in August is shown in Figure 8. The raw data plot (Figure 9) shows much high frequency variation, and a low frequency component superimposed on the record with approximately a 7 km wavelength. In the wavelength region between 1 km

Table 2. Between station (B) and within station (W) variance components for mid-lake and nearshore station nitrate data in July and August 1977, Lake Tahoe. N.S. = no significant difference between percent variation. ** = significant at $\alpha$ = .01.

| | MID-LAKE 25 JULY | | | | | | NEARSHORE 27 JULY | | | | |
|---|---|---|---|---|---|---|---|---|---|---|---|
| | $S_T^2$ | $S_B^2$ | $S_W^2$ | $\%_B$ | $\%_W$ | | $S_T^2$ | $S_B^2$ | $S_W^2$ | $\%_B$ | $\%_W$ |
| 5 m | .00267 | .00141 | .00126 | 53 | $47^{NS}$ | 5 m | .01626 | .01456 | .00170 | 90 | $10^{**}$ |
| 20 m | .00247 | .00142 | .00105 | 58 | $42^{NS}$ | 20 m | .00458 | .00270 | .00188 | 59 | $41^{NS}$ |
| 50 m | .00347 | .00264 | .00083 | 76 | $24^{NS}$ | 50 m | .00619 | .00315 | .00304 | 51 | $49^{NS}$ |
| | MID-LAKE 22 AUGUST | | | | | | NEARSHORE 22 AUGUST | | | | |
| | $S_T^2$ | $S_B^2$ | $S_W^2$ | $\%_B$ | $\%_W$ | | $S_T^2$ | $S_B^2$ | $S_W^2$ | $\%_B$ | $\%_W$ |
| 5 m | .00215 | .00203 | .00012 | 94 | $6^{**}$ | 5 m | .00117 | .00110 | .00007 | 94 | $6^{**}$ |
| 20 m | .00117 | .00109 | .00008 | 93 | $7^{**}$ | 20 m | .00047 | .00043 | .00004 | 92 | $8^{**}$ |
| 50 m | .00276 | .00267 | .00009 | 97 | $3^{**}$ | 50 m | .00110 | .00104 | .00006 | 94 | $6^{**}$ |

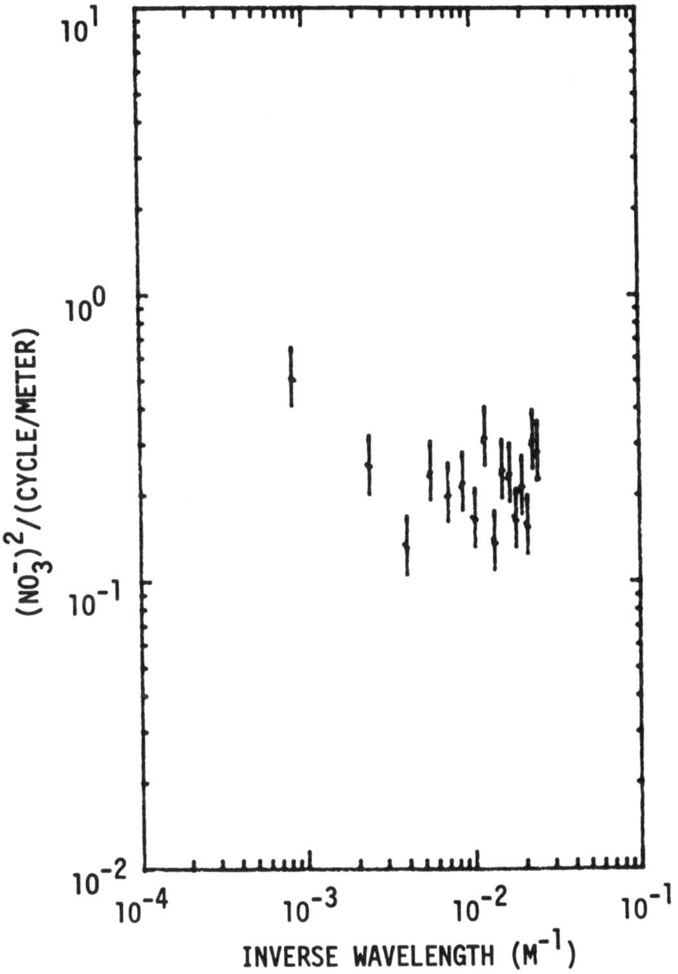


Figure 8. Variance spectrum for nitrate in a mid-lake transect on 29 August 1977 at a depth of 10 m, Lake Tahoe. Error bars are 80% confidence intervals.

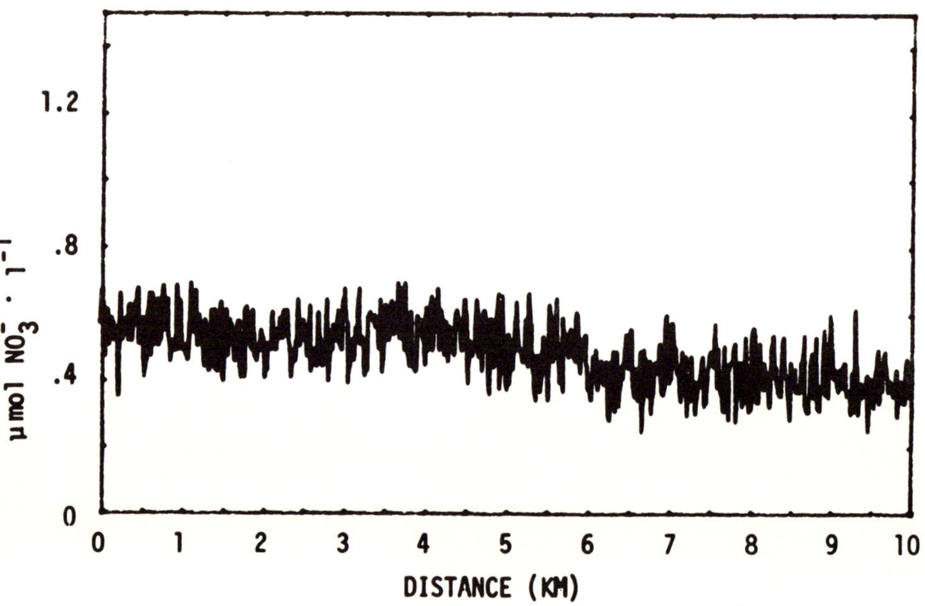


Figure 9. Raw data plot of nitrate concentration in a mid-lake transect on 29 August 1977 at a depth of 10 m, Lake Tahoe.

and 40 m that spectral analysis can resolve from this record, Figure 8 indicates an essentially random distribution.

In October 1977 we collected both nitrate and orthophosphate samples along a continuous transect in the mid-lake region, with the hose inlet towed at 75 m, well below the 30 m thermocline. Figure 10 is a plot of the raw data for nitrate in this tow and Figure 11 is the resulting variance spectrum. The phosphorus data also show a similar pattern. It is clear from the contrast between the epilimnion and hypolimnion transects that nutrients are distributed on different scales, most likely due to the protection from vigorous mixing offered by stratification. These data, considered together, indicate that this effect leads to large scale heterogeneity in the mixed layer and smaller patches in the hypolimnion. This conclusion is consistent with the assumption that the dominant sources of nutrient heterogeneity are inputs from the many streams flowing into the lake that are separated by several kilometers; i.e., large scale.

Since transect studies show much of the spatial variance in all parameters of biological interest in lakes and estuaries to be concentrated at large scales, the less laborious synoptic and station data can be interpreted with new confidence. Even thinly scattered synoptic stations will represent large scale patterns

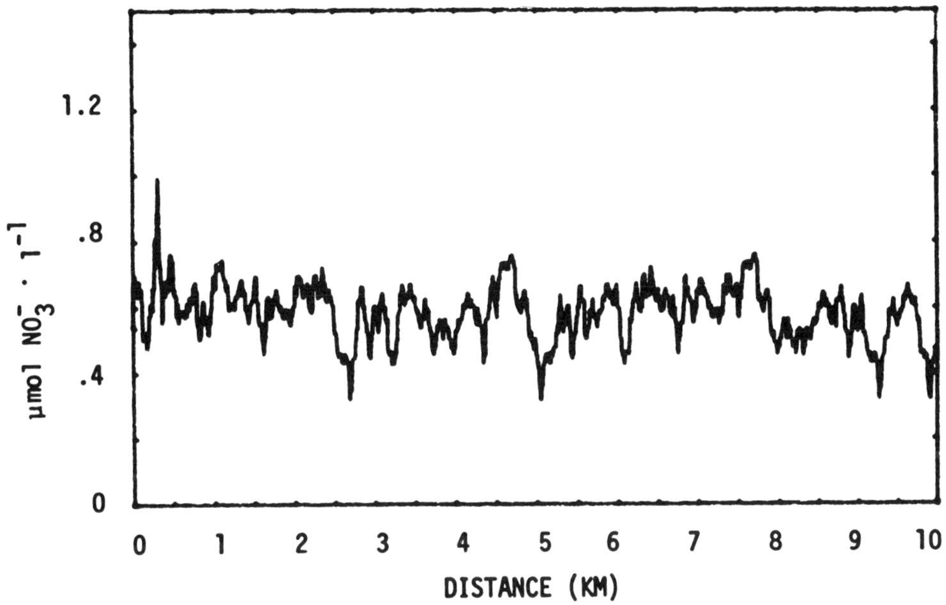


Figure 10. Raw data plot of nitrate concentration in a mid-lake transect on 4 October 1977 at a depth of 75 m, Lake Tahoe.

(e.g. Goldman *et al.*, 1972) without serious danger that the apparent large scale features have variance aliased from smaller scale patterns. A small number of randomly chosen stations, separated by a mean distance designed to correspond to the large scale variance in the water body, will give a statistical picture of spatial heterogeneity adequate for many purposes. For example, Platt and Filion's (1973) study of the spatial variability in P/B ratios showed that between-station differences can be significant, a requirement if spatial differences are important to plankton community organization (Richerson *et al.*, 1970; Margalef, 1969; Lund, 1964). Therriault (1977) has extended these measurements and shown how strong turbulent mixing reduces spatial variability, a confirmation of the Kierstead-Slobodkin theoretical approach to spatial heterogeneity.

As discussed earlier, phytoplankton distributions are highly dependent on physical processes. Thus the differences in physical properties between the mid-lake and coastal regions should create different spatial patterns. In the summer of 1977, a series of horizontal transects was made in Lake Tahoe, measuring chlorophyll and temperature simultaneously (Leigh-Abbott *et al.*, 1978). Figure 12 shows a mid-lake transect done in July at a depth of 12 m, when the thermocline depth was 15 m. The record shows two regions of uniform temperature (indicative of thorough mixing)

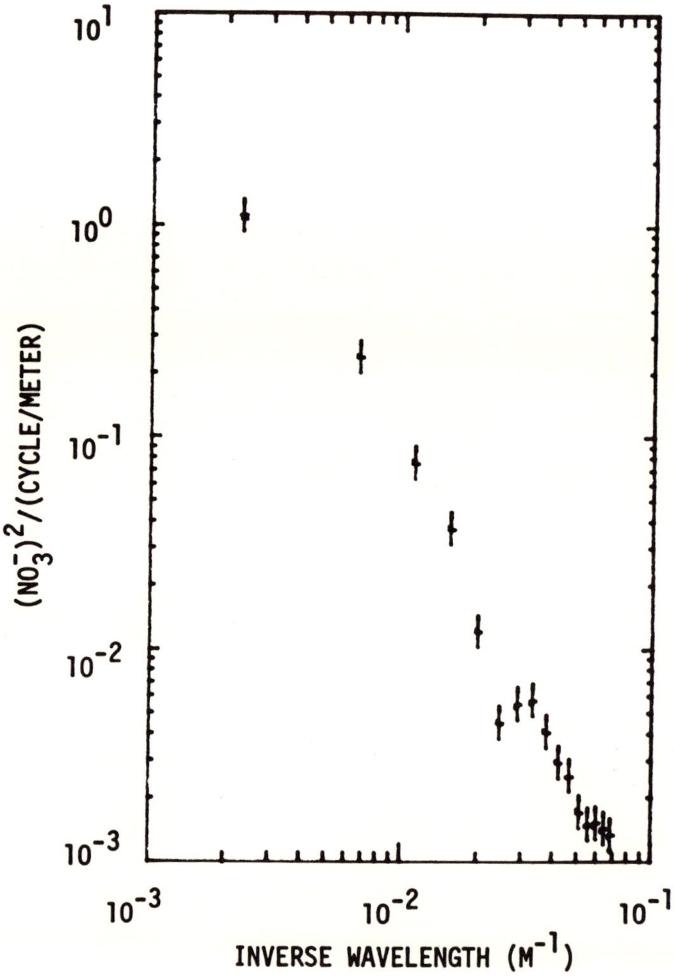


Figure 11. Variance spectrum for nitrate in a mid-lake transect on 4 October 1977 at a depth of 75 m, Lake Tahoe. Error bars are 80% confidence intervals.

separated by an area typical of internal waves. The final region is characterized by a low, variable temperature, and a high, variable chlorophyll content. This is similar to a transect done close to shore on the same day (Figure 13). The latter tow was always within one kilometer of shore while the former started in mid-lake and ended close to shore (see Figure 14). Thus the similarity between the nearshore transect and the end of the mid-lake transect can be explained by the presence of a distinct coastal water mass, extending two to three kilometers offshore.

The formation of the coastal zone is influenced by the nature of the surface wind stress. Lake Tahoe is dominated in early summer by strong afternoon winds from the southwest. This can cause surface drift offshore of the warm upper layer of water in the coastal zone on the western shore, leading to an upwelling of cold water. Though the details of the vertical motion in this region remain to be investigated, significantly lower mean temperatures in the coastal zone (3°C, see Figures 12 and 13) strongly suggest upwelling. However, only limited upwelling can be maintained, due to the strong stratification present and the transitory nature of the surface wind stress. This allows the mid-lake region to retain its normal thermal structure, whereas the coastal zone is influenced by upwelling, thus creating the frontal region between them. Any change in the wind stress and stratification should affect this coastal-mid-lake separation.

Figure 15 shows two chlorophyll variance spectra, one from the mid-lake transect before encountering the coastal front, and the other from the nearshore transect. Both spectra have been normalized to the total variance in each record, facilitating comparison of spectral shapes. At large scales (greater than 100 m), the nearshore region shows a significantly greater proportion of variance than mid-lake. This is consistent with the idea that stream inflow is a large scale source of variance, and its effects should be more evident in the nearshore region than in mid-lake. The mid-lake spectrum shows less variation at large scales, due to the absence of stream effects. At intermediate scales (100 m to 10 m) the nearshore spectrum has a flat region that might be caused by an input of momentum at these scales. The presence of side and bottom boundaries in the nearshore enhances the production of intermediate size eddies, giving a plateau at intermediate scales in the nearshore spectrum. The midlake spectrum, at intermediate scales, shows monotonically decreasing variance without the plateau present in the nearshore spectrum. If mixing is introduced at large scales and chlorophyll is a passive contaminant at intermediate scales, turbulence theory predicts that its spectrum should show a smooth slope at these scales. Both spectra at small scales (less than 10 m) are identical, since tests have shown that mixing within the hose-pump system and electrical noise limits resolution to 10 m (assuming a towing speed of 1 m/s).

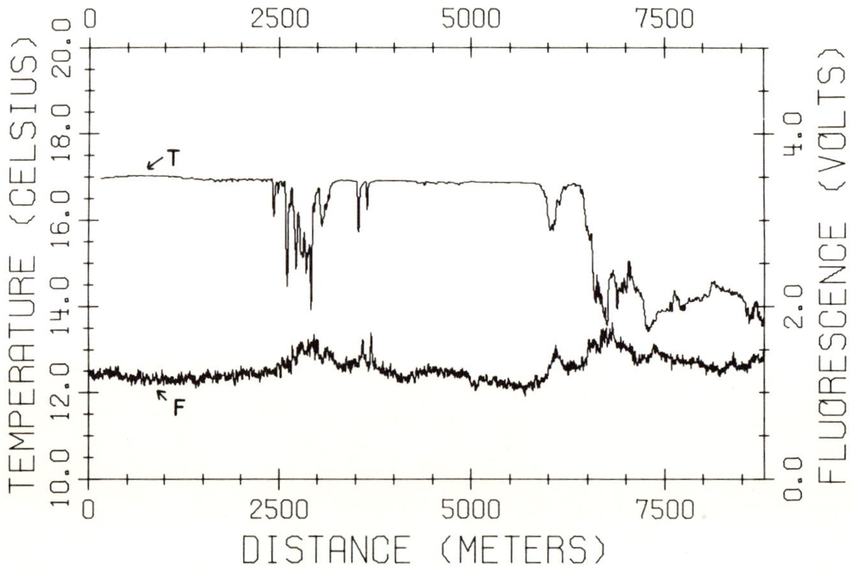


Figure 12. Chlorophyll and temperature transect of 26 July 1977 in mid-lake region (from Leigh-Abbott *et al.* (1978). Copyrighted by American Geophysical Union.

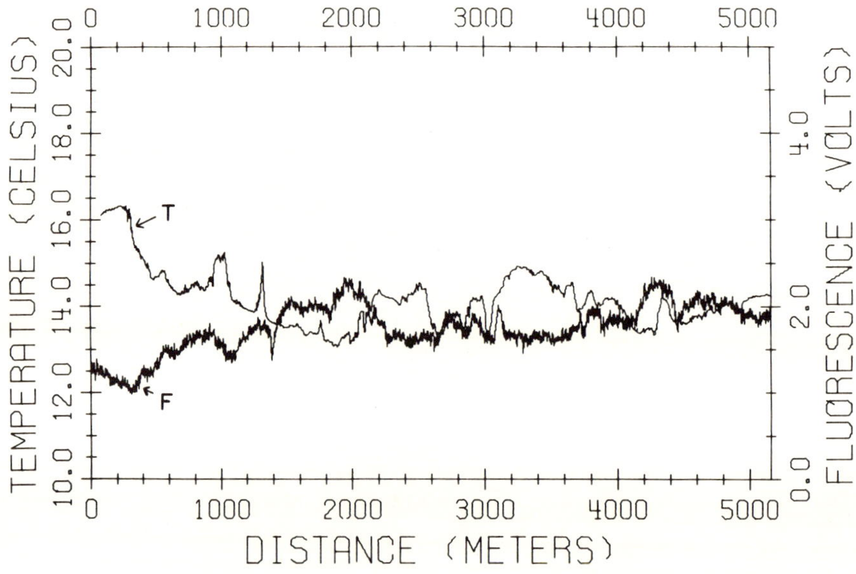


Figure 13. Chlorophyll and temperature transect of 26 July 1977 in nearshore region (from Leigh-Abbott *et al.* (1978). Copyrighted by American Geophysical Union.

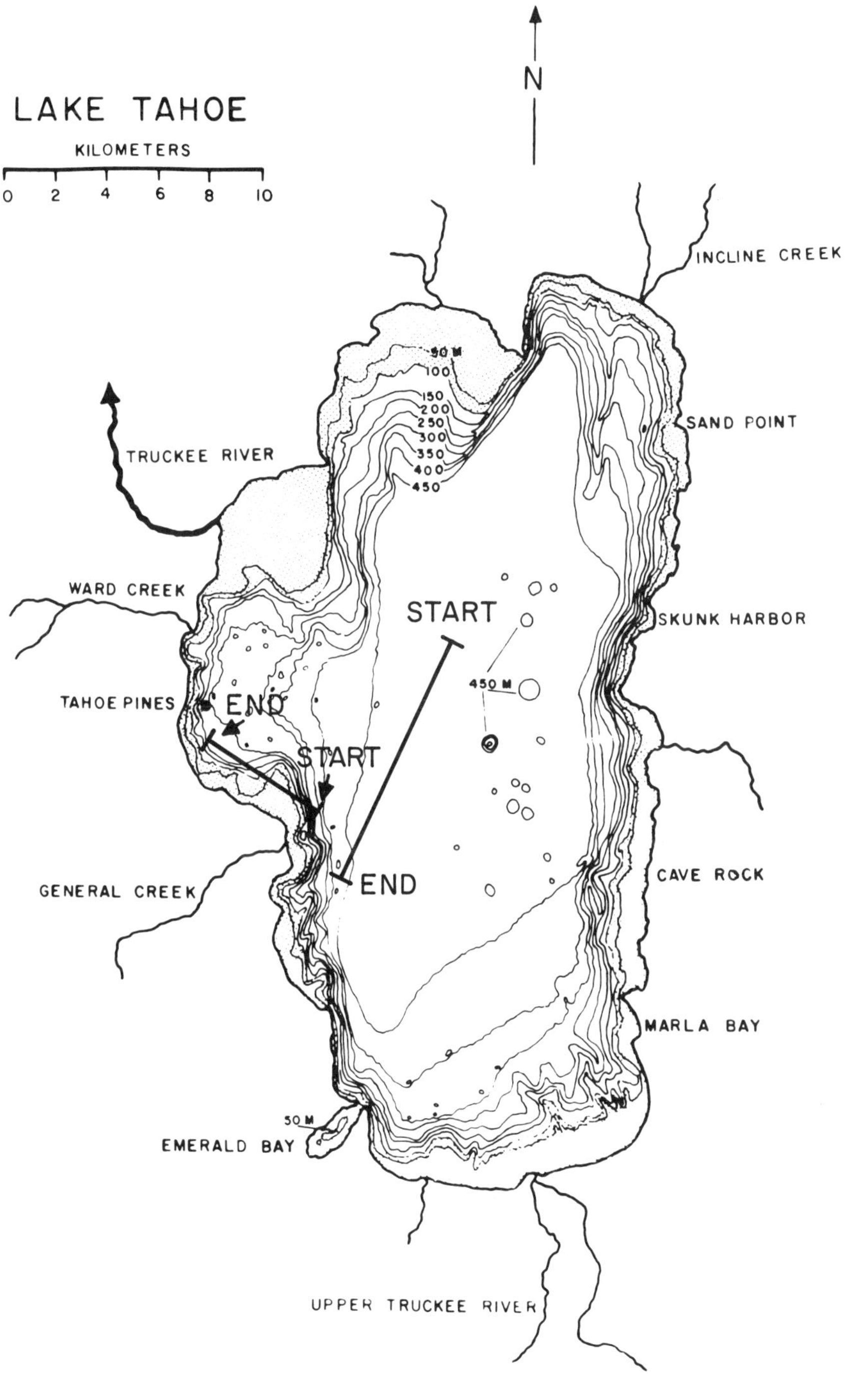


Figure 14. Morphometric map of Lake Tahoe showing transect locations.

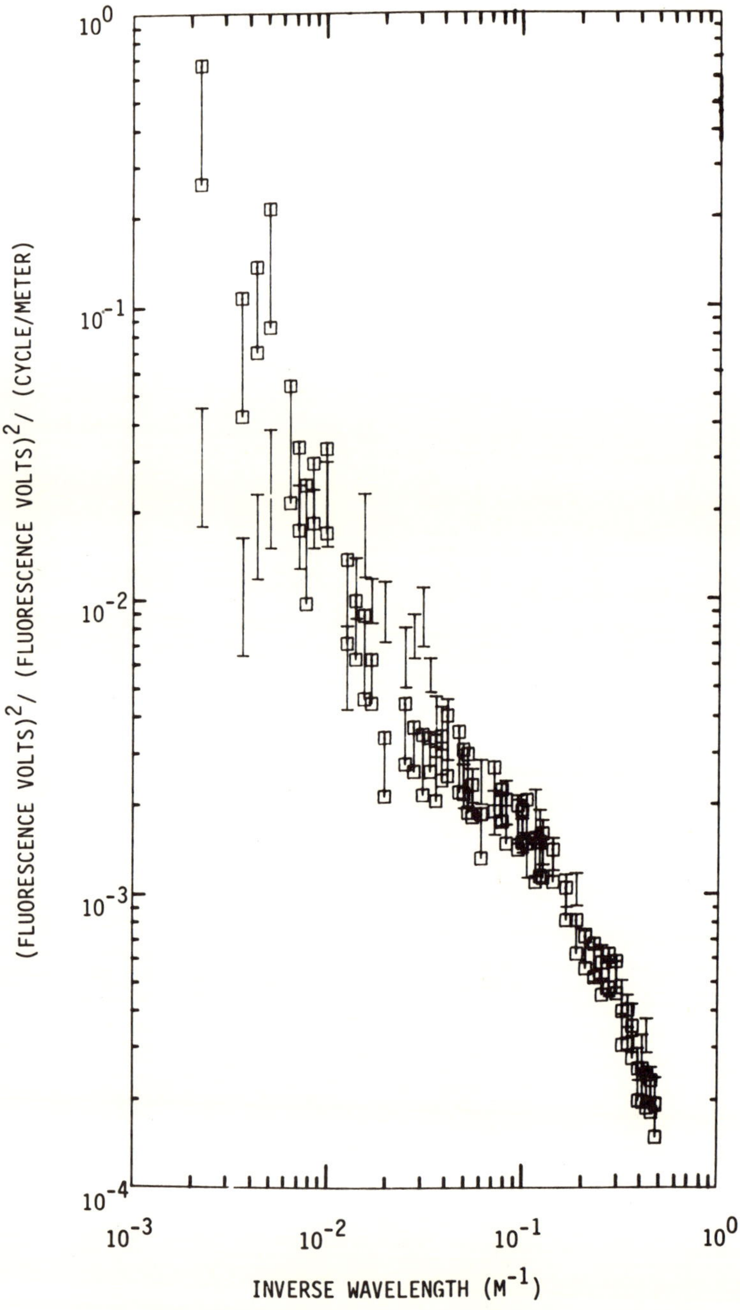


Figure 15. Fluorescence spectra normalized to total variance from 26 July 1977. Error bars reflect 80% confidence interval. —, midlake transect; □ , nearshore transect (from Leigh-Abbott *et al.*, 1978). Copyrighted by American Geophysical Union.

As suggested earlier, changes in wind stress and stratification should affect the coastal-mid-lake separation. After a period of persistent winds, the thermocline had dropped to 23 m. Transects done in August from both mid-lake and nearshore were essentially identical (Figures 16 and 17), with variation in chlorophyll and temperature sharply reduced in comparison to the July transects. The normalized spectra from the chlorophyll transects (Figure 18) also indicate little significant difference between mid-lake and nearshore. The variance at large scales in the coastal zone has decreased, probably due to the cessation of stream inflow and thus nutrient input by mid-August.

In summary, a coastal zone, distinguished by high, variable chlorophyll levels and low, variable temperatures, is present during early summer. The chlorophyll distribution in this zone is influenced by stream inflow of nutrients and possible upwelling events at large scales. This is consistent with the nitrate data presented earlier, that showed higher large scale variance in the nearshore regions at 5 m, where stream inflow should have its greatest effect. After a period of increased mixing, the mid-lake and nearshore regions are virtually identical. Nutrient inputs have ceased, lowering the large scale variance in chlorophyll (again, this agrees with the nitrate data that also showed less large scale variance at the same time).

## Comparison with Oceanic Measurements

The bulk of the recent studies of spatial heterogeneity have been made in estuaries and coastal oceans. Open oceanic measurements are less common, and frequently zooplankton, rather than phytoplankton or other parameters, have been the object of study (Haury *et al.*, this volume). Phytoplankton data exist from the Hardy Plankton Recorder survey of the North Sea and the eastern North Atlantic (e.g. Lucas, 1941; Robinson, 1961), Bainbridge's (1957) review of the early observations, and Venrick's (1972) careful statistical studies in the open Pacific.

There is much similarity between small closed basins (St. Margaret's Bay, Lake Tahoe), semi-enclosed estuaries (Long Island Sound), open coastal waters (North Sea), and oceanic waters. In all situations examined, the existence of heterogeneity and the relatively large scale of such variability has been documented. Despite the major differences in the physical processes that characterize inshore and enclosed habitats as opposed to open water, the organisms appear to respond to the habitats similarly, though the characteristic length scales vary. Venrick's (1972) finding that a minority of phytoplankton populations were heterogeneous on scales of 30 km is reminiscent of the Lake Tahoe plankton at half that scale.

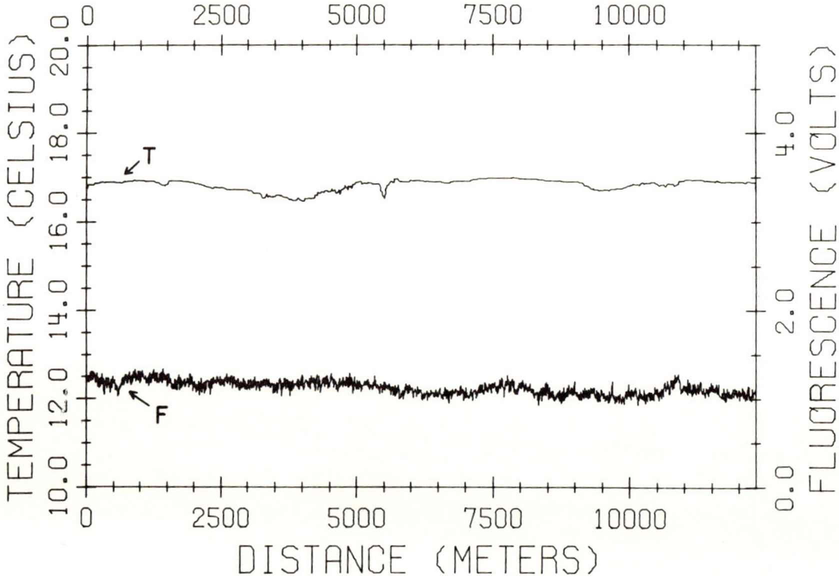


Figure 16. Chlorophyll and temperature transect of 29 August 1977 in mid-lake region (from Leigh-Abbott *et al.*, 1978). Copyrighted by American Geophysical Union.

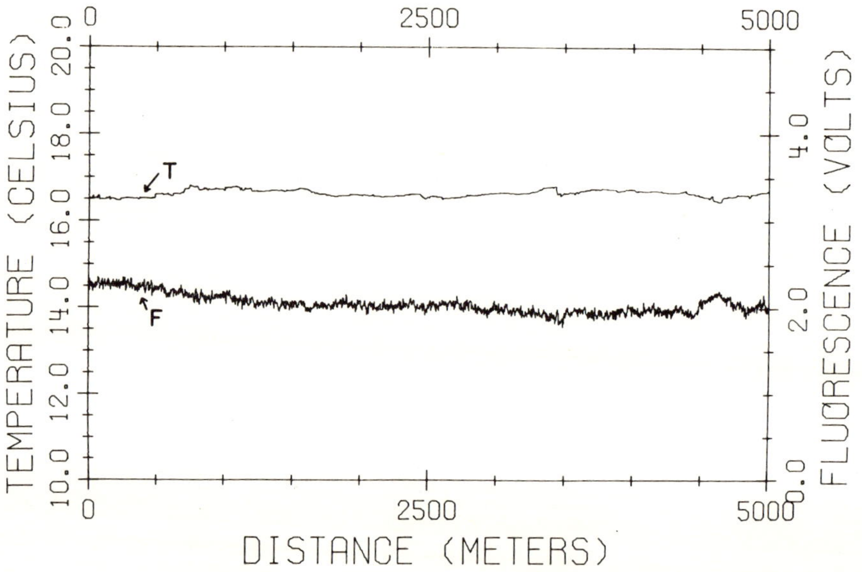


Figure 17. Chlorophyll and temperature transect of 30 August 1977 in nearshore region (from Leigh-Abbott *et al.*, 1978). Copyrighted by American Geophysical Union.

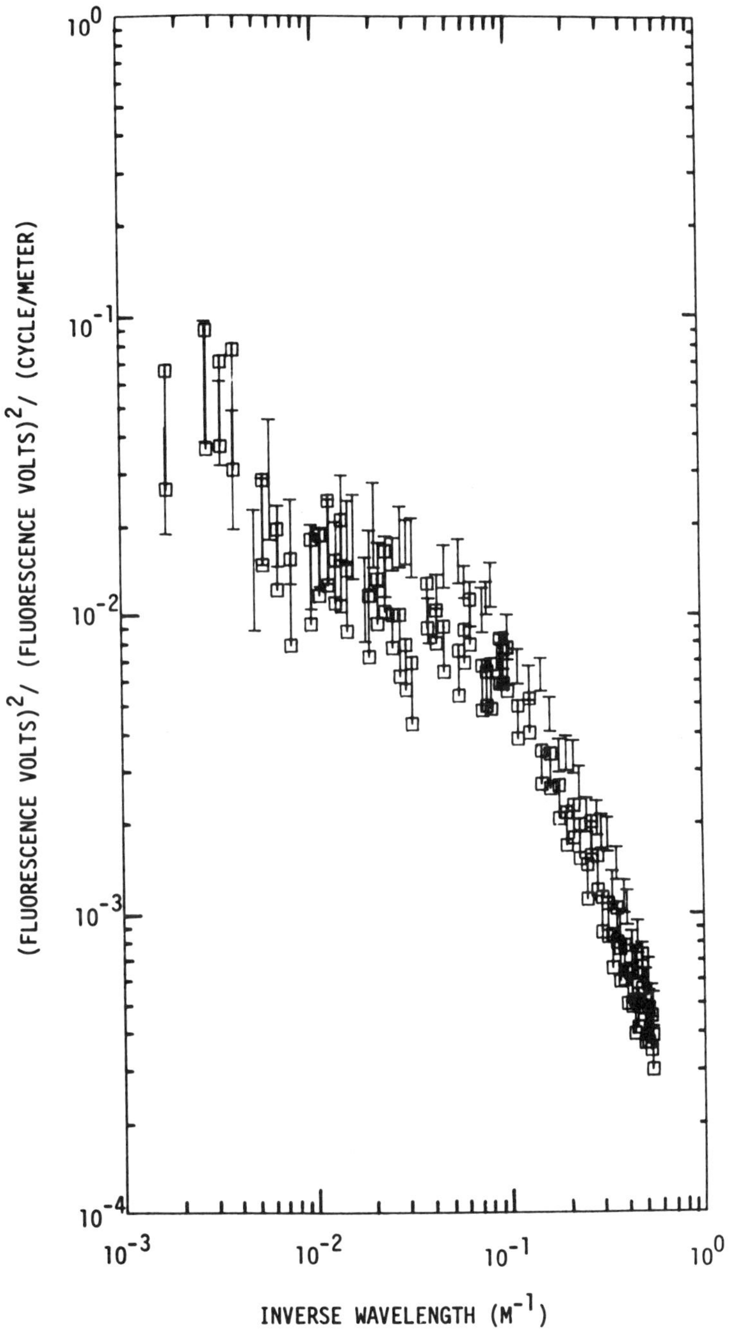


Figure 18. Fluorescence spectra normalized to total variance from 29 August and 30 August 1977. Error bars reflect 80% confidence interval. —, mid-lake transect; □ , nearshore transect (from Leigh-Abbott *et al.*, 1978). Copyrighted by American Geophysical Union.

Zooplankton tend to be more patchy, or at least to have more easily detected heterogeneity, than phytoplankton. There are, of course, some obvious differences between closed and open basins that must have important consequences for the biota. Lakes are isolated habitats, while the oceans are directly connected to distant, and perhaps different, habitats. This fact probably explains the faunistic poverty of lakes compared to the ocean for many groups (Richerson *et al.*, 1977).

## CONCLUSIONS

Spatial and temporal heterogeneity have been hypothesized to have a substantial role in a number of ecological phenomena, from the maintenance of genetic polymorphisms to the regulation of community diversity (Wiens, 1976; Levin, 1976). Pelagic systems are simpler than terrestrial and benthic environments; the best data on spatial heterogeneity comes from planktonic systems. Thus we would expect confirmations of ecological theory to be derived from aquatic environments. We can only make tentative judgments about the importance of these hypotheses from the studies of plankton heterogeneity conducted to date. Our pessimistic conclusion echoes that of Steele in the introduction to this volume.

Despite the complexity of the problem, the achievements are not negligible. The data show that spatial heterogeneity is dominated by large scale, stochastic events. The role given to deterministic processes, such as Langmuir circulations (e.g. Stavn, 1971), is misleading. The measurements of chlorophyll, phytoplankton, and zooplankton suggest that small scale, deterministic heterogeneity is unlikely, except in the vertical, under stratified conditions (Fee, 1976; Richerson *et al.*, 1978). As Margalef (1969) has emphasized, the plankton habitat is influenced by the turbulent physical environment since plankton organisms are impinged upon by a variety of stochastic to semi-organized processes. Biological structure acts like a filter, damping some processes, amplifying others. For example, the exponential growth of phytoplankton amplifies fluctuations larger than the Kierstead-Slobodkin length.

Motions are seldom organized or repeatable, and the classic steady-state communities with strongly interacting populations do not occur. Despite theory (Wiens, 1977), plankton communities may be representative in their lack of pattern, since many populations may coexist without obvious interactions. Abundances change substantially over space scales on the order of one km and time scales the order of a week, but the populations have little statistical relationship to each other. This fact has led to divergent interpretations (Richerson *et al.*, 1970; Hulburt, 1977; Hutchinson, 1961; Tilman, 1977).

Several lines of investigation will be required to understand how the large patches of plankton interact with each other. First, we must continue to search for regularities among the physical and biological variables of interest. For example, the spatial heterogeneity of nutrient ions is poorly understood. Continued measurement of the coherence of chlorophyll and physical factors should bring an increase in our understanding of the relationship between growth and mixing. However, the most interesting measurements are the most difficult. To supplement our structural measurements, we need to know more about the rates at which physical and biological processes operate. Measurements by Platt and Filion (1973) and Stull *et al.* (1973) of the spatial and species-specific variations in photosynthesis are examples. Information on species-specific interactions is also required, such as Tilman's (1977) study of the competition process in phytoplankton. Grazing studies, a staple of biological oceanography and limnology, are introduced into theoretical models with apologies for their lack of realism. Second, as Levin (this volume) shows, we possess a considerable variety of models with spatial and temporal heterogeneity, but their behavior fails to lead to consistent conclusions. The results of any model depend largely on its details and parameter values. Thus, future modeling efforts must include the wealth of data collected from aquatic systems. Only the Kierstead-Slobodkin argument stands out as theoretically (and empirically) robust. The analysis of the stability of grazer-herbivore interactions developed by Steele (1974) is a typical example of our present difficulty. Diffusion may stabilize or destabilize such interactions even in closely related models.

The question of spatial and temporal heterogeneity has been taken seriously in the last few years in ecology because theory without heterogeneity fails to account for many features of ecosystems. At first viewed as a simple solution to the problems posed by simpler deterministic and white noise stochastic theories (e.g., Hutchinson, 1961), spatial heterogeneity can lead to a variety of effects. The task is to find whatever general rules exist among the many possibilities.

## SUMMARY

Spatial heterogeneity of the plankton community in closed basins is caused by a complex interaction of physical and biological processes. Neither deterministic effects nor the stochastic nature of the aquatic environment alone are sufficient to describe the observed spatial (and temporal) variability. A variety of techniques, including direct observation of patches, statistics that summarize large data sets, and time series analysis, confirm that the largest variability is found on the largest scales. This result applies to the coastal ocean, estuaries, and lakes, as well as the mid-ocean. The same basic interactions occur in all of these systems, though at different scales due to the differing sizes of the bodies of water.

Investigations from several lakes are reviewed. Nutrients from inflowing streams seem to be an important source of variability. Spatial variability of phytoplankton appears to be linked primarily to the physical dynamics in lakes of moderate size.

Though spatial heterogeneity has been invoked in numerous theoretical models of ecological processes in both terrestrial and aquatic habitats, its significance is still unclear. More detailed field and theoretical investigations, especially at the species level, are needed.

## ACKNOWLEDGEMENTS

We acknowledge the support of both the Physical Oceanography and Ecology sections of the National Science Foundation (U.S.A.) over the past five years. This support has enabled us to undertake the field investigations reported here. Robert C. Richards of the Institute of Ecology (University of California, Davis) has given generously of his time and expertise in boat operations that were partially supported by an N.S.F. grant to C. R. Goldman. Ms. Cecelia Odelius (University of California, Davis) and Ms. Dottie Bottini and her staff at the Lawrence Berkeley Laboratory conscientiously typed the manuscript during various stages of preparation. Linda Thorpe, Tim Long, and Kenzo Kawahira assisted with computer analysis of the data. The two senior authors (P. J. Richerson and T. M. Powell) are grateful for the hospitality of John Harte and Bob Budnitz of the Energy and Environment Division of the Lawrence Berkeley Laboratory which enabled them to complete this report.

## REFERENCES

Bainbridge, R., 1957: The size, shape and density of marine phytoplankton concentrations. Cambridge Phil. Soc., 32, 91-115.

Bernhard, M. and L. Rampi, 1965: Horizontal microdistribution in the Ligurian Sea. Proc. Mar. Biol. Symp. Goteborg, 5, 13-24.

Blanton, J. O., 1974: Some characteristics of nearshore currents along the northshore of Lake Ontario. J. Phys. Oceanogr., 4, 415-424.

Boyce, F. M., 1974: Some aspects of Great Lakes physics of importance to biological and chemical processes. J. Fish. Res. Bd. Canada, 31, 689-730.

Caldwell, D. R., J. M. Brubaker and V. T. Neal, 1978: Thermal microstructure on a lake slope. Limnol. Oceanogr., 23, 372-374.

Cassie, R. M., 1959: Micro-distribution of plankton. N.Z.J. Sci., 2, 398-409.

Csanady, G. T., 1975: Hydrodynamics of large lakes. Ann. Rev. Fluid Mech., 7, 357-386.

Denman, K. L., 1977: Short term variability in vertical chlorophyll structure. Limnol. Oceanogr., 22, 434-441.

Denman, K. L. and T. Platt, 1975: Coherences in the horizontal distributions of phytoplankton and temperature in the upper ocean. Mem. Soc. R. Sci. Liege, 6e serie, 7, 19-30.

Fee, E. J., 1976: The vertical and seasonal distribution of chlorophyll in lakes of the Experimental Lakes Area, northwestern Ontario: implications for primary production estimates. Limnol. Oceanogr., 21, 767-783.

Goldman, C. R., G. Moshiri and E. deAmezaga, 1972: Synoptic study of accelerated eutrophication in Lake Tahoe, a subalpine lake. In International Symposium on Water Pollution Control in Cold Climates, M. Murphy and D. Nyquist (eds.), pp. 1-21.

Harris, G. P. and R.E.H. Smith, 1977: Observations of small-scale spatial patterns in phytoplankton populations. Limnol. Oceanogr., 22, 887-899.

Haury, L. R., 1973: Sampling bias of a Longhurst-Hardy plankton recorder. Limnol. Oceanogr., 18,500-506.

Horne, A. J., J. R. Sandusky and C.J.W. Carmiggelt, 1978: Nitrogen fixation in Clear Lake, California. III: Super synoptic studies on the spring Aphanizomenon bloom. Limnol. Oceanogr., in press.

Horne, A. J. and R. C. Wrigley, 1975: The use of remote sensing to detect how wind influences planktonic blue-green algal distribution. Verh. Internat. Verein. Limnol., 19, 789-791.

Hulburt, E. M., 1977: Coexistence, equilibrium, and nutrient sharing among phytoplankton species in the Gulf of Maine. Amer. Natur., 111, 967-980.

Hutchinson, G. E., 1957: A treatise on limnology. Vol. I, J. Wiley, N.Y., 1017 pp.

Hutchinson, G. E., 1961: The paradox of the plankton. Amer. Natur., 95, 137-146.

Kierstead, H. and L. B. Slobodkin, 1953: The size of water masses containing plankton blooms. J. Mar. Res., 12, 141-147.

Leigh-Abbott, M. R., J. A. Coil, T. M. Powell and P. J. Richerson, 1978: Effects of a coastal front on the distribution of chlorophyll in Lake Tahoe, California-Nevada. J. Geophys. Res., (Oceans and Atmospheres), in press.

Levin, S. A., 1976: Population dynamic models in heterogeneous environments. Ann. Rev. Ecol. Syst., 7, 287-310.

Lucas, C. E., 1941: Continuous plankton records; phytoplankton in the North Sea, 1938-39, Part I, Diatoms. Hull. Bull. Mar. Ecol., 2, 19-46.

Lund, J.W.E., 1964: Primary production and periodicity of phyto-plankton. Edgar Baldi Mem. Lect., Verh. Internat. Verein. Limnol., 15, 37-56.

Margalef, R., 1969: Estudios sobre la distribucion a pequena escala del fitoplankton marino. Mem. Real Acoda Cienc. Artes Barcelona, 40C(1), 3-22.

McAlice, B. J., 1970: Observations on the small-scale distribution of estuarine phytoplankton. Mar. Biol., 7, 100-111.

Mortimer, C. H., 1974: Lake hydrodynamics. Mitt. Internat. Verein. Limnol., 20, 124-197.

Okubo, A., 1976: The critical size of water masses containing plankton blooms under scale-dependent diffusion. Manuscript.

Pingree, R. D., P. R. Pugh, P. M. Holligan and G. R. Forster, 1975: Summer phytoplankton blooms and red tidal fronts in the approaches to the English Channel. Nature, 258, 672-677.

Platt, T., 1972a: The feasibility of mapping the chlorophyll distribution in the Gulf of St. Lawrence. Fish. Res. Bd. Canada Tech. Rep. 332, 8 pp.

Platt, T., 1972b: Local phytoplankton abundance and turbulence. Deep-Sea Res., 19, 183-187.

Platt, T. and K. L. Denman, 1975: Spectral analysis in ecology. Ann. Rev. Ecol. Syst., 6, 189-210.

Platt, T., L. M. Dickie and R. W. Trites, 1970: Spatial heterogeneity of phytoplankton in a nearshore environment. J. Fish. Res. Bd. Canada, 27, 1453-1473.

Platt, T. and C. Filion, 1973: Spatial variability of productivity: biomass ratio for phytoplankton in a small marine basin. Limnol. Oceanogr., 18, 743-749.

Powell, T. M., P. J. Richerson, T. M. Dillon, B. A. Agee, B. J. Dozier, D. A. Godden and L. O. Myrup, 1975: Spatial scales of current speed and phytoplankton biomass fluctuations in Lake Tahoe. Science, 189, 1088-1090.

Richerson, P. J., R. Armstrong and C. R. Goldman, 1970: Contemporaneous disequilibrium, a new hypothesis to explain the "Paradox of the Plankton." Proc. Nat. Acad. Sci., 67, 1710-1714.

Richerson, P. J., B. J. Dozier and B. T. Maeda, 1975: The structure of phytoplankton associations in Lake Tahoe (California-Nevada). Verh. Internat. Verein. Limnol., 19, 843-849.

Richerson, P. J., M. Lopez and T. Coon, 1977: The deep chlorophyll maximum layer of Lake Tahoe. Verh. Internat. Verein. Limnol., 20, in press.

Richerson, P. J., C. Widmer and T. Kittel, 1977: The limnology of Lake Titicaca (Peru-Bolivia). Institute of Ecology Publ. 14, University of California, Davis, 78 pp.

Ricker, W. E., 1937: Statistical treatment of sampling processes useful in the enumeration of plankton. Arch. Hydrobiol., 31, 68-84.

Robinson, G. A., 1961: Contributions towards a plankton atlas of the northeastern Atlantic and the North Sea. Part I: Phytoplankton. Bull. Mar. Ecol., 5, 81-89.

Sandusky, J. C. and A. J. Horne, 1977: A pattern analysis of Clear Lake phytoplankton. LBL publ. 6302, Lawrence Berkeley Laboratory, University of California, Berkeley, 39 pp.

Simpson, J. H. and J. D. Woods, 1970: Temperature microstructure in a freshwater thermocline. Nature, 226, 832-834.

Stavn, R. H., 1971: The horizontal-vertical distribution hypothesis: Langmuir circulations and Daphnia distributions. Limnol. Oceanogr., 16, 453-466.

Steele, J. H., 1974: Spatial heterogeneity and population stability. Nature, 248, 83-85.

Stull, E. A., E. deAmezaga and C. R. Goldman, 1973: The contribution of individual species of algae to primary productivity of Castle Lake, California. Verh. Internat. Verein. Limnol., 18, 1776-1783.

Therriault, J. C., 1977: Studies of the spatial heterogeneity in the phytoplankton of St. Margaret's Bay, Nova Scotia. Ph.D. Dissertation, Dalhousie University, Nova Scotia, 89 pp.

Thorpe, S. A., A. J. Hall, C. Taylor and J. Allen, 1977: Billows in Loch Ness. Deep-Sea Res., 24, 371-379.

Tilman, D., 1977: Does environmental variability limit the number of coexisting species of algae? or Why aren't there more coexisting species of algae? Abstract, Am. Soc. Limnol. Oceanogr., Annual Mtg.

Venrick, E. L., 1971: The statistics of subsampling. Limnol. Oceanogr., 16, 811-818.

Venrick, E. L., 1972: Small scale distributions of oceanic diatoms. Fish. Bull., 70, 363-372.

Walsh, J. J., 1976: Herbivory as a factor in patterns of nutrient utilization in the sea. Limnol. Oceanogr., 21, 1-13.

Wiens, J. A., 1976: Population responses to patchy environments. Ann. Rev. Ecol. Syst., 7, 81-120.

Wiens, J. A., 1977: On competition in variable environments. Amer. Sci., 65, 590-597.

Wroblewski, J. S., 1977: Vertically migrating herbivorous plankton - their possible role in the creation of small-scale phytoplankton patchiness in the ocean. In Oceanic Sound Scattering Prediction, N. R. Anderson and B. J. Zahuranec (eds.), pp. 817-847, Plenum Press, New York.

Wroblewski, J. S., J. J. O'Brien and T. Platt, 1975: On the physical and biological scales of phytoplankton patchiness in the ocean. Mem. Soc. R. Sci. Liege, 6e serie, 7, 43-57.

# PATTERNS AND PROCESSES IN THE TIME-SPACE SCALES OF PLANKTON DISTRIBUTIONS

L. R. Haury, J. A. McGowan* and P. H. Wiebe

Woods Hole Oceanographic Institution
Woods Hole, Massachusetts 02543

## INTRODUCTION

It is evident that organisms have aggregated, patchy distributions of abundance on a wide variety of space and time scales. This can easily be seen in the terrestrial and littoral environments. It has been more difficult to observe in the pelagic realm simply because we cannot see into the ocean. Thus we must depend upon sampling to gain an impression of the space-time scales of pattern in this habitat. Despite the difficulties, enough sampling of the right sort has now been done so that we can make some very general statements about the nature of pattern in the ocean, particularly with regard to pattern in the distribution and abundance of planktonic organisms. All the evidence indicates that plankton is patchy on a broad spectrum of scales. Because this aggregated spatial pattern is such a general phenomenon, there is little question of its ecological and evolutionary importance. Further, because we assume our samples represent a larger universe, patchiness strongly affects our efforts to obtain estimates of the abundance of organisms and our ability to detect significant spatial and temporal changes in abundance. It is therefore of great importance that we understand its nature, causes, and effects.

The purpose of this paper is to provide a conceptual framework for the consideration of time-space scales of oceanic pattern and within this framework to review the types of patterns observable

*Scripps Institution of Oceanography, La Jolla, California 92037

in the world's oceans, to relate the spectrum of patterns to factors which appear important to their generation and maintenance, and to touch on some of the ecological and evolutionary consequences of patchiness. The framework which will be used we call a Stommel Diagram, a three-dimensional graph of parameters of biological variability plotted against scales of space and time.

## THE STOMMEL DIAGRAM

One of the most significant factors contributing complexity to the interpretation of zooplankton distributional and time-series data is the great variability of abundance estimates on all scales in space and time caused by patchiness. Physical oceanographers, faced with a similar level of complexity in physical variables, long ago turned to time-series analysis techniques to surmount the problems and to determine the relative importance of variability at different scales. This transition in analytical technique was made possible by automated methods of data collection and reduction. Biological oceanographers are turning to these methods (Platt and Denman, 1975) but are hindered by the lack of instrumentation needed for the collection of the large numbers of samples required and, most particularly, by the lack of rapid methods of sample analysis. Present methods allow the biomass of phytoplankton, and in some cases zooplankton, to be measured relatively easily, but the ultimate need is for identification to functional group, if not to the species level.

The limitations mentioned above do not prevent the conceptual use of spectral techniques to aid in planning and interpreting data collections. This was done in physical oceanography almost 15 years ago by Stommel (1963, 1965) to show how to improve the design of sampling programs and how to distinguish achievable goals from the unachievable. Stommel's papers are remarkably pertinent to many aspects of biological oceanography; we have chosen to use some of the ideas contained in them as a core for the discussion of space-time scales of patchiness and their relationship to sample design, data interpretation, and ultimately, our understanding of marine ecosystems.

Stommel presented three-dimensional displays portraying the spectral distribution of physical variables, such as sea level and deep ocean velocities, plotted as a function of logarithmic scales of space and time. The vertical axis was used to represent the amount of fluctuation of the physical variable within a certain range of period and wavelength. Using this same technique, we can apportion the amount of variability of a biological parameter, such as biomass, into the appropriate regions of space and time and plot it as shown in Figure 1. Many problems are inherent in the

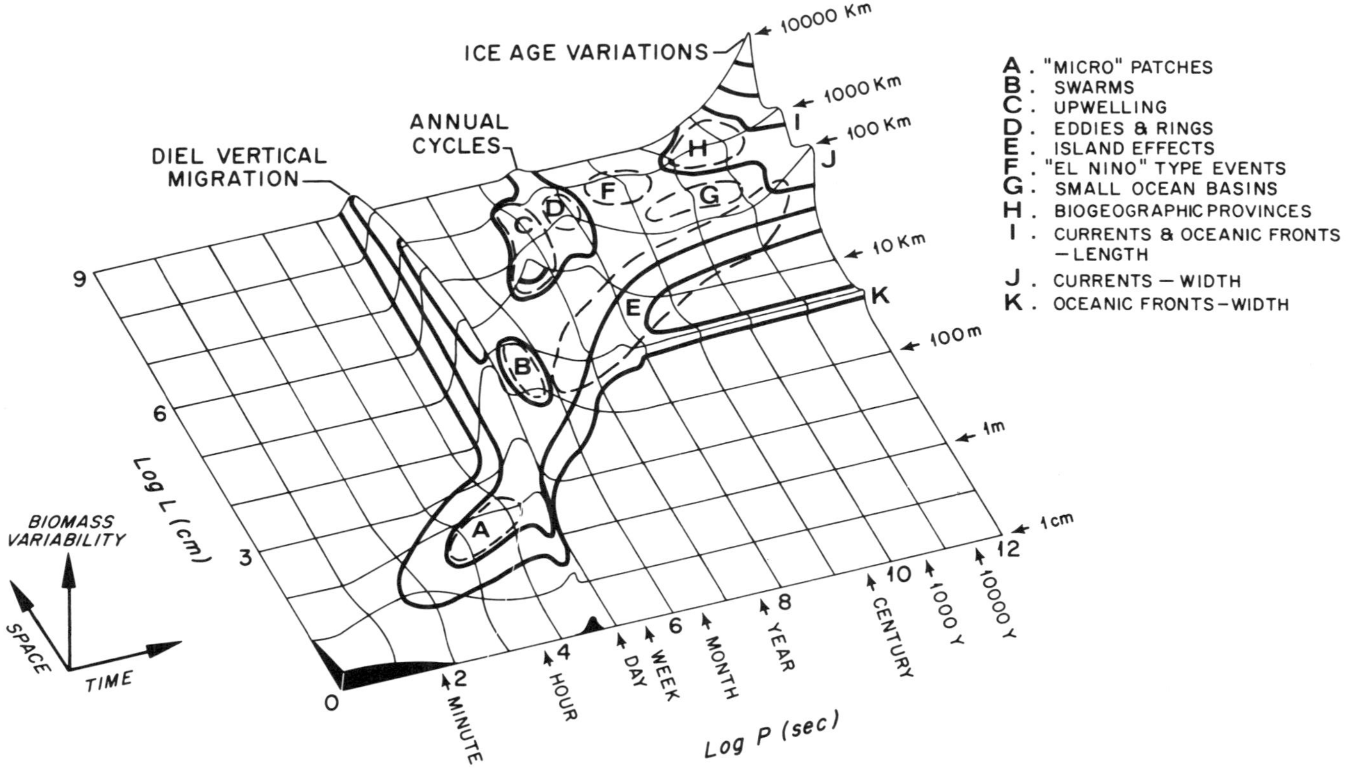


Figure 1. The Stommel Diagram, a conceptual model of the time-space scales of zooplankton biomass variability and the factors contributing to these scales. I, J and K are bands centered about 1000's, 100's and several kilometers in space scales, with time variations between weeks and geological time scales.

construction of such a schematic diagram. We will discuss three of the most important.

The first problem was discussed by Stommel (1965). The shape of the diagram will depend on the geographical area chosen for detailed examination, and the small-scale features may depend on the phase of the larger scale phenomena. Thus, to draw the diagram, we must do a "desperate thing" (Stommel, 1965) -- we must assume that the statistical and spectral properties of the biology of the ocean do not depend upon position or time.

The second difficulty is the lack of sufficient data on biomass (or any other parameter) variability over all scales in space and, especially, over long time periods. We have, therefore, had to use the little data available and a large amount of 'best guesses' to fill in the diagram.

The third problem rests on the fact that we have tried to include in the estimate of variability at any scale a measure of its biological importance. That is, the height of the peaks depend not only on the strict biomass variance associated with that scale, but also the effect this variance may have on the ecosystem. For example, small fluctuations in biomass at short time and space scales may have equal or greater importance in determining ecosystem structure and function than larger scale fluctuations, such as climatic cycles. We are hindered here by the lack of any quantitative measure of what, exactly, is meant by 'importance.'

In a rigorous sense, then, our diagram is not the spectral distribution of biomass in the world's ocean. The assumptions of spectral analysis are not met (stationarity, uniform variance, time independence). The vertical axis is not a strict measure of variance, nor is the diagram a variance conserving plot. Therefore, the 'variability' cannot be integrated onto a set of two-dimensional axes to look at the distribution of variance over all time or length scales.

Despite the shortcomings of the data base and assumptions used for the construction of the Stommel diagram in its present form, it offers certain insights regarding the relationships between scales of variation of zooplankton biomass. We will treat these relationships according to appropriate categories of spatial and temporal structure, and later amplify these with examples of data from the real ocean.

## SCALES OF PATTERN

### Space Scales of Pattern

In the spatial dimension, patchiness refers to the aggregated (non-random) distribution of biomass or number of individuals per unit volume of habitat. Nearly all analyses of spatial pattern start with the determination of the deviation of the variability in a sample set from a random expectation. For plankton, the appropriate theoretical distribution of randomness is the Poisson random wherein the probability of an individual occupying a particular location in space is very small (i.e. the number of spaces is large relative to the number of spaces occupied) and the individuals are considered independent and noninteractive (i.e. the occurrence of one does not influence the occurrence of another).

Beyond the simple statement that a data set deviates from a random expectation, however, there are a growing number of analytical techniques which permit a description of spatial pattern and interpretation of its meaning. The mathematical techniques can be divided into at least three classes: that of comparing the fit of a data set to a theoretical model expectation derived from assumed properties of the underlying spatial distribution, e.g., negative binomial distribution, Thomas-Neyman distribution, Lloyd's patchiness index (Lloyd, 1967); that of quantitatively assessing patchiness in terms of its structural components, i.e. patch size, shape, distribution, number of individuals within versus outside a patch, longevity, etc., as has been done by Wiebe (1970), Ziemann (1970), Fasham, Angel and Roe (1974) and others; and that of analyzing the variance of a data set in terms of its characteristic time and space scales (Platt, 1972; Platt and Denman, 1975; Fasham and Pugh, 1976).

These three mathematical approaches to the interpretation and understanding of spatial structure can be loosely termed model fitting, empirical description and spectral analysis. Model fitting is the least desirable because of certain ambiguities which can arise: models based on conflicting assumptions can often produce the same expectation (Bliss and Fisher, 1953), or can be fit to a data set with the same or similar degree of significance. Empirical description of spatial structure is of critical importance in defining scales of pattern, the scope of intra- and interspecific interactions, and species reactions to physical aspects of the environment. Spectral analysis provides an alternate method of characterizing scales of variability in space or time. Limitations regarding the use of this technique are discussed below.

We think of the range of spatial scales of aggregated patterns from micro-scale patches to ocean-wide distributions as being a continuum. This does not imply absence of locally dominant scales. Nor does it imply the continuum structure is constant over time. It is important to emphasize that spatial pattern is not a cascade in the sense that large scale structures are dissipated by their breaking up into smaller and smaller features. While both small and large scale patterns of plankton distributions can be broken up by turbulence into smaller patches, small patches can contribute to patterns at larger scales as well. Some mechanisms of pattern generation thus do not seem to be entirely unidirectional, apparently due to the nonlinear nature of the interaction between predators and their prey (Steele, 1974c; Okubo, 1974; Levin and Segel, 1976).

Although biomass peaks and plankton species aggregations appear to occur on a continuum of scales, it is convenient to break the continuum into categories and to apply terms to them. Table 1 summarizes these categories and some of their ecological attributes. In the sections which follow, we will amplify this table by presenting examples of pattern appropriate to the categories and relate some of them to the Stommel Diagram.

Mega-scale. Mega-scale patterns of high and low biomass and of species presence and absence are reasonably well documented for most of the world's oceans (Beklemishev, 1969; McGowan, 1971, 1974; Backus *et al.*, 1977; Reid *et al.*, in press). These biogeographic patterns (Region H, Figure 1) extend over very large parts of entire oceans on a scale equal to or greater than 3000 km and closely resemble in shape the major oceanic circulation gyres. For example, the Subarctic cyclonic gyre of the North Pacific exhibits a distinct biomass and species pattern. Another distinct pattern is confined to (and by) the equatorial zonal recirculation systems. These and other mega-scale geographic distributions, together with the associated circulation systems, form the major ecosystems of the world's oceans (McGowan, 1974). Although their sizes are enormous, they do represent an ecologically and evolutionarily important scale of aggregation which appears to be the direct result of the extant circulation patterns.

Macro-scale. Within the large systems, there are macro-scale features on the 1000 km to 3000 km range. The distinction between mega- and macro-scale structures lies not only in the differences in their sizes but also in the degree of distinctness in species composition. Mega-scale pattern involves strong contrast in species composition, while macro-scale pattern involves highs and lows of species abundance throughout areas where differences in species compositional structure is low. The zonal band of high biomass along the central sector of the Pacific within the equatorial circulation system is an example (Figure 2). This is brought about by the quasi-steady state vertical circulation in

Table 1. Categories of scales of pattern.

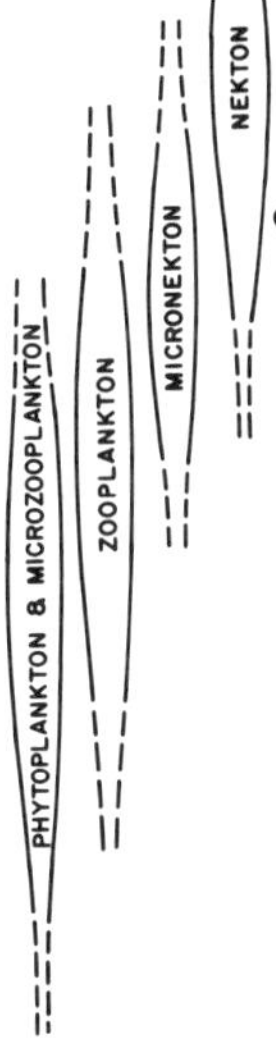


| NAME | SPACE SCALE | DOMINANT PATTERN * | HOW BEST LOOKED AT | WHAT WE LEARN |
|---|---|---|---|---|
| MEGA | $10^4$ km | Vectorial | Communities<br>Biomass<br>Species | Biogeography<br>Evolutionary history |
| MACRO | $10^3$ km | Vectorial<br>Reproductive | Communities<br>Biomass<br>Species | Biogeography<br>Speciation<br>"Best" places to live<br>Ecotones<br>Inter-community competition<br>"Hot spots" within ecosystems |
| MESO | $10^2$ km | Vectorial<br>Reproductive | Biomass<br>Species | Faunal boundaries<br>Invasions<br>Nekton ambit<br>Genetic selection<br>Relationship to environmental parameters |
| COARSE | 10 km<br>1 km | Vectorial<br>Reproductive<br>Coactive<br>Social | Species | Intra-community competition<br>Upwelling responses<br>Micronekton ambit<br>Relationship to environmental parameters |
| FINE | 100 m<br>10 m | Vectorial<br>Reproductive<br>Coactive<br>Social | Species | Coexistence, niche partitioning<br>Inter-and intra-species competition<br>Predation<br>Food densities required<br>Zooplankton ambit<br>Relationship to environmental parameters |
| MICRO | 1 m<br>10 cm<br>1 cm | Vectorial<br>Social | Species<br>Individual | Inter-and intra-species competition<br>Niche partitioning<br>Relationship to environmental parameters |

* Stochastic acts on all scales

this band. The large area surrounding the Costa Rican Dome in the eastern tropical Pacific Ocean is another example where a quasi-steady state circulation system embedded within a mega-scale pattern leads to a productivity and biomass peak (Figure 3). In contrast to regions of high biomass within a given biogeographic province, there are also regions of distinctly lower biomass which occur as macro-scale features in the central gyres of the North Atlantic and North Pacific Oceans.

Meso-scale. There is good recent evidence for meso-scale aggregations on the 100 to 1000 km scale. The cold Slope Water communities isolated in Gulf Stream cold core rings and distributed throughout the Sargasso Sea (Wiebe *et al.*, 1976) are examples of significant patches of both biomass and species as are the meso-scale biomass patches of *Euphausia pacifica* and *Limacina helicina* which occur in the southward extension of the California Current (Brinton, 1967, p. 85; McGowan, 1967, p. 12). Such meso-scale plankton patches are related to the formation and distribution

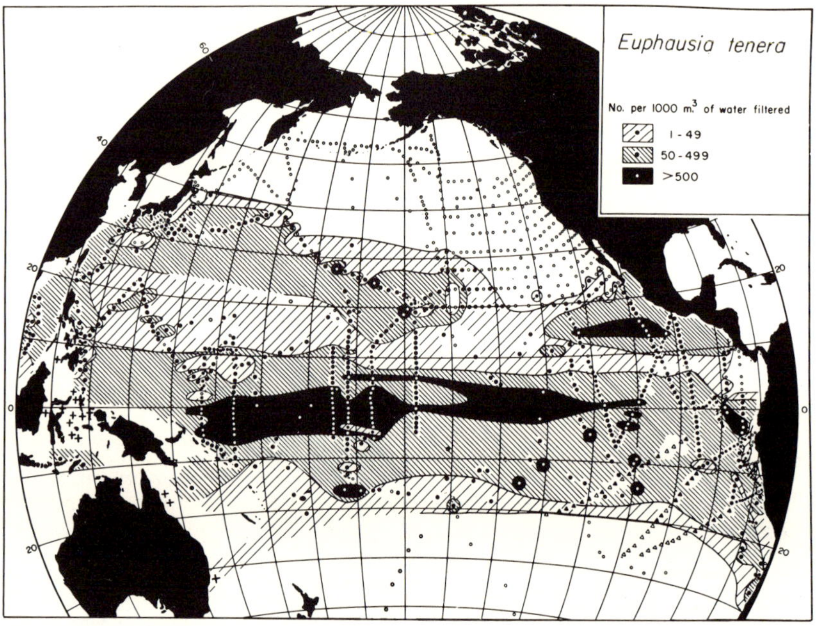


Figure 2. Ocean-wide distribution of a single broadly ranging species, *Euphausia tenera*, showing peaks in abundance along the equatorial zone of upwelling, Peru current extension and the eastern tropical Pacific (from Brinton, 1962).

of meso-scale eddies. These large bodies of water (200-300 km in diameter) are pinched off meanders of sections of eastern and western boundary currents. In the case of western boundary currents, the rings spin off into areas physically and biologically different from the ones of their origin. Although most of the recent biological work on this type of meso-scale structure has been done in the northwestern Atlantic, these features are known to occur in the regions of the Kuroshio, Agulhas, Somalia and east Australia current systems with biological contrasts which undoubtedly parallel those described for the northwestern Atlantic and California Current. Meso-scale patches of this sort may also occur in the fast-moving equatorial recirculation systems as well. For example, it is known that the loop current in the eastern portion of the Gulf of Mexico spins off eddies which then move into the western Gulf of Mexico (Maul *et al.*, 1974). The meso-scale pattern induced by large scale physical phenomena obviously represents a sort of invasion of one ecosystem by a complex of species from another system where the habitat structure and community composition are quite different. As such, they provide us with the opportunity of following definable pelagic communities through time for study of rates of change of population and community parameters with the changing physical conditions as the meso-scale water bodies mix and deteriorate.

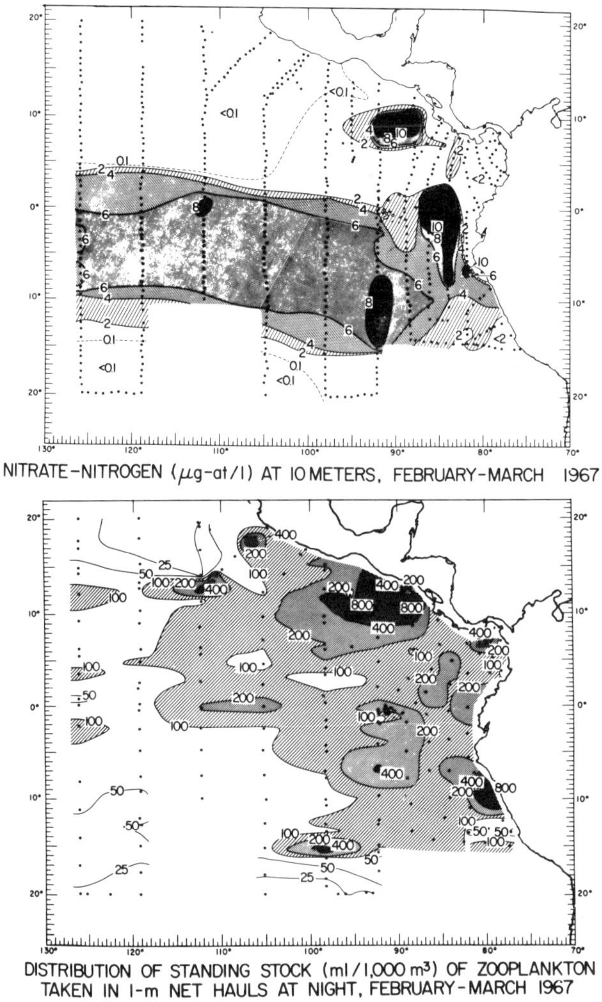


Figure 3. Distribution of nitrate and zooplankton standing stock in the eastern tropical Pacific. The influence of the Costa Rica Dome (9°N; 89°W) is clearly evident. Maps of other properties; phosphate, primary production, chlorophyll and larval myctophid fish also show this feature. Similar data taken in the summer show similar patterns (from Love, 1971).

Broad upwelling events (Smith, 1968), as opposed to plume upwelling (e.g., Beers *et al.*, 1971), in the eastern boundary current regions are another quite different form of meso-scale feature which result in significant contrasts in biomass and species composition, as are the fresh water plumes resulting from outflows from major river systems.

Coarse-scale. At 1-100 km scales, distinct aggregations of both biomass and species occur. Plume-type upwelling events (Region C, Figure 1; Beers *et al.*, 1971; Walsh *et al.*, 1977), widths of oceanic frontal zones (Region K, Figure 1; Pingree *et al.*, 1976; Savidge, 1976), and island wakes (Region E, Figure 1; Sette, 1955; Uda and Ushino, 1958) are areas where patterns on this scale are found. Other, less easily explainable patterns occur on these scales (e.g., Brinton, 1962; Cushing and Tungate, 1963; Wyatt, 1973).

It is on the coarse-scale that one can visualize the ambits[1] of some pelagic organisms becoming an important factor. The ambit of the larger nekton or nektonic schools is on the order of kilometers per day in their search for prey and mates or in attempting to avoid predators. The probability of encountering a predator or finding a prey or mate must depend upon the distance covered and the density of these "targets." If we assume that nektonic animals must, on the average, find food at least every few days and further, that food limitation is a factor in survivorship, then we have a spatial-temporal scale where fundamental ecological interactions occur between populations. The exact dimension of this ambit and the resultant coarse-scale structure will, of course, vary depending on aspects of behavior, size and swimming ability of particular nekton; but we assume, here, that these ambits are between 1-100 km. As important as this scale may be to individuals in the short term, there is a longer term and perhaps larger scale related to the lifetime or generation time ambit that is probably more important in the regulation of populations. The survivorship of a population or year class may well depend on the mean patchiness field that the population experiences per generation.

Fine-scale. The fine-scale (meters to hundreds of meters) patchiness in the distribution of plankton (e.g., Region B, Figure 1) is now widely recognized although few investigators have specifically sought information about dimensions of single or multi-species aggregations, the changes that might take place in this scale or spatial structure over time and space. Until recently, most attention has been paid to the role of fine-scale patchiness as a component of sampling error. Indeed, the work of Wiebe and Holland (1968), Wiebe (1971, 1972), Wiebe *et al.* (1973), McGowan (1971), Chanut (1975) and Smith (1977) all demonstrate the important effect of fine-scale distribution on determining the magnitude of sampling error. The studies which have been designed

---

[1]Ambit means the sphere of action or influence of individuals over days, weeks, lifetimes. An integral part of the concept of ambit is the space inhabited by an individual during the course of movement through its environment.

to look quantitatively at fine-scale aggregations have shown that pronounced variations in the numerical abundance of species occur over small distances (e.g., Barraclough *et al.*, 1969; Wiebe, 1970; George and Edwards, 1973; Fasham *et al.*, 1974; Haury, 1976a,b; Smith *et al.*, 1976). Figure 4 presents recent results from the northwestern Atlantic Slope Water which illustrate both coarse- and fine-scale structure in zooplankton populations at 75 m depth. There are also many qualitative observations of patchiness on this scale (e.g., Bainbridge, 1952; Marr, 1962; Komaki, 1967; Baker, 1970; Kawamura, 1974).

It is on this scale that individual interactions between zooplankton take place. Similar to the coarse-scale patterns for nekton, this interaction scale for zooplankton depends on the relation between the time it takes for an important interaction to occur (such as finding food, meeting a predator) between individuals, on the average timing of these interactions for a population and on the spatial dimension of the habitat that a fine-scale aggregation experiences during this time. It is reasonable to assume for a calanoid copepod that this time period is on the order of minutes to several days and that during this time meters to hundreds of meters or even several kilometers may be traveled. This is especially so if the organism is a vertical migrator. Again, while this scale is important for individuals, the fate of populations may well depend on the field of patchiness, on the spatial scale of populations, and on the temporal scale of generation times.

Micro-scale. There is a still smaller scale of patchiness which we call micro-scale, having dimensions of 1 cm to 1 m. There are a few studies illustrating the pronounced non-randomness which can occur on these scales (Bainbridge, 1952; Cassie, 1959; Owen, 1966; Emery, 1968; Clutter, 1969). There are obvious problems with attempts to describe patchiness on this scale. These result from the sampling devices disrupting (by turbulence) the system being measured. To the extent that biological patchiness on this scale is the product of small-scale physical phenomena, we might expect the patches to be as ephemeral in time as are the physical features. That is, they may be of short duration compared to a significant fraction of the generation time or daily ambit of the organisms in the micropatch. We have reservations, however, concerning the ephemeral nature of such small zooplankton patches. Much of the evidence cited above suggests persistence of fine-scale and micro-scale patches in the face of turbulence much greater than that usual for the open ocean. For example, Emery's (1968) and Clutter's (1969) observations give evidence of patches of copepods and mysids which remain coherent in swash over reefs and near surf zones. This suggests that behavioral factors play an important role in the

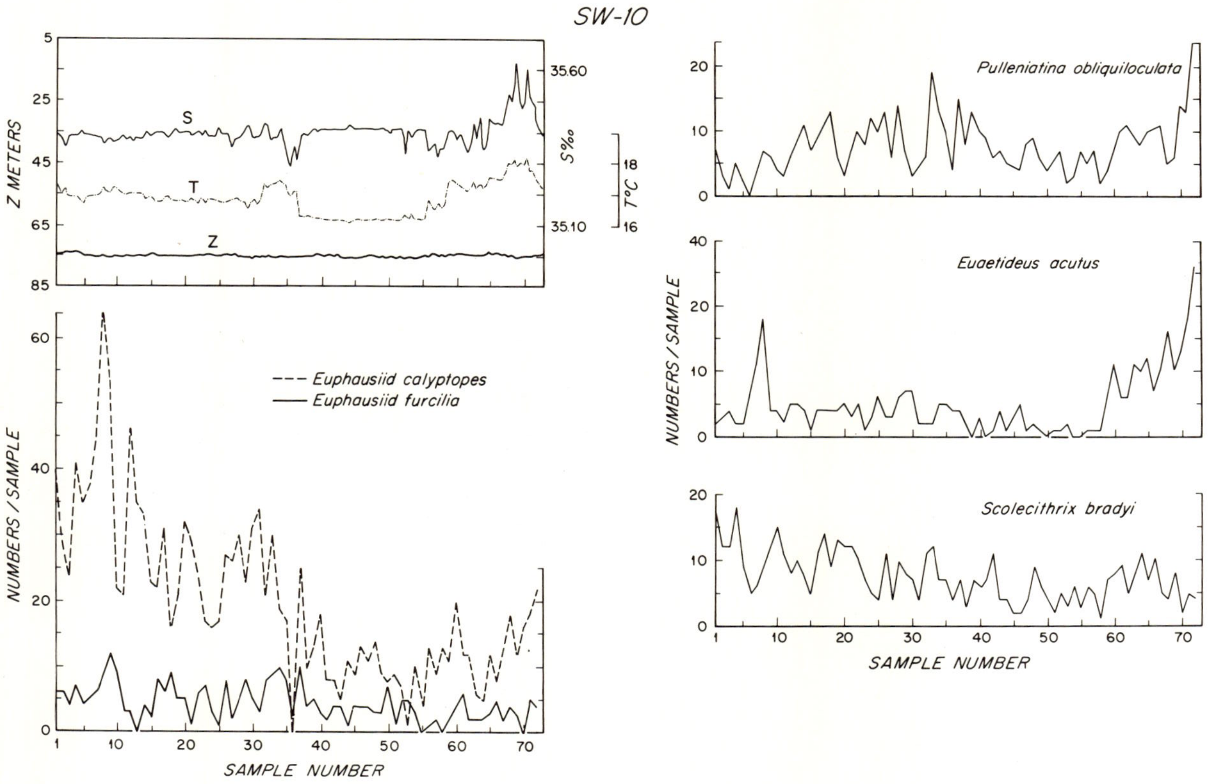


Figure 4. Fine- to coarse-scale patterns of species in the northwestern Atlantic Slope Water at a depth of 75 meters. Longhurst-Hardy Plankton Recorder: tow length 2730 meters; average volume filtered per sample 14.4 m$^3$; mean sample length 38 meters (Haury, unpublished data).

formation and persistence of micropatches, and by their persistence, these patches will have a significant role in ecosystem function. We will discuss these aspects in the last section.

## Time Scales of Pattern

The features discussed above are, as indicated, variable in time, as well as in space, and thus add the additional complexity of a fourth dimension to efforts to understand marine ecosystems. This section is a brief description of some of the time scales on which the space scales appear to vary and how they affect patterns of distribution.

On the longest time scales, pattern effects are ocean-wide and are caused by the cycles of climate, with variability ranging from a few years to many thousands of years (e.g., Ice Age Peak, Figure 1). The geological record in cores of bottom sediments present limited information on the cycles in organization of plankton communities; for plankton with no hard parts, nothing is known. The historical fishing records (Uda, 1961; Cushing and Dickson, 1976), and cores from anoxic basins (Soutar and Isaacs, 1974) give evidence of time variations on shorter intervals. These variations usually occur on smaller scales, e.g., within small ocean basins (Figure 1, Region G). The long-term plankton surveys in the North Atlantic (Colebrook, 1972), North Sea (Williamson, 1961); and California Current (Smith, 1971), the records from some oceanic weather stations (LeBrasseur, 1965) and from isolated, relatively long term sampling efforts by individuals (e.g., Tont, 1976) all give examples of changes occurring over a few years. Random or quasi-cyclic, rare, oceanographic events of large magnitude are sometimes noted and give evidence of significant effects on ecosystem patterns (Figure 1, Region F; Cowles *et al.*, 1977; Colebrook, 1977; Dugdale *et al.*, 1977). Some of the changes may be due to cycles in current strengths (Wickett, 1967; Martin, 1972; Saur, 1972; Wyrtki, 1975).

Between the cycles of climate and season occur the more areally restricted meso-scale oceanic phenomena, such as Gulf Stream rings (noted above), Mode-type eddies (Rhines, 1976), and eddies, meanders and waves in current systems and ocean fronts (e.g., Bernstein *et al.*, 1977; Legeckis, 1977a,b). These phenomena, which appear to have lifetimes of from a few months to several years, impress significant biological variability on oceanic communities through their advection or displacement of differing environmental conditions in the ocean.

Seasonal changes are by far the best documented in terms of ocean-wide to meso-scale changes in abundance and distribution

(Figure 1, Annual Cycles). However, there are no data on how coarse- and fine-scale pattern may vary through time and whether these changes are synchronized with the physical environment on these scales.

Weather, acting at periods from a few days to seasons can significantly alter the structure of the upper water column. Storms cause mixing (e.g., Lasker, 1975; Iverson, 1977); summer heating and relaxation of mixing cause stratification and the formation of seasonal thermoclines (Denman and Miyake, 1973); unusually severe winters cause formation of surface water masses, for example the winter of 1976-1977 forming 18°C water in the northern Sargasso Sea (Leetmaa, 1977; Worthington, 1977). An example of how weather patterns may determine plankton and fish distribution for upwelling regions, both over short time periods and evolutionary time scales, is given by Walsh *et al.* (1975, 1977). In this case, it is not only the length of periods of wind which induce upwelling that are important, but also the length of time between wind events which determines the productivity and availability of food to grazers.

Cyclic events of shorter time scales than seasonal changes are dominated by tidal cycles, both barotropic and baroclinic, and by diel cycles of vertical migration (DVM, see Figure 1). Kamykowski (1974) has described how semi-diurnal internal tidal cycles, both with and without DVM, can intensify coarse-scale phytoplankton patchiness and change productivity. He suggested how these in turn could also affect zooplankton. The time aspects of higher frequency internal waves similarly could affect phytoplankton and zooplankton particularly in the phasing of the internal wave cycles with the diel light cycle. Hardy and Gunther (1935) long ago described how DVM could interact with vertical current gradients to produce a navigation mechanism creating coarse-scale patchiness. Riley (1976) has used this mechanism, together with tides, residual drift, and herbivore grazing in a verbal model to explain coarse-scale phytoplankton and zooplankton patchiness. Wroblewski (1977) used a numerical model to explore the creation of coarse-scale phytoplankton patchiness by vertically migrating grazers and the diel changes this causes in nutrient, phytoplankton and zooplankton distributions.

Diel cycles may exist in intensity and size of fine- and micro-scale patchiness which are dependent on the light and activity regimes of organisms. This may be true for migrating midwater fish (e.g., Barham, 1970) and is probably a general feature of the distribution of invertebrates which form patches through visual cues (e.g., Clutter, 1969; Omori, 1974). The fine-scale patterns observed by Wiebe (1970) and Haury (1976a,b) did not indicate a significant shift in day-night patch structure.

However, recent work using high frequency side-looking sonars has shown marked day-night shifts in pattern scales of acoustic scatterers, presumably zooplankton (R. Pinkel and P. Greenblatt, personal communication).

Random, short term, areally restricted events can lead to marked alteration of distributions. For example, vertical pursuit of food (Hamner and Hamner, 1977) would result in horizontal redistribution of individuals through interaction with vertical current gradients. This is an application of Hardy's navigation hypothesis to fine- and micro-scale pattern which could operate on levels from zooplankton detritivores to nektonic scavengers. Random passage of groups of mobile predators, particularly fish schools, through an area introduces both locally intense predation pressure and nutrient enrichment through excretion (Whitledge and Packard, 1971; McCarthy and Kamykowski, 1972) thus affecting most levels in the food chain.

Perhaps the most important time factor of pattern is the temporal persistence or longevity of individual patches on the micro- to coarse-scale. Much has been inferred, but little is known, about this problem. Since it is of importance both to our understanding of the ecological significance of patchiness and to how sampling programs should be designed, we will discuss patch longevity in more detail in later sections.

## GENERATION, MAINTENANCE AND OBSERVATION OF SPACE-TIME SCALES OF PLANKTON PATCHINESS

It is clear that spatial patterns of plankton distribution illustrated in Figure 1 can arise and be maintained through physical and biological processes acting alone and in concert. These processes vary in time in both a cyclic and random manner, thus the causal factors and their time variation make any categorization by causative mechanism both difficult and artificial. For the convenience of discussion, we will separate the various factors which we believe are important in generating and maintaining the features of the Stommel Diagram of plankton biomass variability. The time factors, which operate simultaneously with the physical and biological factors, have been discussed previously. Observational and analytical procedures, such as sampling methodology and the type of statistical analysis, are factors which modify "true" patterns. A discussion of some of these factors will conclude this section.

We have based our outline of the causative factors upon that of Hutchinson (1953). The five kinds of pattern he described (vectorial, coactive, reproductive, social, stochastic) have been reviewed and in some cases modified (e.g., Stavn, 1971). We prefer to vary Hutchinson's organization only by assuming stochastic mechanisms act as an all-encompassing "blanket" on the other mechanisms, i.e. each of the other major mechanisms contain stochastic elements. Figure 5 illustrates our attempt to summarize these mechanisms in diagrammatic form. The literature dealing with most aspects of the figure is voluminous and well known. We will limit our discussion to those factors which are less well established or are controversial and which have particular relevance to the ecological implications of patchiness.

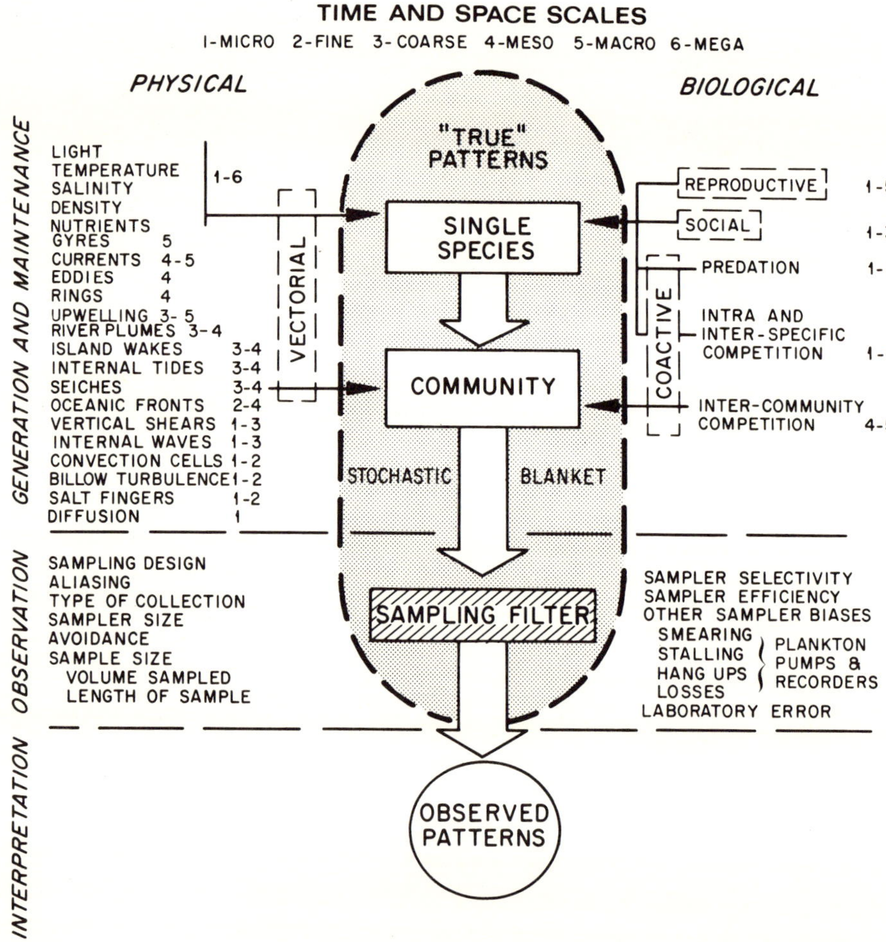


Figure 5. Factors in the generation, maintenance and observation of plankton patterns.

## Biological Factors

Of the three biological kinds of pattern listed by Hutchinson, reproductive and/or social patterns are reasonably well described for some specific areas and organisms (e.g., Bainbridge, 1952; Taft, 1960; Marr, 1962; Komaki, 1967; Clutter, 1969; Baker, 1970; Wyatt, 1973; Kawamura, 1974). In most oceanic areas little is known about the role of these mechanisms in causing micro- to coarse-scale patchiness; this is particularly true for the smaller zooplankton and phytoplankton. The understanding of how the two processes act to generate these scales of patchiness is reasonably clear. The major question of process concerns how the physical dynamics, which tend to destroy heterogeneity, interact with the swimming and energetic abilities of the plankton to form and maintain a patch.

The coactive processes (involving competition, predation and parasitism) are by far the least understood of the biological mechanisms of patch formation on any scale. While it is conceptually quite clear how predator-prey interactions and grazing, and even interspecific competition for a common resource, can create pattern, it is not at all clear what the relative importance of these processes are to the development and maintenance of the patterns observed in the ocean.

Many models (e.g., Dubois, 1975a,b; Wroblewski and O'Brien, 1976; Wroblewski, 1977; Sjöberg, 1977) suggest analytically how coarse-scale coactive pattern can occur, but they say little about persistence of patterns in space and time. Other models (e.g., Okubo, 1974; Steele, 1974; Levin and Segel, 1976; Ziegler, 1977) suggest that the interaction of physical diffusive processes with predators and prey of different mobility can create persistent patchiness and theoretically affect abundance on even larger scales of pattern.

## Physical Factors

The different physical mechanisms which generate and maintain vectorial patterns are in many cases quite obvious intuitively and, for those acting on macro- and mega-scales, reasonably well studied. On the smaller scales, however, only recently has their significance to ecosystem structure and function been recognized. We will here give examples of our recent work on two of these factors: Gulf Stream rings and internal waves.

Gulf Stream cold core rings. Gulf Stream cold core rings are now known to occur throughout the northwestern Sargasso Sea (Figure 6). They constitute a dominant source of meso-scale variability in the plankton communities in this area. At any one

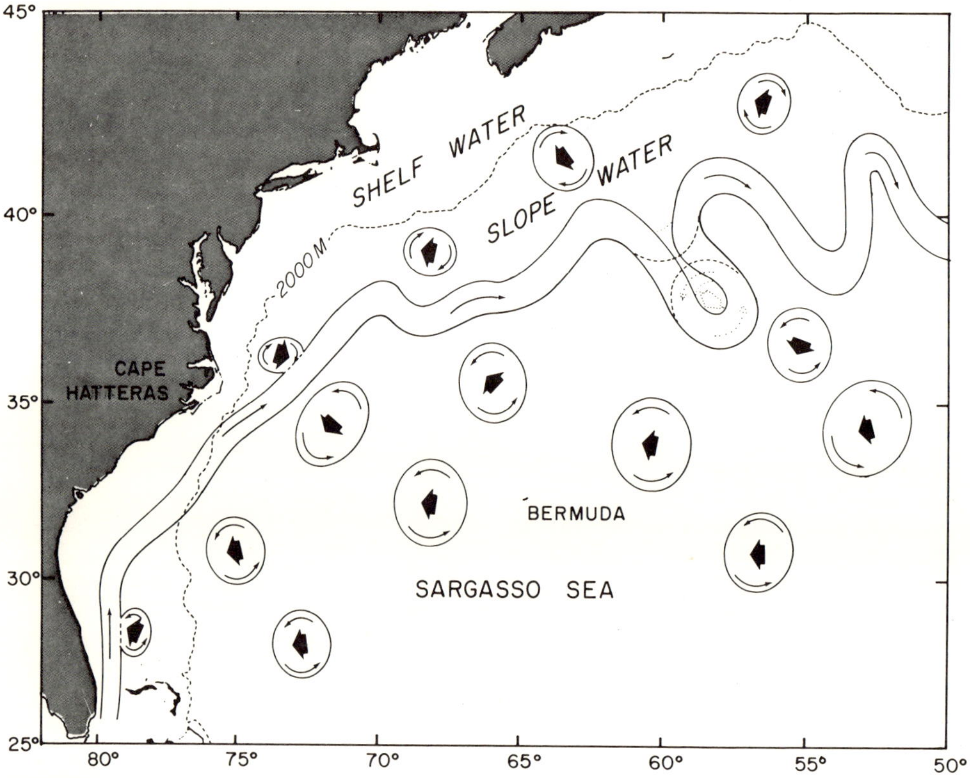


Figure 6. Richardson's (1976) schematic representation of the path of the Gulf Stream and the distribution and movement of rings.

time there may be 10-15 such rings covering between 6 and 13% of the area (Richardson, 1976) and as an advective feature they transport water and biota in a column at least 1000 m deep. During the first year of their existence they are generally more productive than the surrounding Sargasso Sea and may augment the production of the northern portion of this region by 3 to 7% (Ortner *et al.*, 1978). They have a zooplankton biomass in the upper 800 m which varies between being 1.3 and 2.0 times higher than in the Sargasso Sea, depending upon ring age (Wiebe *et al.*, 1976; Ortner *et al.*, 1978; Lai and Richardson, 1977). Highest biomass occurs at the center of a ring and progressively declines on the ring flanks. Compared to either the surrounding Sargasso Sea or the parent Slope Water, an unusually large percentage of the 0-800 m biomass in rings is found at depths greater than 200 m especially in rings between 3 and 12 months old. This distribution is related to the hydrographic and biological changes associated with ring decay.

The strong meso-scale contrast created by the presence of cold core rings in the Sargasso Sea is not limited to gross biological properties such as biomass. For a significant period of time, at least one year, rings are also a major source of meso-scale variability in the species compositional structure of the area because the cold core waters in a ring contain a fauna and flora of Slope Water origin which is distinctly different from that of the Sargasso Sea. Although the vectorial forces responsible for the formation, movement and decay of rings are dominant elements in establishing and maintaining this kind of meso-scale pattern in the Sargasso Sea, the specific character of the ring species composition is biologically regulated to a significant degree. For example, the changes in phytoplankton composition that accompany ring decay do not form a sequence of intermediate compositional forms which parallel the transformation of physical properties such as temperature and salinity from a Slope Water character to a Sargasso Sea character (Ortner *et al.*, submitted). Not only are differences in species composition variable and not clearly related to ring age, but in none of the four cases observed to date has a ring phytoplankton species composition appeared intermediate between that of the Slope Water and Sargasso Sea. This circumstance must in part be due to a combination of the uniqueness of the phytoplankton species reproductive capacities, their response to the changing nutrient conditions, and the changing character of herbivore grazing that accompanies modification of the surface zooplankton structure.

Cold core rings are very likely responsible for the failure of zoogeographic boundaries to clearly coincide with major hydrographic changes in the northwestern Atlantic. For example, in summaries of the zoogeographic distribution of the euphausiid *Nematoscelis megalops* (Mauchline and Fisher, 1969; Gopalakrishnan, 1974; Reid *et al.*, in press) a portion of the Sargasso Sea is included as part of its habitat despite the fact that it is classified by Brinton (1962), Mauchline and Fisher (1969), and Gopalakrishnan (1974) as an inhabitant of the cool transition regions between the tropical-subtropical and the subpolar waters in the Atlantic, Pacific and Indian Oceans. Recent work has shown that the hydrographic limit of *N. megalops* in the northwestern Atlantic coincides with the abrupt changes in water properties across the Gulf Stream (Wiebe and Boyd, 1978). The isolated but related occurrences of this species in the Sargasso Sea were nearly all in cold core rings; exceptions could be linked to the presence of rings (Figure 7). Expatriated individuals have not, in the recent sampling experience, occurred in hydrographically pure Sargasso Sea water. These results suggest that previous Sargasso Sea records of this species were due to the accidental sampling of rings, the presence of which were unknown. Almost certainly the historical data that form the base for most mega-scale distribution

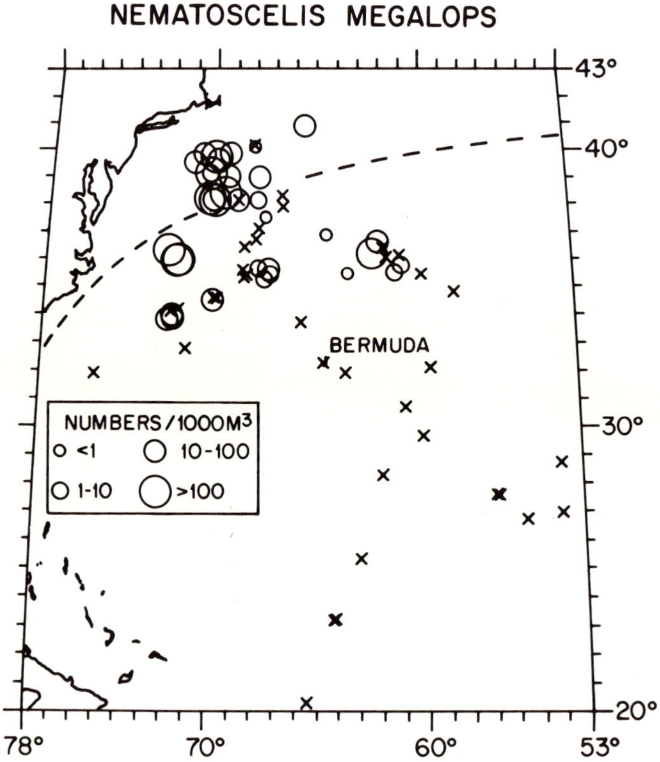


Figure 7. Distribution of *N. megalops* in the upper 750-1000 m. X's indicate samples with no *N. megalops*. The dotted line indicates the mean axis of the Gulf Stream. The samples with *N. megalops* clustered between 68° to 71°W and 37° to 39°N were all taken within the Slope Water. All other positive records to the south of the Gulf Stream axis were from Gulf Stream rings except locations A and B where single individuals were caught in the wake of a ring. Rings are numbered in the order in which they were sampled. Numbers 5a, b and c correspond to the location of ring D at successive sampling times. The X's which appear in ring 5b are actually negative records from collections made when the ring was at location 5a (from Wiebe and Boyd, 1978).

charts for oceanic species is contaminated by expatriates transported from one hydrographic regime into another in meso-scale features such as rings.

High-frequency internal waves. Internal waves, from tidal frequencies to periods of a few minutes, have been invoked as a mechanism for both the generation of plankton patchiness (e.g., Schröder, 1961, 1962; Kamykowski, 1974, 1976; Haury, 1976a), and for the cause of the high variability seen in replicate or repeated net tows (Angel, 1976; Haury, 1976a). As a second example of a physical factor contributing to pattern, we will describe an experiment[2] designed to investigate the interaction of high-frequency internal waves, plankton distribution, and sampling techniques. Specifically, the experiment looked at these interactions in Massachusetts Bay, U.S.A., where internal waves of 6-8 min period, 200 m wavelength, occur on a semi-diurnal cycle controlled by the barotropic tide (Halpern, 1971). The waves occur as discrete sets (packets), one packet per tidal cycle, that propagate westward at about one knot across the Bay from the shallow sill (Stellwagen Bank) marking the eastern boundary. The water column between internal wave packets is relatively quiescent.

The biological objective of the experiment was to look at phytoplankton and zooplankton horizontal and vertical distribution before, during and after the passage of the internal waves through the same water mass. By using different sampling strategies, we hoped to be able to determine the effects of the internal waves on the plankton distributions as well as whether or not the zooplankton reacted actively to the rapid vertical displacement to create new patterns.

The biological sampling was done with a Longhurst-Hardy Plankton Recorder (LHPR) used without a net (Haury *et al.*, 1976) and with a sampling interval of 16 sec. Mounted on the Plankton Recorder was an *in situ* flow-through fluorometer and a Neil Brown CTD (conductivity, temperature, depth) Microprofiler. For most of the biological sampling, simultaneous 200 and 30 kHz echo-sounder records were taken.

Three sampling techniques were used -- "Yo-Yoing" of the CTD-fluorometer combination to give a series of rapid vertical profiles during the wave passage and towing the LHPR-CTD fluorometer either at a fixed depth or along a fixed isotherm.

---

[2]The research discussed here was part of an interdisciplinary experiment conducted with Melbourne G. Briscoe and Marshall H. Orr, Departments of Physical Oceanography and Ocean Engineering, respectively, of the Woods Hole Oceanographic Institution.

Figure 8, based on a Yo-Yo series, illustrates an internal wave packet passing the drifting ship and its effect on temperature and phytoplankton distribution. Note the increase in mean depth of the chlorophyll maximum after the waves arrive. The depth of the maximum gradually decreases through time until the next set of internal waves arrive. This is a graphic illustration on both short (10 min) and long (semi-diurnal) time periods of the conditions modelled by Kamykowski (1974). Also significant are the overturning events (at times 1132 and 1139) which mix that portion of the water column in the region of the breaking wave. Close to shore we have observed the complete breakdown of the internal waves as they reach shoal (∿30 m) depths. This appears to mix the entire water column and possibly resuspends bottom sediments and nutrients. These data suggest that mixing caused by breaking internal waves can be one of the important enrichment mechanisms responsible for the high productivity of many continental shelf areas and around islands.

Figure 9 illustrates temperature, salinity, chlorophyll fluorescence and abundance of three categories of zooplankton taken at 17 m during a tow 960 m long just after the arrival of the leading edge of the internal wave packet. These data, together with that contained in Figure 8, graphically show how vertical distributions are affected and how the thickness of layers of water between isotherms can change. This latter process would make any set of integrating, repeated, tows whether oblique, vertical or horizontal, extremely variable, even if there were no "real" (i.e. changes in numbers/$m^3$) biological patchiness present.

The results of a similar tow taken just below the surface are shown in Figure 10. This tow began before the arrival of the wave packet and extended through several of the wave crests. There are two points to be made from these data: 1) deeper-living species are brought to the surface waters in highly localized regions (e.g., *Calanus* copepodites) and 2) the patchiness of abundance in different species manifests itself in different ways and to different degrees (e.g., appendicularians seem little affected by the internal waves, while *Temora* and *Centropages* show similar trends except at the end of the tow).

These two examples, which illustrate how physical factors provide a dominant force shaping the biological character of pattern, are obviously not unique. A summary of other factors which have been recognized are listed in Table 2 along with selected literature describing the factors and their biological effects. However, as we have tried to emphasize in the layout of the diagram in Figure 5, real pattern is distorted in one way or another by the sampling filter through which it is viewed. Thus

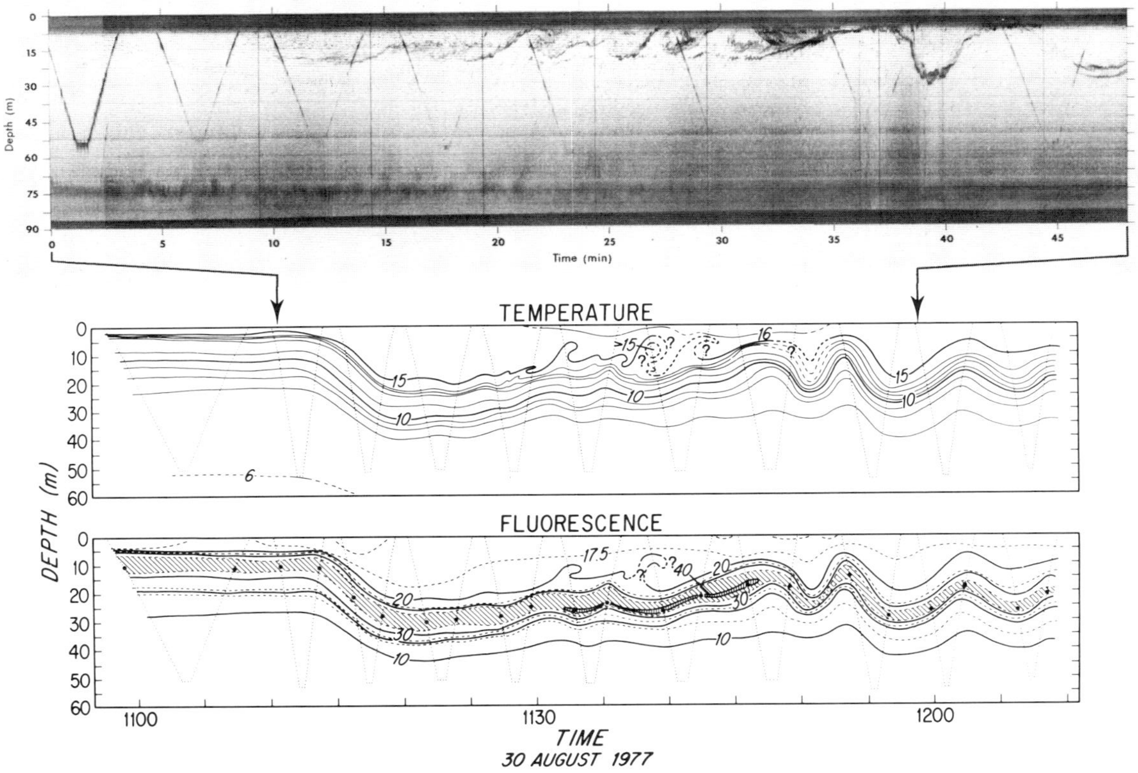


Figure 8. Internal wave packet in Massachusetts Bay. The 200 kHz acoustic back-scattering image (top) shows portions of the "Yo-Yo" profiles of the CTD - *in situ* fluorometer instrument. Vertical exaggeration is about 2x for the temperature and fluorescence profiles; the isotherms and "isoflours" are drawn between profile data points using the acoustic imagery as a guide. The wide hatching denotes fluorescence greater than 37.5 units, the narrow hatching greater than 40.0 units. The large dots within the hatched areas show depth of chlorophyll maximum (Orr, Briscoe and Haury, in prep.).

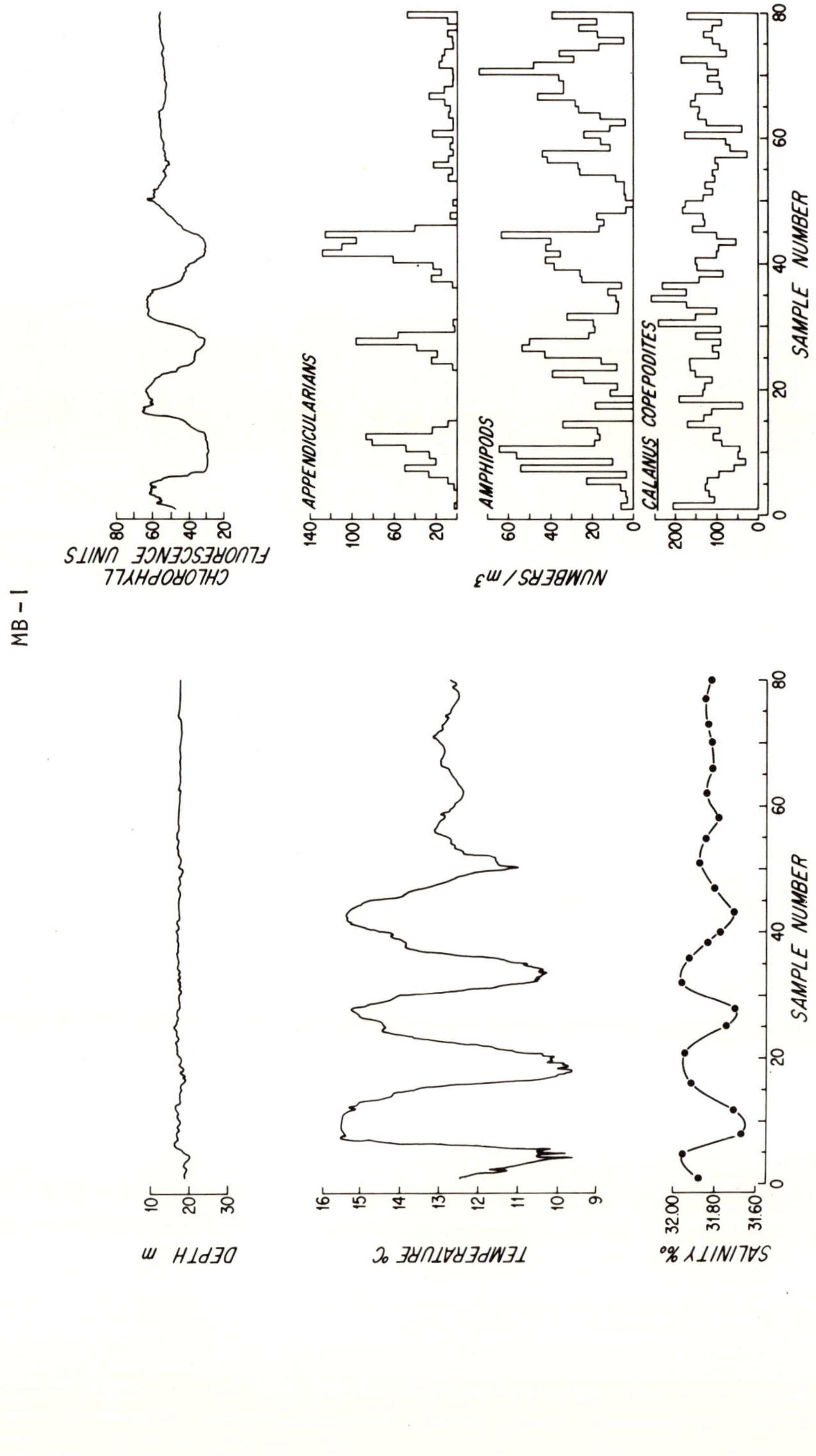


Figure 9. Data from LHPR tow through an internal wave packet; the record begins just after the leading edge (first trough) of the packet passed the ship. The direction of tow was into the wave packet propagation. Tow length 980 meters; mean volume filtered per sample 0.25 $m^3$; mean sample length 12.2 meters.

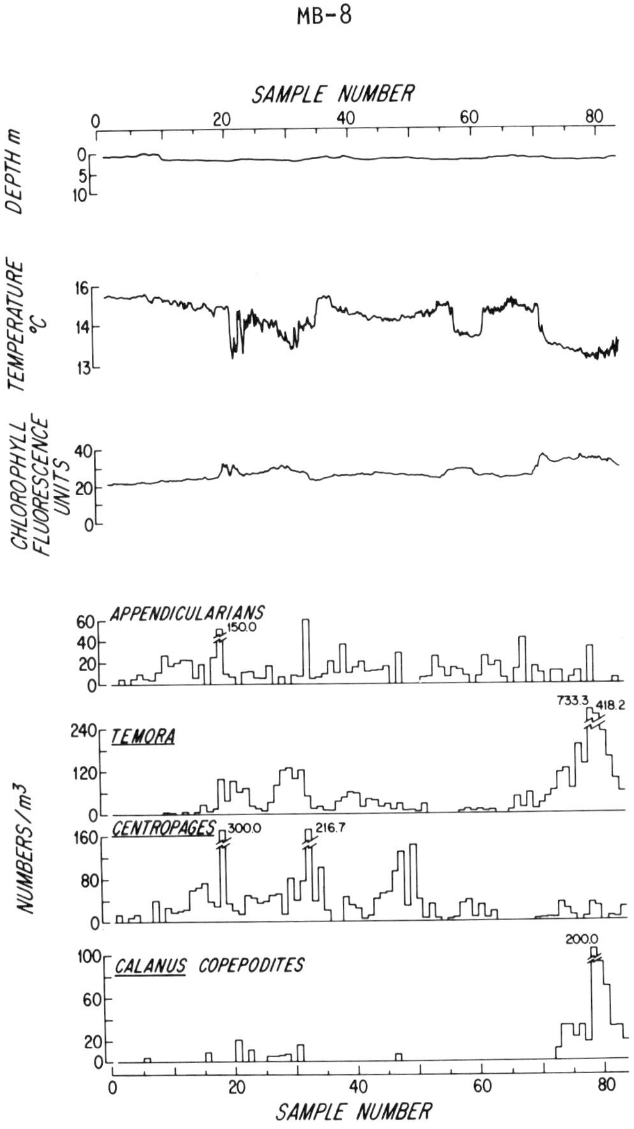


Figure 10. Surface tow (nominal depth 1.5 m) through an internal wave packet; the packet arrival is at sample number 10. Towing direction was into the wave packet propagation. Tow length 750 meters, mean volume filtered per sample 0.17 $m^3$, mean sample length 9.0 meters.

Table 2. Selected references to field observations of the biological effects resulting from dynamic physical processes.

| | |
|---|---|
| Gyres | McGowan, 1971; Backus *et al.*, 1977; Reid *et al.*, in press. |
| Currents | McGowan, 1971, 1974. |
| Eddies | Brinton, 1976; Omori and Gluck, in press. |
| Rings | Wiebe *et al.*, 1976; Wiebe and Boyd, 1978; Chamberlain, 1977. |
| Upwelling | Cowles *et al.*, 1977; Walsh *et al.*, 1977. |
| River Plumes | Calef and Grice, 1967; Kinzer, 1969; Pearcy and Mueller, 1969. |
| Island Wakes | Sette, 1955; Uda and Ishino, 1958. |
| Internal Tides | Kamykowski, 1974, 1976. |
| Billow Turbulence | Schröder, 1961. |
| Seiches | Colebrook, 1960. |
| Oceanic Fronts | Uda, 1959, 1973; Colton *et al.*, 1975; Pingree *et al.*, 1976; Savidge, 1976. |
| Vertical Shears | Angel, 1968. |
| Internal Waves | Schröder, 1962; Fisher and Squire, 1975. |
| Convective Cells | Owen, 1966; Stavn, 1971; George and Edwards, 1973. |

observational factors may play as important a role in determining the nature of the pattern observed as the more fundamental causal factors.

## Observational Factors

We consider here some of the ways in which the collection and interpretation of field data affect pattern. Included in these (Figure 5) are sample program design (the where, when and how often), mechanical aspects dealing with the sampling device, mechanical aspects of laboratory analysis, and the final statistical manipulation of the data.

There are many references dealing with aspects of sampling the plankton (e.g., Cassie, 1971; UNESCO, 1968, 1976; Kelley, 1976; Angel, 1976). All of these works emphasize some factors and neglect others. We will do the same, selecting those which seem to us to be emphasized by the Stommel Diagram and neglected by general works on sampling.

Sampling design - ecological justifications. We have tried to emphasize that all the space and time scales of pattern interact. Thus the choice of scales for investigation is very much dependent upon the question being asked. There are some problems where it is clear which scales are important. For example, if populations (composed of potentially interbreeding individuals (Mayr, 1966)) are the objects of interest then the spatial dimensions of the populations indicate one scale of aggregation which is of importance. However, since many populations exist on scales of 100's to 1000's of kilometers, the temporal scale is very broad, and one might conclude that virtually all phenomena on these scales are of significance. That is, geostrophic eddies, rings, upwelling, annual cycles, diel vertical migration are not only important, but interact in influencing populations. This is not news to most plankton ecologists. Of more interest would be a set of criteria by which to choose the most revealing time-space scales:

1) Since the study of the causes of variation in population size is one of the chief objectives in ecological research, one reasonable approach is to select for study those scales where the variation seems to be the greatest. In our diagram the area between 50 and 1000 km and one half to 80 years is such a scale. This scale is remarkably similar to the one of Stommel's showing the most "energetic" part of the ocean. Further, it is about the size of many populations or stocks of planktivorous fish. This may not be purely coincidental.

2) While many ecological studies have been done on the smaller end of the time scale we show, very little attention has been devoted to the larger (>20 years). The few long term studies mentioned previously (e.g., Williamson, 1961; Colebrook, 1972, 1977; Soutar and Isaacs, 1974; Cushing and Dickson, 1976; Tont, 1976) all indicate important ecological events occur within this range. By selecting the larger scales for study we may learn about aspects of ecology that cannot be understood by studies made on the smaller of the scales.

3) Stommel (1963) has emphasized when variables of different time-space scales interact, it is equally, if not more important

to determine the scales of interaction as well as the scales of the variables themselves. This is a decidedly more difficult task, and it is not immediately obvious what parameters, direct or derived, should be measured or calculated and if, in fact, they can be measured. One can imagine, for example, measures of scales of vertical and horizontal turbulent diffusion rates being combined with scales of local phytoplankton species growth rates to produce a parameter which can be associated with the time and space distribution of the potential for primary production.

4) An additional criterion involves the selection of time and space scales appropriate to the study of the longevity of individual micro- and coarse-scale patches, an attribute of patchiness about which there is little and conflicting evidence. In addition to longevity, the ecological importance of patchiness will depend on the scales of the daily ambit (or the ambit of some larger, but significant fraction of generation time), the distance between patches, the degree of patchiness, and the species structure of patches.

Serious measurement problems arise as a result of the spatial and temporal complexity of open ocean ecosystems. While the detection of micro-scale and larger patches is possible, it is obviously difficult to map the dimensions of these (especially the smaller ones) and to obtain a time-series of measurements of ecological interactions within a particular patch. If patches of these dimensions differ significantly from one another, with respect to the relative proportion of predator and prey or competitors, and if all or most of the interactions take place within patches, then it is obviously important to stake out an individual patch, follow it for an ecologically significant period of time and to make repeated measurements of the populations to detect such things as grazing rates, growth rates, survivorship and measures of successional change. This sort of a study may be impossible for fine- and micro-scale patches. If, on the other hand, the patches in a given area tend to resemble each other with respect to the relative proportions of predator, prey and competitors (i.e. species), we might take this as evidence that each patch has had a similar history and therefore might wish to assume that each will have a similar future. In this case it will be only necessary to take each of our time series of samples in a patch, any patch, to detect the sorts of ecological interactions we wish to study. This is exactly what Savage and Wimpenny (1936), claim to have done in following coarse-scale patches in the English Channel and the North Sea.

Sampling design - sources of error. The partitioning of sampling effort in space and time gives rise to several significant sources of error. The production of incorrect cycles of time or space fluctuations (aliasing) through choosing too large a sampling interval has been discussed by Platt and Denman (1975) and Kelley (1976). This hazard, emphasized by the Stommel Diagram, involves not only the interaction of sampling interval with a known cycle, but also the contamination of the known cycle by additional cycles of unknown period, magnitude, and phase. A simple example of this would be a study of diel vertical migration with sampling times based only on the day-night light cycle at a particular time of year. Neglecting the effects of internal waves of periods from a few minutes to many hours could result in a significant contamination of a desired diel signal.

Cycles with periods or wavelengths longer than the total time or distance over which samples are taken can also bias results. In the above example, the longer cycle of internal tides could extend the potential for interactions with sampling scale to several weeks.

There are two different aspects associated with the significance of contamination of one scale by those above and below. The first and least biologically interesting situation is the case where the various scales are causally unrelated. This affects our observations only through the choice of sampling intervals. The second more biologically important case is where biological and physical interactions between scales create changes in the scales themselves. This is a much more complex situation to sample adequately. It is difficult with our present state of knowledge to distinguish, much less predict, which kinds of interactions are occurring at any place or time, although there are some clear examples, such as Kamykowski's (1974, 1976) studies of the interaction of internal tides with phytoplankton distribution and productivity. The mathematical models of Okubo (1974), Steele (1974) and Levin and Segel (1976) suggest that the latter processes can occur, may be ecologically significant, and indicate a few of the parameters which may be important in causing the interactions between biologically relevant scales.

Stommel (1963) bases two additional comments on the contamination of sampling efforts by space and time scales of variation and the interaction of scales. 1) The noise level present will depend on the kind of variable being measured. Thus the choice of variable to investigate a particular phenomenon may determine whether or not a meaningful signal is observed. The simplest example of this concerns assessment of plankton in terms of biomass or species counts. Biomass measures are usually much less variable than individual species counts. Thus changes in abundance in time or space should be much clearer using biomass measures. However, these are probably less interesting biologically than species variations. 2) Stommel points out that the use of parameters such as correlation coefficients, derived from the observed variables, makes the likelihood of contamination much greater. The use of such parameters effectively reduces the signal-to-noise ratio, with the consequent difficulty in assigning levels of significance to results.

Laboratory analysis. Limitations on interpretation of field data occur at two different stages in the laboratory treatment of collected samples. The first involves the selection of what to count or measure, the second involves the choice of analytical methods to use on the data.

In some cases, the decision as to what will be measured is determined by the sampler; for instance, fluorometers or particle counters which give measures of total biomass. Unless some supplemental samples are taken to augment the crude measures of biomass, these techniques suffer not only from the inherent and obvious problem of how good a measure is the electronic analog signal of what is in the ocean, but also from the problem of not being able to determine the make-up of the biomass. Thus chlorophyll *a* estimates do not determine if the phytoplankton are in the form of coccolithophorids, diatoms or blue-green algae; and estimates of the biomass of organisms 10 cm long do not differentiate between salps and sardines. This lack of information content in biomass data makes it obvious that, if reproductive and social factors are involved in the generation and maintenance of patches, studies of patterns of biomass are inappropriate.

Most other sampling methods give one the choice in the laboratory, however, and the trade-off is between economics, impatience and potential ecological knowledge, usually, impatience and economics win. While biomass knowledge can be useful, it can give a very different picture of what is occurring in the ocean on various time and space scales. The emphasis recently on spectral techniques requires rapid and remote methods of data collection. It is this technique which ultimately will give us an accurate, quantitative version of our Stommel Diagram. Both this ideal diagram, and our crude qualitative one, however, will be very different from one illustrating the biomass variability of a particular species. It seems to us that it is how the various species variabilities interact and fit together to produce the total biomass variability illustrated in the ideal Stommel Diagram that is the essential element to be understood in marine ecology. In a sense, this is what is meant by our comments on measuring scales of interaction of variables.

There is a second motivation for looking at species and not biomass in addition to the acquisition of ecological knowledge about natural populations. This motivation arises from the need to integrate the modern theoretical treatment of ecosystem function with field observations. The recent collection within one volume of a number of reviews of models of population dynamics within patchy environments (Covich, 1976; Levin, 1976; Wiens, 1976; see also for zooplankton, Steele and Mullin, 1977), give ample demonstration of how species information is required and can be used for these models. The application of these models to the oceanic environment will necessitate the collection of environmental and species data for long periods of time (Wiens, 1977).

The nature of data taken to elucidate time and/or space patterns of plankton distributions and their causes requires the analysis of large, not necessarily independent, sets of data made up of many variables. There are many simple methods, as well as complex multivariate techniques that can be used to look at these data, the appropriateness of which are, in many cases, determined by the questions being asked, the frequency with which the data was collected and the particular scales of known (and unknown) pattern in the data. Since the emphasis in this paper has been on the smaller scales of pattern, we will address in particular a few of the techniques which have been used to look for pattern on scales from meters to a few ten's of kilometers. The common element of all we discuss will be how the techniques related to what we believe is the reality of the continuum nature of scales illustrated in the Stommel Diagram.

We have already pointed out some of the undesirable features of fitting data to theoretical model distributions. In terms of gaining information about patches, this technique, by itself, reveals nothing of patch parameters or their position in space; sampling on any scale gives you essentially the same kind of information. The method of Fasham *et al.* (1974), based on the Skellam model (Skellam, 1952), while it provides estimates of three important patch parameters (patch size, numbers of organisms per patch and number of patch centers per sample), has several shortcomings. As Fasham *et al.* (1974) emphasize, several different models fit the data well, thus the confidence limits of "ecological relevance" are wide, even if the statistical confidence limits are not. Second, it reduces to a single number the possible wide range of values a patch parameter could have in a single data set. The method also reveals little of how the animals are organized in space relative to each other.

Limitations of the application of spectral analysis to biological data can be broken down into three broad categories: 1) the assumptions of spectral analysis (normal data, stationarity; see Platt and Denman, 1975) are generally not rigorously met by marine plankton data sets. Non-normal and non-stationary data appear to be the rule on all scales. 2) Spectral analysis requires relatively long time and/or space series of data that are only now becoming feasible to obtain. However, current methods only give information on the gross features of ecosystem components (phytoplankton and zooplankton biomass or particle size) and not on species distributions. 3) Spectral analysis alone does not give information on the physical arrangement in space of patches which may be identified as being characteristic of the data set by means of peaks in a spectrum. If there is no peak in a spectrum, no characteristic or dominant wavelength is presumed to be present; thus the slope remains as the sole information on which to infer cause. Yet there are difficulties in distinguishing significant differences in slopes of spectra and in assigning a causal mechanism for the value of the observed slope (see Fasham, this volume). Even when there are strong spectral peaks in biological data, their cause will be difficult to interpret. In our example, Figure 1, the spectral peaks coincide closely with the spectral peaks in the analogous diagram for physical data presented by Stommel (1963, Figures 1, 3; 1965, Figure 9). This correlation is not sufficient evidence to conclude the cause of the patchiness is solely physical.

The use of cross-spectra (Platt and Denman, 1975; Denman, 1975) offers the potential for indicating possible causal relationships between variables. This seems of limited use, however, until the time when oceanic samples can be obtained and analyzed in sufficient numbers and detail (i.e. to species) to give relationships between individual species and between a species and its physical environment.

The techniques of defining patches of Cushing and Tungate (1963), Wiebe (1970) and Ziemann (1970) appear to offer the best way to get ecological information from sets of data, although they are all subject to such criticisms as those of Fasham *et al.* (1974). Other methods, including those discussed above, all have useful attributes. Certainly each has its place in the analysis of data. In view of the profusion of techniques, it would seem appropriate that someone apply all of them to the same data set to determine which is the most revealing, as well as giving a much-needed estimate of the "analytical" variability possible in what are already highly variable biological data.

## ECOLOGICAL SIGNIFICANCE

In the preceding sections we have been concerned mainly with the spatial dimensions of patchiness structure, the time frames in which it occurs, the biological and physical factors which can be identified as important in the generation and maintenance of patches and some of the observational factors which distort the real patterns. We turn briefly now to aspects of the ecological and evolutionary significance of patchiness.

In all the measures of patchiness, there are a number of variables or attributes of interest in addition to the spatial dimensions of the patches. These include patch intensity, spacing, species composition and longevity. It is especially important that these attributes be considered in relation to some notion of an ecologically important ambit (associated with the distances traveled over days, weeks, lifetimes) of the organisms whose populations are patchy or those that depend on finding patches of food.

Patches may differ from background levels of abundance by factors of 2, 10, 100 times or more so that the degree or intensity of patchiness (Pielou, 1974) may change without a patch scale change. The ecological consequences of degrees of patchiness of 2, 10 or 100 times are, of course, not the same so that the concept of intensity becomes quite important. For example, in an area of 34,000 $km^2$ within the California Current, monthly zooplankton biomass measures taken over a period of three years (1956-1958) show that for as long as three months the indices of dispersion for some years are clearly much greater than other years (Figure 11). Indeed, in one of these years, 1958, the indices were systematically lower for almost the entire year. This index could be influenced by several patch parameters, but it appears that a change in the intensity of coarse-scale patchiness, rather than a scale change, was involved in this case. One may infer this from contoured maps of the areal distribution of biomass from the same cruises (Figures 12 and 13). An examination of these maps gives the impression that the heterogeneity (or patch spatial dimensions) was about the same in July of 1956 as it was in July of 1958. However, the contrast in shading of contour levels was much greater in 1956 than in 1958. It appears that zooplankton were not only generally more abundant in 1956, but the intensity of patchiness was greater. While these data were not taken for and are not well suited to the purpose of examining the temporal changes in patch intensity, at least they indicate that an event of ecological importance to the plankton communities in this region occurred on these scales.

Year-to-year variations in the intensity of patchiness over large areas should have a major impact on the survival of herbivores or carnivores. It can also have significant impact on smaller scales. In a restricted sea area (fine-scale), Mullin and Brooks (1976) showed that a sizable proportion of the population of *Calanus pacificus* was found in water where the standing crops of phytoplankton were insufficient to meet their metabolic requirements. Ultimately, the ecological consequences of this kind of observation depend on whether the degree or intensity of phytoplankton and zooplankton patchiness observed extends over a much larger region, and whether there is net growth of the population as a whole under these circumstances. Mullin and Brooks (1976)

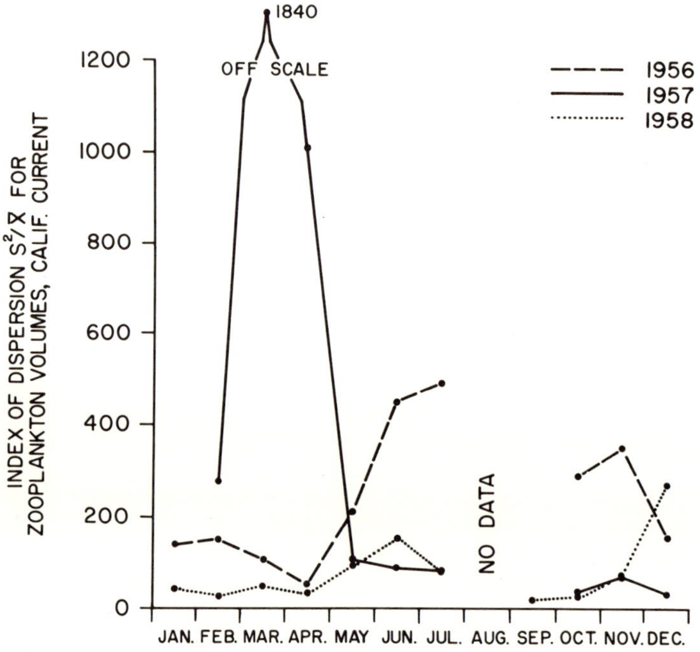


Figure 11. The seasonal indices of dispersion for zooplankton volumes in the central sector of the California Current, an area of about 1.9 x $10^5$ $km^2$ from Pt. Conception to the Mexican border (see Figures 12 and 13). Average number of samples per month = 49.

also point out that it depends on the temporal persistence of the pattern. As this example emphasizes, the time-space dimensions and intensity of patchiness on the scale of what we think may be an encounter ambit of individuals are obviously of importance. But what do we learn of the fate of entire populations from studies done only on the smaller scales? We know that on larger scales of aggregation ecologically important events do happen. For example, the demise of the California sardine and its replacement by the anchovy took place over a large area. Thus while the interactions between individual larvae may have taken place on the coarse- or fine-scale, apparently similar coarse- or fine-scale interactions were going on everywhere over a large area and affected two entire populations.

While it appears that there is an ecological scale of patchiness that involves entire populations, determining these larger scales is uncertain because of the difficulties and ambiguities in defining what we mean by "populations" (Pielou, 1974, pp. 98-99). In general, when we speak of populations in the sea, we are concerned with assemblages of individuals that occupy spatial scales

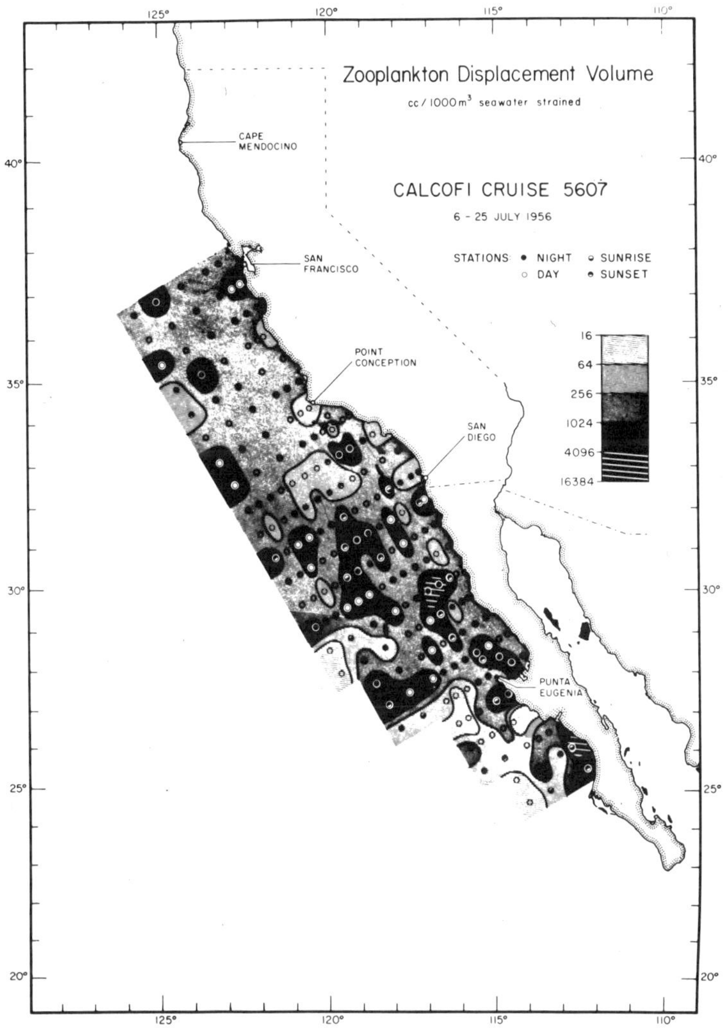


Figure 12. Zooplankton displacement volumes in the California Current in July 1956 (about 8.27 x $10^5$ $km^2$). Oblique, 0-200 m tows were taken with a 1 m diameter net of 0.505 mm mesh.

of at least several hundred kilometers and within which the flow of genetic material is possible over only a few generations.

The distance between patches is also of ecological importance since it very much affects our thinking about the relationship between predators seeking out patches of their prey and the distance apart and degree of patchiness of the prey. If the distances are great and the intensity of patchiness small, then predators

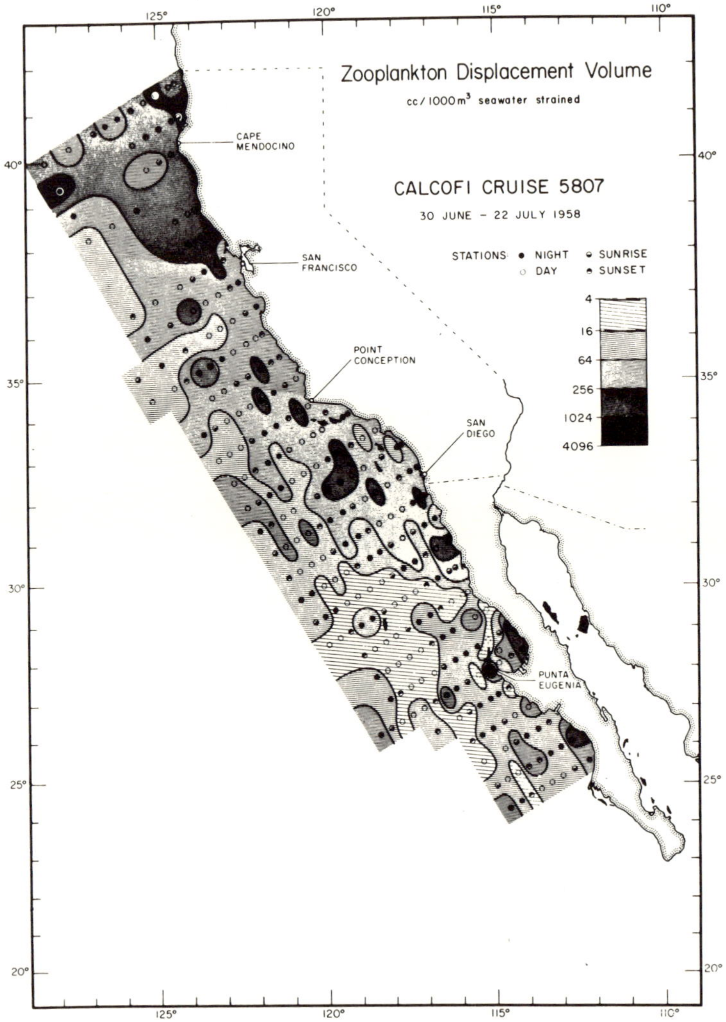


Figure 13. Zooplankton displacement volumes in the California Current in July 1958 (about 1.15 x $10^6$ $km^2$). Oblique, 0-200 m tows were taken with a 1 m diameter net of 0.505 mm mesh.

expend much energy for a relatively small return. Patch spacing also determines the effectiveness of patches in reducing mortality of prey due to predation. Brock and Riffenburgh (1960) have demonstrated theoretically that for patches to be effective against predators, the distance between patches must be larger than the predator sighting distance. For zooplankton and planktivorous fish, both of which are predator and prey at the same time, it is difficult to visualize how the patch spacing requirement associated with being a prey is balanced against the spacing requirement associated with being a predator. Horwood and Cushing (this volume) discuss other aspects of this problem.

Competition and resource utilization theory (e.g., Horn and MacArthur, 1972; Levin, 1974) are affected by whether we conceive of patches as being multispecies or essentially single-species in nature as well as by the intensity and spacing of patchiness. If we believe patches to be multispecies, it is necessary to know the degree to which they resemble one another in terms of species content and species proportions. For example, if one patch has a proportionately greater abundance of chaetognath predators relative to fish larvae prey than another patch of similar dimensions and biomass, then the survival of both organisms might be quite different in the two patches. If there are two or more chaetognath species present per patch, then the consequences may be quite different if one is the better competitor.

Concomitant with the species structure of patches and its importance to species interaction is the temporal persistence or longevity of individual patches. This is not an obvious factor on the land or benthos, for variations in soil type, substrate and microclimate are patchy and relatively long lasting and can be the cause of clumps and patches of attached plants and sessile animals. However, in the pelagic realm there is very little of this sort of physical substrate structure. Eddies or small scale inhomogeneities are, of course, always present but the time constants of eddy turbulence is very much scale dependent. Therefore, while there always are eddies present, the smaller ones do not last very long. If these are the habitat features that mechanically aggregate plankton or are used as cues by plankton to aggregate, then these small patches should be quite ephemeral. Further, if one wishes to describe what sort of average biotic habitat an organism experiences during a significant portion of its life (looking at the ocean from the organism's viewpoint), one must recognize that the smaller scale patches may have a shorter life span than do the constituent organisms and that some, perhaps many, individuals may wander from patch to patch. In this case, what the individual experiences is not merely what is in a certain patch, but rather some sort of integration of the patchiness field it encounters over a significant period of time. Miller (1970) has reported an experiment employing a migrating drogue which was designed to look at a migrating zooplankter's ambit and its biotic experience over a period of days. Miller found the proportions of species encountered through time changed significantly. Assuming the drogue accurately reflected the conditions encountered by a migrating zooplankter, patches on this scale (fine-scale) are variable through time and are multispecies structures. Similarly, Haury (1976a,b) presented results which suggested that community structure varied over distances of 100's to 1000's of meters.

While it was not possible to determine temporal aspects (over days), measures of both distance and day-night differences gave evidence of multispecies, dielly variable structure.

We have been considering the ecological significance of particular aspects of community structure characterized by the components of patchiness (i.e. patch size, intensity, spacing, composition, longevity). In a broader perspective, all these patch characteristics are variable through time as a consequence of random physical and biological perturbations. The interactions of the time scales of these disturbances with the space scales of patchiness will determine the effective environment noise level (e.g., frequency and amplitude of variations in temperature, mixing intensity, light, predator or prey abundance) seen by species populations (Crowley, 1977). As examples of this interaction, we return to the illustrations given previously of physical factors generating patchiness, i.e. Gulf Stream rings and high-frequency internal waves.

To populations of Slope Water and Sargasso Sea species, any single warm or cold core ring is an unpredictable event which by itself would have little effect on the total Sargasso Sea Slope Water population. However, the fact that there are 10 to 15 cold core rings present in the northern Sargasso Sea at any one time suggests that significant effects on productivity and abundance levels of populations would be expected. There should be similar effects on Slope Water populations from multiple incursions of warm core rings.

The division of the Sargasso Sea into northern and southern sections on the basis of the presence and relative abundance of species (Backus *et al.*, 1970; Colton *et al.*, 1975) may be, in part, due to the presence of cold core rings in the northern Sargasso Sea. Moreover, the effects of rings could be time-varying if, as is likely, the number of rings formed per year varies over time scales of years to decades or longer.

The high frequency internal waves occurring in Massachusetts Bay are deterministic in their occurrence in this area due to their formation at a single source. But the intensity of their effects will depend on stochastic events like weather, changes in the density structure of the water column, and also the random nature of the place and time when an internal wave may break. The waves in the Bay probably have little effect on the total plankton community of the Bay because of their highly localized occurrence. On continental shelf regions throughout the world, however, internal waves are a regular feature, impinging on any given area from several directions (Apel *et al.*, 1975). Thus these areally-extensive, interacting waves could be a major factor affecting the distribution and abundance of shelf populations as a whole.

## CLOSING COMMENTS

We have stressed that a knowledge of the scale of patchiness relative to the known or assumed ambits of the target organisms, the degree or intensity of patchiness, the degree to which patch species composition resembles each other, and the temporal longevity of patches are all quite important attributes of the general patchiness problem. These are aspects of patchiness which have not been well studied in the past and it seems evident that studies centered on them will provide useful information in the future. Information on these subjects should allow for better conceptual models of the manner in which important ecological interactions take place in the pelagic realm. Further, it is necessary that consideration be given to which scales of patchiness are of significance to the ecological problem we wish to study.

There are clearly a variety of causes of both physical and biological origin, acting both in concert and independently to shape the scale and temporal longevity of the patch. In the study of the factors which are important in the generation and maintenance of patchiness, one needs to determine the appropriateness of the physical site in which the work is to be done. If the focus of a study is on the patchiness of a particular subpopulation, then regions where larger scale advective processes act to move and mix subpopulations past a geographical location over a time which is short relative to the course of the experiment are obviously not desirable. As we have tried to illustrate, it is possible to choose geographic or hydrographic sites where a particular factor is the dominant element responsible for the overt pattern. Thus studies aimed at the determination of the causes of patchiness and its maintenance can be designed for specific space-time scales. Our concepts of the relationships between spatial and temporal dimensions are therefore very important not only to the design of observational programs but also to our ability to generate testable hypotheses about the ecological significance of plankton patchiness. Finally, there are many aspects of ecosystem structure and function which require knowledge about the ecology of individual species. At this level of complexity, studies of biomass or particle size distributions are inappropriate.

## SUMMARY

The time and space scales of oceanic plankton patterns are discussed using as a conceptual framework a biological modification of the three-dimensional diagrams presented by Stommel (1963). Our diagram plots biomass variability against time (1 sec to $10^4$ years) and space (1 cm to $10^4$ km). The diagram illustrates the overlapping hierarchy of patterns present on the diagonal from small to large time and space scales. There are regions present on this diagonal of particularly large variability; these define focal points for future investigations into the causal processes associated with these patterns. While the spatial patterns in the diagram are based on total plankton biomass, emphasis is placed on the importance of determining patterns at the species (or functional) level. Space scales of pattern are categorized (micro, fine, coarse, meso, macro, mega) and examples given of plankton and nekton distribution for the various scales. The generation, maintenance and observation of plankton patchiness are discussed and related to the ecological significance of scales of pattern (less than 1000 km) associated with the ambits of zooplankton and nekton.

## ACKNOWLEDGEMENTS

We are indebted to P. A. Bernal for pointing out the original Stommel Diagram, for suggesting that a similar one might be made for zooplankton biomass, and for providing the index of dispersion data (Figure 11). Discussions with M. G. Briscoe were helpful in the construction and interpretation of the Stommel Diagram. We thank M. Orr for allowing us to use his acoustic imagery data in Figure 8.

L. R. Haury and P. H. Wiebe were supported by National Science Foundation Grants DES74-02783, OCE77-08682 and OCE77-09132; J. A. McGowan's research has been supported by the Marine Life Research Group of the State of California, the Office of Naval Research and the National Science Foundation. Contribution No. 4051 from the Woods Hole Oceanographic Institution, Woods Hole, Massachusetts 02543.

## REFERENCES

Angel, M. V., 1968: The thermocline as an ecological boundary. Sarsia, 34, 299-312.

Angel, M. V., 1976: Windows into a sea of confusion: sampling limitations to the measurement of ecological parameters in oceanic mid-water environments. In Oceanic Sound Scattering Prediction, Anderson and Zahuranec (eds.), pp. 217-248, Plenum Press.

Apel, J. R., H. M. Bryne, J. R. Proni and R. L. Charnell, 1975: Observations of oceanic internal and surface waters from the Earth Resources Technology Satellite. J. Geophys. Res., 80, 865-881.

Backus, R. H., J. E. Craddock, R. L. Haedrich and D. L. Shores, 1970: The distribution of mesopelagic fishes in the equatorial and western North Atlantic Ocean. J. Mar. Res., 28, 179-201.

Backus, R. H., J. E. Craddock, R. L. Haedrich and B. H. Robinson, 1977: Atlantic mesopelagic zoogeography. In Fishes of the Western North Atlantic, Part 7, Memoir 1, pp. 266-287, Sears Foundation for Marine Research.

Bainbridge, R., 1952: Underwater observations on swimming of marine zooplankton. J. Mar. Biol. Ass. U.K., 31, 107-112.

Baker, A. de C., 1970: The vertical distribution of euphausiids near Fuerteventura, Canary Islands ("Discovery" SOND Cruise, 1965). J. Mar. Biol. Ass. U.K., 50, 301-342.

Barham, E. G., 1970: Deep-sea fishes: lethargy and vertical orientation. In Biological Sound Scattering in the Ocean, G. B. Farquhar (ed.), pp. 100-118, MC Report 005, U.S. Govt. Printing Office.

Barraclough, W. E., R. J. LeBrasseur and O. D. Kennedy, 1969: Shallow scattering layer in the subarctic Pacific Ocean: detection by high-frequency echo sounder. Science, 166, 611-613.

Beers, J. R., M. R. Stevenson, R. W. Eppley and E. R. Brooks, 1971: Plankton populations and upwelling off the coast of Peru, June 1969. Fish. Bull., 69, 859-876.

Beklemishev, K. V., 1969: Ecology and biogeography of the open sea. Naval Oceanogr. Off. Trans. NOOO T-25 1976.

Bernstein, R. L., L. Breaker and R. Whritner, 1977: California Current eddy formation: ship, air, and satellite results. Science, 145, 353-359.

Bliss, C. I. and R. A. Fisher, 1953: Fitting the negative binomial distribution to biological data. Biometrics, 9, 176-200.

Brinton, E., 1962: Variable factors affecting the apparent range and estimated concentration of euphausiids in the North Pacific. Pac. Sci., 14, 374-408.

Brinton, E., 1967: Distributional atlas of Euphausiacea (Crustacea) in the California Current region, Part I. CalCOFI Atlas No. 5, State of California Marine Research Committee.

Brinton, E., 1976: Population biology of Euphausia pacifica off Southern California. Fish. Bull., 74, 733-762.

Brock, V. E. and R. H. Riffenburgh, 1960: Fish schooling: a possible factor in reducing predation. J. Cons. int. Mer., 25, 307-317.

Calef, G. W. and G. D. Grice, 1967: Influence of the Amazon River outflow on the ecology of the western tropical Atlantic. II. Zooplankton abundance, copepod distribution, with remarks on the fauna of low-salinity areas. J. Mar. Res., 25, 84-94.

Cassie, R. M., 1959: Micro-distribution of plankton. N.Z.J. Sci., 2, 398-409.

Cassie, R. M., 1971: Sampling and Statistics. In A Manual on Methods for the Assessment of Secondary Productivity in Fresh Waters, W. T. Edmondson and G. G. Winberg (eds.), pp. 174-209, IBP Handbook No. 17, Blackwell Scientific Publications.

Chamberlin, J. L., 1977: Monitoring effects of Gulf Stream meanders and warm core eddies on the Continental Shelf and Slope. International Commission for the Northwest Atlantic Fisheries. Selected Papers No. 2, 1977, pp. 145-153.

Chanut, Jean-Pierre, 1975: Simulation d'un plan d'experience en oceanographie: effets de la dispersion des organisms sur l'erreur d'echantillonnage. J. Exp. Mar. Biol. Ecol., 17, 239-259.

Clutter, R. I., 1969: The microdistribution and social behavior of some pelagic mysid shrimps. J. Exp. Mar. Biol. Ecol., 3, 125-155.

Colebrook, J. M., 1972: Variability in the distribution and abundance of the plankton. ICNAF Spec. Publ. No. 8, pp. 167-186.

Colebrook, J. M., 1977: Annual fluctuations in biomass of taxonomic groups of zooplankton in the California Current, 1955-1959. Fish. Bull., 75, 357-368.

Colton, J. B., Jr., D. E. Smith and J. W. Jossi, 1975: Further observations on a thermal front in the Sargasso Sea. Deep-Sea Res., 22, 433-439.

Covich, A. P., 1976: Analyzing shapes of foraging areas: some ecological and economic theories. Ann. Rev. Ecol. System, 7, 235-257.

Cowles, T. J., R. T. Barber and O. Guillen, 1977: Biological consequences of the 1975 El Niño. Science, 195, 285-287.

Crowley, P. H., 1977: Spatially distributed stochasticity and the constancy of ecosystems. Bull. Math. Biol., 39, 157-166.

Cushing, D. H. and D. S. Tungate, 1963: Studies of a Calanus patch. I. The identification of a Calanus patch. J. Mar. Biol. Ass. U.K., 43, 327-337.

Cushing, D. H. and R. R. Dickson, 1976: The biological response in the sea to climatic changes. Adv. Mar. Biol., 14, 1-122.

Denman, K. L., 1975: Spectral analysis: a summary of the theory and techniques. Fish. Mar. Serv. Res. Dev. Tech. Rep. 539.

Denman, K. L. and M. Miyake, 1973: Upper layer modification at Ocean Station Papa: observations and simulation. J. Phys. Oceanogr., 3, 185-196.

Dubois, D. M., 1975a: Simulation of the spatial structuration of a patch of prey-predator plankton populations in the Southern Bight of the North Sea. Mem. Soc. Roy des Sci. Liege, 7, 75-82.

Dubois, D. M., 1975b: A model of patchiness for prey-predator plankton populations. Ecological Modelling, 1, 67-80.

Dugdale, R. C., J. J. Goering, R. T. Barber, R. L. Smith and T. T. Packard, 1977: Denitrification and hydrogen sulfide in the Peru upwelling region during 1976. Deep-Sea Res., 24, 601-608.

Emery, A. R., 1968: Preliminary observations on coral reef plankton. Limnol. Oceanogr., 13, 293-303.

Fasham, M.J.R., M. V. Angel and H.S.J. Roe, 1974: An investigation of the spatial pattern of zooplankton using the Longhurst-Hardy plankton recorder. J. Exp. Mar. Biol. Ecol., 16, 93-112.

Fasham, M.J.R. and P. R. Pugh, 1976: Observations on the horizontal coherence of chlorophyll a and temperature. Deep-Sea Res., 23, 527-538.

Fisher, F. H. and E. D. Squier, 1975: Observation of acoustic layering and internal waves with a narrow-beam 87.5 kHz echo sounder. J. Acoust. Soc. Am., 58, 1315-1317.

George, D. G. and R. W. Edwards, 1973: Daphnia distribution with Langmuir calculations. Limnol. Oceanogr., 18, 798-800.

Gopalakrishnan, K., 1974: Zoogeography of the genus Nematoscelis (Crustacea, Euphausiacea). Fish. Bull., 72, 1039-1074.

Halpern, D., 1971: Observations of short period internal waves in Massachusetts Bay. J. Mar. Res., 29, 116-132.

Hamner, W. M., L. P. Madin, A. L. Alldredge, R. W. Gilmer and P. P. Hamner, 1975: Underwater observations of gelatinous zooplankton: sampling problems, feeding biology, and behavior. Limnol. Oceanogr., 20, 907-917.

Hamner, P. and W. M. Hamner, 1977: Chemosensory tracking of scent trails by the planktonic shrimp Acetes sibogae australis. Science, 195, 886-888.

Hardy, A. C. and E. R. Gunther, 1935: The plankton of the South Georgia whaling grounds and adjacent waters, 1926-27. Discovery Rept., 11, 1-456.

Haury, L. R. 1976a: Small-scale pattern of a California Current zooplankton assemblage. Mar. Biol., 37, 137-157.

Haury, L. R., 1976b: A comparison of zooplankton patterns in the California Current and North Pacific central gyre. Mar. Biol., 37, 159-167.

Horn, H. S. and R. H. MacArthur, 1972: Competition among fugitive species in a harlequin environment. Ecology, 53, 749-752.

Hutchinson, G. E., 1953: The concept of pattern in ecology. Proc. Acad. Natur. Sci., Phila., 105, 1-12.

Iverson, R. L., 1977: Mesoscale oceanic phytoplankton patchiness caused by hurricane effects on nutrient distribution in the Gulf of Mexico. In Oceanic Sound Scattering Prediction, Anderson and Zahuranec (eds.), pp. 767-778, Plenum.

Kamykowski, D., 1974: Possible interactions between phytoplankton and semi-diurnal internal tides. J. Mar. Res., 32, 67-89.

Kamykowski, D., 1976: Possible interactions between plankton and semi-diurnal internal tides. II. Deep thermoclines and trophic effects. J. Mar. Res., 34, 499-509.

Kawamura, A., 1974: Food and feeding ecology in the Southern Sei whale. Sci. Rep. Whales Res. Inst., 26, 25-144.

Kelley, J. C., 1976: Sampling the sea. In The Ecology of the Seas, Cushing and Walsh (eds.), pp. 361-378, Saunders.

Kinzer, J., 1969: Quantitative distribution of zooplankton in surface waters of the Gulf of Guinea during August and September 1963. In Proceedings of the Symposium on the Oceanography and Fisheries Resources of the Tropical Atlantic, pp. 231-240, UNESCO.

Komaki, Y., 1967: On the surface swarming of euphausiid crustaceans. Pac. Sci., 21, 433-448.

Lai, D. V. and P. L. Richardson, 1977: Distribution and movement of Gulf Stream rings. J. Phys. Oceanogr., 7, 670-683.

Lasker, R., 1975: Field criteria for survival of anchovy larvae: the relation between inshore chlorophyll maximum layers and successful first feeding. Fish. Bull., 73, 453-462.

LeBrasseur, R. J., 1965: Seasonal and annual variations of net zooplankton at Ocean Station "P", 1956-1964. Fish. Res. Bd. Can., Oceanogr Limnol., M.S. Rep. 202, 163 pp.

Leetmaa, A., 1977: Effects of the winter of 1976-1977 on the Northwestern Sargasso Sea. Science, 198, 188-189.

Legeckis, R., 1977a: Oceanic polar front in the Drake Passage: satellite observations during 1976. Deep-Sea Res., 24, 701-704.

Legeckis, R., 1977b: Long waves in the eastern equatorial Pacific Ocean: a view from a geostationary satellite. Science, 197, 1179-1181.

Levin, S. A., 1976: Population dynamic models in heterogeneous environments. Ann. Rev. Ecol. Syst., 7, 287-310.

Levin, S. A. and L. A. Segel, 1976: Hypothesis for origin of planktonic patchiness. Nature, 259, 659.

Lloyd, M., 1967: Mean crowding. J. Anim. Ecol., 36, 1-30.

Love, C. M., 1971: Eastropac Atlas. Vol. 2, Circular 330, NOAA, Washington, D.C.

McCarthy, J. J. and D. Kamykowski, 1972: Urea and other nitrogenous nutrients in La Jolla Bay during February, March and April, 1970. Fish. Bull., 70, 1261-1274.

McGowan, J. A., 1967: Distributional atlas of pelagic molluscs in the California Current region. CalCOFI Atlas No. 6, State of California Marine Research Committee.

McGowan, J. A., 1971: Oceanic biogeography of the Pacific. In The Micropaleontology of the Oceans, B. M. Funnel and W. R. Riedel (eds.), pp. 3-74, Cambridge University Press.

McGowan, J. A., 1974: The nature of oceanic ecosystems. In The Biology of the Oceanic Pacific, C. B. Miller (ed.), pp. 9-28, Oregon State University Press.

Marr, J.W.S., 1962: The natural history and geography of the Antarctic Krill (Euphausia superba Dana). Discovery Rept., 32, 33-464.

Martin, J.H.A., 1972: Marine climatic changes in the north-east Atlantic, 1900-1966. Rapp. Cons. Explor. Mer., 162, 213-219.

Mauchline, J. and L. R. Fisher, 1969: The biology of euphausiids. Adv. Mar. Biol., 7, 454 pp.

Maul, G. A., D. R. Norris and W. F. Johnson, 1974: Satellite photography of eddies in the Gulf Loop current. Geophys. Res. Ltrs., 1, 256-258.

Mayr, E., 1966: Animal Species and Evolution. Harvard University Press.

Miller C. B., 1970: Some environmental consequences of vertical migration in marine zooplankton. Limnol. Oceanogr., 15, 727-741.

Mullin, M. M. and E. R. Brooks, 1976: Some consequences of distributional heterogeneity of phytoplankton and zooplankton. Limnol. Oceanogr., 21, 784-796.

Okubo, A., 1974: Diffusion-induced instability in model ecosystems: another possible explanation of patchiness. Chesapeake Bay Inst. Tech. Report 86, 16 pp.

Omori, M., 1974: The biology of pelagic shrimps in the ocean. Adv. Mar. Biol., 12, 233-324.

Omori, M. and D. Gluck, 1978: Life history and vertical migration of the pelagic shrimp Sergetes similis Hansen off the Southern California coast. Fish. Bull., in press.

Ortner, P. B., E. M. Hulburt and P. H. Wiebe, 1978: Phytohydrography and herbivore contrasts in the western North Atlantic. Deep-Sea Res., submitted.

Ortner, P. B., P. H. Wiebe, L. Haury and S. Boyd, 1978: Variability in zooplankton biomass distribution in the Northern Sargasso Sea: the contribution of Gulf Stream cold core rings. Fish. Bull., 76, 323-334.

Pearcy, W. G. and J. L. Mueller, 1969: Upwelling, Columbia River plume and albacore tuna. Proc. VI Internat. Symp. on Remote Sensing of the Environment, pp. 1101-1113.

Pielou, E. C., 1974: Population and Community Ecology. Gordon and Breach.

Pingree, R. D., P. M. Holligan, G. T. Mardell and R. N. Head, 1976: The influence of physical stability on spring, summer and autumn phytoplankton blooms in the Celtic Sea. J. Mar. Biol. Ass. U.K., 56, 845-873.

Platt, T., 1972: Local phytoplankton abundance and turbulence. Deep-Sea Res., 19, 183-197.

Platt, T. and K. L. Denman, 1975: Spectral analysis in ecology. A. Rev. Ecol. Syst., 6, 189-210.

Reid, J. L., E. Brinton, A. Fleminger, E. L. Venrick and J. A. McGowan, 1978: Ocean circulation and marine life. In Proc. Gen. Symp. Joint Oceanogr. Ass., Edinburgh, Scotland, 1976. H. Charnock and G. Deacon (eds.), Plenum.

Rhines, P., 1976: Physics of ocean eddies. Oceanus, 19, 26-39.

Richardson, P., 1976: Gulf Stream rings. Oceanus, 19, 65-68.

Riley, G. A., 1976: A model of plankton patchiness. Limnol. Oceanogr., 21, 873-880.

Saur, J.F.T., 1972: Monthly sea level differences between the Hawaiian Islands and the California coast. Fish. Bull., 70, 619-636.

Savage, R. E. and R. S. Wimpenny, 1936: Phytoplankton and the herring. Fish. Invest. Lond. II, xv. 1, 88 pp.

Savidge, G., 1976: A preliminary study of the distribution of chlorophyll a in the vicinity of fronts in the Celtic and Western Irish Seas. Est. Coastal Mar. Sci., 4, 617-625.

Schröder, R., 1961: Untersuchungen über die Plankton-verteilung mit Hilfe der Unterwasser-Fern sehanlage und des Echographen. Arch. Hydrobiol. Suppl., 25, 228-241.

Schröder, R. 1962: Vertikalverteilung des Zooplanktons und Thermokline. Arch. Hydrobiol. Suppl., 25, 401-410.

Sette, O. E., 1955: Consideration of midocean fish production as related to oceanic circulatory systems. J. Mar. Res., 14, 398-414.

Sjöberg, S. 1977: Are pelagic systems inherently unstable? A model study. Ecol. Modelling, 3, 17-37.

Skellam, J. G., 1952: Studies in statistical ecology. I. Spatial pattern. Biometrika, 39, 346-352.

Smith, L. R., C. B. Miller and R. L. Holton, 1976: Small-scale horizontal distribution of coastal copepods. J. Exp. Mar. Biol. Ecol., 23, 241-254.

Smith, P. E., 1971: Distributional atlas of zooplankton volume in the California Current region, 1951 through 1966. CalCOFI Atlas No. 13, State of California Marine Research Committee.

Smith, R. L., 1968: Upwelling. Oceanogr. Mar. Biol. Ann. Rev., 6, 11-46.

Smith, W., 1978: Environmental survey design, a time series approach. Estuarine Coastal Mar. Sci., 5, in press.

Soutar, A. and J. D. Isaacs, 1974: Abundance of pelagic fish during the 19th and 20th centuries as recorded in anaerobic sediments off the Californias. Fish. Bull., 72, 257-273.

Stavn, R. H., 1971: The horizontal-vertical distribution hypothesis: Langmuir circulations and Daphnia distributions. Limnol. Oceanogr., 16, 453-466.

Steele, J. H., 1974: Spatial heterogeneity and population stability. Nature, 248, 83.

Steele, J. H. and M. M. Mullin, 1977: Zooplankton dynamics. In The Sea, E. D. Goldberg et al. (eds.), Vol. 6, pp. 857-890, Wiley.

Stommel, H., 1963: Varieties of oceanographic experience. Science, 139, 572-576.

Stommel, H., 1965: Some thoughts about planning the Kuroshio survey. In Proc. Symp. on the Kuroshio, Tokyo, October 29, 1963, Oceanogr. Soc. Japan and UNESCO.

Taft, B. A., 1960: A statistical study of the estimation of abundance of sardine (Sardinops caerulae) eggs. Limnol. Oceanogr., 5, 245-264.

Tont, S. A., 1976: Short-period climatic fluctuations: effects on diatom biomass. Science, 194, 942-944.

Uda, M. and M. Ishino, 1958: Enrichment pattern resulting from eddy systems in relation to fishing grounds. J. Tokyo Univ. Fish., 44, 105-129.

Uda, M., 1959: Water mass boundaries - "Siome". Frontal theory in oceanography. Fish. Res. Bd. Canada Manuscript Report Series 51, pp. 10-20.

Uda, M., 1961: Fisheries oceanography in Japan, especially on the principles of fish distribution, concentration, dispersal and fluctuation. CalCOFI Progress Report, 8, 25-31.

Uda, M., 1973: Pulsative fluctuation of oceanic fronts in association with the tuna fishing grounds and fisheries. J. Fac. Mar. Sci. Technol., Tokai Univ., 7, 245-265.

UNESCO, 1968: Zooplankton Sampling. Monographs on Oceanographic Methodology 2.

UNESCO, 1976: Zooplankton fixation and preservation. Monographs on Oceanographic Methodology 4. H. F. Steedman (ed.).

Walsh, J. J., J. C. Kelley, T. E. Whitledge, J. J. MacIsaac and S. A. Huntsman, 1974: Spin-up of the Baja California upwelling ecosystem. Limnol. Oceanogr., 19, 553-572.

Walsh, J. J., T. E. Whitledge, J. C. Kelley, S. A. Huntsman and R. D. Pillsbury, 1977: Further transition stages of the Baja California upwelling ecosystem. Limnol. Oceanogr., 22, 264-280.

Whitledge, T. E. and T. T. Packard, 1971: Nutrient excretion by anchovies and zooplankton in Pacific upwelling regions. Inv. Pesq., 35, 243-250.

Wickett, W. P., 1967: Ekman transport and zooplankton concentrations in the North Pacific Ocean. J. Fish. Res. Bd. Can., 24, 581-594.

Wiebe, P. H. and W. R. Holland, 1968: Plankton patchiness: effects on repeated net tows. Limnol. Oceanogr., 13, 315-321.

Wiebe, P. H., 1970: Small-scale spatial distribution in oceanic zooplankton. Limnol. Oceanogr., 15, 205-217.

Wiebe, P. H., 1971: A computer model study of zooplankton patchiness and its effects on sampling error. Limnol. Oceanogr., 16, 29-38.

Wiebe, P. H., 1972: A field investigation of the relationship between length of tow, size of net and sampling error. J. Cons., Cons. Perm. Int. Explor. Mer., 34, 268-275.

Wiebe, P. H., G. D. Grice and E. Hoagland, 1973: Acid-iron waste as a factor affecting the distribution and abundance of zooplankton in the New York Bight. II. Spatial variations in the field and implications for monitoring studies. Estuarine Coastal Mar. Sci., 1, 51-64.

Wiebe, P. H., E. M. Hulburt, E. J. Carpenter, A. E. Jahn, G. P. Knapp, S. H. Boyd, P. B. Ortner and J. L. Cox, 1976: Gulf Stream cold core rings: large-scale interaction sites for open ocean plankton communities. Deep-Sea Res., 23, 685-710.

Wiebe, P. H. and S. H. Boyd, 1978. Limits of Nematoscelis megalops in the northwestern Atlantic in relation to Gulf Stream cold core rings. Part I. Horizontal and vertical distributions. J. Mar. Res., 36, 119-142.

Wiens, J. A., 1976: Population responses to patchy environments. Ann. Rev. Ecol. Syst., 7, 81-120.

Wiens, J. A., 1977: On competition and variable environments. Am. Sci., 65, 590-597.

Williamson, M. H., 1961: An ecological survey of a Scottish herring fishery. Part IV. Changes in the plankton during the period 1949 to 1959. Bull. Mar. Ecol., 5, 207-229.

Worthington, L. V., 1977: Intensification of the Gulf Stream after the winter of 1976-77. Nature, 270, 415-417.

Wroblewski, J. S., 1977: Vertically migrating herbivorous plankton - their possible role in the creation of small scale phytoplankton patchiness in the ocean. In Ocean Sound Scattering Prediction, N. R. Anderson and B. J. Zahuranec (eds.), pp. 817-847, Plenum.

Wroblewski, J. S. and J. J. O'Brien, 1976: A spatial model of phytoplankton patchiness. Mar. Biol., 35, 161-175.

Wyatt, T., 1973: The biology of Oikopleura dioica and Fritillaria borealis in the Southern Bight. Mar. Biol., 22, 137-158.

Wyrtki, K., 1975: El Niño: the dynamic response of the Equatorial Pacific Ocean to atmospheric forcing. J. Phys. Oceanogr., 5, 572-584.

Zeigler, B. P., 1977: Persistence and patchiness of predator-prey systems induced by discrete event population exchange mechanisms. J. Theor. Biol., 67, 687-713.

Ziemann, D. A., 1970: The horizontal distribution of zooplankton in Kaneohe Bay, Oahu, Hawaii. Masters Thesis, Univ. Hawaii, 56 pp.

# GROWTH AND SURVIVAL OF FIRST-FEEDING NORTHERN ANCHOVY LARVAE (*ENGRAULIS MORDAX*) IN PATCHES CONTAINING DIFFERENT PROPORTIONS OF LARGE AND SMALL PREY

Reuben Lasker and James R. Zweifel

National Marine Fisheries Service, NOAA
Southwest Fisheries Center
La Jolla, California 92038

## INTRODUCTION

As early as 1914, Hjort proposed that differential mortality of first-feeding larvae due to variable feeding conditions could be the cause of extreme variations in year class strength of fish. Tests of this simple attractive hypothesis have given equivocal results at best. With regard to clupeid larvae, on the average, there is usually not enough food available for the larvae to survive, considering the amount of water a larva is capable of searching and its inefficiency in capturing food when it first starts to feed (Hunter, 1972). With the sampling gear currently used (pumps and plankton nets*), it is still difficult to establish whether any body of water can support enough larval fish survival to ensure the establishment of a robust year class. To do so, however, would be a first step toward defining criteria for successful recruitment, a recognized need in all fisheries research today (Gulland and Boerema, 1973). The success of pelagic fish in their environment has prompted fishery scientists to postulate that there have to be suitable concentrations of food organisms in space and time (patches) to provide for larval feeding and that Hjort's hypothesis is still viable and should be rigorously tested.

The significance of food aggregations for larval fish can only be assessed if we know important features of larval feeding

---

*Both kinds of gear sample a lot of water and tend to give results which average numbers of organisms over depths or distances and consequently obscure fine-scale patchiness of larval fish food.

behavior. Although we know relatively little about the behavior of most fish larvae, it is possible now, with at least one species, *Engraulis mordax*, to assemble the pertinent behavioral data as a result of some recent intensive laboratory and field investigations. Two recent papers by Lasker (1975 and 1978) have provided evidence for the aggregation of properly sized food particles in the habitat of the first-feeding larval anchovy off California. He found that stability of the water column in the upper 30 meters appeared to be the key condition required for larval anchovy food organisms to aggregate in densities high enough to exceed the threshold for the feeding stimulus of first-feeding Northern anchovy larvae. Disruption of this stability was caused by vigorous storms or by upwelling along the coast, both resulting in a dilution of the microplankton needed for feeding and, by inference, excessive mortality of any larvae which needed to feed immediately after these events occurred. In addition, the work of Hunter (1972, 1976, 1977), Hunter and Sanchez (1976), Arthur (1976, 1978), Lasker (1964), Lasker *et al.* (1970) and Vlymen (1974, 1977) provide us with a background of behavioral and physiological information on the larvae of the northern anchovy, *Engraulis mordax*. The interest in anchovies goes beyond their importance in the California Current ecosystem; anchovies contribute to some of our largest fisheries, and, indeed, when the Peruvian anchovy fishery was at its peak, the anchovy catch made up 25% of the entire world fish catch.

These studies suggest that patchiness, which results in relatively high concentrations of larval anchovy food, is a major factor in the survival of clupeid larvae, although, as we will illustrate, high concentrations alone cannot be the sole criterion for survival. Other factors, such as the nutritive value and digestibility of the food, size of the food particle, food selection by the larva, learning by the larva, and a variety of other behavioral modalities, have to be taken into consideration if one is to predict whether a particular feeding regime will be favorable to larval anchovies. In this investigation, we consider the naive*, first-feeding anchovy larva presented with food on the first day it is capable of feeding. Patchiness obviously provides higher absolute concentrations of food particles than is usually found in the anchovy larva's environment. We attempt to show here that high food concentrations characteristic of patches in the sea, which appear to provide abundant food in the normal foraging distance of the larva, may still not be sufficient to provide for larval

---

*The term naive encompasses both the concept of non-selectivity of prey and a pattern of search based upon the random walk ('the blundering idiot' model, Murdie, 1971).

anchovy growth and survival on the first day of feeding. Because of the rapid development of learning in larvae and the increase in success of capture and selectivity, we have chosen to analyze only the first day of feeding by anchovy where learning has not had the opportunity to develop. However, we believe this modeling technique can be extended to assess growth and survival in older larvae as well.

## LARVAL ANCHOVY BEHAVIOR AND FOOD DISTRIBUTION

A first-feeding larval anchovy is dependent for food on a very local scale. Its swimming and feeding behavior has been studied by Hunter (1972) in great detail. The pertinent results from his study relative to the capture of food is that feeding appears to be entirely visual and that, at first feeding, success of prey capture ranges from 10-30% of the micro-, phyto- and zooplankters attacked by the larva (Figure 1).

From laboratory and field work, Lasker (1975) showed that there is a critical threshold concentration of food particles for gut-filling by first-feeding anchovy larvae. At the prevalent oceanic temperature, 13-14$^{o}$C in the anchovy's spawning ground, Lasker found that about thirty 40-50 μm particles/ml, were needed in the larva's immediate environment to stimulate feeding and gut filling. This is remarkably similar to prey concentrations which are determined from Vlymen's (1977) computer simulation for meeting the larva's energetic needs on the first day of feeding. In the laboratory, however, other factors for successful feeding are apparent and can modify these threshold concentrations. These were listed by Lasker (1975) as follows.

### Particle Size at First Feeding Appears to be Critical

We know from the stomach content work on larval anchovies by Arthur (1976) as well as from investigations by Rojas de Mendiola (1974) on the larva of the Peruvian anchovy *E. ringens*, that these larvae can capture and eat particles as large as 100 μm in smallest dimension when they first begin to feed, but that the usual diet contains smaller particles with a minimum size ingested from 25 μm, and nauplii with an average width of 68 μm. Arthur (1976) mentions that about 20% of the food found in very early-stage anchovy larvae were "unidentified spheres ... probably of plant origin" but mostly, he and other workers describe copepod eggs and nauplii as dominant identifiable foods found in the larval intestine. For example, during the first few days of feeding, stomach contents revealed 43% of ingested particles were copepod nauplii; at 9 days of age, when the larva is 6 mm long, 68% of the food particles found are copepod

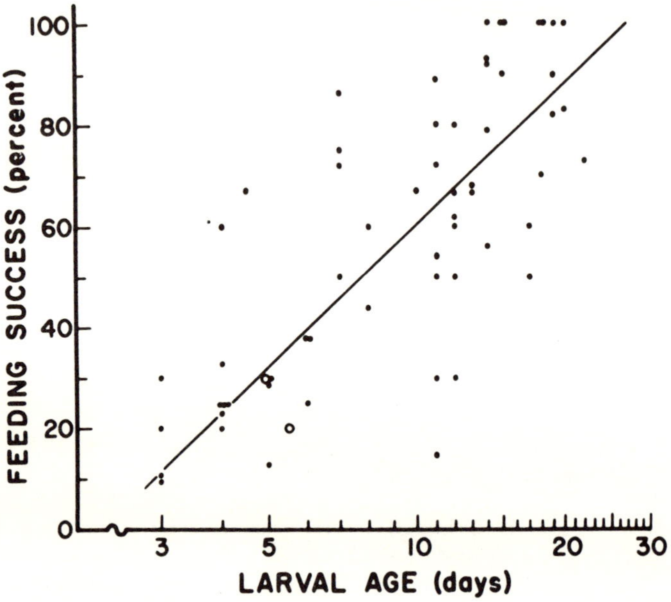


Figure 1. Feeding success (percent of attacked prey captured) of anchovy larvae of various ages fed *Brachionus*. Larval age is plotted on log scale, equation for line is percent success = 93.2 (log age) - 33.30. Two open circles, larvae fed *Gymnodinium* (from Hunter, 1972).

nauplii and at 14 days (ca. 8 mm long), 91% were nauplii*. Doubling the diameter of a spherical particle increases its volume 8 times. In the environment of the anchovy larva, this difference can have a profound effect on the feeding larva. For example, when feeding

---

*Gut content studies suffer from the uncertainty that particles found may be there because they could not be digested. However, the particle sizes actually ingested are pertinent in the following discussion.

on small particles, the anchovy larva must capture (according to Hunter) approximately 230 forty-micron diameter particles a day. If the capture efficiency on the first day of feeding is 10%, the larva must attack 2,300 particles. However, the capture and digestion of a single 80 μm particle provides energy at least equivalent to 8 forty-micron particles. Large particles, however, are not captured with as high an efficiency as small ones and this lower efficiency is related to mouth size (Hunter, 1977). Hunter (1977) also shows that when the anchovy larva reaches 6 mm in size (approximately 9 days of age), its metabolism has reached a level whereby it cannot survive on small particles alone. Thus, in any analysis of the feeding environment, consideration of both particle size and particle concentration seems to be essential.

### The Number of Particles Per Unit Volume in the Anchovy Larva's Environment Must be Above a Minimum Concentration

Hunter (1977) evaluated two methods available for determining the food density requirements for survival of anchovy larvae using a simple searching model as well as laboratory density experiments. His conclusions were that laboratory density experiments yield higher food density requirements for survival of anchovy larvae than simple searching models, but that the former suffer from the possible existence of larval density dependent interactions and that the small experimental tank volume may limit the searching capacity of the larva. While the searching model is not affected by these variables, it is grossly affected by errors in the parameters and the assumptions required by the model.

Using laboratory-spawned, first-feeding anchovy larvae as a bioassay of sea water aboard ship, Lasker (1975) showed that gut filling by a first-feeding anchovy larva in natural water at 13-14°C only occurs when a density of about 30 forty-to-fifty micron particles/ml is present at first feeding. He verified this with laboratory experiments. As the larva grows and becomes a more efficient feeder, the number of particles per unit volume needed for feeding is less because the animal increases the size of the particles it takes and begins to ignore smaller ones. At least for first-feeding larvae, the results obtained by Lasker give a reasonable estimate of the minimum density of 40-50 μm particles required for first feeding by anchovy larvae. Vlymen (1977) has demonstrated, through a prey concentration random walk model and a Markov chain prey attack model, that there is a quantitative connection between the growth of a larva and geometric arrangement of prey in its environment. He concludes from his model that survival after the first day of feeding can be achieved at particle concentrations of between 14 and 32 forty-micron particles/$cm^3$ in a geometrically structured patchy environment with patches having a particle concentration of 32 particles/$cm^3$ and an interpatch region having concentrations 14 particles/$cm^3$. The radius of the patches, and the interpatch region used in Vlymen's simulation, are in centimeters, a scale which is not routinely sampled in the sea.

### The Kind of Food Organism Determines Larval Survival and Growth

Scura and Jerde (1977) showed in a number of laboratory experiments that first-feeding Northern anchovy larvae will ignore and fail to feed on particles less than 20 μm in diameter. These included diatoms, whether single or in chains, and microflagellates, e.g., *Dunaliella* or *Tetraselmis*. Diatoms are rarely reported from gut content studies of anchovy larvae. In the event a particle is large enough to be seen and eaten by an anchovy larva, the nutritional and caloric content of the particle must be sufficiently high in the aggregate to provide for metabolism and growth. Lasker *et al.* (1970) also showed that there was a striking difference between the ability of various microzooplankton to support larval anchovy growth and survival. For example, veliger larvae of some snails supported no anchovy larva growth or survival, while the dinoflagellate *Gymnodinium splendens* and the rotifer *Brachionus plicatilis* could be used to rear the animal (Hunter, 1976). The dinoflagellate *Prorocentrum* does not support growth, as Scura and Jerde (1977) have shown, nor does the dinoflagellate *Gonyaulax polyedra*. *Prorocentrum* may simply be too small and not have enough calories per cell; but *Gonyaulax*, a larger cell, can be captured in great quantity by the larva and, although smaller than *Gymnodinium*, might have been expected to support growth and survival, but did not do so in laboratory experiments.

## PROBABILITIES OF ENCOUNTER AT DIFFERENT DENSITIES

In this section, we suggest an approach which, when coupled with field information on the distribution of proper-sized food particles for first-feeding anchovy larvae, may provide a tool with which to determine the extent and goodness of fit of anchovy larvae feeding grounds. With this information, it may be possible to predict whether recruitment will be successful from any particular spawning.

Important features of anchovy larvae behavior have been pointed out above. The most important of these, for purposes of this discussion, are the threshold concentrations for feeding, the number of strikes a first-feeding larva makes in a day*, the percent success with which it can capture food, and the maximum size particle it can successfully ingest at first feeding. Although usually there is a hyperbolic distribution of particle size in ocean water with the largest particles being fewest in number (Vlymen, 1977), often in the environment of these larvae we find patches containing two or more specific size classes of particles together, e.g., copepod nauplii and dinoflagellates. As a first-feeding anchovy larva swims among these particles, it encounters the most prevalent organism most frequently and captures at least 10% of those at which it strikes. If the animal attacks a nauplius twice the diameter of a dinoflagellate in a hunting series and manages to capture it, the nauplius can provide as much as an order of magnitude more energy than the smaller particle on which it was feeding. Therefore, the feeding behavior of an individual anchovy larva is constrained by a) the concentration and kind of prey which will stimulate a feeding attack; b) the prey distribution; c) the physical and physiological limitation on the number of possible attacks; and d) the gut volume.

In the first analysis shown below, Vlymen's (1977) computer growth model** for Northern anchovy larvae is used with some modification to simulate a random distribution of two size ranges of particles, 45-55 μm and 95-105 μm in diameter. Therefore, on the first day of feeding, the larva is faced with a locally homogeneous environment containing a random selection of particles suitably sized for ingestion. The concentrations of the particles (Table 1) are such that they encompass the range for these sizes found by surveys in the Southern California Bight, the major spawning and feeding ground of the Northern anchovy. When one does this, concentrations of the two particles can be found that provide for metabolism at least, and growth when these concentrations are exceeded. This analysis assumes that the placement of large particles in the feeding queue of small particles is random, with the results given as the average growth rate for the population of first-feeding anchovy larvae on the first day.

---

*Hunter and Thomas (1974) have shown that the rate of larval anchovy feeding increases with increasing food density. Strike rates of anchovy larvae can be as high as 10 per minute, but usually average 4 to 5 per minute. We use 10 strikes per minute in this simulation model.

**Some of the terminology employed (e.g., size specific capture rates) is more general than actually required for this paper. The mathematics and notation used were chosen in order to simplify the adaptation of Vlymen's work and in anticipation of further work on the same topic.

Table 1

Analysis of average growth in millimeters of anchovy larvae for the first day of feeding in varying concentrations of organisms in two size ranges: 45-55 and 95-105 μm at 14.0°C, a 10% capture success rate and random distribution. Energy per each small particle is $4.236 * 10^{-5}$ cal. and for each large particles is $4.523 * 10^{-4}$ cal. based on caloric content of *Gymnodinium splendens* (Vlymen, 1977) and a copepod nauplius, respectively (Lasker, 1965).

| Concentrations of 45-55 μm particles/ml | Concentrations of 95-105 μm particles/ml | | | | | |
|---|---|---|---|---|---|---|
| | 0.00 | 0.001 | 0.01 | 0.1 | 1.0 | 10.0 |
| 0.0 | -0.08 | -0.08 | -0.08 | -0.08 | -0.08 | -0.02 |
| 1.0 | -0.08 | -0.08 | -0.08 | -0.08 | -0.08 | -0.01 |
| 10.0 | -0.08 | -0.08 | -0.08 | -0.08 | -0.07 | -0.01 |
| 20.0 | -0.07 | -0.07 | -0.07 | -0.07 | -0.07 | -0.01 |
| 30.0 | -0.07 | -0.07 | -0.07 | -0.07 | -0.06 | 0.00 |
| 40.0 | -0.06 | -0.06 | -0.06 | -0.06 | -0.06 | 0.00 |
| 50.0 | -0.06 | -0.06 | -0.06 | -0.06 | -0.05 | 0.01 |
| 60.0 | -0.05 | -0.05 | -0.05 | -0.05 | -0.05 | 0.01 |
| 70.0 | -0.05 | -0.05 | -0.05 | -0.05 | -0.04 | 0.01 |
| 80.0 | -0.04 | -0.04 | -0.04 | -0.04 | -0.04 | 0.02 |
| 90.0 | -0.04 | -0.04 | -0.04 | -0.04 | -0.03 | 0.02 |
| 100.0 | -0.04 | -0.04 | -0.04 | -0.03 | -0.03 | 0.02 |
| 200.0 | -0.03 | -0.03 | -0.03 | -0.03 | -0.02 | 0.01 |
| 300.0 | -0.03 | -0.03 | -0.03 | -0.03 | -0.02 | 0.00 |

If the larva could attack all prey in its visual field, the expected number of captures, $N_c$, would be the product of the number $N$ of potential attacks (excursions), the probability of attack $PA$, the probability $PS(S)$ of actually selecting organisms of size $S$ (in this analysis $PS(S)=1$), the relative frequency $R(S)$ of organisms of size $S$ in the prey distribution and the probability of capture $PC(S)$, i.e., where $E(N_c)$ is the average number of prey captured,

$$E(N_c) = N\ PA \sum_{s=\ell}^{u} PS(S)\ R(S)\ PC(S)$$

where $N \leq N_{max}$, and $\sum_{s=\ell}^{u}$ indicates a summation over all sizes of acceptable or available prey.

Vlymen's (1977) model corrects the attack and capture rates for delays between each. In other words, once an attack is made, there is a finite time before the next attack. This is taken into consideration in programming the survival and growth of anchovy larvae feeding on different sized particles; the attack probabilities are calculated as defined in the program diagrammed in Figure 2.

The expected volume $V_c$ of captured organisms will be

$$E(V_c) = N\ PA \sum_{s=\ell}^{u} PS(S)\ R(S)\ PC(S)\ V(S)$$

where $N \leq N_{max}$

$V_c \leq V_{max}$.

When $V(S)$ is the volume of organisms of size $S$, the gross energy return $GE_c$ will be

$$E(GE_c) = N\ PA \sum_{s=\ell}^{u} PS(S)\ R(S)\ PC(S)\ CAL(S)$$

where $CAL(S)$ is the calories per organism of size $S$ and

$N \leq N_{max}$

$V_c \leq V_{max}$

$S_1$ = 45-55 μm; $S_2$ = 95-105 μm

$N_{max}$ = 10 strikes/min = 600/hr. = 7200/12-hr. day

$PC(S)$ = constant for any length

$PS(S_1) = PS(S_2) = 1$

$N\ PA$ is calculated from Vlymen (1977) and is a function of total concentration

$TC = C_1 + C_2$ where $C_1$ = 1, 10, 20, 30, 40, 50, 60, 70, 80, 90, 100, 200, and 300/ml and

$C_2$ = 0.001, 0.01, 0.1, 1.0 and 10.0/ml.

$$R(S_1) = \frac{C_1}{TC},\ R(S_2) = \frac{C_2}{TC}$$

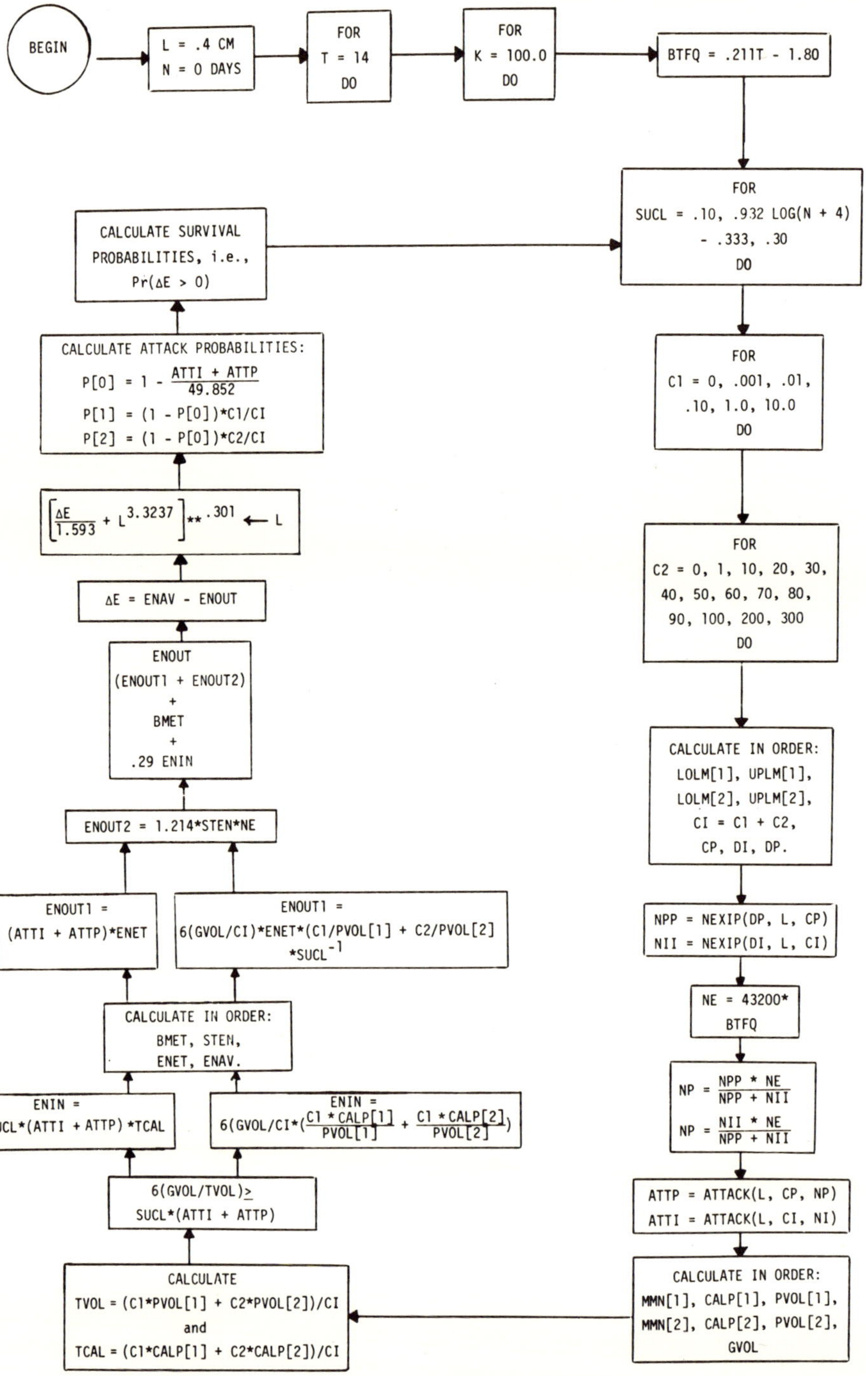


Figure 2. Program modified after Vlymen (1977) for determining survival and growth of first-feeding larval anchovy in various food concentrations. Notation after Vlymen (1977) follows:

L = larval anchovy length, 4 mm
N = 0, first day of feeding
T = 14.0°C
K = dispersion coefficient of 100.00, i.e. random distribution of prey
BTFQ = temperature dependent excursion frequency
SUCL = age dependent success of capture function
C1 = concentration of small prey
C2 = concentration of large prey
LOLM = lower prey concentration limits
UPLM = upper prey concentration limits
CI = interpatch concentration
CP = intrapatch concentration
DP = patch diameter
DI = intrapatch distance
NEXIP (D,L,C) = number of steps required to traverse distance D in concentration of food particles C by larva of length L
NPP = NEXIP (DP,L,CP) = number of excursions a larva of length L will require to traverse a patch of radius DP and concentration CP
NII = NEXIP (DI,L,CI) = number of excursions a larva of length L will require to traverse an interpatch of radius DI and particle concentration CI
NE = number of excursions in 12 hours of foraging
NP = proportion of total daily foraging excursions spent in patches, and NI = in between patches
ATTACK (L,C,N) calculates the number of attacks on prey made by an anchovy larva of length L with N excursions in prey concentration C; NI, attacks between patches; NP, attacks in patches
CALP = caloric value of particles
MMN = size of particles
PVOL = volume of particle
GVOL = maximum gut volume of larva of length L
TVOL = total volume of particles eaten
TCAL = total calories of particles eaten
SUCL*(ATT1+ATTP) = particles successfully attacked
NE = number of excursions; excursions at night = 0.214 NE
STEN = energy per excursion
ENIN = energy in, or daily ration of calories
ENOUT1 = energetic debt of attacks
ENOUT2 = energetic debt of 24 hours of swimming, i.e. 1.214*NE*STEN
BMET = temperature and length dependent basal metabolic debt for 24 hours
ENAV = available energy or 0.48 ENIN
ENET = energy of attack = 3 STEN
0.29 ENIN = energy cost for mechanical processing of food
P[0] = probability of no attack
P[1] = probability of attack on small particle
P[2] = probability of attack on large particle

This simulation implies that the total number of particles in the larva's environment governs the frequency of strikes. The larger particles, even at their highest concentration, are ingested randomly (non-selectivity). Technically, despite the highest concentrations of the largest particles (e.g., nauplii), these must be taken very infrequently if the concentration of small particles is so high that the larger particles rarely show up in the feeding queue. Therefore, it is not the absolute concentration of the large particles that govern their capture by the larva, but rather their concentration relative to the larger number of small particles in the larval environment. Tables 1, 3 and 5 show the results of this simulation. A full discussion of these results is given in the following section.

A second analysis (Tables 2, 4 and 6) is based upon the same random arrangement of large particles among small particles, but

Table 2

Analysis of percent survival of anchovy larvae after the first day of feeding in varying concentrations of organisms in two size ranges: 45-55 and 95-105 μm at 14.0°C and random distribution. Attack rate is 10 per minute and success of capture is 10%. Energy per each small particle is $4.236 * 10^{-5}$ cal. and for each large particle is $4.523 * 10^{-4}$ cal. based on caloric content of *Gymnodinium splendens* (Vlymen, 1977) and a copepod nauplius, respectively (Lasker, 1965).

| Concentrations of 45-55 μm particles/ml | Concentrations of 95-105 μm particles/ml | | | | | |
|---|---|---|---|---|---|---|
| | 0.00 | 0.001 | 0.01 | 0.1 | 1.0 | 10.0 |
| 0.0 | 0.0 | 0.0 | 0.0 | 0.0 | 0.0 | 31.0 |
| 1.0 | 0.0 | 0.0 | 0.0 | 0.0 | 0.0 | 30.0 |
| 10.0 | 0.0 | 0.0 | 0.0 | 0.0 | 0.0 | 34.7 |
| 20.0 | 0.0 | 0.0 | 0.0 | 0.0 | 0.0 | 41.1 |
| 30.0 | 0.0 | 0.0 | 0.0 | 0.0 | 0.0 | 47.5 |
| 40.0 | 0.0 | 0.0 | 0.0 | 0.0 | 0.0 | 54.1 |
| 50.0 | 0.0 | 0.0 | 0.0 | 0.0 | 0.0 | 60.2 |
| 60.0 | 0.0 | 0.0 | 0.0 | 0.0 | 0.0 | 66.2 |
| 70.0 | 0.0 | 0.0 | 0.0 | 0.0 | 0.0 | 71.6 |
| 80.0 | 0.0 | 0.0 | 0.0 | 0.0 | 0.0 | 76.6 |
| 90.0 | 0.0 | 0.0 | 0.0 | 0.0 | 0.1 | 81.0 |
| 100.0 | 0.0 | 0.0 | 0.0 | 0.0 | 0.5 | 84.7 |
| 200.0 | 0.1 | 0.1 | 0.1 | 0.1 | 1.4 | 62.9 |
| 300.0 | 0.1 | 0.1 | 0.1 | 0.1 | 0.7 | 40.5 |

Table 3

Analysis of average growth in millimeters of anchovy larvae for the first day of feeding in varying concentrations of organisms in two size ranges: 45-55 and 95-105 μm at 14.0°C, a 20% capture success rate and random distribution. Energy per each small particle is $4.236 * 10^{-5}$ cal. and for each large particle is $4.523 * 10^{-4}$ cal. based on caloric content of *Gymnodinium splendens* (Vlymen, 1977) and a copepod nauplius, respectively (Lasker, 1965).

| Concentrations of 45-55 μm particles/ml | Concentrations of 95-105 μm particles/ml | | | | | |
|---|---|---|---|---|---|---|
| | 0.00 | 0.001 | 0.01 | 0.1 | 1.0 | 10.0 |
| 0.0 | -0.08 | -0.08 | -0.08 | -0.08 | -0.07 | 0.07 |
| 1.0 | -0.08 | -0.08 | -0.08 | -0.08 | -0.07 | 0.07 |
| 10.0 | -0.07 | -0.07 | -0.07 | -0.07 | -0.05 | 0.08 |
| 20.0 | -0.06 | -0.06 | -0.06 | -0.05 | -0.04 | 0.09 |
| 30.0 | -0.04 | -0.04 | -0.04 | -0.04 | -0.03 | 0.10 |
| 40.0 | -0.03 | -0.03 | -0.03 | -0.03 | -0.02 | 0.11 |
| 50.0 | -0.02 | -0.02 | -0.02 | -0.02 | -0.00 | 0.12 |
| 60.0 | -0.01 | -0.01 | -0.01 | -0.01 | 0.01 | 0.13 |
| 70.0 | 0.00 | 0.00 | 0.00 | 0.01 | 0.02 | 0.14 |
| 80.0 | 0.02 | 0.02 | 0.02 | 0.02 | 0.03 | 0.15 |
| 90.0 | 0.03 | 0.03 | 0.03 | 0.03 | 0.04 | 0.15 |
| 100.0 | 0.04 | 0.04 | 0.04 | 0.04 | 0.05 | 0.16 |
| 200.0 | 0.06 | 0.06 | 0.06 | 0.06 | 0.07 | 0.13 |
| 300.0 | 0.06 | 0.06 | 0.06 | 0.06 | 0.06 | 0.10 |

the results are given in percent survival. Survival probabilities are calculated from the frequencies with which the 49,852 excursions (made by a larva in a 12-hour day) would end in an attack. Vlymen (1977) has assumed that "awareness" or conditions for an attack will occur when the number of particles within the visual field, $K$, is $\geq 1$ and calculated this probability to be:

$$PA = Pr(K \geq 1) = 1 - e^{(-\lambda V_{pr})}$$

where $V_{pr}$ is the volume of the visual field $= 0.0159L^3$ where $L$, the length of the larva, is in centimeters and $\lambda$ is concentration in number of prey per milliliter. The condition necessary for survival is that the energy balance $\Delta E \geq 0$; the available energy is calculated from the actual intake of $N_1$ small particles and $N_2$ large particles when food is limiting or from the maximum digestible volume of food estimated to be 6 * gut volume in a 12-hour day.

Table 4

Analysis of percent survival of anchovy larvae after the first day of feeding in varying concentrations of organisms in two size ranges: 45-55 and 95-105 μm at 14.0°C and random distribution. Attack rate is 10 per minute and success of capture is 20%. Energy per each small particle is 4.236 * $10^{-5}$ cal. and for each large particle is 4.523 * $10^{-4}$ cal. based on caloric content of *Gymnodinium splendens* (Vlymen, 1977) and a copepod nauplius, respectively (Lasker, 1965).

| Concentrations of 45-55 μm particles/ml | Concentrations of 95-105 μm particles/ml | | | | | |
|---|---|---|---|---|---|---|
| | 0.00 | 0.001 | 0.01 | 0.1 | 1.0 | 10.0 |
| 0.0 | 0.0 | 0.0 | 0.00 | 0.0 | 0.0 | 98.6 |
| 1.0 | 0.0 | 0.0 | 0.00 | 0.0 | 0.0 | 98.7 |
| 10.0 | 0.0 | 0.0 | 0.00 | 0.0 | 0.0 | 99.6 |
| 20.0 | 0.0 | 0.0 | 0.00 | 0.0 | 0.4 | 99.9 |
| 30.0 | 0.0 | 0.0 | 0.00 | 0.0 | 5.4 | 99.9 |
| 40.0 | 0.1 | 0.1 | 0.1 | 0.4 | 29.8 | 99.9 |
| 50.0 | 14.1 | 14.1 | 15.3 | 22.1 | 72.2 | 99.9 |
| 60.0 | 80.5 | 80.5 | 80.4 | 83.4 | 95.8 | 99.9 |
| 70.0 | 99.5 | 99.5 | 99.5 | 99.5 | 99.8 | 99.9 |
| 80.0 | 99.9 | 99.9 | 99.9 | 99.9 | 99.9 | 99.9 |
| 90.0 | 99.9 | 99.9 | 99.9 | 99.9 | 99.9 | 99.9 |
| 100.0 | 99.9 | 99.9 | 99.9 | 99.9 | 99.9 | 99.9 |
| 200.0 | 99.9 | 99.9 | 99.9 | 99.9 | 99.9 | 99.9 |
| 300.0 | 99.9 | 99.9 | 99.9 | 99.9 | 99.9 | 99.9 |

To simulate the conditions needed for survival (as opposed to average growth shown in Tables 1, 3 and 5), the following conditions must be met by the model: a) the energy balance is such that $\Delta E \geq 0$ (in Vlymen's notation *ENIN-ENOUT*), and b) the available energy is the smaller of either of two quantities, $ENIN = 0.48\,[N_1\,CAL(S_1) + N_2\,CAL(S_2)]$ (i.e., the actual catch) or when food is not limiting,

$$ENIN = \frac{6*gut\ volume}{C_1 + C_2}\left[\frac{C_1\,CAL(S_1)}{V(S_1)} + \frac{C_2\,CAL(S_2)}{V(S_2)}\right].$$

For any realization $k$ of a trinomial, the relative frequency $f(N_{0,k}, N_{1,k}, N_{2,k})$ of obtaining $N_{0,k}$, $N_{1,k}$ and $N_{2,k}$ attacks on particles is

Table 5

Analysis of average growth in millimeters of anchovy larvae for the first day of feeding in varying concentrations of organisms in two size ranges: 45-55 μm and 95-105 μm at 14.0°C, a 30% capture success rate and random distribution. Energy per each small particle is $4.236 * 10^{-5}$ cal. and for each large particle is $4.523 * 10^{-4}$ cal. based on caloric content of *Gymnodinium splendens* (Vlymen, 1977) and a copepod nauplius, respectively (Lasker, 1965).

| Concentrations of 45-55 μm particles/ml | Concentrations of 95-105 μm particles/ml | | | | | |
|---|---|---|---|---|---|---|
| | 0.00 | 0.001 | 0.010 | 0.1 | 1.0 | 10.0 |
| 0.0 | -0.08 | -0.08 | -0.08 | -0.08 | -0.06 | 0.12 |
| 1.0 | -0.08 | -0.08 | -0.08 | -0.08 | -0.06 | 0.12 |
| 10.0 | -0.07 | -0.07 | -0.07 | -0.06 | -0.04 | 0.13 |
| 20.0 | -0.06 | -0.05 | -0.05 | -0.05 | -0.03 | 0.14 |
| 30.0 | -0.03 | -0.03 | -0.03 | -0.03 | -0.01 | 0.16 |
| 40.0 | -0.01 | -0.01 | -0.01 | -0.01 | 0.01 | 0.17 |
| 50.0 | 0.00 | 0.00 | 0.00 | 0.00 | 0.02 | 0.18 |
| 60.0 | 0.02 | 0.02 | 0.02 | 0.02 | 0.04 | 0.19 |
| 70.0 | 0.03 | 0.03 | 0.03 | 0.04 | 0.05 | 0.20 |
| 80.0 | 0.05 | 0.05 | 0.05 | 0.05 | 0.07 | 0.22 |
| 90.0 | 0.06 | 0.06 | 0.06 | 0.06 | 0.08 | 0.23 |
| 100.0 | 0.08 | 0.08 | 0.08 | 0.08 | 0.09 | 0.24 |
| 200.0 | 0.11 | 0.11 | 0.11 | 0.11 | 0.11 | 0.19 |
| 300.0 | 0.11 | 0.11 | 0.11 | 0.11 | 0.11 | 0.16 |

$$f(N_{0,k}, N_{1,k}, N_{2,k}) = \frac{N!}{N_{0,k}!\, N_{1,k}!\, N_{2,k}!} P_0^{N_{0,k}} P_1^{N_{1,k}} P_2^{N_{2,k}}$$

where $N$ = total number of excursions made by the larva (obtained from tail beat frequencies),

$N_0$ = $N-N_1-N_2$ is the number of excursions in which no attacks have been made,

$P_1$ = probability of attack on small particles,

$P_2$ = probability of attack on large particles, and

$P_0$ = $1-P_1-P_2$ is the probability of no attacks.

Table 6

Analysis of percent survival of anchovy larvae after the first day of feeding in varying concentrations of organisms in two size ranges: 45-55 and 95-105 μm at 14.0°C and random distribution. Attack rate is 10 per minute and success of capture is 30%. Energy per each small particle is $4.236 * 10^{-5}$ cal. and for each large particle is $4.523 * 10^{-4}$ cal. based on caloric content of *Gymnodinium splendens* (Vlymen, 1977) and a copepod nauplius, respectively (Lasker, 1965).

| Concentrations of 45-55 μm particles/ml | Concentrations of 95-105 μm particles/ml | | | | | |
|---|---|---|---|---|---|---|
| | 0.00 | 0.001 | 0.01 | 0.1 | 1.0 | 10.0 |
| 0.0 | 0.0 | 0.0 | 0.0 | 0.0 | 0.0 | 99.9 |
| 1.0 | 0.0 | 0.0 | 0.0 | 0.0 | 0.0 | 99.9 |
| 10.0 | 0.0 | 0.0 | 0.0 | 0.0 | 0.3 | 99.9 |
| 20.0 | 0.0 | 0.0 | 0.0 | 0.0 | 7.2 | 99.9 |
| 30.0 | 0.0 | 0.0 | 0.1 | 0.6 | 43.1 | 99.9 |
| 40.0 | 31.0 | 31.0 | 31.9 | 42.3 | 87.7 | 99.9 |
| 50.0 | 97.1 | 97.1 | 97.1 | 97.4 | 99.5 | 99.9 |
| 60.0 | 99.9 | 99.9 | 99.9 | 99.9 | 99.9 | 99.9 |
| 70.0 | 99.9 | 99.9 | 99.9 | 99.9 | 99.9 | 99.9 |
| 80.0 | 99.9 | 99.9 | 99.9 | 99.9 | 99.9 | 99.9 |
| 90.0 | 99.9 | 99.9 | 99.9 | 99.9 | 99.9 | 99.9 |
| 100.0 | 99.9 | 99.9 | 99.9 | 99.9 | 99.9 | 99.9 |
| 200.0 | 99.9 | 99.9 | 99.9 | 99.9 | 99.9 | 99.9 |
| 300.0 | 99.9 | 99.9 | 99.9 | 99.9 | 99.9 | 99.9 |

The probabilities $P_0+P_1+P_2 = 1$ and the attacks plus no attacks equal total number of excursions, i.e., $N_0+N_1+N_2 = N$.

To estimate survival over the first day of life, all $f(N_{0,k}, N_{1,k}, N_{2,k})$ must be determined to satisfy

$$.48\ (N_{1,k}\ C_1+N_{2,k}C_2) > ENOUT \text{ when } N_{1,k}\ V_1+N_{2,k}\ V_2 \leq 6*gut\ volume$$

or

$$\frac{.48\ (N_{1,k}\ C_1+N_{2,k}C_2)\,6*gut\ volume}{N_{1,k}\ V_1+N_{2,k}\ V_2} > ENOUT$$

where $N_{1,k}\ V_1+N_{2,k}\ V_2 > 6*gut\ volume.$

In addition, since the maximum strike rate is 10/minute, we must impose the finite restriction that $N_1+N_2 \leq 12\ (hrs.) * 60\ (min/hr) * 10\ (strikes/min) = 7200$.

Although all of the $N(N-1)$ possible outcomes have a finite probability of occurrence, the number of calculations can be significantly reduced by considering only outcomes in the ranges:

$$N P_0 - m \sqrt{N P_0 (1-P_0)} < N_{0,k} < N P_0 + m \sqrt{N P_0 (1-P_0)}$$

$$N P_1 - m \sqrt{N P_1 (1-P_1)} < N_{1,k} < N P_1 + m \sqrt{N P_1 (1-P_1)}$$

$$N P_2 - m \sqrt{N P_2 (1-P_2)} < N_{2,k} < N P_2 + m \sqrt{N P_2 (1-P_2)}$$

where $N_{2,k} = N-N_{0,k} - N_{1,k}$ and where $m$ represents the number of standard deviations and can be made as large as necessary. In this calculation, $m=4$ so that the probability of any outcome falling outside the range will be negligible except for the very lowest particle densities; i.e., for estimating survival of say <0.01%.

The probabilities $P_0$, $P_1$ and $P_2$ are determined as follows:

$P_0 = 1 - (ATTI + ATTP)/49{,}852$ (Figure 2)

where $P_0$ is the probability of no attacks and

$1 - P_0$ is the probability of attack on either large or small particles, and $ATTI + ATTP$ is the total number of attacks.

Then $P_1 = (1 - P_0) \left(\dfrac{C_1}{C_1 + C_2}\right)$

and $P_2 = (1 - P_0) \left(\dfrac{C_2}{C_1 + C_2}\right)$

where $C_1$ = concentration of small particles,

$C_2$ = concentration of large particles.

## RESULTS OF SIMULATED FEEDING BY FIRST-FEEDING ANCHOVY LARVAE

Minimum survival by first-feeding anchovy larvae to a second day of feeding depends on threshold concentrations of small particles and not on the presence or absence of large particles. From Tables 2, 4 and 6, large particles can be seen to enhance survival, but even in the highest concentration of large particles ever recorded from the Northern anchovy larva's environment, about 0.3/ml, and with capture success set at the highest level possible (30%), anchovy larvae can neither grow nor survive the first day of feeding on a diet of large particles alone. A probable food environment that the larva encounters is a proportion of small to large food particles of 30:0.1 or 30:0.01 per milliliter. At 30 small particles per milliliter, no anchovy larvae will survive the first day (Table 6). Only when large particles become prevalent at this low concentration of small particles, i.e., when there are between 0.01 and 0.1/ml or 10 to 100 per liter, is there a small increase in survival, although average growth remains negative. However, the difference in growth and survival is striking when the larva has in its immediate environment 40 small particles per milliliter instead of 30. Survival in this instance is increased to 31% and large particles make little difference even in concentrations as high as 0.1 per milliliter. Above 0.1 per milliliter (100/L) of large food particles, survival becomes greatly enhanced as long as 40 small particles per milliliter are also present. At concentrations greater than 10/ml, the distance of a first-feeding larva from its nearest possible food item is within its perceptive field, 2.7 mm (Hunter, 1972). The distance at 30 and 40 particles per milliliter are 1.78 to 1.62 mm, respectively (Gallagher and Burdick, 1970). This may be a critical range for the larva.

It is clear from the simulation of a 10% capture success rate shown in Tables 1 and 2 that under virtually all concentrations and combinations of food, a concentration of nauplii not yet seen in the sea, 10/ml (10,000/L), is needed to insure reasonable survival to the second day of feeding. At 20 and 30% capture rates, there is good agreement with the experimental data of O'Connell and Raymond (1970) who found that there was good survival of young anchovy larvae at naupliar concentrations of 4000/L, also abnormally high when compared to concentrations found in the ocean.

P. E. Smith (personal communication) believes that survival after the first day of feeding, as deduced from field collections, can be as high as 75%. This value is encountered in this simulation in concentrations of prey frequently found in the sea in dinoflagellate blooms (i.e., where small particles are between 40 and 50/ml) and emphasizes the importance of spatial and temporal aggregations of food patches in the larva's environment.

When the capture success rate is low (10%, Tables 1 and 2), this increases the negative aspect of having large numbers of small particles available to the larva. It is evident that the larva is so occupied by attacks on energy-poor particles and that it rarely encounters and cannot capture enough large, energy-rich particles at this low capture success rate to survive or grow through the first day of feeding.

Tables 3 and 4 illustrate the results of an intermediate capture success rate, 20%. Positive growth is only achieved when 70-80 small food particles per milliliter are available (Table 3) and there is very little to be gained by having large particles in the environment unless they are in hundreds per liter. Survival (Table 4) is achieved at much lower concentrations of small particles with the critical value about 50/ml. Survival enhancement by large food particles occurs in this simulation when 10-100/L of these are present coincidentally with the larger number of smaller particles.

## THE RELATION BETWEEN PHYTOPLANKTON AND ZOOPLANKTON SPATIAL DISTRIBUTION AND LARVAL ANCHOVIES IN THE SEA

It is logical to inquire to what extent the concentrations of small and large food particles needed for adequate survival and growth of first-feeding anchovy larvae are found in the Northern anchovy spawning ground during the spawning season, December through March.

### Phytoplankton

In 1974, Lasker (1975) found that chlorophyll maximum layers occasionally contained particle densities which could provide good feeding for first-feeding anchovy larvae. For example, in a series of stations along the California coast near Los Angeles, the densities of the dinoflagellate *Gymnodinium splendens* induced feeding by first-feeding anchovy larvae (on shipboard) at concentrations from 34 to more than 300 per milliliter. *Gymnodinium* was present as an apparently continuous patch which extended along the coast approximately 100 kilometers. An important observation was that stormy conditions were found to disrupt these layers and reduce the minimum densities below threshold levels for larval anchovy first feeding.

In a later study, Lasker (1978) reported on an extensive survey made along the California and Baja California coasts in 1975 where he sought to detect areas of anchovy spawning where particle densities were above minimum thresholds for first-feeding larval anchovy. He found a remarkable patch of the dinoflagellate

*Gonyaulax polyedra* extending from at least Point Conception to the Mexican border, a distance of 300 kilometers. The seaward extent of this very large patch ranged from a few kilometers in December 1974 to 40 kilometers in January 1975; the latter extending across the Southern California Bight from the mainland shore to the Channel Islands.

The vertical distribution of dinoflagellate patches has been studied from time to time. Sharp chlorophyll maxima coincide with dinoflagellate maxima but, with current sampling gear, these patches seem to be many meters thick. However, Kiefer and Lasker (1975) analyzed the vertical migration of *Gymnodinium splendens* populations and got very sharp chlorophyll maxima only 1 to 2 meters thick (Figure 3). Visual sightings of dinoflagellate blooms at the surface often give the familiar red or rust color of a "red tide." In 1976, the rich Peruvian upwelling system experienced a vast bloom of *Gymnodinium splendens* which became so dense in some places that hydrogen sulfide was produced when the organisms died. In Peru this phenomenon is known as "aguaje" and extended from Pimental to Matarani, a distance of 2,000 kilometers, in a more or less continuous patch (B. Rojas de Mendiola and J. Valdivia of the Instituto del Mar del Peru, personal communication).

In 1975, when *Gonyaulax polyedra* was the dominant particle in the feeding size range of first-feeding anchovy larvae and *Gymnodinium splendens* similarly dominated the Peruvian upwelling system, it is tempting to speculate that *Engraulis mordax* first-feeding larvae ate *Gonyaulax* in numbers off California while the Peruvian anchovy first-feeding larvae, *Engraulis ringens*, were eating *Gymnodinium*. In the former, because of the poor nutritional value of *Gonyaulax*, a poor year class would have been expected while in the latter we would have expected to see a good year class. Complicating the *Gonyaulax* situation in 1975, however, was the fact that a massive upwelling in the midst of the anchovy spawning season destroyed the patches of *Gonyaulax* and in fact severely reduced particle concentrations below the threshold needed for gut filling and metabolic needs of first-feeding larvae. Indeed, the 1975 year class of the Northern anchovy was very poor compared to other years (Sunada, California Department of Fish and Game, personal communication) while the Peruvian anchovy showed a resurgence from previous low levels in its 1975 year class. While these are at present only single observations, they suggest that it may be worthwhile to investigate both particle densities and distribution and nutritional content of larval anchovy food to forecast whether a recruitment will be successful or not. Microzooplankton concentrations never exceeded 30 per liter in hundreds of samples taken during 1974 and 1975 in the California studies described above.

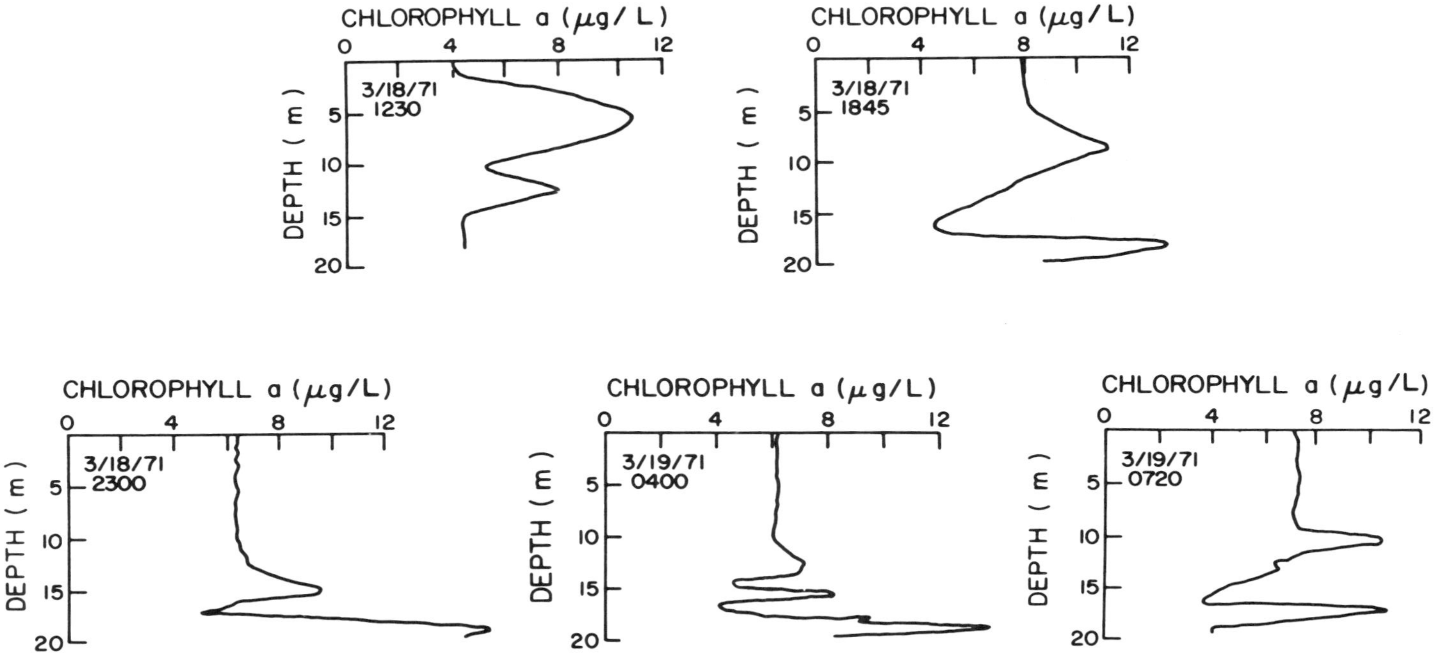


Figure 3. Profiles of the concentration of chlorophyll *a* based upon fluorescence profiles. The upper layer of *Gymnodinium splendens* at 1230 h was concentrated at 6 m; by 1845 it had moved to 8 m and reached 15 m by 2300. Movement upward commenced at 0200 reaching 10 m by 0720. The lower layer remained relatively close to the bottom during this time. Profiles were obtained in Coyote Bay, Bahia Concepcion, Mexico (lat. 26°43.0'N, long. 111°53.0'W) (from Kiefer and Lasker, 1975).

## Zooplankton

When food is found in anchovy larvae intestines, it usually consists of copepod nauplii, copepod eggs and a variety of similarly small microzooplankters, regardless of where the anchovy population is found*. Yet, as we indicated before, the sampled densities of these potential food organisms have always seemed to be too low to maintain engraulids. This conflicting evidence leads one to believe that patches of microzooplankton may be prevalent in productive areas where anchovies are found, but present sampling gear underestimates the concentration factor. On occasion, when a sample of water is discovered to have a large number of nauplii or other microzooplankters, the sample, when retaken in the same place, does not duplicate the results. However, on one occasion, Lasker (1978) showed that in samples containing over 100 microzooplankters per liter taken 0.8 km apart, i.e., 2 and 2.8 km from the California shore, there was a remarkable coherence in the numbers and size distribution of copepod nauplii and post-naupliar copepods (Figure 4). This duplication indicated that a patch of at least 0.8 km in linear extent offshore was sampled despite a prevailing along-shore current. It should also be noted that motile phytoplankton food reproduce on an hourly or daily scale while nauplii are a stage in a copepod life cycle of several weeks. The predation on copepods by a major fish population like the Northern anchovy could cumulatively have a marked depressant effect on the local abundance of nauplii.

The point we wish to make is that temporal and spatial stability of the ocean is a prerequisite for successful feeding by clupeid larvae. Food organisms must have time to aggregate and, once patches are formed, they must have a sufficiently long life to permit larval fish feeding. However, it is much more likely that there will be aggregations of small particles than large ones and this in turn favors first day larval survival and growth.

The model described here to determine adequate feeding of various size particles for first-feeding anchovy larvae can be extended to older larvae as Vlymen (1977) has shown in his analysis of the effect of prey microdistribution to larval anchovy growth.

---

*Most of the anchovy larvae examined have no food in their intestines. Whether this is due to rough treatment in plankton nets, to the severity of the usual Formalin preservation, or is the normal situation in the sea is not known. Kjelson *et al.* (1975) have shown, however, that menhaden larvae lose much of their gut contents with rough laboratory handling equivalent to field capture.

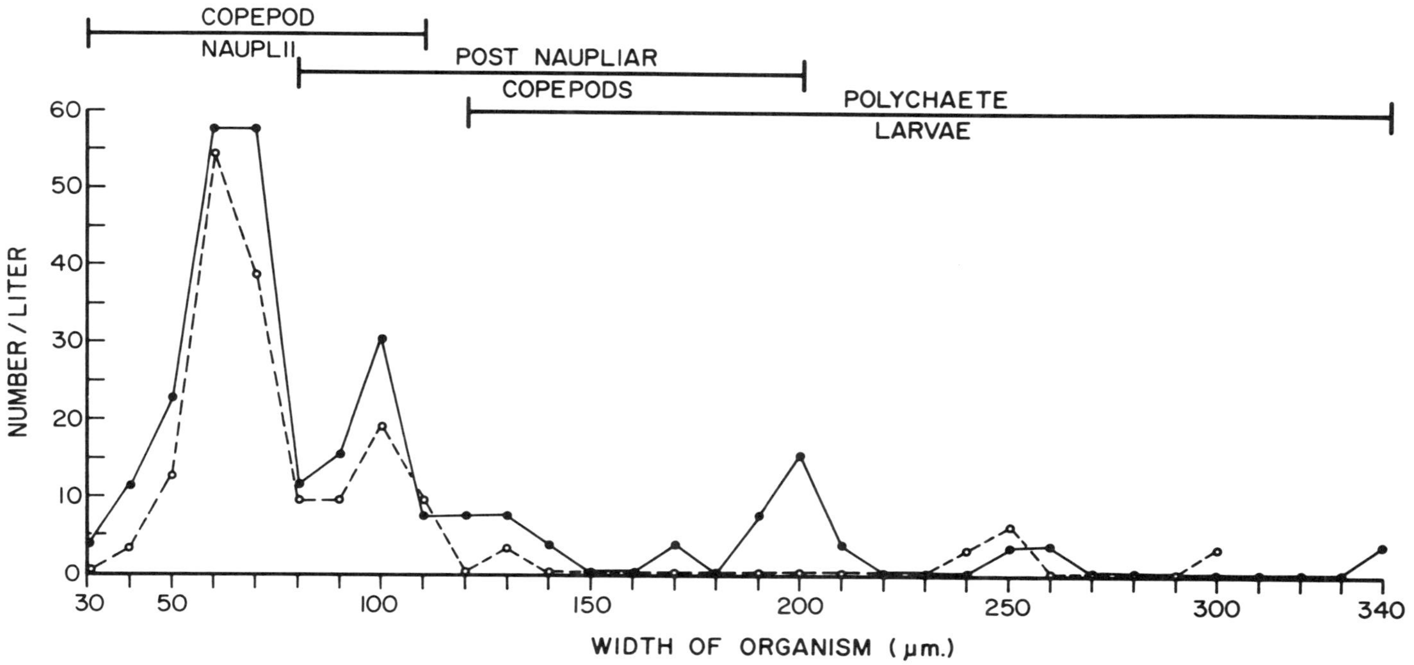


Figure 4. Width frequency distribution of two samples of microzooplankton taken from the chlorophyll maximum layer in March 1976, 0.8 km apart within 2 km of the shore off San Onofre, California. Closed circles indicate the sample taken nearest the shore which contained 280 microzooplankters per liter. The other sample contained 117 microzooplankters per liter. These were the only two of hundreds of pump samples taken in the Southern California Bight over a 3-year period which contained more than 100 microzooplankters per liter (from Lasker, 1978).

We know from the simulation presented in this paper that the enigma presented by the lack of reasonably large numbers of nauplii in the sea for larval anchovy food may be explained because the larvae don't need them, nor do the larvae encounter them with a sufficiently high frequency for these large particles to make more than a token contribution to survival and growth on the first day of feeding.

## SUMMARY

Based on laboratory and field data, a simulation model of first-feeding anchovy larvae, *Engraulis mordax*, was constructed to investigate the effect of different prey size on survival and growth. First-feeding, naive anchovy larvae must depend almost entirely on small particles (ca. 45 μm diameter) for early survival, despite their ability to capture large particles. The absolute concentration of large particles (ca. 90 μm diameter) does not govern their capture by the larvae; rather, the more numerous small particles distract the larvae from capturing large particles. Simulation shows that a 10% prey capture efficiency will not suffice for survival or growth and that, at higher capture efficiencies (20-30%), there is a threshold of 30-50 small particles per milliliter needed for substantial survival and growth. Large particles, e.g., copepod nauplii, can enhance survival when they reach concentrations between 10 and 100 per liter in the same environment containing above-threshold numbers of small particles. An important distinction is made between survival and average growth; it is shown that larvae may survive in significant numbers in feeding regimes in which the average growth for the population is negative. These results are discussed in relation to actual field data from the spawning ground of the Northern anchovy.

## ACKNOWLEDGEMENTS

We thank Dr. William J. Vlymen, radiologist-bioengineer, Stanford University Medical Center, for his generous help. Partial funding for this work was provided by the Oceanography Program of the Brookhaven National Laboratory, Energy Research and Development Administration (ERDA), Upton, Long Island, New York.

## REFERENCES

Arthur, D. K., 1976: Food and feeding of larvae of three fishes occurring in the California Current, Sardinops sagax, Engraulis mordax, and Trachurus symmetricus. Fish. Bull. U.S., 74, 517-530.

Arthur, D. K., 1978: Differences in heart size between ocean-caught and laboratory-grown larvae of the northern anchovy, Engraulis mordax: implications for larval behavior and survival. J. Exp. Mar. Biol. Ecol., in press.

Gallagher, B. S. and J. E. Burdick, 1970: Mean separation of organisms in three dimensions. Ecology, 51, 538-540.

Gulland, J. A. and L. K. Boerema, 1973: Scientific advice on catch levels. Fish. Bull. U.S., 71, 325-335.

Hjort, J., 1914: Fluctuations in the great fisheries of northern Europe viewed in the light of biological research. Rapp. P.-v. Reun. Cons. Perm. Int. Explor. Mer., 20, 1-228.

Hunter, J. R., 1972: Swimming and feeding behavior of larval anchovy, Engraulis mordax. Fish. Bull. U.S., 70, 821-838.

Hunter, J. R., 1976: Culture and growth of northern anchovy, Engraulis mordax, larvae. Fish. Bull. U.S., 74, 81-88.

Hunter, J. R., 1977: Behavior and survival of northern anchovy, Engraulis mordax, larvae. Cal. Coop. Fish. Invest. Rep., 19, in press.

Hunter, J. R. and C. Sanchez, 1976: Diel changes in swim bladder inflation of the larvae of the northern anchovy, Engraulis mordax. Fish. Bull. U.S., 74, 847-855.

Hunter, J. R. and G. Thomas, 1974: Effect of prey distribution and density on the searching and feeding behavior of larval anchovy, Engraulis mordax Girard. In The Early Life History of Fish, J.H.S. Blaxter (ed.), pp. 559-574, Springer-Verlag, Berlin.

Kiefer, D. A. and R. Lasker, 1975: Two blooms of Gymnodinium splendens, an unarmored dinoflagellate. Fish. Bull. U.S., 73, 675-678.

Kjelson, M. A., D. S. Peters, G. W. Thayer and G. N. Johnson, 1975: The general feeding ecology of postlarval fishes in the Newport River estuary. Fish. Bull. U.S., 73, 137-144.

Lasker, R., 1964: An experimental study of the effect of temperature on the incubation time, development and growth of Pacific sardine embryos and larvae. Copeia 1964, 399-405.

Lasker, R., 1965: The physiology of Pacific sardine embryos and larvae. Calif. Coop. Ocean. Fish. Invest. Rep., 10, 96-101.

Lasker, R., 1975: Field criteria for survival of anchovy larvae: the relation between inshore chlorophyll maximum layers and successful first feeding. Fish. Bull. U.S., 73, 453-462.

Lasker, R., 1978: The relation between oceanographic conditions and larval anchovy food in the California Current: identification of factors contributing to recruitment failure. Rapp. P.-v. Reun. Cons. Int. Explor. Mer., 173, 212-230.

Lasker, R., H. M. Feder, G. H. Theilacker and R. C. May, 1970: Feeding, growth and survival of Engraulis mordax larvae reared in the laboratory. Mar. Biol., 5, 345-353.

Murdie, G., 1971: Simulation of the effects of predator/parasite models on prey/host spatial distribution. In Statistical Ecology, G. P. Patil, E. G. Pielou and W. E. Waters (eds.), Vol. 1, pp. 215-233, The Pennsylvania State University Press.

O'Connell, C. P. and L. P. Raymond, 1970: The effect of food density on survival and growth of early post-yolk-sac larvae of the northern anchovy (Engraulis mordax Girard) in the laboratory. J. exp. mar. Biol., 5, 187-197.

R. de Mendiola, B., 1974: Food of the larval anchoveta Engraulis ringens J. In The Early Life History of Fish, J.H.S. Blaxter (ed.), pp. 277-285, Springer-Verlag, Berlin.

Scura, E. D. and C. W. Jerde, 1977: Various species of phytoplankton as food for larval northern anchovy, Engraulis mordax, and relative nutritional value of the dinoflagellates Gymnodinium splendens and Gonyaulax polyedra. Fish. Bull. U.S., 75, 577-583.

Vlymen, W. J., 1974: Swimming energetics of the larval anchovy, Engraulis mordax. Fish. Bull. U.S., 72, 885-899.

Vlymen, W. J., 1977: A mathematical model of the relationship between larval anchovy (Engraulis mordax) growth, prey microdistribution, and larval behavior. Environ. Biol. Fishes, 2, 211-213.

# SPATIAL DISTRIBUTIONS AND ECOLOGY OF PELAGIC FISH

J. W. Horwood and D. H. Cushing

MAFF
Fisheries Laboratory
Lowestoft, NR33 OHT, England

## INTRODUCTION

Homogeneity in nature is rare. Even in the regular structure of crystals it is impurities which provide color. The distribution of fish provides no exception to this rule and variations in their distribution are observed on almost all time and space scales. During the 1970's the annual world catch of fish was about 70 million tons with an estimated maximum potential sustained catch of fish taken by traditional methods of 100 million tons (Gulland, 1974). With this degree of sampling, it is reasonable to assume it is known where most of the fish live. The early arguments on the degree of variation took place before the statistics of small samples were understood and later such variability became interesting in itself, partly to establish differences between samples and partly to determine the physical or biological origin of such differences and their functional significance. The scale of patches was examined as an end in itself and the study of large patches played a part in the development of the technology of marine science. The use made by animals of the large patches observed is fairly obvious, but in recent years attempts have been made to find what use is made by animals of the small scale variability available to their senses.

## GEOGRAPHICAL DISTRIBUTIONS

Figure 1 shows the global distribution of coastal pelagic fish catches but pelagic catches also include the oceanic tuna-like fish. This distribution is primarily governed by the productivity in different areas. Most prominent in Figure 1 is the stock of Peruvian

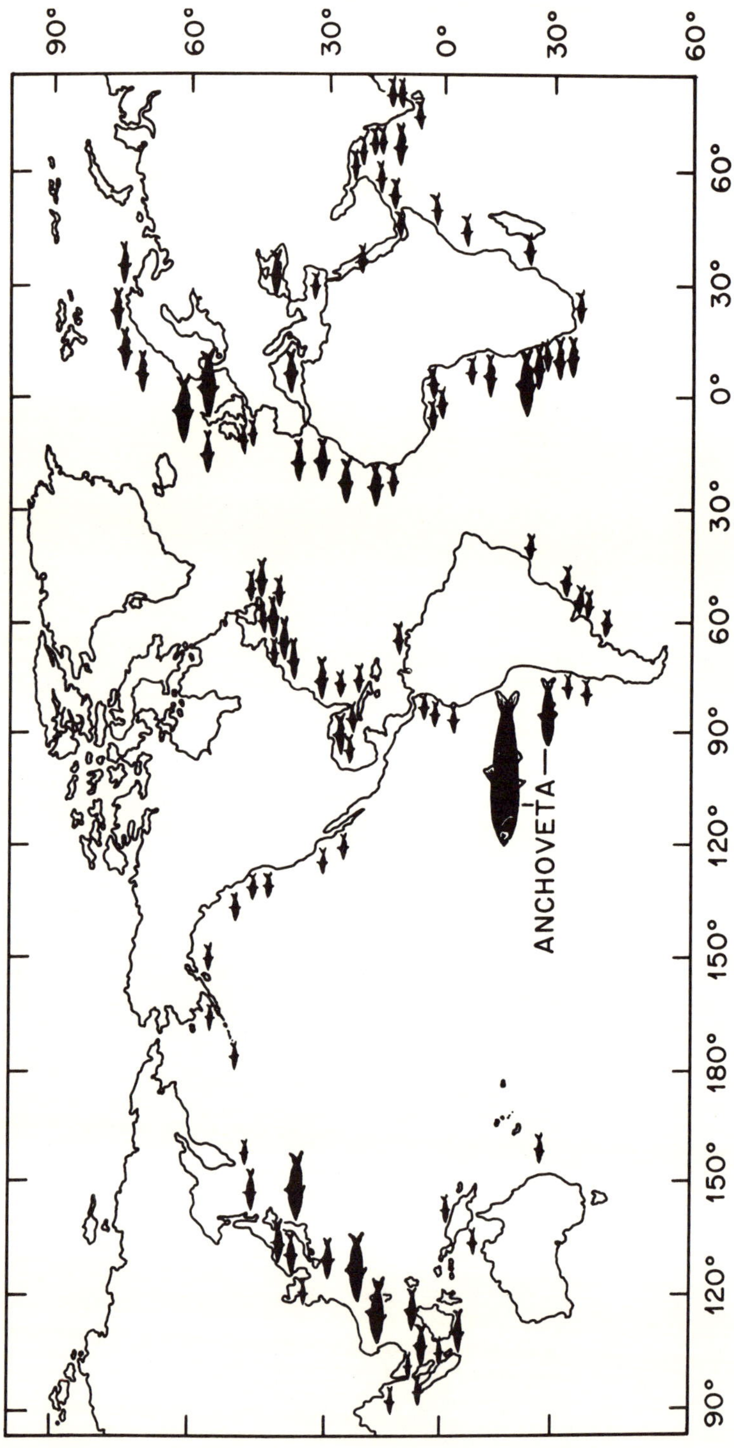


Figure 1. Distribution of the world coastal pelagic catches of fish in 1968 illustrating the regions of greatest fish production (from Gulland, 1971).

anchoveta which in 1968 yielded over 10 million tons, about a third of the world pelagic catch. This stock occurs in an upwelling area where nutrients are fed to the surface waters from below to maintain primary productivity at a much higher level than would otherwise occur. The distribution and biology of global upwelling systems are reviewed by Cushing (1971). Other large catches of pelagic fish come from the shallow seas of the continental shelfs such as the North Sea and Barents Sea where high nutrient levels and coastal production allow a complicated and rich food web (Wyatt, 1976). The oceanic tunas do not live in such areas of high productivity but in general take advantage of the supply of food generated in offshore divergences and advected and diffused through ocean currents as illustrated by Reid (1962).

Within the broad limits of food availability the stocks of fish are constrained or associated with physical discontinuities. Cushing (1968) cites three types of boundaries. Temperature isotherms show that some species have a preference for a given temperature field and that movement of this field can result in dispersion or concentration of the fish. The interface between ocean currents has also been found to attract fish (the "siome" of Japanese fishermen) and so do the regions of convergence and divergence in the oceans. The temperature constraint in the Arctic cod may be physiological but the associations of fish with the other two phenomena may also be connected with food supply. This has been observed in the case of North Sea herring (Steele, 1961) where aggregations occur on the edge of the Baltic outflow into the North Sea during early spring when the Baltic water was relatively rich in phytoplankton and zooplankton. However, the factors determining this gross distribution of fish in space also affect the temporal distribution of fish.

## MIGRATIONS

The most spectacular fish migrations are for spawning and feeding. The biology of larval, young and mature fish is sufficiently diverse that frequently the different groups need to be in different areas to feed. The North Sea plaice is a good example with the mature fish spawning in the central southern North Sea. The eggs and larvae are carried by the residual currents to the north-east and as they metamorphose they stay near the bottom and are carried to the nursery grounds of the Dutch coast or Waddensea. The fish grow and move out towards the central North Sea but there is a graduation by size with the larger fish in the deeper waters. After spawning the mature fish return to these areas. Harden-Jones (1968) has schematized this migration as in Figure 2, so it can be concluded that the occurrence of fish in any area is likely to be temporal.

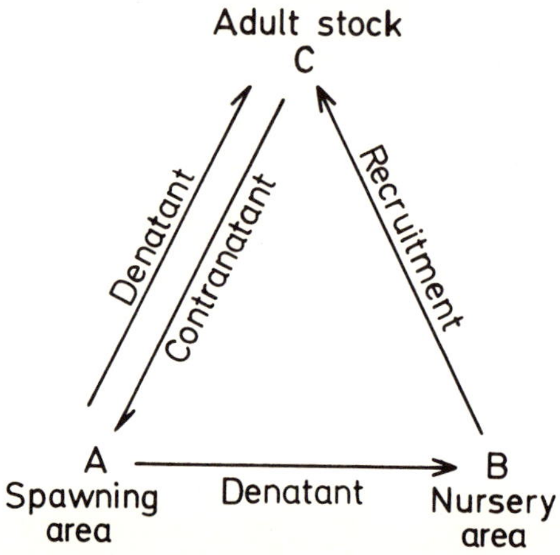


Figure 2. A schematic migration (from Harden-Jones, 1968).

The life history and migration of the North Sea herring stocks are described by Harden-Jones (1968) and Cushing (1968a). Harden-Jones' illustration of the migration of the Bank stock is shown in Figure 3 and it again shows the movement between spawning and feeding grounds. The whitebait drift along the continental coast, living quite close inshore, and at age one they move to the Bløden ground to the east of the Dogger. These young herring will either move north to join the adult stock or else move around the Dogger to join up in the central North Sea.

The mechanisms which enable the fish to undertake such large journeys are poorly understood but recent work on North Sea plaice and cod indicate that fish in shallow seas can make extensive and efficient use of tidal streams to migrate. Greer-Walker, Harden-Jones and Arnold (in press), using a sector-scanning sonar, have shown that fish can remain on the bottom during one part of the tidal cycle and then rise when the tide changes direction to be carried passively. In this way substantial progress can be made even against a residual tidal flow. This illustrates that there is an active use of the third dimension and that the vertical distribution of fish is by no means uniform.

Diurnal migrations have been observed in many fish species, pelagic as well as those normally defined as demersal. The herring usually stay in deep water or near the bottom during the day, rise to the surface during dusk and return to the deep water at dawn,

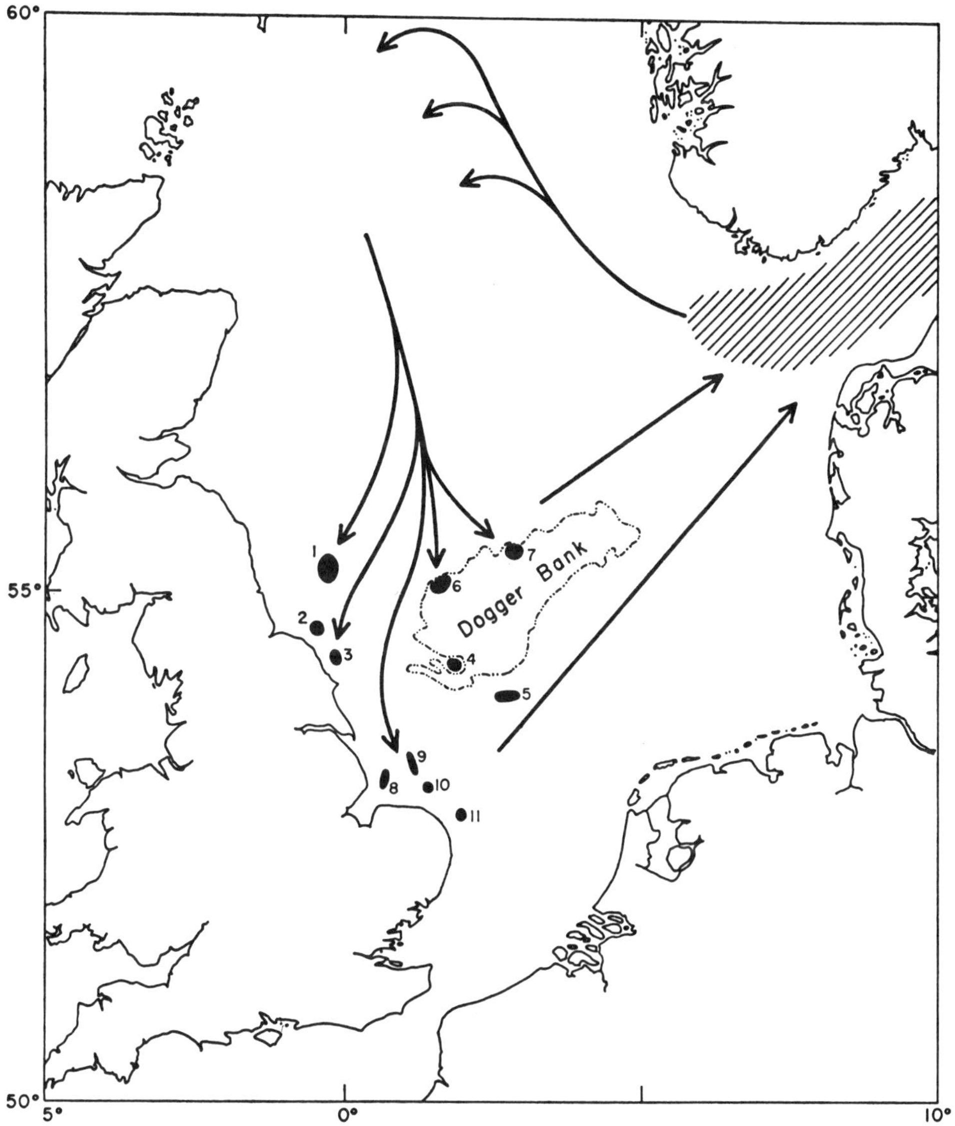


Figure 3. The migrations of adult Dogger or Bank spawning herring. The migration is anti-clockwise in the direction of the residual currents. The herring spawn to the west (spawning grounds black and numbered), and winter to the east in the Skagerrak (diagonal shading). (From Harden-Jones, 1968). Spawning ground key: 1 NE Bank; 2 Whitby; 3 Scarborough; 4 Dogger Bight; 5 South-side Great Silver Pit; 6 NW Rough; 7 Middle Rough; 8 Inner Dowsing; 9 Outer Dowsing; 10 Cromer Knoll; 11 Winterton Ridge.

although a slight midnight decline has been detected. The extent of migration may be related to the state of maturity of the animal, but in most cases light is cited as a consistent casual factor and its biological significance is usually assumed to be in relation to feeding (Harden-Jones, 1968). However, Richardson (1952) observed shoals of both sprat and herring undertaking diurnal migrations when not feeding. One problem is that most records are from echo-sounders and it is often unclear what species is being observed. In deeper waters the blue whiting migrate (Zilanov, 1968); during winter they remain between 100 m and 300 m depth in the day rising to 30 m or shallower at twilight, although in the summer the range is much less, between 30 m to 100 m. Similarly, the North Atlantic herring remain in the top 30 m to 40 m during summer but undertake diurnal migrations of 300-400 m during the winter. This differential seasonal pattern is also observed in the Arctic cod. The phase of their diurnal migrations differs amongst areas and is superimposed on an annual vertical migration with fish in the upper waters during the summer (e.g. Trout, 1957). Consequently, observations and analyses of fish distributions must take into account the time of year and time of day for each species.

## FISHERIES DISTRIBUTIONS

Those distributions of fish, which have been charted primarily to describe the geographical extent and quantity of a specific fish species, we have termed fisheries distributions and they provide the basic fisheries data. Usually these data are collected with the prime aim of estimating numbers, although internal spatial patterns may be observed incidentally. Various techniques have been employed in this process and the most modern summary is given by Cushing (1973).

Aerial mapping of fish distributions was first done in 1919 by Welch of the USA Bureau of Fisheries, in 1921 in France, and in 1921 by the English Ministry of Agriculture and Fisheries searching for herring shoals off Plymouth. In 1924 similar experiments were carried out in Scottish waters by Wood and McGee (1925). Patterns of herring shoals are recorded and appeared to be of the order of 10 miles diameter with a greater distance than this between shoals. However, the approach did not attract much attention. More recently, aircraft have been used to photograph and record the bioluminescent intensity of pilchards off south-west Africa (Cram and Hampton, 1976). The technique has the advantage that it is near-synoptic and is not biased by the fact that movement of the fish may involve their being counted twice or not at all. Cram and Hampton further note that shoals have been observed to actively avoid ships approaching them and that this effect may thus cause a bias in acoustic measurements. Visual aerial surveys of tuna have also been undertaken off Australia but it is not clear whether the tuna are as efficiently observed in deep water as in shallow water and this reflects the major disadvantage of this technique. Cram and Hampton note that the behavior of the African pilchard results in a good observational record and do not consider that many shoals are missed.

As described by Cushing (1973) the majority of the extensive stock assessment work uses acoustic techniques with accompanying trawl hauls for calibration and back-up. Tungate (1958) reported on a series of North Sea surveys using a Kelvin-Hughes echo-sounder to describe the movement and distribution of herring. Catches by local drifters were compared to ensure that the traces were of herring. Tungate reports they were able to direct drifters into large aggregations of herring that then rated some of the best catches of the season. From his figures it can be seen that herring aggregations were again of the order of 10 miles diameter with perhaps double that as the distance between groups.

More practical surveys can be done in regions where the target species is more isolated. Combined with trawl hauls these provide a good estimate of abundance. An example of this type of survey is given by Pawson *et al.* (1975, 1976) who undertook an acoustic survey of the spawning aggregation of blue whiting to the west of Britain. The equipment used was a 30 kHz Kelvin-Hughes Humber equipment with a Simrad QM integrator, as described by Burridge *et al.* (1977) and the signal was integrated to give a density value each mile. Pawson *et al.* (1975, 1976) reported that the fish lay to the west of the 200 m contour (Figure 4), typically at depths between 200 and 500 m and in a band between 20 and 50 m thick. During March and early April the blue whiting were probably not feeding, but in late April food was observed in their stomachs. During May, when the fish had begun to move to the north, over 95 percent were feeding. Later the fish are too dispersed to be usefully surveyed. Similar surveys (reported at a symposium on hydro-acoustics in fisheries research, Margetts, 1977) have been undertaken for Californian anchovies, sockeye salmon, Pacific hake and herring, Barents Sea 0-group fish and others.

There are instances, however, when acoustics are of little assistance. Blacker (1974) undertook an 0-group survey at Faroe. The main gear used was a Norwegian capelin trawl and samples were taken every 15 miles along a rectangular grid. Similar acoustic equipment to that used by Pawson was also employed. Two adjacent stations (81 and 82) yielded the following trawl catch.

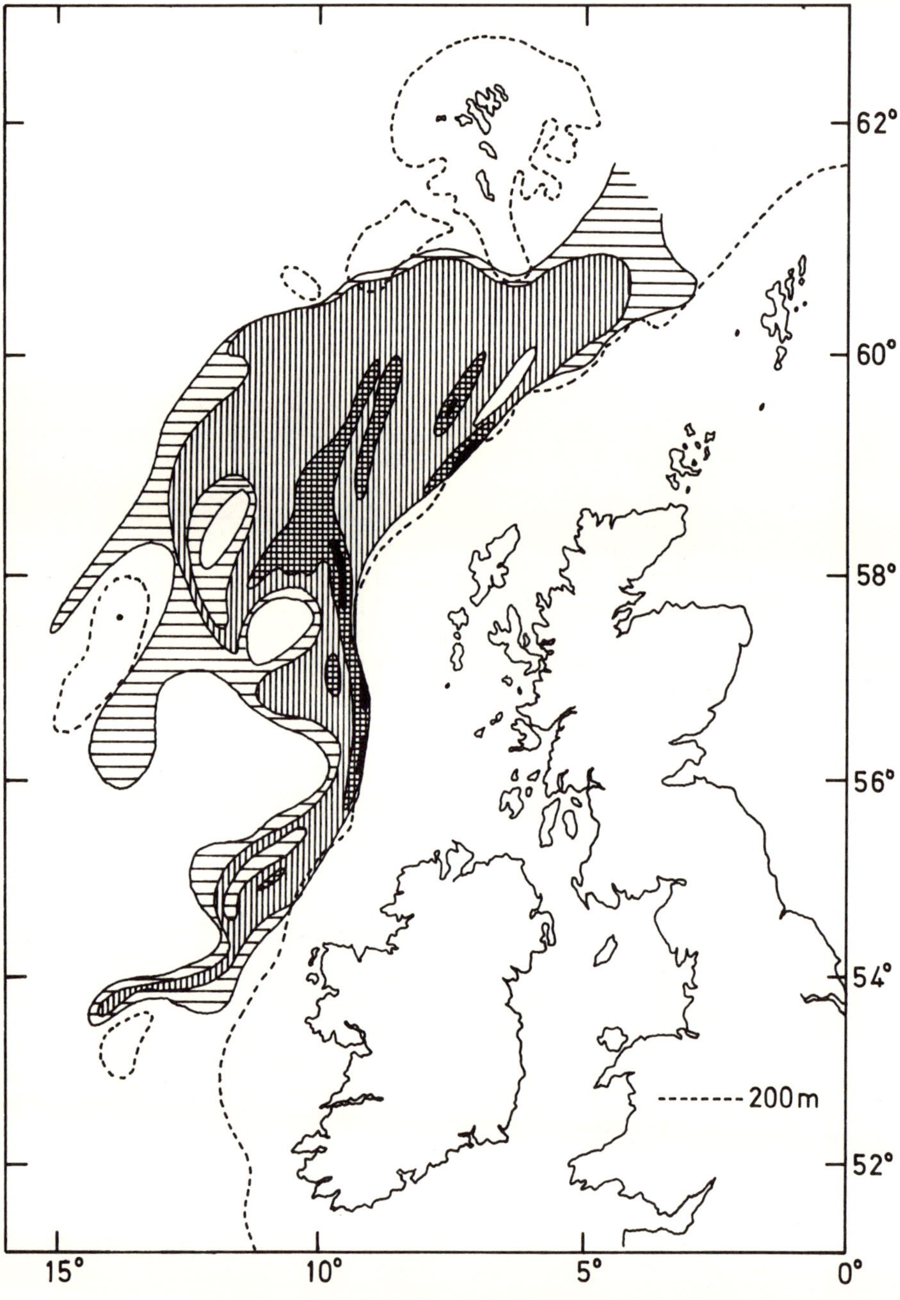


Figure 4. Distribution of blue whiting along the western continental shelf of Europe in April (from Pawson *et al.*, 1975).

| | 81 | 82 |
|---|---|---|
| Sandeel | 11148 | 5758 |
| Norway pout | 24768 | 18029 |
| Cod | 48 | 977 |
| Haddock | 756 | 183 |
| Blue whiting | 36 | - |
| Argentine | - | 3 |
| Herring | - | 3 |
| Anglerfish | - | 3 |
| Jellyfish | 60 liters | 9 liters |

It is clear that with such diversity, and particularly acoustic diversity, trawl hauls are necessary to evaluate the fish distribution in such a location; no method of acoustic identification of targets is yet available.

Taylor (1953) discussed fitting single and compound distributions to such trawl data. In general, it was found that the number of species followed a Poisson distribution but the numbers per species were clumped. These techniques were applied by Roessler (1965) for fish populations off Florida. However, fitting compound distributions to fisheries data is not common since virtually any compound model will yield adequate fits to the data and little, if any, insight into the mechanisms of fish distribution is achieved. Nevertheless, a general knowledge of the nature of fish distributions enables surveys to be designed to minimize the variance of stock estimates (e.g. Taylor, 1953; Jones and Pope, 1973).

## DISTRIBUTIONS WITHIN MAJOR AGGREGATIONS

If the major aggregations are defined as the extent of distribution of a species at any time in its life history or seasonal pattern, then within that aggregation specific spatial distributions are observed. An example already discussed is that of the North Sea herring where shoals occur at about 10 miles diameter with a somewhat greater distance between shoals. Many such examples can be found in the fisheries literature but usually as an aside to the main points under discussion. For example, Lockwood and Johnson (1976) described an acoustic and trawl survey of the mackerel population off the south-west of England and explained that this mackerel population is found throughout the whole European shelf in the summer months when it is in small shoals searching for prey, such as whitebait and copepods. As winter approaches, they return to their overwintering grounds where they do not feed. However, during the daytime they form large identifiable shoals which disperse during the night. They comment on two such shoals observed; one of dimensions 5 miles by 1.5 miles by 12 m thick containing 750 million fish and a second of 24 miles by 7 miles containing 990 million fish. A collation of these isolated pieces of information has not yet been undertaken.

The acoustic data provide an opportunity to study in more detail the spatial distribution of fish within its main aggregation. Horwood (1976) described the internal distribution of blue whiting observed by Pawson *et al.* (1975, 1976) off the west coast of Britain. Such data are amenable to treatment by spectral analysis as described by Platt and Denman (1975). Fourier transforms of some of the results of the 1976 survey are illustrated in Figure 5 and Figure 6. If the blue whiting had aggregated in shoals about 10 miles apart, a peak in the Fourier spectrum would have been observed at such a frequency. However, the shape shows

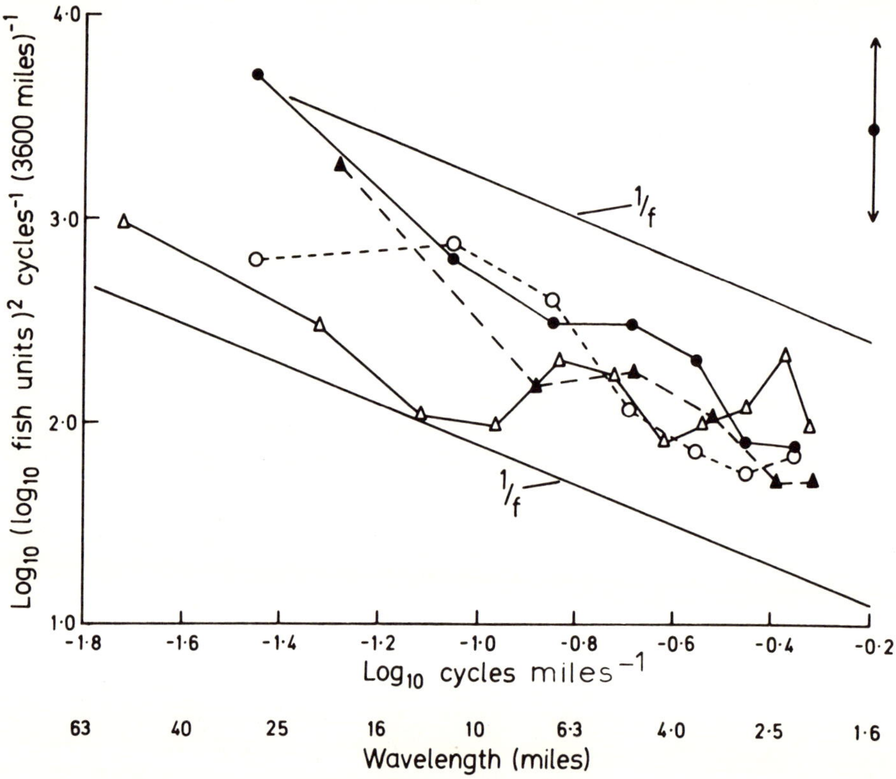


Figure 5. The power spectra of the distribution of blue whiting. An envelope of two lines of slope -1 is included. The approximate 95% confidence interval is in the right-hand corner.

no peaks at all and the transform can be described by a "1/f" or slightly steeper distribution (on the log-log plot this appears as a straight line of slope -1). Voss and Clarke (1975) explain that this implies a degree of correlation in the fish distribution over all length scales for which the spectrum is 1/f-like. Analysis of the 1975 survey gave similar results, even during May when most of the fish caught were feeding.

This is not a unique application of the techniques of spectral analysis to fish distributions. Mishima *et al.* (1966a) considered the spatial distribution of pink and chum salmon in the catches by gill nets in the North Pacific Ocean. Spectra totally unlike that for the blue whiting were obtained. The general distribution was flat but with prominent peaks. Mishima *et al.* concluded that pink and chum salmon were distributed similarly but that at the beginning of the season a short periodic component was observed and that later in the season a medium periodic component appeared also. The frequencies were at about 2½ and 10 $tan^{-1}$ (1 tan is about 12 m). Further experiments were undertaken to study this distribution in

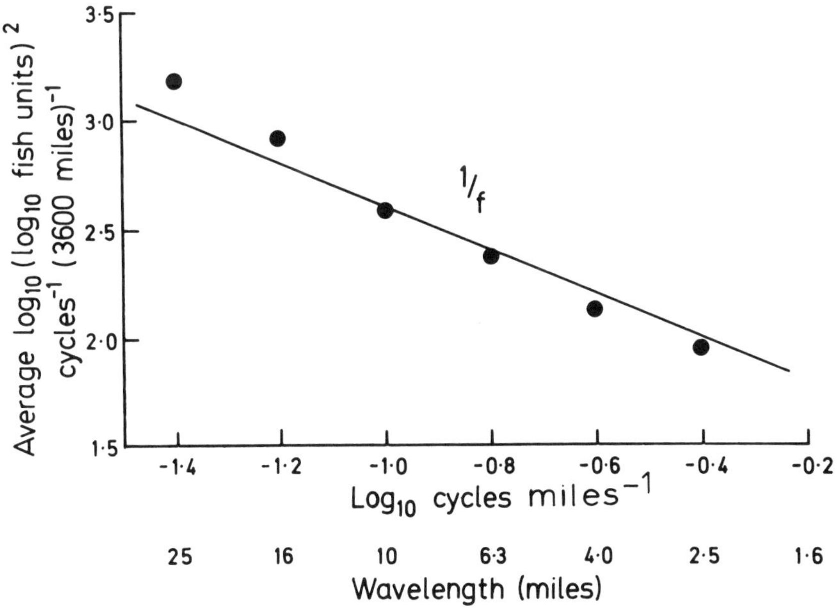


Figure 6. The average of the power spectra shown on Figure 5. The lines were averaged at specific frequencies and a line of slope -1 has been fitted by eye.

relation to diurnal movement (Mishima *et al.*, 1966b). Again the experiments were carried out with gill nets. When the nets were set at sunset and sunrise, periodic components of fish schools of 2-3 tans and about 4 tans were conspicuous in the spectra while at night the 4 tans component was still present but in a more noisy signal. Earlier tuna distributions caught by long-line have been described by Maeda (1959) although using statistical distributions.

Distributions of fish within a major aggregation have been described by Smith (1970) who reported on shoal sizes of anchovy with data obtained with a 30 kHz sonar. The distributions appear to have been very different to the blue whiting aggregations with shoals predominating. In the 200,000 square mile study area off California about one million schooled sonar targets were found with most shoals between 10 m and 30 m in diameter. Less than five percent exceeded 60 m. Fish shoals and plankton patches differ in that a shoal coheres by behavioral factors but a plankton patch is the result only of the population dynamics within a local hydrodynamic frame. Again, fish assemble at particular places at particular times having made positive migrations; for example, the Arcto-Norwegian cod spawns in a layer about 20 miles long on the northern edge of the Vestfjord in northern Norway. A shoal of spawning herring extended uptide for 27 km from Sandettié Bank in the first week of November. The Atlanto-Scandian herring formed a large overwintering shoal in the east Icelandic current in late autumn where they were in the end caught by Norwegian and Icelandic purse seiners. Figure 7 shows a record of one shoal of herring about 8 km in length which moves with the tide; the record was made with a large sidescan sonar (Rusby, 1977).

The density of fish within a shoal is summarized in the following table:

| | $n/m^3$ | m/fish | |
|---|---|---|---|
| Sardine | 2 | 0.79 | (Cushing, 1957) |
| Herring | 0.7-1.0 | 1.00 | (Truskanov and Scherbino, 1966) |
| Gadoids | $1\text{-}8.10^{-5}$ | 44-91 | (Cushing, 1968b) |
| Tuna | $1.10^{-4}$ to $1.10^{-6}$ | 22-100 | |
| Pilchards | 0.003-0.05 | 3 - 7 | (Weston *et al.*, 1969) |

The sizes of shoals vary considerably and those measured tend to be the large and spectacular ones. Like the Sandettie shoal of herring referred to above, one of Pacific hake at Port Susan in Puget Sound extended for 12 km along the depth contours; it was 6-20 m thick and 0.3-0.5 km wide (Alverson, 1967). Shoals of hairtails in the east China Sea were 40 miles by 5.5 miles and were 10 m thick; three such shoals were found about 50 miles apart.

Shoals of Atlanto-Scandian herring in the east Icelandic current in autumn were about 2-5 miles across (Truskanov and Scherbino, 1966). The commonest shoal of pelagic fish makes the well-known "plume" trace on the paper record of an echo sounder and it is 20-50 m across, depending on depth, and its vertical extent is 10-30 m; it occupies roughly 15,000-50,000 $m^3$.

The differences in shoal pattern can be elucidated by studying shoals of the same species at different times of the year. Spent

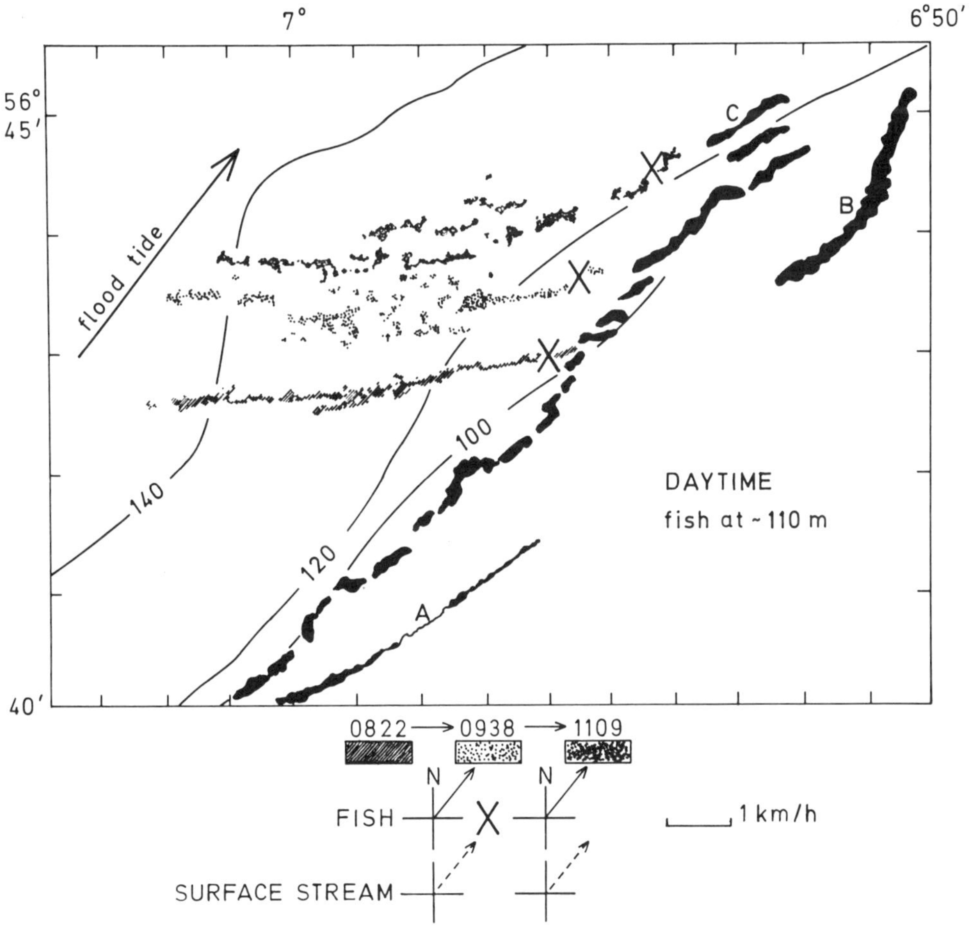


Figure 7. Large fish shoals: herring off western Scotland (Rusby *et al.*, 1977); this long range record was taken with GLORIA, the long range side scan sonar; the patch of fish, about 8 km in length, moved northeasterly on the flood tide (as shown by the crosses). The other marks represent rocks.

herring overwintering in the Norwegian Deep were observed as single fish each recorded separately on the paper record. Prespawners formed a diffuse extensive layer of small shoals in the southern North Sea. Migrating prespawning Atlanto-Scandian herring formed a dense layer at 138 m as they crossed the Norwegian Sea. When spawning, the herring formed dense tower-like traces on the bottom in the southern North Sea, as they laid their eggs on the seabed. In the feeding areas herring formed small plumes. Feeding shoals are small and scattered yet they can be shown to aggregate on patches of their food; spawning shoals are dense and large and spent herring disperse to considerable distances between fish. This account is taken from Cushing (1973).

Detailed behavioral patterns over a range of 20 km have been detected by long range sonar working at a frequency of 1 kHz (Weston and Revie, 1971). A pulsed beam is received back through 4 degrees in the horizontal and covers the entire depth (Figure 8). The sinusoidal waves are shoals of pilchard in the Bristol Channel, confirmed as such by fishing, and the waves are due to passive movement of these shoals with the tide. Results over a period of 23 days are described by Revie *et al.* (1974). A diurnal pattern can be seen with the shoals dispersing at dusk and aggregating at dawn; the rates of such dispersion can be estimated from these data. Even the noise pattern during the night can identify the distribution of the fish and indicates that the fish are homogeneously distributed in either the top, or the bottom of the water column, but not both. Weston and Revie gave an average of 2 shoals per $km^2$. The typical shoal was about 10-15 m across with about $3 \times 10^4$ pilchards in it. Weston (personal communication) has said that the number of shoals per linear mile varies seasonally with a peak in July to October, and also from year to year. The shoal number on any day does not appear to vary by more than a factor of two from mile to mile. Shoals other than pilchard have been detected. Some sprat shoals were observed further offshore than the pilchards and held their position against the tide. These sprat shoals, at 30-40 km, have been observed to make pronounced shoreward movements at dusk at rates of up to 2 knots (Weston and Revie, 1975). (The opposite reaction has been observed in skipjack tuna (Yuen, 1970), when the tuna are thought to return to the banks during daytime to feed.) Finally, this method allows for size discrimination and high frequency noise has been attributed to fish clustering around the hydrophone (Ching and Weston, 1975).

## FISH DISTRIBUTION AND FEEDING

Geographical and seasonal scales of feeding have already been mentioned and give a gross strategy of fish behavior; the diurnal behavior of the fish is more tactical. Generally, pelagic fish disperse at night and shoal during the day, feeding predominantly

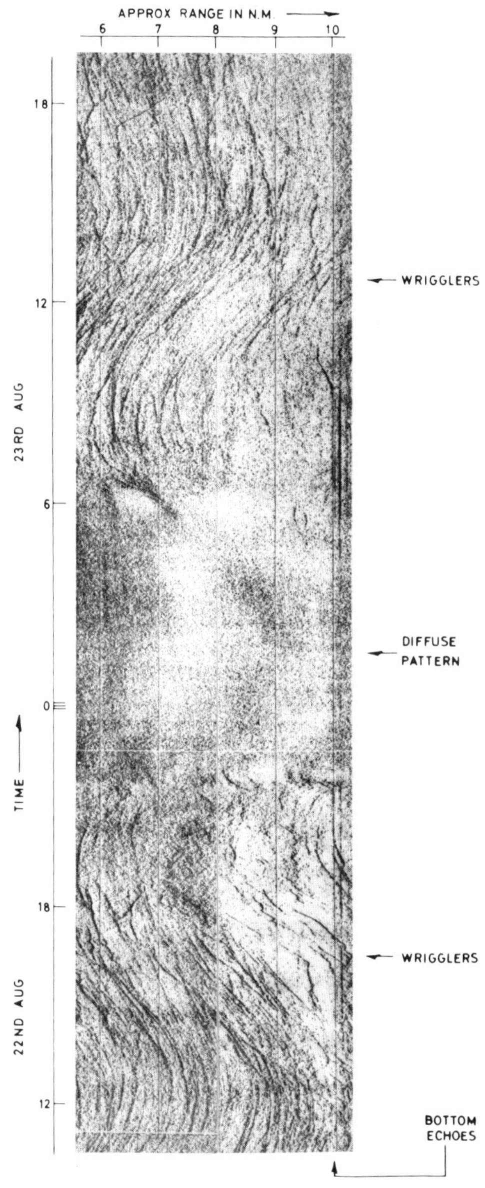


DISPLAY FOR 22ND/23RD AUGUST 1962 (MONTAGE)

Figure 8. Shoals, probably pilchards, observed for 2 days over long range; this record was taken with a long range, low frequency sonar. A range of 4-16 nm is given along the abscissa and the ordinate represents time for a period of 30 hours. The lines of "wrigglers" in daytime represent fish shoals (probably pilchards) moving back and forth in range with the tidal cycle; at night, however, the lines become diffuse.

at dusk and dawn. The reasons for this must be complex and intimately related to the diurnal behavior of zooplankton which are usually nearer the surface during the night. There must be an evolutionary battle between fish and zooplankton with the zooplankton being least active during the daytime when they could be most easily observed. Similarly, the fish are forced to be inactive during the day when they could be most efficient. At night there is perhaps less fish activity than at dusk and dawn. The zooplankton may also be more efficient at the dusk and dawn light levels. Applying search theory, Saila and Flowers (1969) indicated that predators should be most efficient if they worked at periods when their prey was most active and it has been established (Olson, 1964) that the advantage of shoaling, to a predator, reduces as the detection range of the shoal is increased. At nightfall the increased activity for efficient hunting is predictable but the break-up of shoals implies a communication system such as that described by Magnuson (1964) for tuna.

Feeding rates of fish vary seasonally. Thorpe (1977) reported that the daily ration of perch in Loch Leven varied between 3.2 percent wet body weight in December to 6.5 percent in July. Cushing (1964) analyzed the feeding relationships of the North Sea herring and suggested that a daily ration of about 3 to 3.3 percent would be expected. Examination of gut contents of herring, however, indicated maximum rations of copepods of about 1000 per day, yielding about 0.7 percent. In both calculations it was necessary to accept a digestion time of 14 hours. The weight of copepods found in the herring guts at any one time was much less than the typical fraction of body weight found for a range of fish by Blegvad (1917). The two separate sources of information may be reconciled by the observed growth needing some 3 percent per day and by the digestion rate of copepods by herring being much faster than 14 hours. Of equal interest, however, is the shape of the feeding curves of the herring and two features are apparent. First, under a range of natural food densities there is no obvious saturation level of feeding, and below prey densities of 100 $mm^3$ per $m^3$ very few prey are taken. Secondly, in some years it was observed that at high prey densities, smaller copepods were taken. Satiation levels have been observed on shoals of small fish kept in laboratories. Ishiwata (1968) fed schools of jack mackerel, puffer, grouper, filefish and yellowtails to satiation levels; except for the mackerel, which took 20 percent of its body weight, the others stopped at 10 percent. Zusser (1961) carried out feeding experiments on fish under different light conditions and found that satiated fish avoided light whereas hungry fish would ignore the light. It might therefore be assumed that fish will feed until satiated if it is economical for them to do so.

The food of different species is usually well described and Hardy's food web is probably the most cited example but this approach has been criticized by Wyatt (1976) who explained that in any area at any time the food web is best described by a simple food chain, and the fish move from one chain to the next. Associated with this idea must be the concept of optimum food size for any animal. Cushing (1964) concluded that the older copepodite stages of *Calanus finmarchicus* were probably the optimum size for herring, and the general size-orientated approach to marine ecology has been advocated by Parsons and Le Brasseur (1970) who fed juvenile fish on size sorted particles. There are no data to quantify the densities at which a predator species will switch to an alternative but less desirable prey but, as pointed out by Wyatt, the catholic tastes of fish make it highly probable that such switching will occur. Such changes in feeding will depend on the spatial variation in plankton concentration.

The existence of large patches of plankton in the sea was first established by Savage and Wimpenny (1936) who showed that extensive populations of the diatom *Rhizosolenia styliformis* lived in the southern North Sea for a period of four months; indeed they were able to draw a curve in time of the slow rise of the population to a peak before it collapsed. It is not known how such patches originated, but they first appeared off the Yorkshire coasts where thermocline fronts (Hughes, 1976) are found in summer. It is possible that the patches of *Rhizosolenia* were shed from the front and moved in a roughly southeasterly direction towards the tidally dominated regions of the southern North Sea.

In 1954 a patch of *Calanus* was followed for about three months (Cushing and Tungate, 1963); as one cruise was completed, the next was started and there was no doubt that the patch was the same from cruise to cruise. The time series of numbers at each stage is self-consistent and mortality was low during the younger and middle copepodite stages. Heavier mortality was inflicted on the adults and older copepodites by the aggregating herring (Cushing, 1955). The large patch of *Calanus* was a technical convenience but the work raises an interesting point. In the waters round the British Isles two forms of production cycles have been observed: first, the production cycle reaches its peak before the nutrients start to decline (Harvey *et al.*, 1935; Cushing and Vucetic, 1963), and in the second, the nutrients had fallen by half before the production cycle had risen to half its peak (Steele and Henderson, 1977). There are two forms of explanation. The first is a dynamic one: in the second example, the rate of change of production was probably greater than the rate of change of grazing capacity, as if there had been a sudden period of calm weather. In the first example, the rates of change of production and grazing capacity were perhaps more closely matched and the decline of nutrients only occurred when the major transfer of material to herbivore flesh occurred, after the peak of production. The second is a statistical explanation which summarizes many production cycles but the first example may represent that of a single population.

A second type of patch observed at sea is exemplified by fish eggs or larvae. Patches of plaice eggs are spawned between the Thames and the Rhine and drift northeastwards towards Texel Island on the northwest coast of Holland. Here, at certain states of the tide, plaice larvae in the lower half of the water column drift across a boundary between the Dutch coastal water and the open Southern Bight. Having crossed the boundary, the larvae, which are metamorphosing, reach their nursery ground on the flats of the Waddensea. Changes in such large patches provide estimates of mortality rate as they drift from spawning ground to nursery ground (Harding and Talbot, 1973).

Large patches were also used to study the aggregation and disengagement of pelagic fish (estimated acoustically) on to patches of *Calanus* off the northeast coast of England (Figure 9). Such patches were 50 to 100 km across and the fish, which were probably herring, gathered on the patches for about three weeks. They aggregated predominantly on copepodite stages V, which (as noted above) suffered the greatest mortality, and as numbers reduced, they disengaged; the processes of aggregation and disengagement were estimated by correlation coefficients. The same phenomenon was observed by Jakobsson (1963) in the feeding fishery for herring north of Iceland. Such a mechanism of aggregation and disengagement requires, first, a "reservoir" of fish from which the aggregation can be drawn, and, second, a sufficient mechanism to generate aggregation, for example, by fish decelerating to feed.

Little is known about the ability of pelagic fish to aggregate on their prey. The bioeconomics of searching and the desirability of leaving a poor feeding area in search of a better one (that might not exist) have been studied by Royama (1971) and the theory has been shown to hold for birds (Goss-Custard, 1977) and university students subjected to a similar problem. The theory predicts a range of predator density with prey density. The aggregation of herring onto patches of copepods (Cushing, 1955) were confirmed by herring catches and the results show unequivocally a general increase in echo-level with density of copepods. It was noticeable, however, that the two very high levels of echo readings were at intermediate copepod densities. Over the season, as a consequence of grazing, the larger copepods decline and aggregation ceases. With the same data, Cushing noted the succession of grazing from older stages of copepods to smaller stages and in poor years to smaller copepods. His results amplify the early claims of the success of the plankton "indicators" given to drifter skippers (Hardy, 1936). The skippers fished in areas of high *Calanus* abundance and frequently claimed greater success than their colleagues not using the indicator. The interpretation was that herring were aggregating onto *Calanus*.

If the marine ecosystem dynamics are to be understood, then much more information needs to be obtained on the distribution of fish in relation to their food, their ration and feeding strategies.

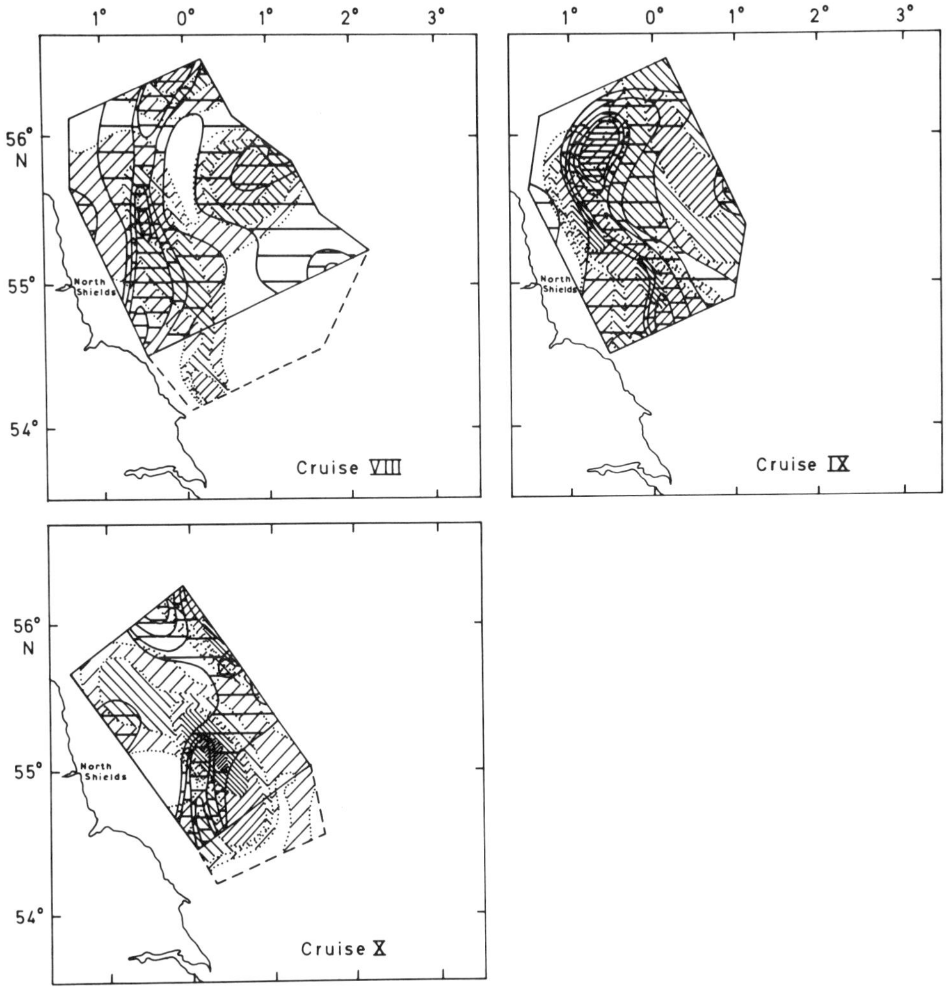


Figure 9. Aggregation of pelagic fish (estimated acoustically) on a patch of *Calanus* (Cushing, 1955); *Calanus* patches are shown within dotted contours and pelagic fish within full line contours.

The importance of such information for the construction of ecosystem models has been described by Steele (1976). The form that the predation rate takes affects the dynamics of all the trophic levels. Such questions as -- do fish feed for a certain number of hours then stop or do they feed until satiated -- or what densities of food induce movement away to other areas or other prey; all still need to be answered.

The natural mortality of fish (and possibly invertebrates) declines with age (Cushing, 1973). Figure 10 shows hypothetically how a patch of rich food might modify such mortality of larval fish. If the death rate depends on the availability of food, then mortality and its rate of change both decrease during the period of rich food. When the patch of food disappears after a short period, mortality is greater than would have been expected if the patch had not been encountered. If such mortality is modified by the availa-

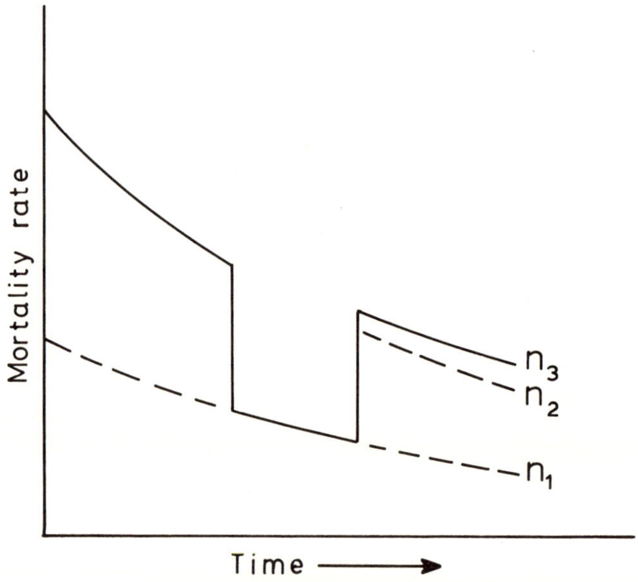


Figure 10. Modification of mortality rate by a patch of food; mortality rate is assumed to decline with time and to be density dependent but determined by food. A patch of food appears and the mortality rate drops from the line yielding numbers $n_3$ to that yielding $n_1$. Because the rate of change of mortality rate differs, the final number is $n_2$.

bility of food, it is also density dependent. If density independent mortality is modulated in a density dependent way, the environment is being exploited continuously by this sampling procedure. In this way the exploitation of the available food is maximized merely by its variability in time.

One of the remarkable points about those populations which pass through the plankton during their life histories is that most grow from the egg by three to seven orders of magnitude and are fecund to the same degree. As only two in each generation survive from a parental egg production, mortality (M) and growth (G) within that cohort are of the same order as the fecundity. Because the eggs die and do not grow, biomass must increase by perhaps two orders of magnitude during the life cycle which means that $(G/M) > 1$ until a critical age after the cohort has reproduced.

In the patch of rich food referred to in Figure 10, larval fish would grow more quickly and mortality would be reduced. The ratio (G/M) must increase in the patch of rich food and because the survivors are bigger they demand more food, and as a consequence have an inbuilt advantage in taking it. It is obvious that the specific growth rate will increase in a patch of rich food, but it is less obvious that the specific mortality rate should decrease. Then it is possible that the patches of rich food maintain the ratio (G/M) at values greater than unity. Further, the increment of biomass during the life cycle may in part be due to the variability of food in time.

The exploitation of food by adult herring depends on the mechanism of aggregation on to *Calanus* patches described above. Because the fish find their food more readily in patches, they slow down to feed and use the acquired energy more economically. As a consequence, the specific growth rate in the population must be variable. If a gain in weight makes a further gain more probable, the distribution of weight is probably skewed; perhaps more eggs might be produced from such a distribution than from a normal one. Patches of food may be exploited by larvae to maximize the food available and to determine the ultimate biomass that recruit to the adult population, whereas patches exploited by the adults may establish the numbers of eggs laid for the subsequent generation.

The distribution of newly laid plaice eggs in the Southern Bight of the North Sea between the Thames and the Rhine is a large diffused patch generated by females each about 80 m apart on average. Their spawning produces a large patch of eggs and subsequently a large patch of larvae that is some tens of kilometers in diameter. These patches retain their identity against the diffusive processes in the sea so that the majority of the surviving larvae reach the nursery ground on the Waddensea inside Texel Island (Cushing, 1972); indeed a patch of metamorphosing or metamorphosed larvae has been recorded off Texel Island in the North Sea (Harding *et al.*, 1975). Further, because the water is somewhat shallower towards the end of the larval drift, the baby fish are carried into more productive water at the time when their need for food is greatest.

This example is really part of a more general argument which states that in temperate waters, at least, spring spawning fish lay their eggs at a fixed season just before the spring outburst so that the larvae have the greatest chance of finding food. Spawning grounds are usually at a fixed position and so are the nursery grounds. Thus the larval drift forms the geographical base to the migration circuit. In an evolutionary sense it is preordained that in early spring a large patch of plaice larvae in the Southern Bight of the North Sea lives in water originating from the English Channel. In this water large patches of their preferred food, *Oikopleura* (or *Pseudocalanus* in 1963), appear and the degree of superposition of such patches may govern the generation of biomass during the larval drift.

Large patches of diatoms exist and it is difficult to know if the larvae gain advantage from them. Nevertheless, Lasker (this volume) and Shelbourne (1957) have observed such features in anchovy and plaice larvae. If the algae were evenly distributed in a temperate sea, they would exhaust the local nutrients with little transfer of material to the herbivores. In the large patches numbers are high enough to generate grazing capacity and return nutrients to the water and sustain production far longer.

Grazing introduces a dynamic character into the production cycle, the first step towards the steady state system, and this system really depends on the generation of patches.

## DISCUSSION

Although the original interest in patchiness was a technical one, the recent interest stems from the conviction that animals might make use of patches to their advantage. The advantages differ, but any advantage to an animal population is expressed in its growth, mortality and reproduction. An animal in the sea avoids death by growing and it grows to reproduce. Differences in the ratio of growth to mortality may express the capacity of animals to explore and exploit the physical and biological environment through its variability. By such processes the magnitude of recruiting year classes in fish populations may well be established and by extension the population may be controlled by them.

There are two obvious points about patches. The first is that it is hard to imagine animals not exploiting the high densities particularly if they recur in any evolutionary sense. The second is that it is hard to imagine a world of evenly distributed characteristics, so deeply is the variability ingrained in our senses. Perhaps the sea is particularly valuable for such a study because the variations in time and space are intermingled, whereas those on the land have a quasi-permanent character.

The advantages and features that accrue from variability in space or time may be:

1. it is cheaper to feed on patches than on evenly distributed individuals;

2. if larval and juvenile growth should maximize biomass, and if mortality is density dependent, variability in food ensures that the biological and physical environments are exploited with greatest efficiency;

3. if the growth of adults should maximize egg production, the variability in specific growth rate may skew the weight distribution enough to achieve this end;

4. large patches of eggs are laid in the sea to obtain geographical continuity in the life cycle. Thus, if the spawning ground is always in the same place, so is the nursery ground;

5. fishes may shoal to minimize predation, and

6. diurnal behavior may give a pulsed predation with selection of food items, aggregation and possibly satiation occurring.

## SUMMARY

It has been shown that fish distributions occupy all space and time scales. The spatial features occur from global scales determining major productivity sites to those of the dimensions of the fish themselves when feeding individually. Other major intermediate scales are those of the shoal and the migration circuit. The time scales exhibit a similar pattern with stocks disappearing and reappearing over historical times. Scales of seasons or the lifetime of the fish are reflected in migrations, and diurnal movements are usually associated with feeding. The feeding ecology is poorly known although basic rations can be easily calculated. Evidence suggests that pelagic fish take advantage of patchy food through aggregation and that satiation can be achieved in the laboratory although it is not clear whether satiations occur naturally. Feeding, affecting growth and consequently modifying mortality, has significant effects on the population dynamics of the stocks.

## REFERENCES


Alverson, D. L., 1967: Distribution and behaviour of Pacific hake as related to design of fishing strategy and harvest rationale. FAO Conf. Fish Behaviour. FAO Fish. Rep. 62, 2, 361-376.

Blacker, R. W., 1974: Report of the English 0-group fish survey at Faroe, July 1974. I.C.E.S. C.M. 1974/F:28, 8 pp.

Blegvad, H., 1917: On the food fish in the Danish waters within the Skaw. Rep. Dan. Biol. Sta., XXIV, 17-72, 1916.

Burridge, A. J., P. G. Griffiths, J.W.R. Griffiths and D. W. Hoare, 1977: Time varied gain echo sounder receiver systems for use with fish echo signal processing equipment. Rapp. P.-v. Reun. Cons. int. Explor. Mer., 170, 83-87.

Ching, P. A. and D. E. Weston, 1971: Wideband studies of shallow-water acoustic attenuation due to fish. J. Sound Vibr., 18, 499-510.

Ching, P. A. and D. E. Weston, 1975: Fast acoustic fluctuations caused by fish. J. Sound Vibr., 39, 287-292.

Cram, D. L. and I. Hampton, 1976: A proposed aerial/acoustic strategy for pelagic fish stock assessment. J. Cons. int. Explor. Mer., 37, 91-97.

Cushing, D. H., 1955: Production and a pelagic fishery. Fishery Invest., Lond., Ser. 2, 18, 104 pp.

Cushing, D. H., 1957: The number of pilchards in the Channel. Fish. Invest. Lond., 2, 21, 27 pp.

Cushing, D. H., 1964: The work of grazing in the sea. In Grazing in Terrestrial and Marine Environments, D. J. Crisp (ed.), pp. 207-225, Blackwell Scientific Publications, Oxford.

Cushing, D. H., 1968a: Fisheries Biology. A Study in Population Dynamics. University of Wisconsin Press, 200 pp.

Cushing, D. H., 1968b: Direct estimation of a fish population acoustically. J. Fish. Res. Bd. Can., 25, 2349-2364.

Cushing, D. H., 1971: Upwelling and the production of fish. In Adv. Mar. Biol., F. S. Russell and M. Yonge (eds.), 9, 255-334.

Cushing, D. H., 1972: The production cycle and numbers of marine fish. Symp. Zool. Soc. Lond. (1972), 29, 213-232.

Cushing, D. H., 1973: The Detection of Fish. Pergamon Press Ltd., Oxford, 200 pp.

Cushing, D. H. and D. S. Tungate, 1963: Studies on a Calanus patch. I. The identification of a Calanus patch. J. Mar. Biol. Ass. U.K., 43, 327-337.

Cushing, D. H. and T. Vucetic, 1963: Studies on a Calanus patch. III. The quantity of food eaten by Calanus finmarchicus. J. Mar. Biol. Ass. U.K., 43, 349-371.

Goss-Custard, J. D., 1977: Predator responses and prey mortality in redshank, Tringa totanus (L.) and a preferred prey, Corophium volutator (Pallas). J. Anim. Ecol., 46, 21-35.

Greer Walker, M., F. R. Harden-Jones and G. P. Arnold, 1978: The movements of plaice (Pleuronectes platessa) tracked in the open sea. J. Cons. Int. Explor. Mer., in press.

Gulland, J. A., 1971: The Fish Resources of the Ocean. Fishing News (Books) Ltd., West Byfleet, England, 255 pp.

Gulland, J. A., 1974: The Management of Marine Fisheries. Scientechnica, Bristol, 198 pp.

Harden-Jones, F. R., 1968: Fish Migration. Edward Arnold, London, 325 pp.

Harding, D. and J. W. Talbot, 1973: Recent studies on the eggs and larvae of the plaice (Pleuronectes platessa L) in the Southern Bight. Rapp. P.-v., Cons. int. Explor. Mer., 164, 261-269.

Harding, D., D. S. Tungate and J. H. Nichols, 1975: The spawning of plaice in the southern North Sea and English Channel. Symp. Changes in North Sea Fish Stocks. I.C.E.S. Paper 42, 9 pp., mimeo.

Hardy, A. C., 1936: The ecological relations between the herring and the plankton investigated with the Plankton Indicator, Part 1. J. Mar. Biol. Ass. U.K., 21, 147-177.

Harvey, H. W., L.H.N. Cooper, M. V. LeBour and F. S. Russell, 1935: The control of production. J. Mar. Biol. Ass. U.K., 20, 407-441.

Horwood, J. W., 1976: The spatial distribution of the blue whiting (Micromesistius poutassou) by power spectral analysis. I.C.E.S. C.M. 1976/H:13, 5 pp.

Hughes, D. G., 1976: A simple method for predicting the occurrence of sea sonal stratification and fronts in the North Sea and around the British Isles. I.C.E.S. C.M. 1976/C:1, 10 pp.

Ishiwata, N., 1968: Ecological studies on the feeding of fishes. II. Acclimatization of a school of fish and saturation amounts. Bull. Jap. Soc. Scient. Fish., 34, 498-502.

Jakobsson, J., 1963: Some remarks on the distribution and availability of the Iceland north coast herring. Rapp. P.-v. Cons. int. Explor. Mer., 154, 73-83.

Jones, B. W. and J. G. Pope, 1973: A groundfish survey of Faroe Bank. Int. Comm. Northw. Atlant. Fish. Res. Bull., 10, 53-61.

Lockwood, S. J. and P. O. Johnson, 1976: Mackerel research in the southwest. Lab. Leafl. MAFF Direct. Fish. Res., Lowestoft (32), 26 pp.

Maeda, H., 1959: A tentative analysis of the distribution pattern of tuna projected in the long-line. J. Shimonoseki Coll. Fish., 293, 1-218.

Magnuson, J. J., 1964: Activity patterns of scombrids. Proc. Hawaiian Acad. Science, 39th Annual Meeting, 26 pp.

Margetts, A. R., 1977: Hydroacoustics in fisheries research. A symposium held in Bergen, June 1973. A. R. Margetts (ed.). Rapp. P.-v. Cons. int. Explor. Mer., 170, 327 pp.

Mishima, S., M. Ueno and R. Kawashima, 1966a: On the characteristics of spatial distribution of salmon in the sea areas at the fishing season. Bull. Jap. Soc. Scient. Fish., 32, 837-842.

Mishima, S., M. Ueno and R. Kawashima, 1966b: On the spatial distribution of pink salmon with the lapse of time during a night. Bull. Jap. Soc. Scient. Fish., 32, 843-846.

Olson, F.C.W., 1964: The survival value of fish schooling. J. Cons. int. Explor. Mer., 29, 115-116.

Parsons, T. R. and R. J. LeBrasseur, 1970: The availability of food to different trophic levels in the marine food chain. In Marine Food Chains, J. H. Steele (ed.), pp. 325-343, Oliver and Boyd, Edinburgh.

Pawson, M. G., S. E. Blanchflower, S. T. Forbes and J. Richards, 1976: Preliminary results of the 1976 English and Scottish blue whiting surveys. I.C.E.S. C.M. 1976/H:15, 12 pp.

Pawson, M. G., S. T. Forbes and J. Richards, 1975: Results of the 1975 acoustic surveys of blue whiting to the west of Britain. I.C.E.S. C.M. 1975/H:15, 13 pp.

Platt, T. and K. L. Denman, 1975: Spectral analysis in ecology. Ann. Rev. Ecol. Syst., 6, 189-210.

Reid, J. L., 1962: On circulation, phosphate phosphorous content and zooplankton volumes in the upper part of the Pacific Ocean. Limnol. Oceanogr., 7. 287-306.

Revie, J., J. Wearden, P. D. Hocking and D. E. Weston, 1974: A twenty-three day twenty mile echo record of fish behaviour. J. Cons. int. Explor. Mer., 36, 82-86.

Richardson, I. D., 1952: Some reactions of the pelagic fish to light as recorded by echo sounding. Fishery Invest. Lond., Ser. 2, 18, 20 pp.

Roessler, M., 1965: An analysis of the variability of fish populations taken by otter trawl in Biscayne Bay, Florida. Trans. Amer. Fish. Soc., 94, 311-318.

Royama, T., 1971: Evolutionary significance of predator's response to local differences in prey density: a theoretical study. In Dynamics of Populations, P. J. den Boer and G. R. Gradwell (eds.), pp. 344-357, Proc. Adv. Study Inst. Oosterbeck, 1970. Centre for Agri. Publ. and Documentation, Wageningen.

Rusby, J.S.M., 1977: Long-range survey of a herring fishery by side-scan sonar. Rapp. P.-v. Reun. Cons. int. Explor. Mer., 170, 7-14.

Saila, S. and J. M. Flowers, 1969: Elementary applications of search theory to fishing tactics as related to some aspects of fish behaviour. FAO Conf. Fish Behaviour, FAO Fish. Rep., 62.2, 343-356.

Savage, R. E. and R. S. Wimpenny, 1936: Phytoplankton and the herring. Fish. Invest. Lond., XV, 1, 88 pp.

Shelbourne, J. E., 1957: The feeding and condition of plaice larvae in good and bad plankton patches. J. Mar. Biol. Ass. U.K., 3, 539-552.

Smith, P. E., 1970: The horizontal dimensions and abundance of fish schools in the upper mixed layer as measured by sonar. In Proc. of an Int. Symp. on Biological Sound Scattering in the Ocean, G. Brooke Farquhar, pp. 563-560, Maury Centre for Ocean Science.

Steele, J. H., 1961: The environment of a herring fishery. Mar. Res. Scot., 6, 1-19.

Steele, J. H., 1976: The role of predation in ecosystem models. Mar. Biol., 35, 9-11.

Steele, J. H. and E. W. Henderson, 1977: Plankton patches in the North Sea. In Fisheries Mathematics, J. H. Steele (ed.), pp. 1-20, Academic Press, New York.

Taylor, C. F., 1953: Nature of variability in trawl hauls. Fish. Bull. U.S. Fish and Wildlife Service, 54, 145-165.

Thorpe, J. E., 1977: Daily ration of adult perch during summer in Loch Leven, Scotland. J. Fish. Biol., 11, 55-69.

Trout, G. C., 1957: The Bear Island cod: migrations and movement. Fishery Invest., Lond., Ser. 2, 21, 51 pp.

Truskanov, M. and M. Scherbino, 1966: Methods of direct calculation of fish concentrations by means of hydroacoustic apparatus. ICNAF Res. Bull., 3, 70-80.

Tungate, D. S., 1958: Echo sounder surveys in the autumn of 1956. Fishery Invest., Lond., Ser. 2, 22, 17 pp.

Voss, R. V. and J. Clarke, 1975: "1/f" noise in music and speech. Nature, 258, 317-318.

Weston, D. E., A. A. Horrigan, S.J.L. Thomas and J. Review, 1969: Studies of sound transmission in shallow water. Phil. Trans. Roy. Soc., 265, 567-608.

Weston, D. E. and J. Revie, 1971: Fish echoes on a long-range sonar display. J. Sound Vibr., 17, 105-112.

Weston, D. E. and J. Revie, 1975: Long-range sonar studies - a five-day record of an extensive concentration of fish. In Proc. of Eighth Int. Congr. on Acoustics, London, 1974, 467 pp.

Wood, H. and G. McGee, 1925: Aircraft experiments for the location of herring shoals in Scottish waters. Scientific Investigations, 1, Sea Fishery Board for Scotland, 20 pp.

Wyatt, T., 1976: Food chains in the sea. In The Ecology of the Seas, D. H. Cushing and J. J. Walsh (eds.), pp. 341-358, W. B. Saunders Company, Philadelphia.

Yuen, H.S.H., 1970: Behaviour of skipjack tuna (Katsuwonus pelainis) as determined by tracking with ultrasonic devices. J. Fish. Res. Bd. Canada, 27, 2071-2079.

Zilanov, V. K., 1968: Some data on the biology of Micromesistius poutassou (Risso) in the North East Atlantic. Rapp. P.-v. Reun. Cons. int. Explor. Mer., 158, 116-122.

Zusser, S. G., 1961: The daily vertical migrations of planktophagic fish. Trudy VNIRO, 44, 177-186.

# PATCHES, MOVEMENTS, AND POPULATION DYNAMICS IN ECOLOGICAL SYSTEMS: A TERRESTRIAL PERSPECTIVE

William C. Clark, Dixon D. Jones and C. S. Holling

Institute of Resource Ecology
University of British Columbia
Vancouver, B.C. V6T 1W5, Canada

## INTRODUCTION

Animals have always coped with more or less patchy environments and cannot be held responsible for the fact that ecologists have not. Some species discover or affect a degree of constancy in their surroundings and evolve accordingly. More commonly, they develop adaptive responses to the general experience that "here" and "now" is most unlikely to be the same as either "elsewhere" or "later" (Southwood, 1977).

The key word here is "adaptive." Fifty years ago terrestrial ecologists characterized animal movement as "on the whole a rather quiet, humdrum process...taking place all the time as a result of the normal life of animals..." (Elton, 1972, pp. 147). Further, they argued that it was "almost impossible to spend any time profitably in a deliberate study of animal dispersal as such" (Ibid., p. 155).

At least in terrestrial ecology, research perspectives have long since swung back from this extreme initial position. Dispersal has indeed been studied, and profitably so. The "humdrum" activity described by Elton does of course occur, often through the majority of an animal's life. But at critical stages it is often overridden by a highly differentiated and adaptive movement response to variable habitat pattern. In contrast to what Levin (1976a) has usefully characterized as the "null hypothesis of random diffusion," this later form of movement reflects explicit animal behaviors which in large part determine under what conditions, and with what redistribution effect, a change of habitats will occur.

Fifty years of Lotka-Volterra analyses have shown that theoretical studies can make little real progress until provided with a foundation of realistic functional forms based upon observable ecological relationships (*sensu* Gilbert *et al.*, 1976). Animal movement observed in nature ranges from the ordered, repetitive, migrations of Capistrano's swallows to the apparently ubiquitous "rain" of Hutchinson's fugative species. In this paper, we explore some representative intermediate forms of the adaptive movement response in terrestrial animals, with a view to suggesting how these might be cast as specific, realistic variants on Levin's "null hypothesis." In so doing, we seek to determine how these adaptations can be related to their owner's local population dynamics, and how the resulting interactions are combined in the determination of broader spatial patterns and long term fitness.

The structural components of the movement response appear to exhibit remarkable similarities across a wide range of terrestrial organisms. The richest understanding by far, however, has been developed through the study of insect systems (see reviews in Southwood, 1962; Johnson, 1966, 1969; Dingle, 1972; Kennedy, 1961, 1975). To keep the discussion manageable, we largely constrain the present arguments to these insect systems, and focus on our own studies of space-time dynamics in the spruce budworm/balsam fir ecosystem. For those inclined to generalizations, this may be viewed as a typical three-level food chain consisting of a predator guild (birds), a herbivore (budworm), and a primary producer (the forest). In another context, Steele (1972) has shown us that trees are just immobile fish, so the comparison to marine systems is bound to be illuminating.

Our treatment is arranged as follows: the next section on Patterns in Space and Time briefly describes the budworm system and its observed space-time behavior. Local Interactions analyzes local population interactions of the system, setting the stage on which animal movement is carried out and its adaptive value determined. The Spatial Dimensions introduces physical dimensions to the local analysis, exploring the meaning of traditional concepts of "critical patch size." Adaptive Movement explores the component structure of the movement response, contrasting the "null hypothesis" of random diffusion with the highly directional, often aggregative behavior observed in a wide array of natural systems. The final section draws the previous bits and pieces together in the analysis of emergent space-time patterns in the budworm system, and explores some broader implications of the study.

## PATTERNS IN SPACE AND TIME: THE SPRUCE BUDWORM/CONIFEROUS FOREST ECOSYSTEM

A major obstacle to better understanding of animal movement -- population dynamics -- habitat interactions is the scarcity of higher resolution space-time data on patterns of animal abundance. The development of continuous sampling devices has recently begun to improve this situation for marine systems as evidenced by the existence of the present volume, but even there the situation is far from ideal. In terrestrial studies the sampling problems are different (generally worse) and, except for specialized successional studies on primary producers (see Whittaker, 1974), very few data sets of the desired sort are available now or likely to appear in the near future.

Not unexpectedly, much of the best work concerns species with potential health and/or economic impact on man. A number of dynamic space-time patterns are documented in the epidemiological literature (see Bailey, 1975). Probably the most immediately relevant for questions of plankton patchiness is that of human measles (Bartlett, 1957; Cliff *et al.*, 1975) to which we will return in <u>The Spatial Dimensions</u>.

Studies on insect pest populations have resulted in numerous records of abundance at various places and times, but rarely with sufficient attention to questions resolution or accompanying population dynamics to make the results more than suggestive. The Rothamstead light trapping program for nocturnal flyers and the Canadian Forest Insect and Disease Survey have spawned numerous progency and data. The strengths and weaknesses of such survey approaches to analysis of space-time patterns are well illustrated by Taylor and Taylor (1977) in a recent synthesis of some of the Rothamstead work.

Long term research programs focussed on particular species have occasionally managed to both document and explain complex space-time patterns, thereby providing rich bases for the development and testing of theory. Perhaps the classic case is the *Cactoblastis-Opuntia* interaction originally documented by Dodd (1940, 1959) and later extensively analyzed by Monro (1967) and Caughley (1976). Representative of the best work on other insect systems is that of Broadhead on *Mesopsocus* (Broadhead and Cheke, 1975); R. E. Jones on the butterfly *Pieris rapae* (R. E. Jones, 1977a,b); Hughes and his coworkers on the Australian bushfly (*Musca vetustissima*; Hughes and Nicholas, 1974); Myers on the cinnabar moth, *Tyria jacobaeae* (Myers, 1976; Myers and Campbell, 1976) and, especially, Wellington on the western tent caterpillar, *Malacosoma californicum pluvale* (Wellington, 1957, 1964, 1965; Iwao and Wellington, 1970).

## The Spruce Budworm Ecosystem

Our own work has focussed on the spruce budworm/coniferous forest ecosystem of eastern Canada. The spruce budworm *(Choristoneura fumiferana)* is a defoliating lepidopteran insect of the North American boreal forest. Throughout most of its range it is usually rare, held in check by a combination of natural enemies, inadequate resource levels, and weather. These controls periodically break down, however, and the system experiences budworm outbreaks inflicting extensive mortality on the preferred host trees, balsam fir *(Abies balsamea)* and white spruce *(Picea glauca)* (Davidson and Prentice, 1967). Outbreaks spread rapidly in space since budworm moths may disperse tens or even hundreds of kilometers in a single generation. Detailed studies on the system have been carried out since the late 1940's, especially in eastern Canada. Early results are synthesized in Morris (1963), and more recent work summarized in Miller and Ketella (1975).

Data on temporal and spatial behavior of the budworm/forest system are unusually complete for a terrestrial study. Tree-ring measurements, backed by contemporary records, allow documentation of outbreaks back to 1704 (Blais, 1968). The general temporal picture is one of irregular outbreaks recurring at 30-70 year intervals and lasting from 12-15 years in any given area (Figure 1). Synoptic maps of outbreaks occurring east of the Great Lakes since the early part of this century have been compiled from the files of the Canadian Forest Insect Survey (Brown, 1970), and show a complex but distinctly wave-like pattern of spatial spread. A typical synoptic sequence is shown in Figure 2.

The Canadian Province of New Brunswick has experienced some of the worst outbreaks, undergone the most intensive management, and compiled the most complete records. Since extensive management began in the 1950's, yearly estimates of insect density have been taken at nearly a thousand locations distributed over a seven million hectare area. Each estimate results from a standardized sequential sampling procedure which is continued until the standard error is reduced to a predetermined proportion of the mean (Morris, 1955). These data are summarized in Figure 3 (see also Webb, Blais and Nash, 1961) and represent one of the most detailed descriptions of time-space population dynamics in the terrestrial literature. The confused temporal and spatial behaviors of the management-perturbed system shown in Figure 3 are radically different from the highly periodic patterns of the natural situation. Although a fairly simple statistical model might account for either pattern alone, the challenge is to develop a single, spatially dimensioned theory of the system's dynamics which can account for both sets of observations.

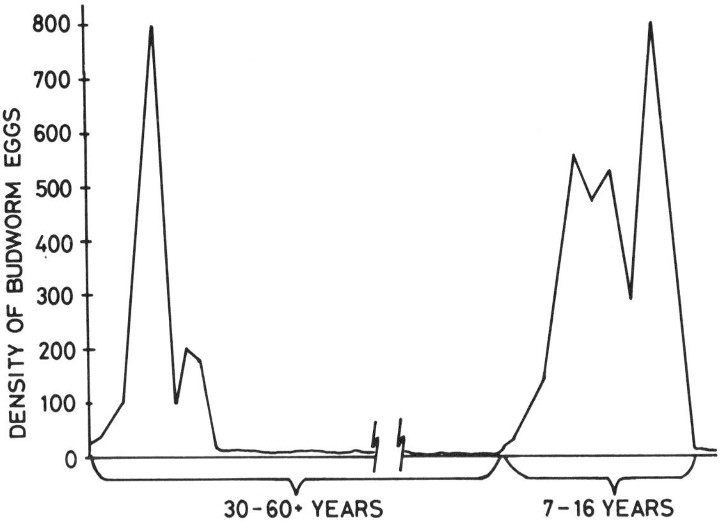


Figure 1. The pattern in time. Representative historical pattern of spruce budworm outbreaks, synthesized from the work of Blais (1968). There have been four major outbreaks since 1770. The density measure of budworm is per typical balsam fir branch (one square meter of foliage).

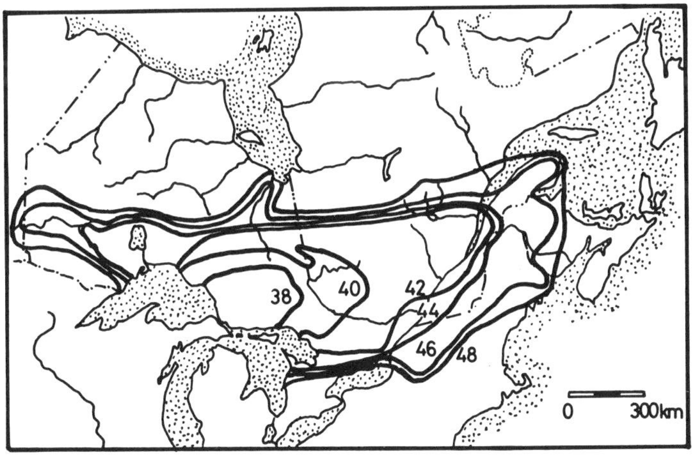


Figure 2. The pattern in space. Maximal extent of budworm infestations in eastern North America for 1938-1948, the last outbreak sequence before extensive insecticide spraying was introduced. Redrawn from Brown (1970).

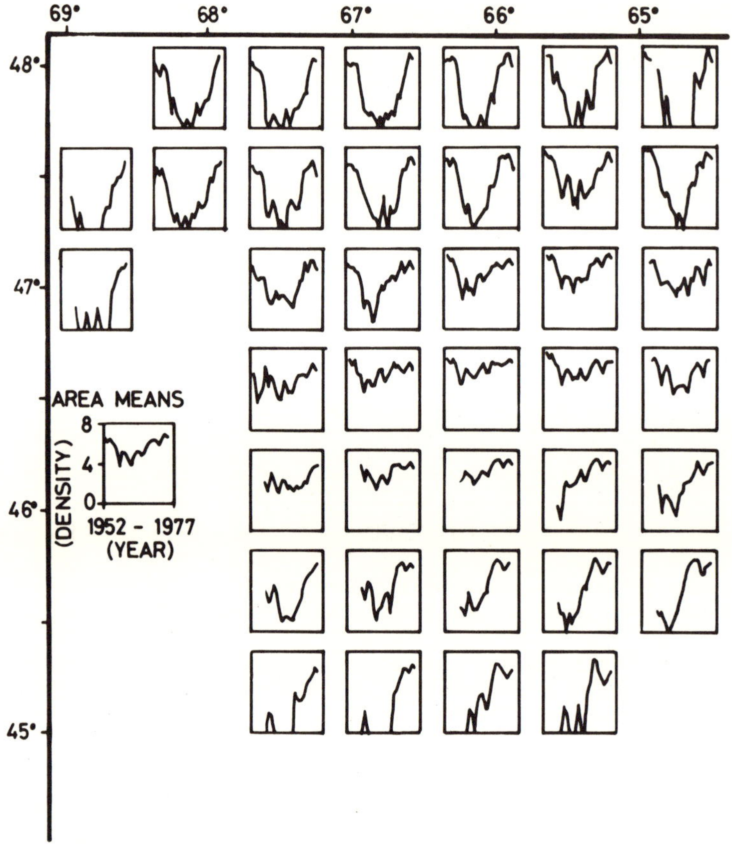


Figure 3. Budworm densities in New Brunswick under historical management conditions. Each grid represents a 50 x 50 km square. Mean egg densities (n≈20) are shown on the ordinate as the natural logarithm of (one plus egg density per square meter of foliage). Time is given on the abscissa for years 1952 to 1977. Mean densities for all areas combined are given in the isolated grid.

Our initial approach to the problem involved development of a complex simulation model representing a "dynamic life table" (Gilbert *et al.*, 1976) for the budworm and its local interaction with trees, weather, natural enemies, and human management. The seven million hectares of New Brunswick were treated as a mosaic of 393 patches governed by these local interactions, and individual patches were linked via insect dispersal. Data were drawn from the fifteen years of analysis summarized in Morris (1963) and subsequent cooperative studies with researchers of the Canadian Forest Service. The detailed results of this work are reported elsewhere (Jones, 1977; Clark, Jones and Holling, in press) and an extensive monograph on the work is under preparation. As we will show in the last section, the resulting model predicts certain qualitative aspects of both observed behavior patterns in an aesthetically pleasing way. Nonetheless, the historical data are sufficiently detailed to highlight significant shortcomings in our present understanding -- a luxury not often encountered in spatial population studies. Equally important for our present purposes, however, are some of the issues raised in the analysis regarding the interrelationships of habitat character, local dynamics, and animal movement.

## LOCAL INTERACTIONS: WHY MOVEMENT MATTERS

The effects of animal movement may be as simple as the colonization of a previously barren island, or as complex as the destabilization of a hitherto balanced interspecies relationship. In any event, it is the local (or proximal, *sensu* Levin, 1976) interaction properties of the ecosystem -- intrinsic growth functions, prey-predator or competitors equilibria, extinction rates, and so on -- which set the stage on which movements are acted out and attain their significance.

Early theoretical work on dispersal in spatially dimensioned environments tended to explore only the simplest of local interaction situations. Recently it has become apparent that multiple equilibria in local dynamics may seriously affect the conclusions of dispersal studies (Holling, 1973; Levin, 1974 and this volume; Whittaker and Levin, 1977). We illustrate some of the implications, and an approach to their elucidation, through reference to the budworm system described earlier.

Equations governing local interactions of the budworm, its food host, and its natural enemies have been drawn from the extensive studies of Morris (1963) and his coworkers, and are summarized in Jones (1977). These are too complex to be solved analytically. Numerical simulation models are the only tools available for treating such situations, but they are no more attractive for their inevitability. The necessary exploration of numerous initial conditions and parameter values is always expensive, frequently ambiguous, and potentially misleading. We need understanding as well as solutions, and space-time streams of simulation output are notoriously inadequate for this purpose.

A great deal of understanding can be gained, however, if we use the detailed simulation model to calculate equilibrium values for insect densities under a range of ecologically meaningful conditions, and then plot these as equilibrium manifolds of the sort found in bifurcation analysis and topology (Levin, this volume; Thom, 1975). Jones (1975) introduced this approach into the ecological literature with a manifold analysis of the budworm situation, and his original paper should be consulted for further details of the work reported here. The general approach has since been used to good effect in the analysis of other ecological situations (Southwood and Comins, 1976; Jones and Walters, 1976; May, 1977).

Figure 4 is the equilibrium manifold for local budworm dynamics as a function of forest density (i.e., habitat quality). For forest densities less than $F'$ there is one stable budworm equilibrium corresponding to low endemic levels of the insect. Any initial density of budworm will move to this level. Above $F''$ there is only one budworm equilibrium, this time at high densities associated with insect outbreak. All initial densities greater than zero will here rise to the upper level. Between $F'$ and $F''$ three budworm equilibria exist: the stable endemic level noted earlier (e.g. $N^-$), the stable upper outbreak level (e.g. $N^+$), and an unstable intermediate level (e.g. $N^o$). Initial densities above $N^o$ will be attracted to $N^+$ generating an outbreak, while those below $N^o$ will return to the endemic equilibrium at $N^-$.

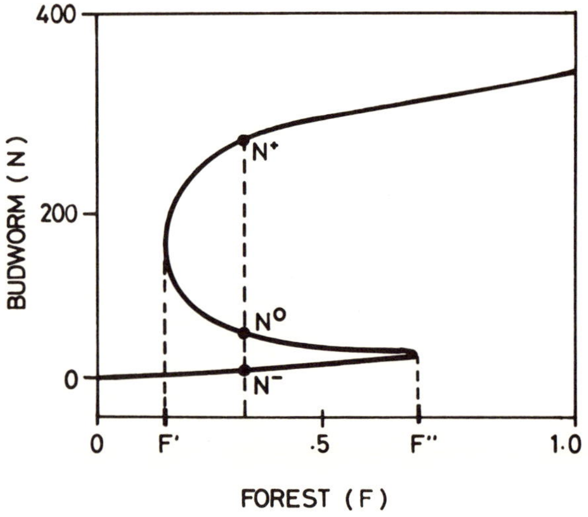


Figure 4. Equilibrium manifold for the budworm simulation model. The surface represents egg densities which show no tendency to change when branch density and foliage quantity are held constant at their specified values. Weather conditions used in these calculations are averages for New Brunswick (see text).

The general shape of the budworm manifold is typical of a wide range of other ecological systems, ranging from man-malaria to wolf-caribou interactions, and including a large number of well-studied insect systems (May, 1977; Haber, Walters and Cowan, 1976; Southwood, 1975). For budworm and many of the insect cases the upper equilibrium surface is defined by competition for food at high densities. The existence of an intermediate unstable surface is critical to the arguments about spatial pattern and animal movement arguments we will advance. Indeed, a similar structural property is implicated in many of the interesting theoretical results on spatial dispersion dynamics in ecology (Levin, 1978). The phenomenon can arise from any of several auto-catalytic or Allee-type mechanisms which impart an increasing survivorship rate with increasing population size over a range of densities (Okubo, 1974, this volume; Segel and Jackson, 1972; Segel and Levin, 1976). In budworm and many other situations cited above it is due to saturation of the functional response of parasites and/or vertebrate predators by rising prey densities (Holling, 1959, 1965, 1973; Oaten and Murdoch, 1975).

A lower equilibrium will always exist at zero density. In the present case, a non-zero lower stable equilibrium also arises due to search inefficiencies and/or switching in budworm predators at low densities (Hollings' "Type-III" functional response; see reference above). Finally, the merging of the lower stable and unstable surfaces at $F''$ represents the fact that basic habitat conditions have become so good that predation is no longer able to compensate for increasing growth rate of the budworm population. The prey "escapes" (Takahashi, 1964).

Refer now to the budworm model manifold of Figure 5. As the forest grows following fire, logging, or previous budworm destruction, $F$ rises -- i.e. the habitat quality increases. Budworm densities remain on the low endemic surface and any random fluctuations are drawn down to that surface. As the improving forest passes density $F'$, budworm remain locked on the lower stable surface. Now, however, random fluctuations or immigration of a quantity of budworm greater than $\Delta(F)$ is sufficient to raise densities above the unstable surface. Even if the disturbance is not repeated, the system now naturally evolves to the upper equilibrium and an outbreak ensues. The necessary perturbation, $\Delta(F)$, decreases as $F$ increases, until above $F''$ an outbreak occurs spontaneously, independent of fluctuation or immigration. To complete the picture, note that though the upper manifold surface is stable for budworm, it is not for the forest. High insect densities inflict mortality on the forest, decreasing $F$, starving budworm, and finally cycling the system back to its initial state. These manifold interpretations have been verified by recourse to the dynamic simulation.

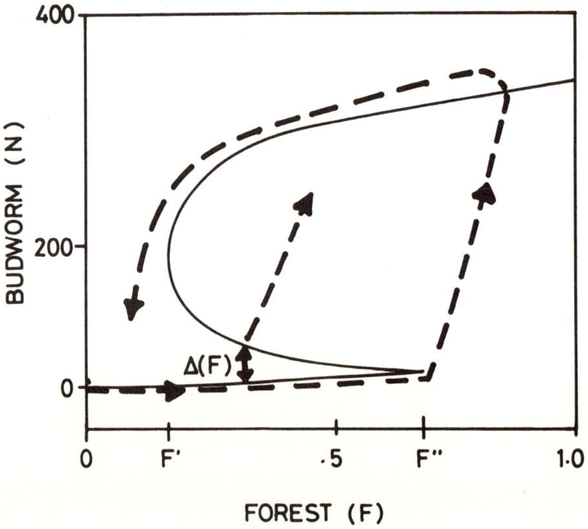


Figure 5. Equilibrium manifold for the budworm simulation model showing typical outbreak trajectory (see text).

This equilibrium manifold compression of the full simulation model immediately highlights the two potentially conflicting functions of animal movement which concern us in this paper. High insect densities limit the duration of favorable habitat conditions. A "family" remaining on the site of its ancestors through several years of outbreak is more than likely doomed. There is an obvious and strong selective pressure to migrate in search of better habitat under such conditions. This circumstance has been remarked upon from the earliest studies on the adaptive role of terrestrial dispersal (e.g., Andrewartha and Birch, 1954).

Compare, however, the situation in a forest of intermediate density (between $F'$ and $F''$). Here the forest habitat is capable of supporting a high rate of increase, but only if local population densities can be boosted above the unstable lower equilibrium. Local dynamics cannot accomplish this. And the increased tendency to disperse which is adaptive at high densities would now have the opposite effect. What is adaptively "needed" here instead is plainly a local anti-dispersal or even aggregative phenomenon: one which collects organisms from a large area of low density infestation and concentrates them sufficiently to trigger a local outbreak with it attendant selective advantage for the outbreak founders. Such aggregative adaptations are, in fact, common in terrestrial systems and often seem to be key determinants of their space-time behavior (Taylor and Taylor, 1977). We will return to consider some specific examples shortly. It would be misleading to go further, however, without explicitly considering the spatial dimensions of the "local" model and its manifold discussed above. This brings us to the classical questions of refuges and critical patch sizes, both of which are discussed from a budworm perspective in the following section.

## THE SPATIAL DIMENSIONS: HABITAT PATCHES, CRITICAL SCALES, AND THE RANGE OF ADAPTIVE OPTIONS

Mathematical and laboratory studies have clearly demonstrated that merely introducing significant spatial scale to the environment within which local interactions occur can have dramatic effects on mean density and expected persistence times of interacting populations (see Steele, 1974a; Roff, 1974a,b; Maynard Smith, 1974, for representative mathematical results, and Huffaker, 1958; Pimentel *et al.*, 1963; Luckinbill, 1974 for a sampler of the lab studies). The addition of spatial heterogeneities, even under isotropic dispersal (Jones, 1974; review in May, 1974), or differential dispersal rates (Hastings, 1977; Hilborn, 1975; Okubo, 1974; and related papers noted by both Okubo and Levin, this volume), or complex dispersal functions (Gadgil, 1971; Jones, 1974; Roff, 1974a, b) further enhances the range of possible space-time patterns, but only at the expense of introducing a large number of loosely specified parameters into an already intractable theory.

To keep our analysis of spatial budworm as comprehensible as possible, we focus in this section on the simplest interactions among habitat structure, movement, and local dynamics, following the "critical scale" approach as originally applied to ecological problems by Skellam (1951) and Kierstead and Slobodkin (1953), and reviewed elsewhere in this volume by Okubo. We will then use the results of this simplified study to suggest which of the many potential complicating factors merit attention in the fuller analysis of the next two sections.

### Critical Scales in the Budworm Study

The natural habitat of budworm is the more or less continuous boreal forest of eastern North America. A small amount of patchiness in habitat quality occurs due to areas of nonsusceptible forest, farmland, and such. More generally significant, however, is that resulting from the variable age composition and thus density ($F$) of the forest. Sizable patches of fairly homogeneous age arise due to effects of fire, logging, and previous budworm damage, and it is this dynamic age/quality mosaic which confronts the budworm (Baskerville, 1975).

The local budworm-forest interactions discussed in the previous section suggest how an outbreak initiating in a patch of favorable habitat ($F > F''$) can propagate throughout areas of lower habitat suitability ($F > F'$) via dispersal. Just such dispersal-triggered outbreaks have been repeatedly observed in the field (Greenbank, 1973). In addition, field experience has long identified extensive patches of high quality habitat as the highest risk areas for outbreak initiation. The same experience shows, however, that small, isolated patches of even the highest quality habitat are unlikely to give rise to outbreaks or sustain them (van Raalte, 1972). Observers have long implicated dispersal losses across patch boundaries in this phenomenon (Morris, 1963), and the analogy with "critical patch size" notions of marine ecology is obvious. More generally, the question arises of how the spatial scale of habitat quality (resource patchiness) interacts with the scale of animal movement to affect the picture of local dynamics presented earlier.

Our starting point, once again, is the detailed simulation model of budworm-forest dynamics. The complexities which led us to develop an equilibrium manifold compression for the local dynamics of the model are compounded to an extraordinary degree in the full spatial case. Further, the potential problems of numerical instabilities in the solution of such gridded spatial models are profound. As part of our efforts to develop a hierarchy of simplifications to cope with complexity in the local case, Ludwig, Jones and Holling (1978) had created a set of coupled differential equations which effectively captured certain aspects of budworm outbreak behavior when tested against the fuller characterizations of the simulation model and its manifold compression. In an attempt to extend this simplify-and-compare strategy to the spatial case, Drs. Don Ludwig of the University of British Columbia and H. F. Weinberger and D. Aronson of the University of Minnesota have added a spatial diffusion term to the local differential equation model. Their fascinating exploration of the resulting patch scale effects is as yet unpublished, and we are indebted to them for permission to quote some initial results here.

The local budworm dynamics of the original Ludwig, Jones and Holling model (hereafter LJH) can be summarized as follows:

$$\frac{dB}{dT} = rB\left(1 - \frac{B}{KS}\right) - \beta\frac{B^2}{(\alpha S)^2 + B^2} \qquad (1)$$

$B$ represents local budworm density. The first expression on the right-hand side is a logistic term with the carrying capacity proportional to forest density ($S$), a term analogous to the ($F$) in the previous section. The second term on the right gives the rate of consumption of budworm by a predator guild with a sigmoid functional response. Scaled for mathematical convenience and coupled to equations governing resource growth, the differential equation system yields the equilibrium manifold for budworm shown schematically in Figure 6. Here $n$ and $f$ are scaled versions of budworm and forest density respectively. Subject to parameter scaling (see LJH), this is the same manifold, with the same interpretation, as that derived from compression of the simulation model described earlier (Figures 4 and 5).

Ludwig, Weinberger and Aronson (LWA) introduced space into equation (1) by adding a random spatial diffusion term with mean zero and variance $\sigma^2\, dt$ (see, for example, Skellam, 1973). The analogous equation for budworm growth then becomes

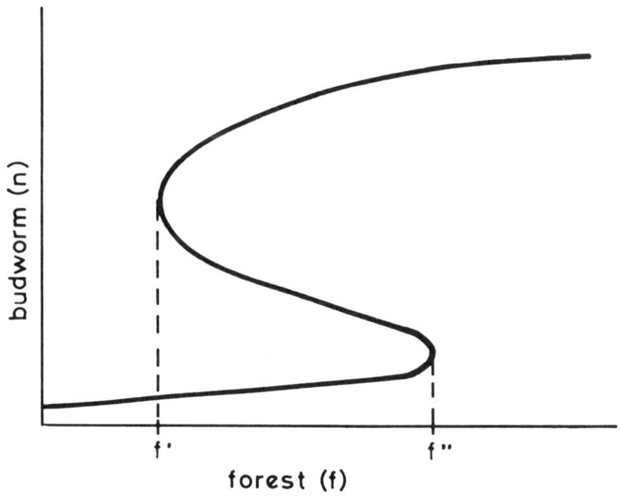


Figure 6. Equilibrium manifold for the Ludwig-Jones-Holling differential equation model. Axes are distorted for ease of interpretation (see text).

$$\frac{\partial B}{\partial T} = g(B) + \frac{\sigma^2}{2} \left( \frac{\partial^2 B}{\partial x^2} + \frac{\partial^2 B}{\partial y^2} \right) \tag{2}$$

The term $g(B)$ is equivalent to the right-hand side of equation (1), and $x$ and $y$ are spatial coordinates. LWA then proceeded to study the number of time independent solutions of equation (2) as a function of the scaled length ($\ell$) of a suitable habitat patch with dimension $x = y = L$, where

$$\ell = L/L_d = L\,(\sigma^2/2r)^{-\frac{1}{2}} \tag{3}$$

Their conclusions, only the simplest of which are outlined here, suggest some critical patch size notions for multiple equilibria systems which are related to but rather more complicated than those developed in the marine literature (Okubo, this volume).

Refer now to Figure 7 which gives the maximum values for budworm density at three fixed habitat qualities ($f$) as a function of the (scaled) size of the habitat patch ($\ell$). Consider first Figure 7a, the situation for poor quality patches ($f$ small). As expected from the local analysis, large patches ($\ell$ large) of such habitat exhibit only endemic budworm equilibria at $n^-(\ell)$. ($n$ is here treated as the density at the center of the patch.) For patch sizes sufficiently small, however, the sustainable endemic equilibrium density decreases rapidly, and for $\ell < \ell^0$ budworm go (locally) extinct. $\ell^0$ is thus a critical minimum patch size for endemic budworm; a budworm "refuge" must be larger than $\ell^0$.

Consider next the medium quality habitat ($f$ moderate) represented in Figure 7b. Again, for large patch sizes the results are essentially identical to those of the local analysis. Three budworm equilibria exist, a stable endemic one at $n^-(\ell)$, a stable outbreak one at $n^+(\ell)$, and an unstable intermediate one between them at $n^0(\ell)$. An outbreak here can be triggered by addition of sufficient immigrants to bring the density above $n^0(\ell)$. As the size of the habitat patch decreases, however, the necessary $\Delta$ grows, and at $\ell < \ell'$ it becomes impossible for the patch to support an outbreak, regardless of the quantity of immigrants. This is a qualitative and potentially significant change from the local interpretation alone. $\ell'$ characterizes a critical minimum patch size for epidemic budworm. Perhaps equally important, it raises the prospect of a small scale ($\ell < \ell'$) "refuge" for patches of the budworms' forest "prey." Further reductions in the habitat patch size beyond $\ell^0$ eliminate budworm entirely.

Finally, consider the case where habitat quality is high (Figure 7c; $f$ large). As expected, large patches exhibit the

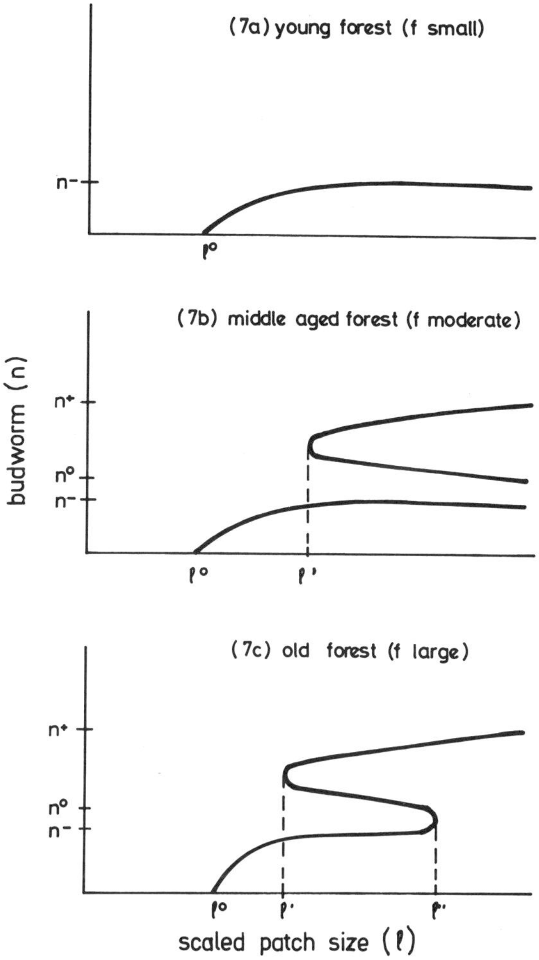


Figure 7. Maximum density solutions for the Ludwig-Weinberger-Aronson model. Abscissa is scaled patch size; three figures are for (a) young (poor quality) forest in the patch; (b) middle age (moderate quality) forest, and (c) old (high quality) forest (see text).

single upper budworm equilibrium: all initial densities lead to outbreak. As the patch size is decreased past $\ell''$, the mature forest begins to behave in a way which the local analysis associated only with intermediate quality ($F' < F < F''$ in Figure 4) patches. Three potential equilibria emerge and immigration-induced outbreaks again become a relevant consideration. Theory is now in accord with the field experience cited earlier. Outbreaks are no longer inevitable in a mature forest patch, but only in a mature patch of sufficient size. Further reductions in patch size bring about the further qualitative changes already noted at $\ell'$ and $\ell^0$.

The results of Figure 7 show important qualitative relationships between patch size and patch quality in the determination of "critical patch" phenomena in the modeled budworm system. Before considering the broader implications of these relationships, one additional feature of the Ludwig-Weinberger-Aronson findings should be noted. Recall (equation 3) that the length parameter $\ell$ in the above analysis was a scaled, dimensionless function of the physical patch size ($L$) and divided by the "diffusive length" of budworm $L_d = (\sigma^2/2r)^{\frac{1}{2}}$. Maintaining suitable reservations with regard to the assumption of random diffusion, this suggests that the physical dimensions of the various "critical" scales suggested in Figure 7 will, in fact, also be functions of an animal's tendency and ability to disperse. Not surprisingly, the physical dimension of the critical patch sizes $\ell$, $\ell'$ and $\ell''$ will generally be larger for animals with high diffusion and/or low growth parameters. And just as changes in patch quality could alter the value of the scaled length $\ell$ for which outbreaks became possible and eventually inevitable (Figure 7), so changes in animal dispersion rates can bring about similar changes in the unscaled length $L$, even for a fixed level of patch quality. This suggests (though now the LWA analysis is being stretched rather grossly beyond its assumptions) that an organism capable of altering its dispersal tendency in relation to habitat conditions would often have a selective advantage over his constant Fickian brothers. In fact, animals have developed a great variety of adaptations for exerting active control over the dimensions of their critical scales, some of which will provide the focus in the next section. For the moment, it is sufficient to summarize the critical scale analysis on budworm as revealing important complementarities among resource quality, patch size, and animal movement in the determination of local population dynamics.

## Critical Scales in Epidemiology

The potential for complex patch size-resource quality-animal movement relationships of the sort suggested here for budworm should exist in all of those systems referred to earlier which exhibit both multiple equilibria in their local dynamics and dispersal by one or more of the organisms involved. A particularly well-documented situation which almost certainly reflects such phenomena is that of measles outbreaks in human populations (Bartlett, 1957). As in budworm, there is a lower population density (resource quality) below which the disease simply does not manifest itself. Above this first critical level, man-measles dynamics are a joint function of community (patch) size, population density, and exchange with other communities (the immigration $\Delta$

of budworm). Below a critical community size (typically 300,000 people under western European or North American conditions) epidemics do not naturally recur, but must be triggered by "immigration" of sufficient infectives from outside. In a pattern which may be followed nicely via Figure 7, larger communities require fewer immigrants to trigger measles outbreak and consequently exhibit a higher outbreak frequency. Above the critical community size, outbreaks trigger naturally and regularly as a result of growth in number of susceptibles (increase of resource quality) and an ever present endemic "lower equilibrium" of infectives. Deterministic multiequilibria models have been used to explain the budworm patterns, while structurally simpler but stochastic ones have achieved most success in the measles analysis (see Bailey, 1975). The similarities underlying the ecology of the two situations suggest that a more rigorous comparison of the two approaches to analysis would be likely to increase the insights of both.

## Scale Interactions and the Problem of Analysis

In one sense, the budworm, measles, and other "epidemic" cases represent the easiest class of space-time patterns to analyze. When the time scales of the interacting species are close together (Levin, this volume) and/or more than one is involved in active movement, the investigation becomes much more difficult and ambiguous. Analytical solutions exhibit accute sensitivity to small structural or parametric variations (e.g. Steele, 1974b; Okubo, this volume). And preliminary simulation studies yield a bewildering array of complex space-time behaviors under even the simplest such assumptions (Jones, 1974; Holling, Jones and Clark, 1977; Hilborn, 1977; Evans, this volume).

One response to this complexity has been to focus on the patterns themselves -- to subject them to various statistical analyses in hopes of pointing towards a more basic understanding of the underlying generating processes. Such descriptive work has its functions, particularly in telling us at which scales the "action" is (Platt; Richerson and Powell; Haury *et al.*, all this volume), in testing specific hypotheses (see last section on Putting the Pieces Together), and in helping to define appropriate scales for more critical observation and experimentation (Steele, this volume). But it is crucial to keep in mind the fundamental impossibility of deducing underlying biological relationships and adaptations from their resulting space-time distributions, no matter how the latter are measured. This is as true for spectral and auto-correlation descriptions of spatial variance (Platt, this volume; Fasham, this volume) as it is for more traditional fits to theoretical distributions (Southwood, 1966; Taylor and Taylor, 1977). A concise proof is given for the random walk case by Gilbert in R. E. Jones (1977a), and Hilborn (1978) demonstrates the problem for a more realistic ecological case. It is significant that most of the impetus for employing descriptive statistical approaches to the analysis of spatial dynamics comes from either the physical (Wiener, 1956) or social sciences (Cliff *et al.*, 1975). In physics success has been achieved primarily where the underlying processes were well known (and largely linear) and the problem was to discover which were more or less important in particular circumstances. In human geography and, to a lesser extent, regional economics the descriptive focus has recently developed a truly impressive array of statistical tools (Cliff *et al.*, 1975), justified on the contention that there is "no theory -- simply no theory at all" (Curry and Bannister, 1974; but see Gould, 1970) available through which to cast alternative causal models.

Ecology, it seems to us, lies between these two extremes. The papers by Fasham and Evans in this volume illustrate convincingly that we have hardly reached a state of theoretical sophistication where observed spectral slopes can safely be used as critical tests of conflicting hypotheses. On the other hand, we do know a fair bit about the underlying process -- the ecological relationships -- which are involved in the generation of observed space-time patterns. Something has been said of local interactions in previous sections of this paper. There remains, however, the process of animal movement itself. Most theoretical studies have treated movement implicitly or in mathematically tractible but biologically simplistic formulations. As a result, we now know a good deal about the properties of the random diffusion "null hypothesis," but precious little about any of the implied alternatives. Skellam (1973), Okubo (this volume) and others have argued convincingly that it is to such alternatives, based on actual movement relationships observed in nature, that theoretical studies must now turn. As Strickler (1977) has noted, movement relationships in the aquatic plankton have been little studied in comparison to the so-called "aerial plankton" of the insects. It is thus to the latter group that we now turn to illustrate some important components of the movement process which might profitably be investigated in the next generation of spatial population dynamic models.

## ADAPTIVE MOVEMENT: A TERRESTRIAL SAMPLER

C. B. Williams has characterized animal movement as a three-stage process involving exodus, displacement and settling (Williams, 1965). The random diffusion hypothesis treats these by assuming a constant proportion of the population in exodus at all times; displacement which is random and uncorrelated with respect to both distance and direction; and (indirectly) settling location as a regular decreasing function of distance displaced. A catalog of specific exceptions to diffusive movement would be easy but pointless. Of more interest are some of the most common extreme departures actually encountered in nature, together with an appreciation -- however tentative -- of their adaptive significance to the animal. Drawing again primarily from insect studies, and most selectively even there, we present in this section a sampler of the terrestrial movement process, organizing it around the behavioral classification of Williams.

## The Exodus Response

Our discussion of local population dynamics in the previous section on Patterns in Space and Time implies that the most adaptive movement response for the budworm might require an increased tendency to remain on the feeding site at low densities and to leave the site at high densities. The differential equation analysis described earlier posed the same trade-off from another perspective: animals with a constant low "diffusion" rate can persist at endemic densities, become eligible for immigrant-triggered outbreak, and undergo spontaneous outbreak, in smaller (or poorer quality) patches than animals with high diffusion rates. However, more of the slow dispersing animals will starve as the outbreak proceeds and fewer will be likely to emigrate and "found" new outbreaks.

The obvious requirement is for an exodus rate responsive to local population dynamics, and this is commonly found in nature. Reviewing a vast number of cases, Southwood (1962) concluded that insects in moderately transient or unpredictable environments usually exhibit an exodus response well tuned to the deterioration of local conditions. Gadgil (1971) has conducted one of the few theoretical studies which explicitly separates the magnitude of the exodus rate from its responsiveness to local conditions. Using a hierarchy of models in a manner similar to that described for our budworm studies, he concluded that selection should favor high sensitivity to local conditions, but a low maximum exodus rate in spatially heterogeneous environments, as opposed to lower sensitivity but higher maxima in temporally heterogeneous ones. More attention must be paid to the relative time scales in Gadgil's analysis (perhaps following Southwood, 1977) before these general conclusions are specifically applicable, but his predictions are certainly much more in accord with observation than random diffusion theory. A somewhat different approach to the issue has been explored by Roff (1975).

The likely importance of a response to habitat condition in exodus is further emphasized by the range of time scales over which it is expressed in nature. Immediate behavioral responses are best documented at the individual level under the heading of "search behavior' (e.g. Fleschner, 1950; Dixon, 1959; Chandler, 1969; Murdie and Hassel, 1973; Brunner and Burts, 1975; R. Jones, 1977b).

In these and a variety of other cases it has been shown that turning rates are positively related to the rate of resource encounter, with the result that animals tend to stay in areas of high resource quality and to leave those with low quality. A similar adaptation is found throughout the animal kingdom; for instance, in the thrushes studied by Smith (1974a,b). Wratten (1973) provides an unusually pure example in which adult coccinellid beetles simply took their eggs elsewhere whenever local prey densities became too low for larvae to be reared successfully.

Ontogenetic changes in populations over the course of one or a few generations also give rise to habitat-dependent exodus responses. The production of migrant alate forms in aphids and of "hoppers" in locusts is well documented (see Kennedy, 1975). Leonard (1970) has also shown such changes to be of crucial importance in determining space-time dynamics of the gypsy moth *(Porthetria dispar)*. Here, well-fed larvae eventually produce eggs with large food reserves. In emergence, these rapidly settle nearby, feed well, and repeat the cycle. Poorly fed larvae yield stressed females which in turn lay eggs with low food reserves. The emerging larvae are small, active, and unlikely to settle to feed. The first-instar phase is thereby prolonged, with an increased tendency for these larvae to disperse via windblown silken threads. New areas with presumably better chances of adequate food supply are thereby colonized, and local resource pressures relieved. Similar results, if different mechanisms, have been observed in birds, microtine mammals (Krebs and Myers, 1974), and the Australian rabbit (Meyers and Parker, 1975).

Finally, sustained population polymorphisms for exodus behavior have been observed. The best documentation is provided by Wellington's work on the western tent caterpillar, *Malacosoma pluviale* (Wellington, 1957, 1964, 1965; Iwao and Wellington, 1970). One morph of these populations is generally inactive, oviposits near its emergence sites, and produces offspring with generally similar characteristics. These best exploit high quality habitat and are killed by stressful conditions. The second morph is more active, leaves it emergence site and travels before egg deposition, and reproduces a high proportion of similar progeny. These organisms are adapted to harsher habitat patches and tend to colonize peripheral areas. Natural habitat and climatic fluctuations are apparently sufficient to maintain the observed polymorphisms over very long periods indeed, a conclusion sustained by extensive modeling studies of the population (Wellington *et al.*, 1975; Thompson *et al.*, 1976).

## Displacement and Settling: The Refractory Period

Ideally, these two phases of movement behavior should be discussed separately; certainly their experimental analysis requires very different approaches. Unfortunately, the field evidence generally is not sufficiently detailed to allow the proper treatment. It is comparatively easy to observe the conditions under which exodus occurs, but a good deal more difficult to follow what happens next.

A critical departure of observed displacement/settling behavior from that hypothesized by random diffusion concerns the existence of a lag or refractory period during which settling does not occur. In a great variety of cases exodus is followed by a period during which the dispersing organism will not stop moving regardless of the stimuli encountered. Studies on the aphids are typical in this respect (Kennedy *et al.*, 1961). As the adult emerges, its response to host plants is inhibited, it exhibits a positive phototaxis, and the combination leads to exodus. The plant inhibition is progressively removed by flight, eventually becoming a positive taxis to host-reflected wavelengths which overrides the remaining sky taxis and induces settling. The host inhibition is reversibly reimposed by landings on non-host plants, and the process repeats itself. Similar refractory effects with different mechanisms have been documented for ambrosia beetles (Graham, 1961; Francia and Graham, 1967), pierid butterflies (Nielsen, 1961), and many other insects for which the phenomenon has been checked.

The existence of the refractory period of movement is at odds with the random diffusion hypothesis. Resulting displacements will not have a mean zero distance, but rather will be offset (often substantially) from the origin, resulting in highly skewed redistributions. The consequences of the refractory period have been observed for the spruce budworm. Radar studies of airborne densities of dispersing moths and their resettlement rates have been conducted in Canada by Schaefer (1976) and his collaborators. Dr. Schaefer kindly made his data available to us, and our analysis suggests the displacement distributions shown in Figure 8. The displacement of the modal dispersal distance from the origin is robust to all forms of analysis which do not impose random diffusion assumptions on the underlying process. It is, however, consistent with the refractory period concepts described above. Simulation experiments with our spatial population model have shown the qualitative existence of any sizable such modal distance displacement to have profound consequences for the system's space/time dynamics (see the concluding section).

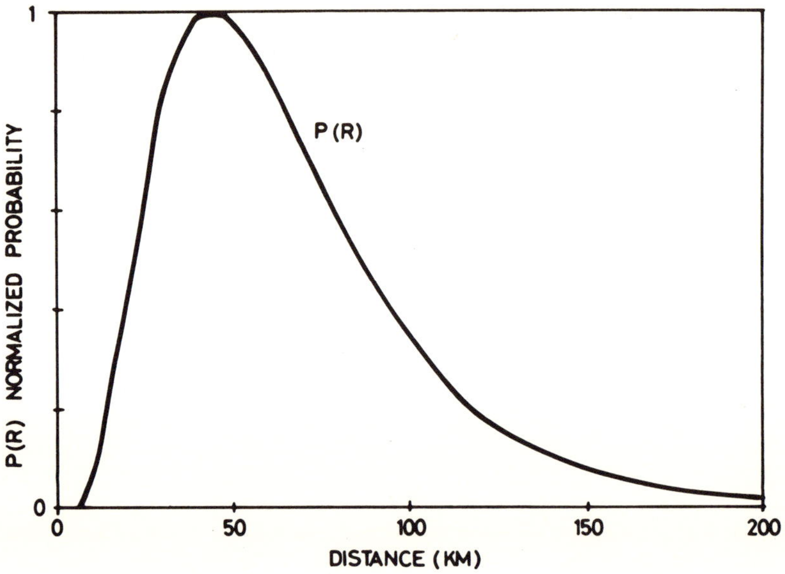


Figure 8. Normalized probability distribution for single flight displacement in budworm.

## Displacement and Settling: "Straightening Out"

A second important departure of natural displacement/settling behavior from random diffusion theory concerns so-called "straightening out" phenomena (Kennedy, 1961). In our discussion of the exodus response, we noted the general finding that rates-of-turn decrease with decreasing rates of resource contact in searching animals. In its limiting form, this tends to become directed displacement behavior with high directional autocorrelation (Nielsen, 1961; R. Jones, 1977a,b). Similarly, the repression of resource-directed responses during the refractory period referred to above eliminates many potential sources of variation in animals' displacement direction. Nearly all flight dispersers in the insects are now known to orient downwind (Johnson, 1969; Schaefer, 1976) after rising above the boundary layer, and this further imposes a high degree of directionality to individual movements. Finally, a growing number of cases exist in which insects are found to orient directionally over long periods and displacements using visual, chemical, and other cues (Schneirla, 1957; Kennedy, 1975; Graham, 1959). Corresponding phenomena in "migratory" birds, fish, and mammals are well known. Once again, the net result is that instead of wandering randomly about the landscape, moving animals tend to be highly directional, achieving maximal net displacement from the original habitat for given time or effort spent in dispersal activities.

## Displacement and Settling: Animal Aggregation

The discussion of local dynamics in the third and fourth sections suggested that an ability to aggregate dispersing animals in high local concentrations would be adaptive in a number of circumstances (see Taylor and Taylor, 1977, for a further development of the topic). Predator protection through concentration of defense secretions in the colonial sawflies (*Neodiprion* sp., Tostowaryk, 1972) and habitat modifications of tent caterpillars (*Malacosoma*, Wellington, 1969) are two of many examples (Ford, 1962). More subtle, but probably even more important, is the common observation from parasite-host theory that a contagious distribution of attacks -- however generated -- will exert a strong stabilizing influence on the overall system (Griffiths and Holling, 1969; Hassell and May, 1974, and references therein). No explicit spatial dimension is invoked in this theory, but the stabilizing mechanism is ultimately the same implicated in such diverse instances as the Huffaker (1958), Utida (1957), and Smith (1972) experiments; i.e., a certain proportion of the prey population is rendered less subject to attack than others. As might be expected, the stabilizing effects of contagious attack apply not only to host-parasite systems but to resource-exploitation systems in general. Myers (1974; Myers and Campbell, 1974), for instance, has shown how a clumped pattern of egg distribution enhances persistence in tansy ragwort (*Senecia jacobaea*)-cinnabar moth (*Tyria jacobaeae*) associations, again through protecting a portion of the food population from consumption. Such contagious egg distributions seem to be the rule among *Lepidoptera* (see references in Myers, 1974) and, indeed, in the insects in general (Southwood, 1966). Their role in the famous *Cactoblastis-Opuntia* saga has been emphasized by Caughley (1976). Aggregative spatial phenomena are thus seen to be essential determinants of ecological stability and persistence in a wide variety of circumstances. They are also, in a sense, the antithesis of the diffusion processes discussed earlier, and a few specific examples follow to suggest the range of mechanisms which may bring them about in nature.

A direct example of aggregative displacement/settling behavior is shown by the various scolytid beetles which prey on coniferous forests in western North America (see reviews by Renwick and Vité, 1970; Bennett and Borden, 1971). Although the details differ from species to species, the following general picture has emerged: female beetles locate a host, often homing on specific hydrocarbon odors emitted from those in suitable condition for attack. If only a few beetles attack the tree, however, they are usually overwhelmed by its defense mechanisms. An increased number of attacks leads to an increased chance of successful reproduction (again, the ubiquitous autocatalytic density effect). These initial females emit powerful pheromones which attract other beetles, first of both sexes and then predominantly males, from great distances.

(Interestingly, these males exhibit a very strong initial refractory period and may have to fly for an hour or more before becoming responsive to the female pheromones.) Males, however, secrete an additional pheromone which in sufficient quantities repels further colonizers. The net result is to achieve the necessary aggregation for escape from prey "control," while limiting densities below those which would reduce fitness through overcrowding. The dilemma we initially posed earlier (in Local Interactions) is resolved through a complex adaptive system of factors inhibiting and eliciting aggregation. (It is not, perhaps, surprising that the bark beetles' predators have also keyed into the complex system of odors, using it to home on their own hosts (Wood *et al.*, 1968).) For other examples, see Johnson (1969).

Another behavioral source of aggregations was mentioned earlier in discussion of the exodus response. The common tendency of searching animals to increase their turning rates under the stimulus of high target encounter rates leads immediately to a clumped distribution of "attacks" (see documentation above). The more prosaic adaptation of laying egg masses (common, for instance, in moths) rather than single eggs tends to produce a similar effect.

Aggregation of dispersers can also be brought about by predominantly physical factors. In insects, it is common to find that the threshold temperatures for development are a few degrees lower than thresholds for exodus. Thus even the unsynchronized development patterns which might be expected due to variations in resource quality or individual physiology may result in a "pile up" of potential migrants. As the season progresses and temperatures rise, all of these may encounter their first exodus window at the same time. The "straightening-out" displacement activity described earlier will then keep the migrants reasonably aggregated until settlement occurs. Other forms of exodus synchronization include light and weather cues, and a great deal of "follow the leader" activity also seems evident. A roughly parallel case can probably be made for many zooplankters.

Physical factors also act to bring about non-random, often aggregative settling of dispersers. Convective air currents, for instance, are well known to build up high concentrations of aerial dispersers, and to deposit them *en masse* when air flow instabilities occur (Johnson, 1969). When topographic features of the physical landscape precipitate such discontinuities, they may become systematic "sinks" of dispersing organisms. Just this phenomenon has been studied in the budworm system, and hypothesized to result in local "hot spots" where outbreaks are initiated (Greenbank, 1973). Some insects are evidently adapted to utilize natural aggregating properties of air flow, as shown by Rainey and his colleagues in analysis of locust concentrations in the intertropical convergence zone (Rainey, 1963).

We summarize this sampler of terrestrial movement somewhat tongue-in-cheek, with an alternative null hypothesis of selective advection, a notion which we propose is both more interesting and more realistic than that of random diffusion:

1) The proportion of animals leaving a location is a function of per capita habitat quality.

2) The exodus population determined by (1) moves as a patch, not as independent individuals.

3) The population of (2) will go off in a single direction and settle together.

4) Migrating populations will not settle within a sizable refractory distance of their original location.

5) Settlement of different populations will be contagious in space, due to biological and physical aggregative factors.

There are doubtless as many exceptions to this extreme position as to its extreme alternative. What we would hope to see in future theoretical work is an exploration of the spectrum of possibilities lying between. An extension of Gadgil's (1971), Roff's (1974a,b) and Jones' (1974) studies, set against the habitat templets of Southwood (1977), would almost certainly be productive of perspective and insight. The prospects for applying such perspectives to yield truly predictive understanding of habitat-movement-population dynamics relationships in the field are less bright, and certainly more distant. The studies of R. E. Jones (1977a,b) on *Pieris* and those of Myers (1974, Myers and Campbell, 1974) on *Tyria* show that such understanding can sometimes be achieved through careful elucidation of the components of movement outlined above. But the effort required is great and the experimental requirements extremely exacting. Whether general progress can be made in this area remains to be seen.

## PUTTING THE PIECES TOGETHER: PROBLEMS IN THE MODELING AND ANALYSIS OF SPATIAL POPULATION DYNAMICS

Analysis of local interactions and movement relationships along the lines discussed in the preceding section can provide the basic components for spatial theories of population dynamics. But theory remains unguided speculation unless it can generate specific predictions, and those predictions can be critically evaluated against empirical observation. Theories of patchiness/movement/population dynamics interrelationships confront serious problems on both counts.

First of all, the theory itself is inherently complex. At best it implies an awkward set of second order differential equations; more likely, a thoroughly intractable combination of nonlinearities, thresholds, and stochastic phenomena. In general, it will be impractical to obtain explicit solutions of the resulting systems of equations at time and space scales relevant to observed properties of the organisms. This forces us to adopt numerical simulation approaches, with all of the limitations, inelegance and pitfalls which that implies.

The other difficulty concerns the observations themselves. Just which spatial properties of ecosystem dynamics should a theory reasonably be expected to predict? More specifically, which of those properties are relevant to the organisms being studied?

## A Spatial Simulation of Budworm

Our own analysis of spatial population dynamics is still incomplete. Preliminary results have been described elsewhere (Clark, Jones and Holling, in press; Jones, 1977), and it will be sufficient here to briefly illustrate the approach, its potential, and some of its problems.

Dispersal by adult spruce budworm involves elements of each of the five factors listed in the "selective advection" hypothesis with which we concluded the preceding section. Most importantly:

1) The liklihood of a given adult emigrating is inversely proportional to the amount of food consumed as a larva.

2) Local emigration is fairly well synchronized due to temperature controls.

3) At any given time, there is a directional bias dictated by the prevailing winds.

4) Displacement reflects a refractory period as shown in Figure 8.

5) Local aggregations may be caused by at least three factors: subflight temperatures associated with high altitudes; turbulent air cells associated with cold front passage; and local convergence associated with land-sea breeze phenomena.

Each of these aspects of dispersal can be explicitly included or implicitly mimicked in the full spatial simulation model of budworm/forest interactions which was mentioned earlier. Recall that we divided the 7,000,000 hectares of New Brunswick into 393 local patches, each of which was independently governed by local population equations (interpretable through the equilibrium manifolds) and all of which were potentially linked via dispersal. Absolute spatial differences in the patches are determined by their respective locations relative to borders, altitude, and proximity to the ocean. It was also necessary to specify the proportion of each patch covered in the budworm's host species and, more importantly, the age-class distribution of these species. The latter two factors combine to determine the forest density ($F$) which manifold analysis showed so critical to outbreak development. These data were gathered from field surveys and forest inventories by research scientists of the Canadian Forestry Service Maritimes Forest Research Centre. Coupled with the egg density surveys shown in Figure 3, these let us characterize with varying degrees of accuracy the time-space pattern of budworm and forest densities from 1953 to 1975.

We then specified the 1953 historical pattern as an "initial condition" for the model and simulated 100 years of subsequent history in the absence of managerial intervention by man. The yearly results are most directly viewed as a series of annual maps of predicted budworm density and forest conditions. A sample is shown in Figure 9. The general asymmetrical wave pattern initially encountered in the historical maps of Figure 2 is evident as are the out-of-phase "hot spots" reported in a number of historical accounts.

When a realistic management model is coupled to the purely biological one, the results are radically different as shown in Figure 10. The integrated spatial waves characteristic of the natural system have vanished, replaced by the more chaotic spatial picture. Individual elements of the spatial mosaic operate more or less independently of each other and exhibit a comparatively fast oscillation period (5-8 years as opposed to 30-35 years in the natural case). Significantly, it is just such "seething outbreak," fast oscillating conditions which developed in central New Brunswick after extensive management activities were initiated in the early 1950's (Figure 3). But it is also clear from even a casual comparison of observed (Figure 3) and predicted maps (Figure 10) that the theory has some problems. In particular, the real world outbreak failed to spread into the northern portion of the province in the mid-1960's. The model does not reproduce this blockage. We assume this involves some as yet unidentified differences in basic patch quality between the highland north and lowland south, but have thus far been unable to realistically mimic the observed phenomenon in even a *post hoc* sense. Further studies on this aspect of spatial heterogeneity-dispersal interactions are continuing.

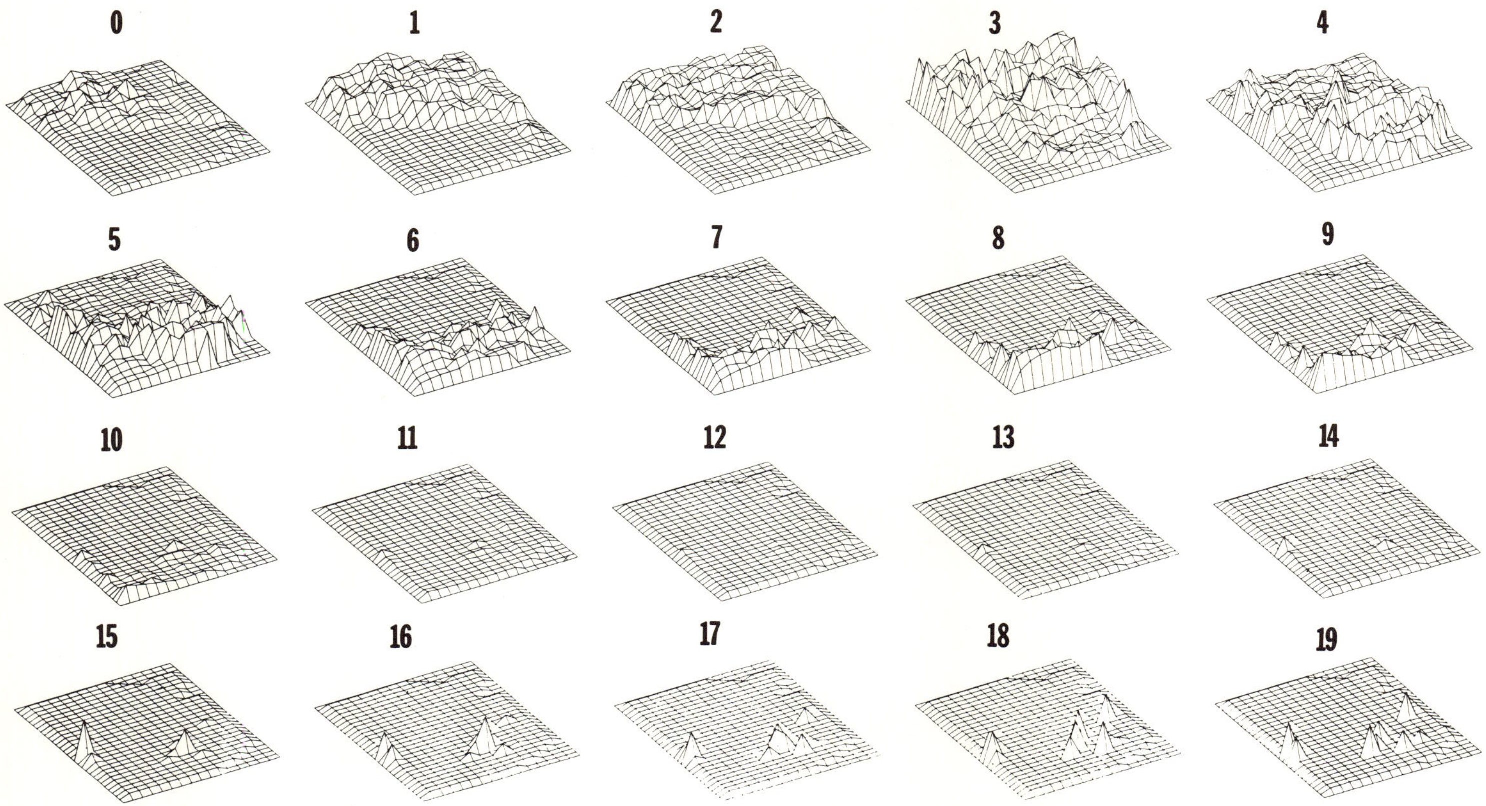


Figure 9. Twenty-year sequence of spatial patterns predicted by the model when no management or harvesting is imposed. Horizontal $(x,y)$ coordinates define a north-south by east-west spatial grid of 393 individual patches, each representing about 17,000 hectares of New Brunswick forest. Initial conditions are those for corresponding locations in New Brunswick in 1953. The vertical $(z)$ axis depicts relative budworm egg densities on a linear scale.

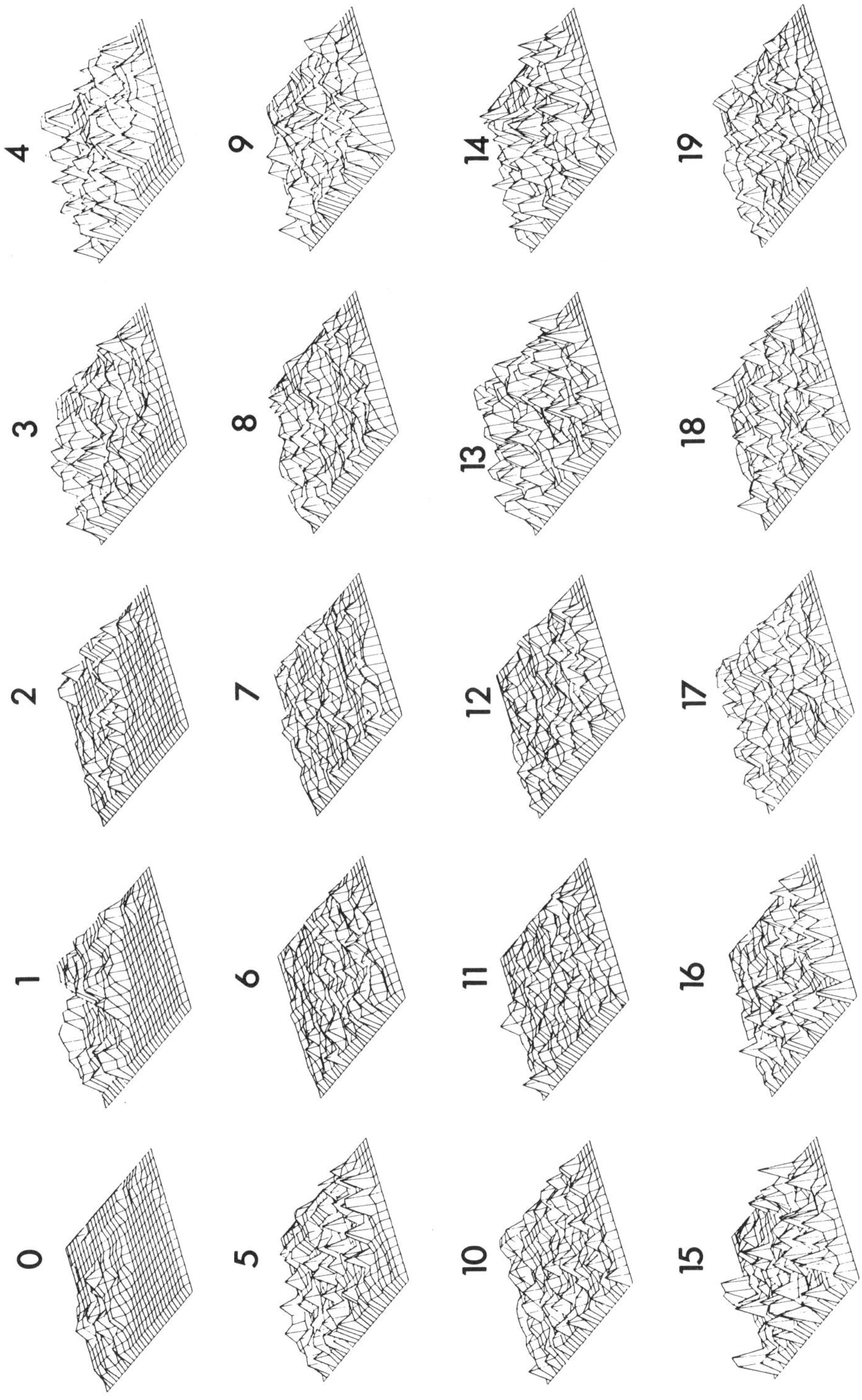


Figure 10. Twenty-year sequence of spatial patterns predicted by the model under historical rules for insecticide spraying and tree harvest. Conventions and initial conditions as in Figure 9. Note the loss of spatial cohesion and temporal pattern relative to Figure 9.

## Spatial Autocorrelation Analyses

The northern outbreak discrepancy illuminates one shortcoming of the present theory, but leaves the broader issues of exploring alternative formulations and assessing their limitations unresolved. The direct comparison of predicted and observed maps on a location-by-location basis fails to provide an adequate basis for rigorous tests of the theory. This problem has received a great deal of attention in geographical analysis but, as reviewed by Brown and Moore (1969), there are still no satisfactory statistical tests developed for the space-time case when dynamic spatial linkage and stochastic effects are involved.

This validation dilemma would be at least partially resolved if we could identify properties of observed patterns which were independent of absolute time and location, but still ecologically meaningful. In some other situations, notably those dominated by physical processes, these criteria seem to be met by the scale-variance relationships revealed by spectral or autocorrelation analysis (Platt and Denman, 1975; Cliff *et al.*, 1975; Platt, this volume; Richerson and Powell, this volume). Despite the fact that the theoretically necessary stationarity assumptions are almost certainly not met in terrestrial systems, we have explored the spatial autocorrelation structure of the budworm data and theories. Preliminary results have been only moderately encouraging.

We adopted the crossproduct correlation coefficient studied by Cliff and Ord (1973), and have expressed it as a standard normal deviate, using expected moments developed by those authors for the assumption of randomly distributed sample values. (Some such significance assessment is mandatory in any autocorrelation or spectral study which is to be remotely credible. The difficulties in the two-dimensional spectral case are well illustrated by Robinson (1975).) A maximum lag of 50 km was defined to reduce spatial trend effects, and the rate of change of density, rather than density itself, used to lessen the effects of a serious temporal trend in the data.

Figure 11 shows the resulting spatial autocorrelation structure for 12 years of the observational data which were summarized in Figure 3. The two features of interest are the absence of any strong pattern of correlation peaks and the lack of credible significance levels in the few peaks which do occur. There is apparently no significant small scale autocorrelation in the observed data; i.e., there is substantial variance at all scales examined. The theoretical predictions shown in the maps of Figure 10 were similarly analyzed, yielding even flatter (and zero valued) autocorrelation functions at scales of 20-50 km.

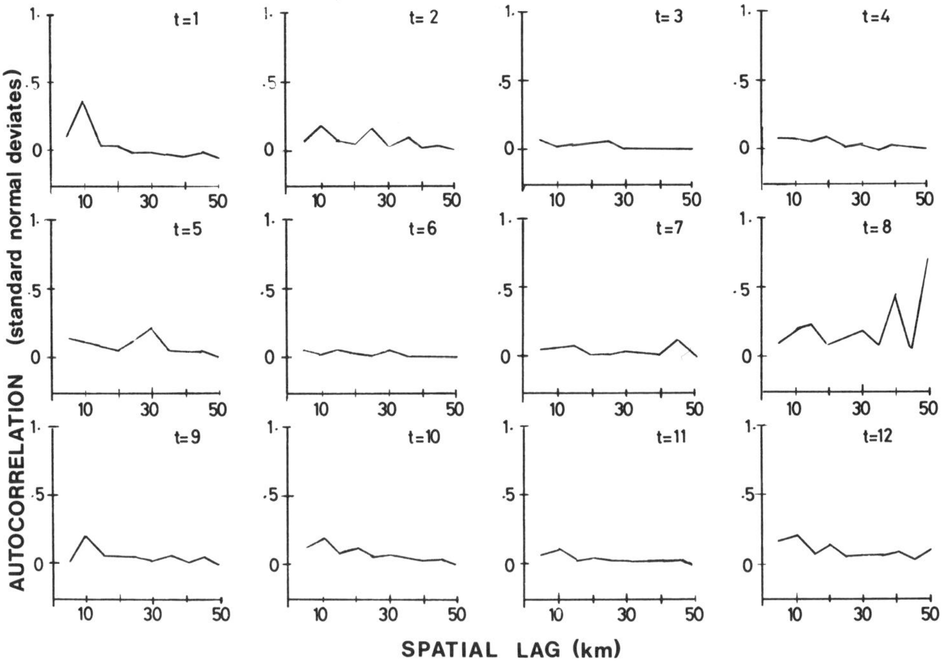


Figure 11. Spatial autocorrelation function for the real world managed pattern of Figure 3. Ordinate is scaled in standard normal deviates of Cliff and Ord's (1973) crossproduct statistic, evaluated under their randomization assumption. Abscissa represents lag in kilometers, summed over all directions. Sequence covers data for 1962 ($t$=1) to 1973 ($t$=12).

There are not sufficient quantitative data available to calculate spatial autocorrelations for the observed natural system outbreak patterns shown in Figures 1 and 2. When the theoretical predictions of Figure 9 are analyzed, however, the autocorrelation functions of Figure 12 are obtained. Here there appears to be a bit more small scale structure during years of active outbreak, with a certain tendency to peak at 20-30 km scales. This may be an artifact of the nonstationarity inherent in the traveling wave front evident in the map series. But in any event, the measured autocorrelation pattern is generally weak and not statistically inconsistent with an hypothesis of random spatial structure at the scales examined.

This result is somewhat disconcerting in view of the "obvious" structure apparent on inspection of the maps themselves. When formal statistics give such apparently silly answers, it is usually the case that we are either asking the wrong questions or we are entertaining silly hypotheses. In this case, we suspect a little

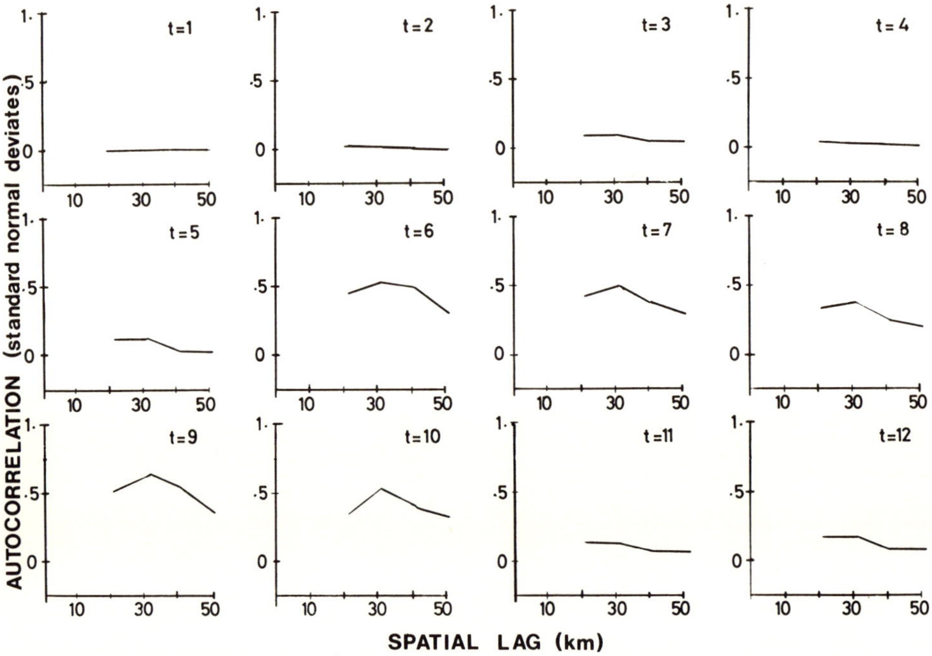


Figure 12. Spatial autocorrelation function for model predictions of the unmanaged system as shown in Figure 9. Sequence covers data for modeled years 38 ($t$=1) to 49 ($t$=12); (see Figure 14d for analogous space-time patterns). Other conventions as for Figure 11.

of both. Our budworm hypothesis involves dispersers from a patch already in outbreak triggering outbreaks in patches of suitable quality some kilometers away. These newly triggered patches will not themselves be in full outbreak for several years, however, so it is perhaps not surprising that spatially lagged, within-year patch comparisons show little correlation structure. But if this realization is not to send us on a mindless hunt through space-time in search of matching correlation peaks in the observed and predicted data series, it requires that our theory specify in advance where those peaks should occur. That is precisely what we are unable, for the moment, to do, and our explorations of the theory and its predictions have consequently proceeded in a somewhat more heuristic manner. The general approach has been to predict space-time patterns for limiting cases of the dispersal process theory, and use these to focus attention on critical areas for research and policy analysis. A few examples will serve to illustrate the nature of the results obtained.

Perturbing the Basic Theory

The maps predicted by the basic theory (Figure 9) provide reference points for further explorations. These data have been condensed in Figure 13 by sampling egg densities along a NW-SE transect of each annual map, and plotting these transects against time. The results show clearly the 30-odd year periodicity, the general tendency to spatial cohesion, and the occasional aperiodic "hot spots" characteristic of the unmanaged system.

The manifold analysis described earlier suggested that in the absence of dispersal, local outbreak timing would be determined largely by the spatially heterogeneous distribution of local patch quality. Setting the exodus rate for female moths to zero in the model allows us to investigate this situation, with the results shown in Figure 14a. Not surprisingly, dispersal is shown to play a powerful role as a spatial homogenizer in the system.

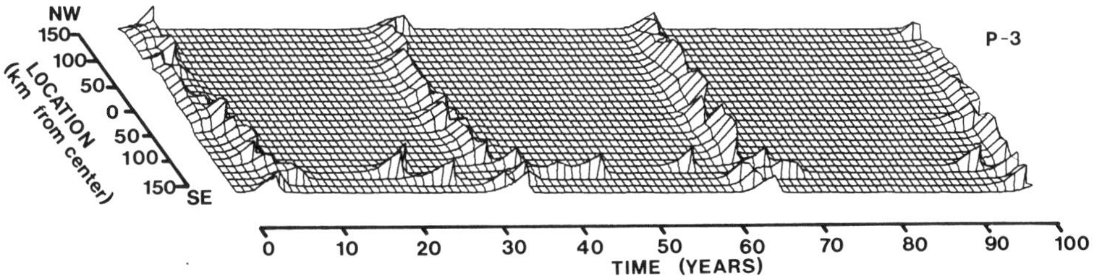


Figure 13. Space-time plot for the basic model. This figure shows a time series plot of predicted egg densities on a one-dimensional spatial transect of the full map series for the unmanaged basic model as originally shown in Figure 9. Each transect runs from NW to SE through the center of the modeled region and shows the relative budworm egg density on the vertical ($z$) axis. A separate transect sample is taken for each of the 100 simulated years, and arranged sequentially on the time ($x$) axis. This figure shows clearly the 30-odd year outbreak periodicity characteristic of the province as a whole, as well as the tendency for a few patches (usually near the SE end of the transect) to oscillate out of phase and at a higher rate. Note the evident spread of the initial ($t$=0) outbreak from NE to SW; the origin of the second outbreak sequence ($t \approx 30$) in the center and its spread NW and SE; and the progression of the third outbreak ($t \approx 60$) from SE to NW. Compare with the complete data of Figure 9. 'P-3' is an index code for the computer run.

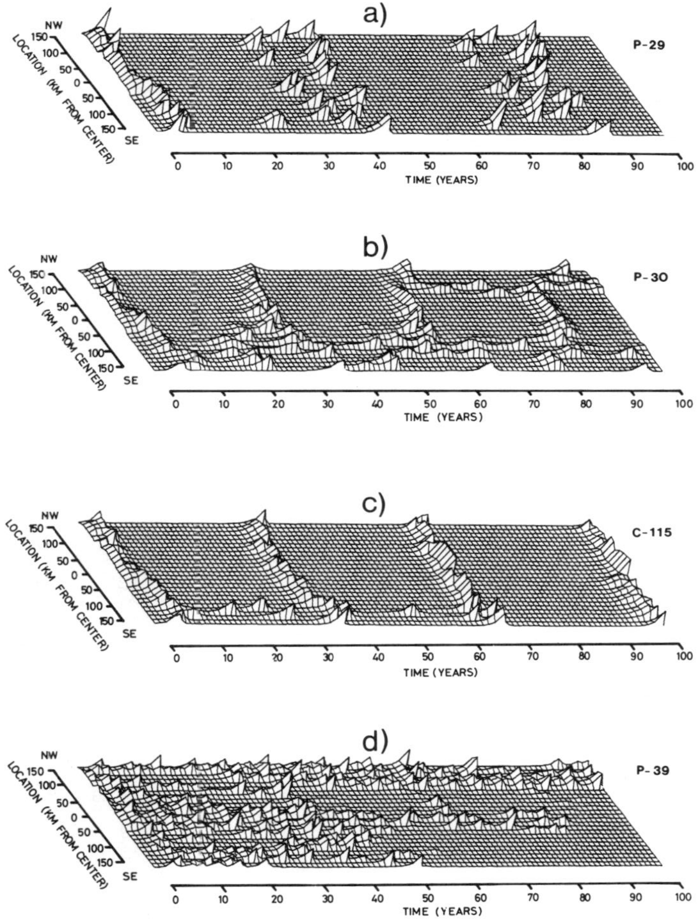


Figure 14. Space-time plots for perturbations to the basic model. The basic format here is the same as for Figure 13. Figure 14a (run P-29) shows the space-time behavior of the basic model when no dispersal occurs. Note the breakdown of spatial homogeneity and temporal synchronization relative to the basic case of Figure 13. Figure 14b (run P-30) shows the results of imposing a constant as opposed to patch quality dependent, exodus rate on budworm. The peculiar space-time pattern is of a sort unprecedented in history or other model experiments. Figure 14c (run C-115) results when multiple dispersal flights are allowed to each female budworm moth. The results are qualitatively similar to the basic single flight case of Figure 13. Finally, Figure 14d (run P-39) replots the results of the management perturbation (Figure 10) in the space-time format. The dearth of eggs in the southeast in later years is due to the fact that the forest there has been overharvested by man, and no budworm food remains. In general, the reduction of temporal and spatial cohesion as a result of management activities is greater than that caused by any realistic biological perturbation we have yet investigated.

Both the theoretical analysis and the empirical review described earlier in this paper point to the likely adaptive significance of a patch quality response in the exodus rate. Such a response is found in budworm (Jones, 1977) and was included in the basic theory. We conducted a series of experiments in which the quality dependence term was set to zero, and the emigration rate set at a constant value. Representative results for a constant emigration rate equal to the mean value of the basic case (FMIG = 0.5 in Jones, 1977) are shown in Figure 14b. The peculiar space-time behavior shows a degree of spatial cohesion intermediate to that of Figures 13 and 14a. A fast oscillating, somewhat discontinuous outbreak pattern in the southeast flows north in the early years, then brackets the central regions and finally moves into the northwest alone. Analysis of the full map series confirms this impression. Results of this sort confirm our earlier prejudices regarding a patch quality response in the emigration rate, focus subsequent research on the elucidation of its precise form, and make us leery of analyses which uncritically adopt constant terms in its stead.

A similar approach has been used to study the effect of variations on the displacement and settling responses originally incorporated in the basic theory. We have found the space-time behavior of the system to be relatively insensitive to plausible changes in these components of the dispersal process. For example, scaling the abscissa of the original displacement distribution function (Figure 8) by a factor of ±25% yields results essentially indistinguishable from those of the basic theory (Figure 13). Forty-five degree shifts in the mean vector of the prevailing windrose distribution likewise have minimal effect. Even allowing for several flights on successive nights by dispersing females gives approximately the same pattern as the basic model, as shown in Figure 14c. This result is particularly gratifying in that it suggests we can avoid the tremendous increases in computational time and costs associated with multiple flight analysis, not to mention the almost insurmountable difficulties of accurately parameterizing it in the field.

Finally, it is instructive to replot the results of our earlier management experiment (Figure 10) in the space-time format of Figure 14d. The general impression of a loosely structured set of semi-independent patches is emphasized, as is the dramatic impact of management as compared to any of the purely biological perturbations discussed earlier. At any given time, several patches are in outbreak condition, generating sufficient emigrants to trigger neighboring patches as food quality and imperfect insecticide targeting permit. Since forest quality is partially protected by management activities, each patch becomes susceptible to outbreak more frequently, and at any given place the interoutbreak period decreases. The problems of designing effective control strategies in such a heterogeneous, disperser-rich system are profound. But approaches along the lines illustrated here have provided an effective starting point. It has been possible, for instance, to show that a number of apparently plausible management policies are simply ineffective within the framework of biological interactions explored by the models. More importantly, it has been possible to proceed with the design and evaluation of alternative policies which explicitly take those interactions into account and offer new hopes for successful management (Baskerville, 1976; Clark, Jones and Holling, in press).

## SUMMARY

The profound difficulties of analyzing spatial heterogeneity-dispersal-population dynamics relationships in ecology are all too effectively illustrated by the studies discussed in this paper. A summary of major problem areas and research needs might include the following:

Biological dispersal is probably more complex and certainly qualitatively different than the spatial displacement processes encountered in physical systems. Accurate predictions of space-time patterns are still rare in ecology (but see, for example, R. Jones, 1977a,b), and elegant undertakings such as the spring bloom analysis reported by Pingree (this volume) are simply beyond the state of the art when significant biological components are involved. This situation is unlikely to improve until the physical analogies of our present dispersal theories are replaced by experimentally derived representatives of ecological reality.

We still do not understand what level of space-time predictability can be expected from even a good ecological theory. Severe *a priori* limitations are now known to exist for spatial meteorological models (Lorenz, 1969; Robinson, 1971), and Platt, Denman and Jassby (1977) have suggested that similar difficulties may hold for even simple phytoplankton productivity models. Terrestrial and complex marine cases remain unstudied. This leaves questions of validation and testing for spatial models in an even more ambiguous situation than we have come to expect in ecology.

More broadly, there remains a dearth of biologically meaningful questions concerning even the qualitative properties of space-time solutions to ecological theories. "Stabilizing" versus "destabilizing" criteria have been employed to some effect in theoretical studies, but their applicability to field studies is not at all straightforward. And despite our own emphasis on the adaptive significance of biological dispersal, we agree with Taylor (1974) and Gilbert *et al.* (1976) that there exist no convincing practical means for assessing the relative "fitness" or even "resilience" (Holling, 1973) implied by nonequilibrium space-time patterns and the models which generate them.

Finally, and encompassing all of the preceding points, data and theory in population studies remain very poorly matched. Scale-variance (e.g. spectral) approaches represent the most widespread attempt to cast predictions and observations in a common metric. But even these techniques, as emphasized by Denman and Platt (in press), owe popularity more to their monopoly of the field than any ability to cleanly bridge the data-theory gap. Specific theoretical models tied to detailed field studies seem to us a most immediate and pressing need. The marine work reported by Evans (this volume) and the terrestrial studies noted earlier suggest both the potential and the difficulties involved.

Despite the manifest problems of studying spatial heterogeneity-dispersal-population dynamics relationships in ecology, it is not inappropriate to close this discussion of the subject on a guardedly optimistic note. Much, in fact, has been accomplished, and the prospects for future progress are brighter than ever before.

Animal movement is no longer considered a "humdrum process" to be safely ignored by sober ecologists. Rather, along with patchiness and habitat questions, it has moved to center stage in the Ecological Theater, and begun to exert a major influence on our interpretation of the Evolutionary Play (Pimentel *et al.*, 1963; Gilpin, 1975).

Further, a growing number of sophisticated studies show that, while surely difficult, it no longer need be considered "almost impossible to spend any time profitably in the deliberate study of animal dispersal as such." In terrestrial work at least, these dispersal studies have progressed from their early preoccupation with statistical problems of pattern description to the present treatment of movement as a component-structured ecological process of the same order as predation or reproduction (Gilbert *et al.*, 1976). When the analogous shift in perspective occurred more than a decade ago in predation research, it led to substantial improvements in both predictive theory and experimental design. We see no reason not to expect a similar trend in dispersal research in the 1980's.

Lastly, it is clear that technical computational problems now present no insurmountable obstacles to the evaluation of realistic spatial theory. It is encouraging, for instance, that the controlled laboratory experiments of Huffaker (1958) have now been adequately treated by an approximate Markov model (Maynard Smith, 1974), a spatially gridded numerical simulation (Hilborn, 1975), and an analytical solution of a differential equation formulation (Hastings, 1977), all with essentially equivalent results. This surely obviates the need for further sterile debates on which tools are "best" for the task at hand. In another positive development, the early critical scale studies, now expanded into general bifurcation analyses (Levin, this volume) and specific parameter space explorations (Ludwig, Weinberger and Aronson, this paper) have shown how good mathematical analysis can be used to simply complex models, highlighting key issues for further numerical study or field investigation.

The close and extended collaboration of mathematicians, laboratory workers, and field ecologists will almost certainly be necessary if the nascent opportunities suggested in this review (and volume) are to be developed and the established difficulties overcome. Students of marine patchiness have perhaps been quicker to appreciate this than their terrestrial counterparts, but the thought is hardly new. Nearly 200 years ago the British ornithologist, Thomas Bewick, summed up the present dilemma and prospect in a way which intervening experience gives us no reason to change:

> *"To investigate, with any tolerable degree of success, the more retired and distant parts of the animal economy, is a task of no small difficulty. An inquiry so desirable and so eminently useful would require the united efforts of many to give it the desired success."*
>
> (Bewick, 1797)

ACKNOWLEDGEMENTS

As suggested in our closing quote from Mr. Bewick, a study such as reported here represents the united efforts of a great many people. We have drawn heavily on the data and ideas of David Greenbank and Charles Miller of the Canadian Forestry Service's Maritimes Forest Research Centre; Chris Sanders of the C.F.S.' Great Lakes Forest Research Centre; Gordon Baskerville and Robert Dickison of the Faculty of Forestry, University of New Brunswick; Glen Schaefer and Keith Allsopp of the Cranfield Institute of Technology; R. C. Rainey of the Centre for Overseas Pest Research; and Bill Wellington of the Institute of Research Ecology, University of British Columbia. Our debt to Don Ludwig, H. F. Weinberger and Dr. Aronson for use of their unpublished results is acknowledged in the text. The simulation model has developed from an early prototype designed at our Institute by Carl Walters, Ray Hilborn, Randall Peterman, Sandra Buckingham, and Jeff Stander. Zafar Rashid has been responsible for programming and maintaining the present version. To all, our gratitude and thanks.

## REFERENCES

Andrewartha, H. G. and L. C. Birch, 1954: The distribution and abundance of animals. University of Chicago Press, Chicago.

Bailey, N.T.J., 1975: The mathematical theory of infectious diseases and its applications. Griffen, London.

Bartlett, M. S., 1957: Measles periocity and community size. J. Roy. Soc. Stat. Soc., 120A, 48-70.

Baskerville, G., 1975: Spruce budworm: super silviculturist. For. Chron., 51, 138-140.

Baskerville, G. (ed.), 1976: Report of the task force for evaluation of budworm control alternatives. Dept. of Natural Resources, Fredericton, New Brunswick, Canada.

Bennett, R. B. and J. H. Borden, 1971: Flight arrestment of tethered Denroctonus pseudotsugae and Trypodenron lineatum in response to olfactory stimuli. Ann. Ent. Soc. Amer., 64, 1273-1286.

Bewick, T., 1797: History of British birds. Vol. 1, Beilby and Bewick, Newcastle.

Blais, J. R., 1968: Regional variation in susceptibility of eastern North American forests to budworm attack based on history of outbreaks. For. Chron., 44, 17-23.

Broadhead, E. and R. C. Cheke, 1975: Host spatial pattern, parasitoid interference and the modelling of the dynamics of Alaptus fusculus (Hym: Myrmaridae), a parasitoid of two Mesocpsocus species (Psocoptera). J. Anim. Ecol., 44, 767-793.

Brown, C. E., 1970: A cartographic representation of spruce budworm infestation in eastern Canada 1909-1966. Can. For. Serv. Publ. No. 1263.

Brown, L. A. and E. G. Moore, 1969: Diffusion research in geography: a perspective. Prog. in Geogr., 1, 119-157.

Brunner, J. and Bruts, E. C., 1975: Searching behaviour and growth rates of Anthocoris nemoralis (Hemiptera: Anthocoridae), a predator of the pear psylla Psylla pyricola. Ann. Ent. Soc. Am., 68, 311-315.

Caughley, G., 1976: Plant-herbivore systems. In Theoretical Ecology, R. M. May (ed.), pp. 94-113, Saunders, Philadelphia.

Chandler, A.E.F., 1969: Locomotory behaviour of first instar larvae of aphidophagous Syrphidae (Diptera) after contact with aphids. Anim. Behav., 17, 673-678.

Clark, W. C., D. D. Jones and C. S. Holling, 1978: Lessons for ecological policy design: a case study of forest ecosystem management. Ecological Modelling, in press.

Cliff, A. D. and J. K. Ord, 1973: Spatial autocorrelation. Pion, London.

Cliff, A. D. et al., 1975: Elements of spatial structure. Cambridge University Press.

Curry, L. and G. Bannister, 1974: Township populations of Ontario 1971-2001 from time-space covariances. In Towards an Urban Future for Central Canada, L. S. Bourne et al. (eds.), pp. 34-59, University of Toronto Press.

Davidson, A. G. and R. M. Prentice (eds.), 1967. Important forest insects and diseases of mutual concern to Canada, the United States and Mexico. Department of Forestry, Canada.

Denman, K. and T. Platt, 1978: Time series analysis of marine ecosystems. In Proc. SIMS Conf. on Time Series and Ecological Processes, H. H. Shugart (ed.), SIAM, Philadelphia.

Dingle, H., 1972: Migration strategies of insects. Science, 175, 1327-1335.

Dixon, A.F.G., 1959: An experimental study of the searching behaviour of the predatory coccinellid beetle Adalia bipunctata (L.). J. Anim. Ecol., 28, 259-281.

Dodd, A. P., 1940: The biological campaign against prickly pear. Comm. Prickly Pear Board, Brisbane, Queensland, 177 pp.

Dodd, A. P., 1959: The biological control of prickly pear in Australia. Monog. Biologicae, 8, 565-577.

Elton, C. S., 1927: Animal Ecology. Methuen and Company, London.

Fleschner, C. A., 1950: Studies of the searching capacity of larvae of three species of predators of the citrus red mite. Hilgardia, 20, 233-265.

Ford, E. B., 1962: Butterflies. Collins, London.

Francia, F. C. and K. Graham, 1967: Aspects of orientation behavior in the ambrosia beetle Trypodendeon lineatum, Can. J. Zool., 45, 985-1002,

Gadgil, M., 1971: Dispersal: population consequences and evolution. Ecol., 52, 253-261.

Gilbert, N., A. P. Gutierrez, B. D. Frazer and R. E. Jones, 1976: Ecological Relationships. Freeman, San Francisco.

Gilpin, M. E., 1975: Group selection in predator-prey communities. Princeton University Press, Princeton.

Gould, P., 1970: Is statistix inferens the geographical name for a wild goose? Econ. Geogr., 46, 439-448.

Graham, K., 1959: Release by flight exercise of a chemotrophic response from photopositive domination in a solytid beetle. Nature, 184, 283-284.

Graham, K., 1961: Air swallowing: a mechanism for photic reversal of the beetle Trypodendron. Nature, 191, 519-520.

Greenbank, D. O., 1973: The role of climate and dispersal in the initiation of outbreaks of spruce budworm. Can. J. Zool., 35, 385-403.

Griffiths, K. J. and C. S. Holling, 1969: A competition submodel for parasites and predators. Can. Ent., 101, 785-818.

Haber, G., C. J. Walters and I. McT. Cowan, 1976: Stability properties of a wolf-ungulate system in Alaska and management implications. IRE R-5-R.*

Hassel, M. P. and R. M. May, 1974: Aggregation of predators and insect parasites and its effects on stability. J. Anim. Ecol., 43, 567-594.

Hastings, A. J., 1977: Spatial heterogeneity and the stability of predator-prey systems. Theor. Pop. Biol., 12, 37-48.

Hilborn, R., 1975: The effect of spatial heterogeneity on the persistence of prey-predator interactions. Theor. Pop. Biol., 8, 346-355.

Hilborn, R., 1978: Some long term dynamics of predator-prey models with diffusion. Ecol. Modelling, 5, in press.

Holling, C. S., 1959: The components of predation as revealed by a study of small mammal predation of the European pine sawfly. Can. Entomol., 91, 293-320.

Holling, C. S., 1965: The functional response of predators to prey density and its role in mimicry and population regulation. Mem. Entomol. Soc. Can., 45, 1-60.

Holling, C. S., 1973: Resilience and stability of ecological systems. Ann. Rev. Ecol. Syst., 4, 1-23.

Holling, C. S., D. D. Jones and W. C. Clark, 1977: Ecological policy design: a case study of forest and pest management. In Proc. Conf. Pest Management, October 1976, G. A. Norton and C. S. Holling (eds.), pp. 13-90, IIASA CP-77-6.*

Huffaker, C. B., 1958: Experimental studies on predation: dispersion factors and predator-prey oscillations. Hilgardia, 27, 343-383.

Hughes, R. D. and W. L. Nicholas, 1974: The spring migration of the bushfly (Musca vetustissima Walk.): evidence of displacement provided by natural population markers including parasitism. J. Anim. Ecol., 43, 411-428.

Iwao, S. and W. G. Wellington, 1970: The western tent caterpillar. Qualitative differences and the action of natural enemies. Res. Popul. Ecol., 12, 81-99.

Johnson, C. G., 1966: A functional system of adaptive dispersal in flight. Ann. Rev. Ent., 11, 233-260.

Johnson, C. G., 1969: Migration and dispersal of insects by flight. Methuen, London.

Jones, D. D., 1974: Stability implications of dispersal linked ecological models. IIASA RM-75-44.*

Jones, D. D., 1975: The application of catastrophe theory to ecological systems. In New Directions in the Analysis of Ecological Systems. Part II: Simulation Councils Proc. 5(2), 133-148, G. S. Innis (ed.). (Reprinted in Simulation, 29, 1-15, 1977.)

Jones, D. D., 1977: The budworm site model. In Proc. Conf. Pest Management, October 1976, G. A. Norton and C. S. Hollings (eds.), pp. 91-115, IIASA CP-77-6.*

Jones, D. D. and C. J. Walters, 1976: Catastrophe theory and fisheries regulation. J. Fish. Res. Bd. Can., 33, 2829-2833.

Jones, R. E., 1977a: Movement patterns and egg distribution in cabbage butterflies. J. Anim. Ecol., 46, 195-212.

Jones, R. E., 1977b: Search behavior: a study of three caterpillar species. Behaviour, 60, 237-259.

Kennedy, J. S., 1961: A turning point in the study of insect migration. Nature, 189, 785-791.

Kennedy, J. S., 1975: Insect dispersal. In Insects, Science and Society, D. Pimentel (ed.), pp. 103-119, Academic Press, New York.

Kennedy, J. S., C. O. Booth and W.J.S. Kershaw, 1961: Host finding by aphids in the field. Ann. Appl. Biol., 49, 1-21.

Kierstead, H. and L. B. Slobodkin, 1953: The size of water masses containing plankton blooms. J. Mar. Res., 12, 141-147.

Krebs, C. J. and J. H. Myers, 1974: Population cycles in small animals. Adv. Ecol. Res., 267-399.

Leonard, D. E., 1970: Intrinsic factors causing qualitative changes in populations of Porthetria dispar (Lepidoptera). Can. Entomol., 102, 239-249.

Levin, S. A., 1974: Dispersion and population interactions. Am. Nat., 108, 207-228.

Levin, S. A., 1976a: Spatial patterning and the structure of ecological communities. Lectures on Math. in Life Sci., 8, 1-35.

Levin, S. A., 1976b: Population dynamic models in heterogeneous environments. Ann. Rev. Ecol. Syst., 7, 287-310.

Levin, S. A., 1978: Population models and community structure in heterogeneous environments. In Math. Assoc., Amer. Studies in Mathematical Biology, Vol. II: Populations and communities, S. A. Levin (ed.), pp. 439-476.

Lorenz, E. N., 1969: The predictability of a flow which possesses many scales of motion. Tellus, 21(3), 389-407.

Luckinbill, L. S., 1974: The effects of space and enrichment on a predator-prey system. Ecol., 55, 1142-1147.

Ludwig, D., D. D. Jones and C. S. Holling, 1978: Qualitative analysis of insect outbreak systems: the spruce budworm and the forest. J. Anim. Ecol., 47, 315-332.

May, R. M., 1974: Ecosystem pattern in randomly fluctuating environments. In Progress in Theoretical Biology, R. Rosen and F. Snell (eds.), pp. 1-50, Academic Press, New York.

May, R. M., 1977: Thresholds and breakpoints in ecosystems with a multiplicity of stable states. Nature, 269, 471-477.

Maynard Smith, J., 1974: Models in Ecology. Cambridge University Press.

Meyers, K. and B. S. Parker, 1975: A study of the biology of the wild rabbit in climatically different regions of eastern Australia. Aust. Wildl. Res., 2, 11-32.

Miller, C. A. and E. G. Ketella, 1975: Aerial control operations against the spruce budworm in New Brunswick 1952-1973. In Aerial Control of Forest Insects in Canada, M. L. Prebble (ed.), Department of Environment, Canada.

Monro, J., 1967: The exploitation and conservation of resources by populations of insects. J. Anim. Ecol., 36, 531-547.

Morris, R. F., 1955: The development of sampling techniques for forest insect defoliators, with particular reference to the spruce budworm. Can. J. Zool., 33, 225-294.

Morris, R. F. (ed.), 1963: The dynamics of epidemic spruce budworm populations. Mem. Ent. Soc. Can., No. 31.

Murdie, G. and Hassell, M. P., 1973: Food distribution, searching success, and predator-prey models. In The Mathematical Theory of the Dynamics of Biological Populations, M. S. Bartlett and R. W. Hiorns (eds.), pp. 87-101, Academic Press, London and New York.

Myers, J. H., 1976: Distribution and dispersal in populations capable of resource depletion: a simulation model. Oecol., 23, 255-269.

Myers, J. H. and B. J. Campbell, 1976: Distribution and dispersal in populations capable of resource depletion: a field study on the cinnabar moth. Oecol., 24, 7-20.

Nielsen, E. T., 1961: On the habits of the migratory butterfly *Ascia monuste*. Biol. Medd., 23, 1-81.

Oaten, A. and W. W. Murdoch, 1975: Functional response and stability in predator-prey systems. Amer. Nat., 109, 289-298.

Okubo, A., 1974: Diffusion-induced instability in model ecosystems: another possible explanation of patchiness. Tech. Rept. 86, Chesapeake Bay Inst., Johns Hopkins University.

Pimentel, D., W. P. Nagel and J. L. Madden, 1963: Space-time structure of the environment and the survival of parasite-host systems. Amer. Nat., 97, 141-167.

Platt, T. and K. L. Denman, 1975: Spectral analysis in ecology. Ann. Rev. Ecol. Syst., 6, 189-210.

Platt, T., K. L. Denman and A. D. Jassby, 1977: Modeling the productivity of phytoplankton. In The Sea, Vol. 6, E. D. Goldberg et al. (eds.), pp. 807-856, Wiley, New York.

Rainey, R. C., 1963: Meteorology and the migration of desert locusts. Locusts, Tech. Notes, WMO, 54, 1-115.

Renwick, J. A. and J. P. Vite, 1970: Systems of chemical communication in Dendroctonus. Contr. Boyce Thompson Inst., 24, 283-292.

Robinson, G. B., 1971: The predictability of a dissipative flow. Quant. J. Roy. Meteorol. Soc., 97, 300-312.

Robinson, J. E., 1975: Frequency analysis, sampling, and errors in spatial data. In Display and Analysis of Spatial Data, J. C. Davis and M. J. McCullaugh (eds.), pp. 78-95, Wiley, London.

Roff, D. A., 1974a: Spatial heterogeneity and the persistence of populations. Oecol., 15, 245-258.

Roff, D. A., 1974b: The analysis of a population model demonstrating the importance of dispersal in a heterogeneous environment. Oecol., 15, 259-275.

Roff, D. A., 1975: Population stability and evolution of dispersal in a heterogeneous environment. Oecol., 19, 217-237.

Schaefer, G., 1976: Radar observations of insect flight. In Insect Flight, R. Rainey (ed.), Blackwells, Oxford.

Schneirla, T. C., 1957: Theoretical consideration of cyclic processes in Doryline ants. Proc. Amer. Phil. Soc., 101, 106-133.

Segel, L. A. and J. L. Jackson, 1972: Dissipative structure: an explanation and an ecological example. J. Theor. Biol., 37, 545-559.

Segel, L. A. and S. A. Levin, 1976: Application of nonlinear stability theory to the study of the effects of diffusion on predator-prey interactions. In Topics in Statistical Mechanics and Biophysics: A Memorial to J. L. Jackson, R. A. Piccirelli (ed.), Proc. AIP Conf., 27, 123-152.

Skellam, J. G., 1951: Random dispersal in theoretical populations. Biometrika, 38, 196-218.

Skellam, J. G., 1973: The formulation and interpretation of mathematical models of diffusionary processes in population biology. In The Mathematical Theory of the Dynamics of Biological Populations, M. S. Bartlett and R. W. Hiorns (eds)., pp. 63-85, Academic Press, London.

Smith, F. E., 1972: Spatial heterogeneity, stability and diversity in ecosystems. In Growth by Intussusception: Ecological Essays in Honor of G. E. Hutchinson, E. S. Deevey (ed.). Trans. Conn. Acad. Arts Sci., 44, 309-335.

Smith, J.N.M., 1974a: The food-searching behaviour of two European thrushes. I. The description and analysis of search paths. Behaviour, 48, 276-302.

Smith, J.N.M., 1974b: The food-searching behaviour of two European thrushes. II. The adaptiveness of the search patterns. Behaviour, 49, 1-61.

Southwood, T.R.E., 1962: Migration of terrestrial arthropods in relation to habitat. Biol. Rev., 37, 171-214.

Southwood, T.R.E., 1966: Ecological Methods. Methuen, London.

Southwood, T.R.E., 1975: The dynamics of insect populations. In Insects, Science and Society, D. Pimentel (ed.), pp. 151-199, Academic Press, New York.

Southwood, T.R.E., 1977: Habitat, the templet for ecological strategies? J. Anim. Ecol., 46, 337-366.

Southwood, T.R.E. and H. N. Comins, 1976: A synoptic population model. J. Anim. Ecol., 45, 949-965.

Steele, J. H., 1972: Factors controlling marine ecosystems. In The Changing Chemistry of the Oceans, D. Dryssen (ed.), Nobel Symp. 20, Wiley, New York.

Steele, J. H., 1974a: Structure of Marine Ecosystems. Harvard University Press, Cambridge.

Steele, J. H., 1974b: Stability of plankton ecosystems. In Ecological Stability, M. B. Usher and M. H. Williamson (eds.), pp. 179-191, Chapman and Hall, London.

Strickler, J. R., 1977: Observation of swimming performances of planktonic copepods. Limnol. Oceanogr., 22, 165-170.

Takahashi, F. 1964: Reproduction curve with two equilibrium points: a consideration on the fluctuation of insect population. Res. Popul. Ecol., 6, 28-36.

Taylor, L. R., 1975: Longevity, fecundity, and size: control of reproductive potential in a polymorphic migrant, Aphis fabae Scop. J. Anim. Ecol., 44, 135-163.

Taylor, L. R. and R.A.J. Taylor, 1977: Aggregation, migration, and population mechanics. Nature, 265, 415-421.

Thom, R., 1975: Structural stability and morphogenesis. Benjamin, Reading.

Thompson, W. A., P. J. Cameron, W. G. Wellington and I. B. Vertinsky, 1976: Degrees of heterogeneity and the survival of an insect population. Res. Popul. Ecol., 18, 1-13.

Tostowaryk, W., 1972: The effect of prey defense on the functional response of Podisus modestus (Hemiptera: Pentatomidae) to densities of the sawflies Neodiprion swainei and N. pratti banksianae (Hymenoptera: Neodiprionidae). Can. Ent., 104, 61-69.

Utida, A., 1957: Population fluctuation: an experimental and theoretical approach. Cold Spring Harbor Symp. Quant. Biol., 22, 139-152.

van Raalte, G. D., 1972: Do I have a budworm-susceptible forest? For. Chron., 48, 190-192.

Webb, F. E., J. R. Blais and R. N. Nash, 1961: A cartographic history of spruce budworm outbreaks and aerial forest spraying in the Atlantic region of North America, 1948-1959. Can. Ent., 93, 360-379.

Wellington, W. G., 1957: Individual differences as a factor in population dynamics: the development of a problem. Can. J. Zool., 35, 293-323.

Wellington, W. G., 1964: Qualitative changes in unstable environments. Can. Ent., 96, 436-451.

Wellington, W. G., 1965: The use of cloud patterns to outline areas with different climates during population studies. Can. Ent., 97, 617-631.

Wellington, W. G., 1969: Effects of three hormonal mimics on mortality, metamorphosis and reproduction of the western tent caterpillar, Malacosoma californicum pluviale. Can. Ent., 101, 1163-1172.

Wellington, W. G., P. J. Cameron, W. A. Thompson, I. B. Vertinsky and A. S. Landsberg, 1975: A stochastic model for assessing the effects of external and internal heterogeneity on an insect population. Res. Popul. Ecol., 17, 1-28.

Whittaker, R. H., 1974: Climax concepts and recognition. In Vegetation Dynamics, Hand. Veg. Sci., 8, 137-154.

Whittaker, R. H. and S. A. Levin, 1977: The role of mosaic phenomena in natural communities. Theor. Pop. Biol., 12, 117-139.

Wiener, N., 1956: Nonlinear prediction and dynamics. Third Berk. Symp. Math. Stat. Prob., 3, 247-252.

Williams, C. B., 1965: Insect Migration. Collins, London.

Wood, D. J., L. E. Brown, W. D. Bedard, P. E. Tilden, R. M. Silverstein and J. O. Rodin, 1968: Response of Ips confusus to synthetic sex pheromones in nature. Science, 159, 1373-1374.

Wrattan, S. D., 1973. The effectiveness of the coccinellid beetle, Adalia Bipunctata (L.) as a predator of the lime aphid Eucallipterus tiliae L. J. Anim. Ecol., 42, 785-802.

*In an effort to make the bibliography as current as possible, we have cited some works available at present only as publications of the International Institute for Applied Systems Analysis (IIASA) and our own Institute of Resource Ecology (IRE). These may be obtained from the following addresses:

Documents and Publications
International Institute for Applied Systems Analysis
Schloss Laxenburg
A-2361 Laxenburg
AUSTRIA

Publications (Attn: Ralf Yorque)
Institute of Resource Ecology
University of British Columbia
Vancouver, B.C., Canada V6T 1W5

# PATTERN FORMATION IN ECOLOGICAL COMMUNITIES

Simon A. Levin

Section of Ecology and Systematics
Cornell University
Ithaca, New York 14853

## INTRODUCTION

The primary concern of this volume is the study of spatial pattern in plankton communities, but the basic issues are fundamental ones in all ecosystems (Levin, 1976b). The plankton has emerged as perhaps the prototypical system for the discussion of spatial pattern principally because of the ubiquitous nature of patchiness, independent of the scale of observation (Platt and Denman, 1975). Population densities are sufficiently high that pattern is easily detected, and planktonic systems have been intensively studied from a variety of aspects.

However, spatial heterogeneity is a readily apparent and important feature of virtually every ecosystem, and this has profound implications with respect to the adaptive syndromes of the component species. Population dynamics and dispersal and foraging strategies depend critically upon the spatio-temporal characteristics of the environment. To understand these effects, one must know the nature of pattern. How does it arise? How stable is it over time? What factors govern the details? How important is chance? Only when these issues are confronted may one begin to ask in a precise way how species adapt to such heterogeneous environments.

Mathematical models may be useful in addressing such questions but need to be modified from the usual fare in theoretical ecology. The classical mathematical approaches, while they have influenced heavily our conceptions concerning ecosystem organization, are inadequate and indeed misleading in their assumption of homogeneity and emphasis upon asymptotics. New approaches are mandated which treat systems as spatially heterogeneous and which recognize the importance of localized transients.

Although there are other choices available, the paradigm for modelling spatially-distributed populations has been based on the diffusion-reaction system (Skellam, 1951; Kierstead and Slobodkin, 1953) and its generalizations. The history of this approach is detailed somewhat more fully in Levin (1976b, 1978). Its use in biology dates back to Brownlee's (1911) work on epidemics, and has become well-established in genetics (Fisher, 1937; Haldane, 1948), in theoretical developmental biology (Turing, 1952), and, of course, in more classical physical and engineering applications. There is a rich literature associated with the diffusion-reaction approach, both for the continuum models which seem most appropriate for the aquatic environment and for the discrete (cellular) versions. The latter may be better suited for many terrestrial situations, especially where patches are clearly defined units with fixed boundaries; examples may be found in agriculture, in forests, and in the rocky intertidal. A more complete discussion of these formalisms may be found in Levin (1976a,b, 1978).

It must be recognized that the diffusion-reaction approach is strictly correct for only very few systems. Most generally, transport is not by passive diffusion and the description must be generalized to include other mechanisms; e.g. advective, chemotactic or interspecific effects. Okubo (1975) has discussed some modifications, and other recent work has considered long-range transport, for which the equations take a very different mathematical form. Further, Fasham (this volume) discusses the introduction of stochastic elements Finally, I shall discuss in later sections a quite different approach to the entire problem of patchiness.

However, despite these qualifications, the generic diffusion-reaction system provides an exceedingly useful metaphor. Such systems have received tremendous attention recently in the mathematical community, and possess a rich behavioral repertoire. Most of the phenomena observed for continuous diffusion-reaction systems also are characteristic of discretized versions in space or time, or of the related integro-difference or integro-differential systems necessary when more general transport mechanisms are considered (Fife, 1978a). This is important because although it is recognized that the assumption of diffusive motion is not correct, it is also the case that we cannot strictly justify any alternative model from first principles. Thus the most prudent form of action is to focus attention on those behaviors which are common to a variety of possible models, and for this purpose diffusion-reaction systems provide a convenient illustration.

## PATTERN

Discussion of community pattern requires a recognition that such pattern occurs simultaneously on many scales: spatial, temporal and structural (Levin, 1976b). For example, populations have characteristic life histories: communities, typical successional sequences. Space and time integrate in an inseparable way, so it is often impossible to restrict attention to either one alone.

What makes patterns interesting to us is that different communities have different ones. Thus, any discussion of pattern begins with the conceptualization of a number of possible worlds, perhaps for convenience indexed by parameters. Each world has a characteristic behavior or pattern, and the problem of primary interest is the relation of pattern to parameters. For example, in the classical view (Clements, 1916), each vegetational succession in a given area will tend toward a unique climax community, the climatic climax, and the problem of interest is the correlation of that climatic climax to the climate of the region. Although modern views of succession differ markedly from that of Clements (see discussion in Drury and Nisbet, 1973; Horn, 1974; Pickett, 1976; Connell and Slatyer, 1977; Whittaker and Levin, 1977), the original concept remains of interest. Many more general notions of succession are modifications or complications of the Clementsian view.

The correlation of communities and environmental parameters has contributed much to our understanding of community structure. Among the most important of these studies has been that of species distributions along environmental gradients (Whittaker, 1967, 1975). Once this quantification of the relationship of community to environment has been completed, the next step is often a classification into broad equivalence classes, community-types, generally according to the climax pattern. As in developmental biology, where cells become organized into distinct organs, populations become molded over time into distinguishable community-types (Whittaker, 1953, 1974) more similar to each other than they are to members of other such equivalence classes. This is more than a statistical clustering; it implies the possibility of discontinuity in response to environment. Within an equivalence class there may be variation, but generally on a scale much different than between classes.

Once these equivalence classes have been identified, three related problems become important: (1) the characterization of corresponding equivalence classes in the "parameter-space", and of the boundaries between them; (2) the "boundary-layer" behavior in the transition zones, where parameters change from one equivalence class to the next; and (3) the developmental or evolutionary problem (Whittaker and Woodwell, 1971): "How do community-types come to be different?"

Problem (3) (and to some extent (2)) is usually a "slow-time" scale problem, assuming parametric change is slow. As parameters vary on this scale, special interest is then focused upon threshold or bifurcation points, where critical determinations in community development occur that restrict the range of ultimate possible community-types. The language of mathematics can be useful here, as it has been in the consideration of related problems in the past (Waddington, 1957; Thom, 1969; Holling, 1973). Suppose initially, for simplicity of illustration, that the local (within-community) dynamics may be described in terms of the rates of change of a set of state variables $u_1,\ldots,u_n$. Setting $\underset{\sim}{u} = (u_1,\ldots,u_n)$, I write

$$\frac{d\underset{\sim}{u}}{dt} = \underset{\sim}{F}(\underset{\sim}{u}, \underset{\sim}{v}) \tag{1}$$

where $\underset{\sim}{v}$ is a vector of parameters. Time $t$ is not included explicitly among the arguments of $\underset{\sim}{F}$, since it may be disguised as one or more of the parameters.

The parameters $\underset{\sim}{v}$ change according to a complementary system of equations,

$$\frac{d\underset{\sim}{v}}{dt} = \underset{\sim}{G}(\underset{\sim}{u}, \underset{\sim}{v}). \tag{2}$$

The treatment of such systems (1)-(2) is discussed to some extent in Levin (1976b), and here I simply paraphrase the discussion given there. Particular interest is attached to the case when the parameters $\underset{\sim}{v}$ may be thought of as varying on a much slower time scale than $\underset{\sim}{u}$; often this can be realized simply by an appropriate labelling of the parameters. Then, assuming certain additional conditions are met (Tikhonov, 1948, 1952; Göbber and Seelig, 1975), the equations (1) and (2) may be dealt with in sequence, rather than in parallel. For example, under some conditions (1) may be solved for equilibrium values under the assumption that $\underset{\sim}{v}$ is constant, and the equilibrium values $\underset{\sim}{u}(\underset{\sim}{v})$ then substituted into the slow-time equations. These latter take the autonomous form

$$\frac{d\underset{\sim}{v}}{dt} = \underset{\sim}{G}(\underset{\sim}{u}(\underset{\sim}{v}), \underset{\sim}{v}). \tag{3}$$

In this way, results are obtained for both time scales.

More generally, $\underset{\sim}{u}$ may not tend to an equilibrium on the fast time scale. In this case the hope is still that the system will have a characteristic "fast-time" behavior for each vector $\underset{\sim}{v}$, described perhaps both in terms of the possible (asymptotic) limit point sets and their adherent "basins of attraction." (A basin of attraction of a limit point set is the set of initial state vectors which will converge ultimately to that set.) If such a characterization is possible, one may then study, on the slow-time scale, how the characteristic fast-time behavior tracks parametric change.

A very simple example may be useful here. Consider the differential equation

$$\frac{du}{dt} = au-u^3, \tag{4}$$

which will for all values of $a$ have an equilibrium point at $u = 0$. If $a \leq 0$, this equilibrium is globally stable in the sense that every solution of (4) will eventually tend to $u = 0$. This stable equilibrium may be viewed as analogous to the climatic climax associated with the environmental parameter $a$. (This is not proposed as a literal model since I have not even discussed what the components of the vector $\underset{\sim}{u}$ are. Thus no importance should be attached to the fact that variables may take negative values.) If $a > 0$, the system (4) has a much more complicated characteristic behavior (phase portrait). There are now three equilibrium points: an unstable one at $u = 0$, and two stable ones at $u = \pm a^{\frac{1}{2}}$. Any initially positive solution will tend to $u_\infty = + a^{\frac{1}{2}}$; any initially negative one to $u_\infty = - a^{\frac{1}{2}}$; and one initially zero will remain so. The two states $\pm a^{\frac{1}{2}}$ represent alternative stable equilibria; the importance of such alternative equilibria for ecological communities has been pointed out elsewhere (Levin, 1974, 1976b; Sutherland, 1974) and will be discussed further later.

If the parameter $a$ is now thought of as being varied slowly from negative values through zero to positive ones, then a bifurcation occurs at the boundary point $a = 0$. Three asymptotic solutions branch from one (Figure 1), and the fast dynamics of the system assume a very different form. Away from $a = 0$, the portrait of the system may change quantitatively as $a$ is varied, but the qualitative picture remains the same. Only at $a = 0$ is there a discontinuity.

In this simple example, there is only one free parameter, and hence the boundary between different qualitative pictures is a single point in parameter space. More generally, however, when several parameters are involved, the boundary will take the form of a curve or a higher-dimensional surface. Indeed, even in the case of equation (4), it is convenient to think of things in terms of two parameters, $a$ and $u_0$ (the initial value of $u$), since there are really three equivalence classes of behavior (two non-trivial) associated with each positive value of $a$ (Figure 2). The transition boundary now takes the form of a pair of orthogonal lines.

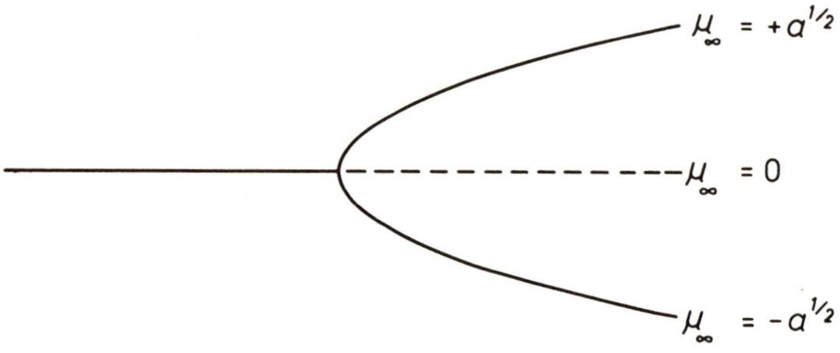


Figure 1. Possible asymptotic states $u_\infty$ as a function of $a$. Bifurcation occurs at $a = 0$.

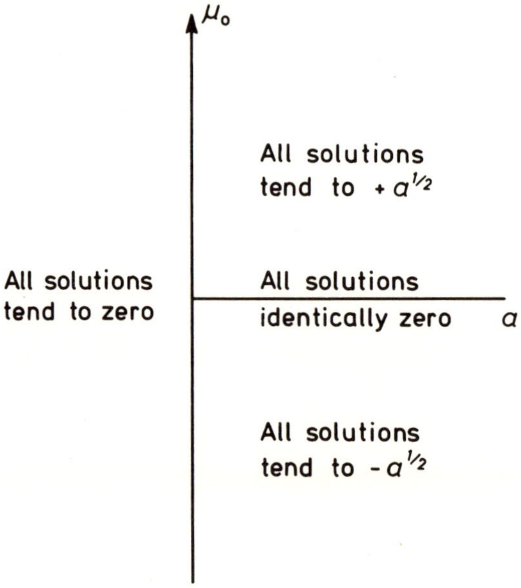


Figure 2. Classification of solutions by asymptotic behavior as $a \to \infty$. Axes are $a$ and initial condition ($u_0$).

The solutions of systems of ordinary differential equations may take a variety of general forms (Smale, 1976) of which the most familiar and simplest are those which tend for large time to stable equilibrium points or stable limit cycles. A classification of solutions according to such asymptotic (long-term) properties is analogous to the Clementsian classification suggested earlier for community-types, since one typically focuses there on the asymptotic behavior or climax. Similarly, the kinds of questions usually asked in the mathematical analysis of dynamical systems parallel those raised earlier in an ecological context: Can the solutions be classified into distinct equivalence classes? How is parameter-space subdivided in relation to these equivalence classes? What is the behavior of the system near the boundaries between these classes? What is the relation between initial and asymptotic state? Some of these questions have been popularized recently in the context of catastrophe theory (Guckenheimer, 1978a,b provides two excellent critiques), but represent much broader issues in the study of dynamical systems.

When communities are modeled by systems of the general form (1)-(2), habitat variables (Whittaker, Levin, Root, 1973, 1975), especially those with spatial extent, are often treated as parameters. Interest may center upon species distributional patterns along environmental gradients of these parameters, and significance is then attached to the question of whether discontinuous behavior (e.g. ecotones) may be observed along continuous gradients. The questions which are raised parallel those I have just discussed, but the modeling paradigm rests upon the assumption that communities at different locations interact only weakly with each other. Such an assumption is a spatial analogue to the earlier one that parameters change on a slow-time scale.

However, in many systems the assumption of weak interaction of neighboring "communities" is unjustified. Nutrients and other chemicals flow freely between systems, and individuals disperse widely. A way around this is to regard the environment as heterogeneous, to characterize a species not by its mean density over a region but by its distribution with respect to environment, and to allow for transport as a means of communication. If the environment is treated as a continuum, this communication is achieved by labelling each point in the environment with its coordinates $\underset{\sim}{r}$, and dealing with the distribution $u(\underset{\sim}{r}, t)$ of a species at time $t$ rather than a simple function of time. A typical way to describe the dynamics is then through the diffusion-reaction system

$$\frac{\partial \underset{\sim}{u}}{\partial t} = f(\underset{\sim}{u}, \underset{\sim}{v}) + \nabla(\underset{\sim}{D}\nabla \underset{\sim}{u}), \tag{5}$$

which may be seen to be the direct generalization of (1). Here $\underset{\sim}{u} = \underset{\sim}{u}(\underset{\sim}{r}, t)$ is the vector of species densities at the point $\underset{\sim}{r}$ at time $t$, $\underset{\sim}{v}(\underset{\sim}{r}, t)$ is the vector embodying local parameters, and $\underset{\sim}{D}$ is the matrix of diffusion parameters. Since the vector $\underset{\sim}{v}(\underset{\sim}{r}, t)$ of parameters now also depends on $\underset{\sim}{r}$, the system (2) must also be modified as was (1).

The specific form (5) is presented here only as an example. As discussed earlier and in Levin (1976b, 1978), there are numerous alternatives. Yet (5) provides a convenient framework for discussion and, as mentioned previously, its behavioral repertoire includes most of the ways pattern forms in natural systems. Hence the questions which I shall discuss in terms of (5) do not depend upon the particular formulation, and are the same ones raised several times previously. What are the characteristic behavioral patterns? How do they depend on parameters? How does behavior vary near threshold or bifurcation boundaries? How is asymptotic behavior related to initial conditions? In the remaining sections, I shall address these questions with special reference to pattern formation in ecological communities.

## CLASSIFICATION

The initial question I address is that of the classification of behaviors for diffusion-reaction systems; this has become a topic of considerable recent mathematical interest (Fife, 1978a,c). Superficially, diffusion is an homogenizing process; and indeed, under many conditions, the only asymptotic states for systems of the form (5) are homogeneous ones (see for example Conway and Smoller, 1978; Conway, Hoff and Smoller, 1978); but this is not always the case.

Clearly, however, even in the face of diffusion, spatial variation in environmental parameters can maintain parallel variation in the species which are responding to them. Biogeographical comparisons of community and species patterns are dependent upon this fact, and it hardly requires discussion here. But as has been pointed out for planktonic systems (Steele, 1973, 1974b), variation in physical (i.e. environmental) parameters is not sufficient to explain all pattern. What is of interest then is how pattern can be initiated and maintained in an apparently homogeneous environment. This does not deny the importance of parametric variation; it simply focuses attention upon those factors which are interacting with that parametric variation in the determination of pattern.

One important mechanism for the maintenance of spatial heterogeneity is the existence of multiple (alternative) stable states in the fast dynamics of the system without diffusion. This was mentioned in the previous section, and several authors (Lewontin, 1969; Holling, 1973; Levin, 1974, 1976b; Sutherland, 1974) have discussed the importance of the phenomenon in natural systems. The experimental literature is replete with examples (e.g. Park, 1962; Sutherland, 1974; Whittaker and Levin, 1977), and further the multiple stable state case is one of the classical feasible outcomes of even such an oversimplified reaction scheme as that described by the Lotka-Volterra competition equations. May (1977) in a recent review provides an in-depth discussion of several important cases of multiple stable states in ecology. One of the most impressive of these is the spruce budworm systems discussed elsewhere in this volume by Clark, and in the paper of Ludwig, Jones and Holling (1978).

My concern here is not with the factors which determine which of the several possible asymptotic states is achieved in a particular succession, since these will involve both history and chance; rather I simply wish to emphasize the fact that several possible outcomes may exist. Moreover, with each possible asymptotic state is associated a basin of attraction, and so the existence of multiple asymptotic states implies the existence of multiple qualitatively distinct successional pathways.

The above discussion applies to the local environments considered separately, and must be reexamined in the face of diffusion. This situation has been studied for cellular environments by Levin (1974). For these, not only may the environment as a whole tend to multiple uniform steady states; but, more interestingly, mosaic patterns may develop in which individual local environments arrive at different steady-states. This possibility is obvious if such local environments operate independently; but Levin (1974) shows that even in the presence of movement between environments, the complex mosaic pattern does not break down provided that movement is not too frequent. That is to say, the pattern varies continuously as the diffusion coefficients are increased until those coefficients become sufficiently large. Then, as the diffusion parameters cross a manifold of critical values, the spatial structure breaks down and homogeneity results. The situation for the continuum version (5) is less well-understood (Fife, 1978a), but similar phenomena seem to occur.

In the above example, such historical factors as initial colonization episodes and even events in geological time (Davis, 1976) will play a role in determining the local trajectory. Different microcommunities thus embark upon different successions of species, and the ensemble of such microcommunities converges to one of the many possible nonuniform equilibrium configurations. Interest may then shift to the slower-time scale, that is the one on which the determining parameters (e.g. extinction rates) vary. On this scale the mosaic pattern will change, sometimes continuously and sometimes abruptly, thereby introducing another class of threshold problems.

Often, the determining parameters vary on a time scale not significantly slower than that of the local dynamics. Typical of this is the situation in which local extinctions and recolonizations are common. In this case, even were there a unique feasible equilibrium state, a snapshot of the system at any moment of time would show a spatial mosaic of various successional states. As time varied, this succession of snapshots would produce a motion picture showing the landscape dynamically undulating and the local environments experiencing over and over again the various successional situations.

Local environments may be virtually asynchronous with one another in their dynamics or only accidentally related by virtue of being subjected to parallel variations in climate and other forcing variables. This type of situation will be the focus of the section on slower time scales in this paper. However, local environments may also experience varying degrees of synchrony through interaction, either in phase (analogously to the Belousov-Zhabotinskii chemical reaction scheme (Zaikin and Zhabotinskii, 1970; Kopell and Howard, 1974; Tyson, 1976) when it is well-stirred); or out of phase (as when that chemical system is not continuously stirred). It is well-known that in the latter case a variety of complex spatio-temporal patterns can result; e.g. targets, spirals, and travelling waves. All of these possible behaviors have ecological analogues. For example, certain species of the harvester ant *Pogonomyrmex* may form spirals as they disperse (Bernstein, 1975). Secondly, carnivorous gastropods of the genus *Thais*, which under most conditions are uniformly distributed over their invertebrate resource, can be observed for certain values of the parameters to form broad marauding bands in their intertidal habitat. This situation has been studied by B. Menge (personal communication), but no completely satisfactory explanation exists and certainly no one has attempted the parametric classification problem. Finally, in balsam fir forests that typically exhibit "asynchronous" patterns as the result of tree falls (local extinctions), under certain conditions of slope and exposure travelling waves of death and regeneration have been observed. This situation has been studied thoroughly by Sprugel (1976).

Patterns such as those above result even assuming that locally there is a single asymptotic state. When the multiple stable states discussed earlier are part of the local description along with their associated multiple successional pathways, the picture becomes even more complex and diversity is magnified accordingly (Levin, 1974, 1976b).

A dynamic, changing and heterogeneous spatial picture with individual micro-environments in different states is fundamental to survival for most species. Species disperse to capitalize on this variability, employing to varying degrees fugitive strategies (Levin, 1976b). The importance of this local temporal variation in phytoplankton communities and specifically in relation to the "paradox of the plankton" has been discussed frequently in the literature. Huffaker's (1958) classic study of mites on oranges showed that a complex, inhomogeneous environment with relatively low predator dispersal rates is a critical requisite for the survival of certain predator-prey systems, and mathematical analogues of this have been studied recently by Smoller and others (Fife, personal communication). The entomological literature is rich in discussion of dispersal (e.g. Wynne-Edwards, 1962; Southwood, 1962a,b; Johnson, 1969; Kennedy, 1975), and interest in the importance of dispersal in marine environments has received increasing recent attention. Moreover, dispersal strategies have become a popular subject for theoretical examination (e.g. denBoer, 1968, 1971; Gadgil, 1971; Janzen, 1968; Reddingius 1971; Reddingius and denBoer, 1970; Roff, 1974; Strathman, 1974).

Refugia, a special form of environmental heterogeneity, can provide stabilization of such systems. Examples are legion. In the rocky intertidal, where the dominant community-type is a spatio-temporal mosaic of algae and sessile invertebrates, many if not all of the principal competitors have refuge areas. Nyblade (1974) has also observed such refugia to be important in the stabilization of hermit crab "communities," and a similar phenomenon may be important in the stability of the coral reef fish communities studied by Sale (1976). Mech (1977) has recently studied the interaction between wolves and white-tailed deer in northern Minnesota where again the stability if not the very persistence of the system is dependent upon the existence of a refuge area for the prey.

As pointed out in Levin (1976b), pattern formed either as the result of environmental heterogeneity or of multiple stable states is in some sense pattern in spite of dispersal. There are, however, other mechanisms for stable pattern formation which depend upon dispersal; these include some of the examples of the preceding few paragraphs. Other examples have been suggested elsewhere (Levin and Segel, 1976; Levin, 1976b; Steele, 1973, 1974a,b), and will to some extent be developed in the next section. In addition, it needs to be emphasized that asymptotically systems need not tend to either stable equilibria or stable spatio-temporal patterns (see, for example, Ruelle and Takens, 1971). As observed recently for much simpler systems (May and Oster, 1976), the behavior of solutions to dynamic models is, in general, so complex and varied that it is not even possible to infer mechanisms from observed spatio-temporal patterns; this is a point which has been made in other ways many times previously in the literature.

I have concentrated attention on long-term behavior. However, in many cases, it is the transient behavior which is of most interest. The study of transient behavior relates closely to the problem of threshold and bifurcation structure, since we are usually interested in those transients which follow a qualitative change in the system. I thus now turn to a general discussion of threshold and bifurcation problems.

## THRESHOLD AND BIFURCATION

Threshold and bifurcation problems are concerned with the behavior of systems near critical values of parameters. Such parameters may appear as we have encountered them to date, that is explicitly in the dynamical equations; or they may be part of initial or boundary conditions. These are just variations on the same theme. Problems of these various kinds have probably provided the most usual entrée to discussions of pattern, and this is especially true of those studies which focus on the initiation of pattern. Often consideration of threshold problems precedes the kind of analysis suggested in the previous section, and is a useful way to begin to classify behaviors.

Many initial value problems are essentially threshold problems, such as those based on threshold models in the theory of epidemics (Waltman, 1974). In population biology, geneticists have been concerned since Fisher's (1937) paper with the description of the wave of advance of a favorable allele along a spatial gradient (see Fife, 1978b). The examination of this problem has been predicated on a general model of the form (5), and concerns itself with the wave front which locally converts the system from one equilibrium state into another. In this case, the introduction of the advantageous allele at a non-trivial level, either through mutation or immigration, elicits the threshold response; zero density of that allele represents the threshold boundary. Similar transient problems in neurophysiology (Rinzel, 1978) treat not wave fronts such as those above, but rather pulses which travel down axons. The threshold character of the response of neurons to stimulation is, of course, well-known. The excellent papers of Clark (this volume) and Ludwig, Jones and Holling (1978) vividly demonstrate the importance of thresholds in spruce budworm outbreaks. Such outbreaks generate travelling waves, much like those referred to elsewhere in this paper.

A somewhat different problem which is certainly familiar to all planktologists is that considered by Kierstead and Slobodkin

(1953) in relation to the initiation of red tides. (The identical mathematical development may be found in the pioneering paper of Skellam (1951), but in a more general setting.) In their simplest treatment, Kierstead and Slobodkin model phytoplankton dynamics by an equation of the form

$$\frac{\partial u}{\partial t} = D\frac{\partial^2 u}{\partial x^2} + ru \tag{6}$$

in which $u$ is phytoplankton concentration, $x$ is the single spatial coordinate, and $r$ is the intrinsic rate of natural increase. The threshold problem arises when boundary conditions are introduced. Kierstead and Slobodkin assume a local accumulation of requisite growth substances restricted to a region of length L. When L is less than the critical value $\pi(D/r)^{\frac{1}{2}}$, all solutions are damped to zero and no blooms can form. However, when L exceeds $\pi(D/r)^{\frac{1}{2}}$, certain modes may grow and phytoplankton growth may be initiated. This approach corresponds to the classical critical mass problem in nuclear reactor theory. The critical value $\pi(D/r)^{\frac{1}{2}}$ represents the threshold point for the parameter L, which, in this case, is contained within the boundary conditions rather than the equations themselves. Variations of the simple model (6), some considered by Kierstead and Slobodkin and others summarized by Okubo (1976), arrive at different quantitative answers but make the same qualitative point.

Following various suggestions of Turing (1952), of Steele (1973, 1974a,b) and of Okubo (1974), Lee Segel and I have considered yet another sort of bifurcation as contributing to the initiation and maintenance of pattern in the plankton (Levin and Segel, 1976). Our discussion, which follows related developments in papers of Segel and Jackson (1972), Levin (1974), and Okubo (1974), begins from a consideration of the "predator-prey" system

$$\begin{aligned} \frac{\partial P}{\partial t} &= P(a + eP - bH) + D_1\nabla^2 P \\ \frac{\partial H}{\partial t} &= H(-d + cP - gH) + D_2\nabla^2 H, \end{aligned} \tag{7}$$

in which all parameters are non-negative. $P$ represents phytoplankton concentration, $H$ herbivores (zooplankton). The positivity of $e$, which introduces an autocatalytic response into the phytoplankton, is a consequence of an assumed Type-2 functional response (Holling, 1965), essentially due to a grazing function with a "diminishing returns" character. The system (7) without diffusion may be regarded as obtained by retention of linear terms in Taylor expansions for $(dP/dt)/P$ and $(dH/dt)/H$ near a possible equilibrium (Levin, 1978).

Without diffusion ($D_1 = D_2 = 0$), the system (7) has a uniform equilibrium which can be, depending on the parameter values, both non-trivial and stable. This spatially uniform solution is still a possible equilibrium position of the system when $D_1$ and $D_2$ are not zero. What is surprising, however, and was first noticed by Turing (1952), is that for $D_2/D_1$ sufficiently large, this uniform equilibrium is unstable. A heuristic explanation of what transpires is that, if zooplankton diffuse much more rapidly than phytoplankton, they do not damp local phytoplankton blooms as they otherwise would; as a result, spatial pattern begins to form. The parameter $D_2/D_1$ has a critical value (see Levin, 1978) beyond which a new nonuniform solution bifurcates from the uniform one. Such a response to variation in the diffusion parameters is often referred to as a diffusive instability (Segel and Jackson, 1972). Segel and Levin (1976) developed the nonlinear analysis of such instabilities to show that under certain conditions the bifurcating solution will eventually stabilize, asymptotically tending to a new and nonuniform steady-state. Thus, due to differential dispersal abilities of phytoplankton and zooplankton, an initially uniform distribution of individuals transforms into a stable nonuniform one.

The kinds of bifurcations described above are only a few of a large number of possibilities, and recent attempts to classify the possible bifurcations for even very restricted classes of systems have met with major difficulties. May (1975) and May and Oster (1976) and, much earlier, Lorenz (1964) and Sarkovskii (1964) (see also Stefan, 1976; Hoppensteadt and Hyman, 1977), show that even for such a simple equation as the discrete logistic

$$\Delta u = ru\ (1-u), \tag{8}$$

a tremendous diversity of behaviors are possible, with an infinite number of bifurcations occurring as $r$ is varied. More general equations, such as those confronted in catastrophe theory, present major obstacles to analysis (Guckenheimer, 1978a).

One important type of bifurcation, the Hopf bifurcation, has considerable relevance to predator-prey systems. For most predator-prey systems common in the theoretical literature, every solution will tend to either a stable equilibrium point or a stable limit cycle (Kolmogorov, 1936; May, 1972; Albrecht *et al.*, 1973, 1974). Which of these two possibilities will result depends on the parameters of the model. For example, depending on the specifics of the model, there will, in general, be a set of parameter values for which a stable equilibrium point represents the only non-trivial asymptotic state of the system. As the parameters are varied, that stable point will continuously track them until some finite point at which a stable limit cycle bifurcates from the equilibrium point and becomes the asymptotic limit of almost every set of starting conditions. If parameter variation is reversed and the parameters change back toward the bifurcation boundary, the limit cycle will shrink back to the equilibrium point. The Hopf bifurcation has been the subject of intensive mathematical investigation, and recently an entire book (Marsden and McCracken, 1976) has been devoted to it. Of special interest to our discussion of pattern formation is the work of Kopell and Howard (1973, 1974), who consider an activator-inhibitor system of the type described above. They transform the predator-prey system with diffusion to a form for which a Hopf bifurcation results, and use this to show the existence of plane-wave patterns. These plane-waves are unstable, but Fife (1978a) points out that stable small amplitude wave trains do exist in some cases. Other patterns may also form under such conditions; for example, the spirals generated by *Pogonomyrmex* as it disperses (see previous section).

Finally, yet another class of bifurcation problems is identified with the multiple-stable state situation discussed earlier, and these are developed more fully in Levin (1978). In particular, the nonuniform patterns which exist due to multiple stable states may do so only for values of the diffusion coefficients up to some finite limit. A boundary exists in the space of diffusion parameters across which a discontinuous response occurs; the nonuniform pattern trades stability type with the uniform one, which is to say that for diffusion coefficients sufficiently large the only asymptotic states are homogeneous. More powerful results are developed in Conway and Smoller (1978) and Conway, Hoff, and Smoller (1978). Other bifurcations of this system occur on the slow time scale of colonization and extinction, and these will be explored somewhat in the next section.

In summary, no one mechanism accounts for all of the pattern we observe in natural systems. Environmental heterogeneity (such as in the Kierstead-Slobodkin example), multiple stable states, diffusive instabilities, and other factors may all play a role; and we are certain to be thwarted in any attempt to find a universal explanation for pattern in the plankton or elsewhere.

## SLOWER TIME SCALES

To this point, emphasis has been on the "fast" time scale. Parameters have been imagined as varying slowly, and attention focused on transient responses and stabilization processes following destabilization. Before discussing the other (slower) time scale, I point out that there may be more than two natural time scales; in particular, what I have just referred to as the fast time scale may, itself, be subdivided. For example, in their treatment of diffusive instability and the *dissipative structures* which result, Segel and Levin (1976) recognize two fast time scales, a fast-fast one on which destabilization takes place and a slow-fast one on which the new pattern becomes stabilized (see also Fife, 1976). An even slower scale, that on which parametric change occurs, is implicit in their treatment but not explicitly discussed.

Segel and Levin (1976) utilize "multiple scale" techniques; these are of considerable potential interest because, in the same way as techniques of spectral analysis, they do not require *a priori* assumptions concerning what the natural scales are. It is to be expected that both spectral techniques and multiple scale methods will become increasingly important tools of pattern analysis in the future. However, for the remainder of this paper I will restrict attention to the simpler situations, in which relative rates of response of variables may be specified in advance and classified according to scale.

There are in nature many situations in which a system may be thought of as having a (reference) state of rest, from which it may be perturbed only as the result of some external stimulus. Let me take these situations as a starting point for discussion, recognizing that the evaluation of a state as a rest state is, in many cases, a subjective judgment. One of the most familiar examples is that of the action of a neuron, which does have an obvious state of rest. Subject the neuron to a sufficient stimulus, and a signal will be initiated which travels down the axon. This signal is a transient pulse, and is followed first by a refractory period and eventually by a return to the rest state; the latter is an asymptotically stable state.

As a second example, consider a biotic infection; for example, a virus. An individual's rest state, "healthy," is one in which the infecting organism is not present or is present at low concentrations. A change in parameters, either to create a more favorable environment for growth or to raise the infecting population above a threshold, triggers a transient increase in the level of infection followed either by the return to the original rest state or by death (a somewhat more stable rest state). Analogous examples are the spread of an epidemic in a population of individuals or the spruce budworm outbreaks discussed by Clark (this volume).

The latter examples are special cases of colonization problems; either an infection or an epidemic involves the colonization of one life form by another. More general colonization processes may involve inanimate hosts, but the basic process is the same. In general, colonization may be triggered either by onset of favorable conditions, for example, due to seasonal climatic changes, or by the arrival of propagules sufficient to initiate population growth. Usually a variety of species will have evolved the capability to colonize any particular target, and these will arrive in a succession, the details of which include chance, temporal factors, history, location, and interspecific interactions. This successional process has been studied for mangrove islands (Simberloff, 1976; Simberloff and Wilson, 1969, 1970), for badger mounds (Platt, 1975), and for algae on intertidal boulders (Sousa, 1977) and high intertidal clearings (Paine: see Paine and Levin, 1978). Other experiments have been carried out by Janzen (1968) and by Root (1973) for plant-arthropod associations, by Shure and Ragsdale (1977) on granite outcroppings, by Paine and Levin (unpublished) in mid-intertidal clearings, and by Sutherland (1974) for subtidal hard surfaces. To this list could be added many others, including, of course, succession in higher plants (see reviews in Whittaker, 1975; Connell and Slatyer, 1977) and among phytoplankton.

In each of these examples, as in the model of Kierstead and Slobodkin, a favorable site for colonization is created, which then experiences a sequence of colonists. Depending on the size of the "island," each successional stage will involve several species, although there may be a fall-off in diversity in the late stages as competitive dominants extend their ranges.

This discussion has been somewhat oversimplified. For example, in a forest succession one might more naturally regard the climax stage as the rest stage, with disturbances such as tree falls, hurricanes, or fire initiating the colonization process. This, however, is a matter of convention. Indeed, many systems may be thought of as having at least two "rest" states: a disturbance may convert a climax "rest" state into a vacant one; successful colonization may then need to await a second parameter change; e.g. a favorable shift in climate. Eventually, the climax "rest" state may be reachieved; but this cyclical process need not be the universal rule. The possibility of multiple stable states adds another dimension, since there can be "transfers of energy" between distinct successional microcycles. Vulnerability to these transfers is highest after a major disturbance. Moreover, as with death, some rest states may be permanently absorbing, stable to all perturbations except on a much longer (messianic?) time scale. The notion of a rest state is simply a convenience: a rest state is a limit state of a process on a fast time scale, and there may be an arbitrary number for any system. A disturbance may initiate a transient which will eventually return to the same rest state, or it may result in the transfer of the system to another such state.

The study of successional processes involves the fast-time scale. Slow-time scale analysis is concerned with somewhat different questions: How do parameters change, and why? How does the fast-time behavior track slow parameter change? These questions may be related. For example, after the onset of an infection, an individual may be treated with antibiotics. By treating symptoms, one is, in effect, changing system parameters in response to the way fast-time behavior varies on the slow-time scale. This classical control problem introduces a feedback into the system. The change in parameters is no longer independent of what's going on in the system; i.e., it is not "external" in the sense the term was used earlier. Recognition of this is important, but does not change our fundamental viewpoint.

In the consideration of how fast-time behavior tracks slow parameter change, a variety of models are possible since the basic forces which govern pattern are not specific to any one time scale. Models can acknowledge environmental heterogeneity of varying degrees, including spatial variation in parameters other than those being varied temporally. I begin by ignoring this additional level of complexity, and thus initially do not consider such differences between local environments as relative accessibility to colonists, exposure, or mean climatic or edaphic parameters.

The simplest models which address slow-time behavior were suggested by Levins and Culver (1971), who hypothesized an environment subdivided into a mosaic of identical patches, each governed by the same local dynamics but possibly out of phase with the others. They did not consider within-patch dynamics (fast-time scale behavior) explicitly, assuming only that each patch could exist in one of several possible states. These states are equivalent to the rest states I have just described. Levins and Culver suggest writing a set of equations describing the frequencies of the various states and the transitions between them; the parameters of such a description are extinction and colonization rates. Those parameters change on a yet slower time scale than the processes of colonization and extinction themselves, so there are implicitly four time scales, those of: change in colonization and extinction rates; the colonization and extinction processes themselves; lateral extension by population growth, competitive interactions and secondary succession within patches; and, finally, attainment of a climax (rest) state. Levins and Culver basically consider two climax states, but allow also for one intermediate stage (a competitive phase). They also allow for invasion of already occupied patches. Although there are some technical problems with their model (Levin, 1974; Slatkin, 1974), the basic idea is sound and the rates potentially measurable. Such an approach allows for a determination of the balance between multiple stable states in terms of a few simple parameters. Horn (1975) suggests a basically equivalent approach for forest succession, using a somewhat different formalism. Horn and MacArthur (1972) also consider an extension of the Levins and Culver (1971) model, to allow for some habitat heterogeneity. Unfortunately, their model retains the same flaw as the Levins and Culver model, but again the basic idea is sound.

Any model is restricted by its assumptions, and so naturally there are a variety of directions in which models of the type just described may be extended. First, in some systems the distinction between slow and fast time scales is not clear, and various processes may interrupt the successional sequence before it is complete. One way in which this can happen was suggested by Hutchinson (1961, 1965) in relation to the paradox of the plankton: environmental parameters change so rapidly as to shift the competitive balance from one species to another and prevent full dominance from ever being achieved. Essentially, the equations (1) are then no longer autonomous. A related possibility is that disturbance may interrupt a successional sequence before its completion and basically reset it; that is, change its phase. Suppose, for example, that a tree fall initiates a successional recovery process, but before this is completed a second tree falls and breaks off the saplings which have begun to grow. The successional process begins again. Consideration of these complications represents one mode of extension of the modeling paradigm of Levins and Culver (1971).

Another possible extension is to systems in which patch boundaries are not fixed over time. These represent an extremely important class of problems. Inoculation of almost any species into an island habitat will be followed by a period of lateral extension in which that species regarded as patch enlarges. Bursts of phytoplankton begin as fairly localized clumps, and extend their ranges outward as they grow; at the same time, the patches of favorable water which serve as substrate for them may be spreading by diffusive and other processes and shifting centers due to advection. A related notion, contemporaneous disequilibrium, was suggested by Richerson *et al.* (1970) to explain the "paradox of the plankton." Similarly, gaps which form in forests or in intertidal mussel beds (Levin and Paine, 1974, 1975) will shrink due to perimeter encroachment respectively by branches or by border mussels. To deal with such possibilities, Paine and I (Levin and Paine, 1974, 1975) suggested that an additional set of slow-time scale equations is necessary to deal with changes in the characteristics of the patch structure of the environment. Regarding the environment as basically a sea within which floats a population of patches, we concentrate attention on the demographics of the patch population: birth, death, and change in size. Following the approach of von Foerster (1959; see also Rubinow, 1973, 1978; Oster and Takahashi, 1974; Oster, 1978), we let $n(a, \xi, t)$ define the distribution of patches with respect to age $a$ and size $\xi$ at time $t$. The continuity equation which results is

$$\frac{\partial n}{\partial t} + \frac{\partial n}{\partial a} + \frac{\partial (gn)}{\partial \xi} = -\mu n \qquad (9)$$

in which $g$ is the mean rate of growth (positive or negative, depending on the context) of a patch and $\mu n$ is the instantaneous "death" rate of patches due to new disturbance. Equation (9) may be explicitly solved (Levin and Paine, 1974) to give the distribution $n(a, \xi, t)$ in terms of the initial distribution $n(a, \xi, 0)$ and the pattern of birth rates of patches $n(0, \xi, t)$. Since, in many cases, the species composition of a patch may (by secondary successional studies of the type previously described) be related directly to $a$, $\xi$, and $t$ with a high degree of predictability, (9) combined with those studies of secondary succession leads to a complete description of system dynamics.

There are still a number of extensions possible. Noise terms can be added; for example, taking into account the variance in the growth rate of a patch (Rubinow, 1973). Second, age and size may not be the only descriptors of interest. For example, if disturbance has come in the form of fire (Heinselman, 1974; Heinselman and Wright, 1973; Wright, 1974), litter level and nutrient status may be important. For such cases, the model (9) may be easily extended to include several "physiological" variables (Oster and Takahashi, 1974; Levin and Paine, 1975; Levin, 1976b). Among these additional variables may be species levels, particularly if they do not adjust on a faster time scale (Whittaker and Levin, 1977).

Third, feedback birth conditions dependent on present composition of the population may be necessary; Oster and Takahashi (1974) discuss the analysis when such feedback relationships are introduced. Fourth, habitat heterogeneity may be included in several ways (Levin, 1976b), most simply by considering several interacting populations of patches with somewhat different general characteristics. One such extension would consider basically similar patch types but for various values of climatic and edaphic parameters. A more extreme version is discussed in the next paragraph.

Hall *et al.* (1978) describes the interaction between phytoplankton populations, fish larvae, and oil spills. All are patchily distributed, and for each the dynamics could be described by an equation of the form (9). Interactions between patch types need then be introduced, and this will mean that the critical parameters $g$ and $\mu$ for one population of patches will depend on the properties of the other two distributions. Such a modeling framework is a logical first approach to the problem confronting Hall *et al.* (1978), the effect of oil spills on fish populations.

The model just proposed is a mass-action (random collision) approach to interactions, and ignores the details of patch dispersion. By dealing with spatial averages, it ignores most of the interesting pattern phenomena described in the preceding sections. While recognizing the importance of spatial variability, its assumption that location is not important is a step backwards towards (1)-(2). It is still the case, as with the system (1)-(2), that interesting threshold problems will appear. For example, Marks (1974), in a study of the dormancy strategies of pin cherry (*Prunus pensylvanica* L.), found that seeds of this species may commonly remain buried in the soil for a half-century or more, germinating only when the formation of a gap creates an opportunity. If the frequency of gap formation is too low, so that seeds must commonly wait considerably longer for a chance to germinate, seed viability goes down and the population will not be able to survive. To a first approximation then, a threshold level of disturbance of the climax is necessary to the local survival of *Prunus*. Similar examples may be drawn from other systems of species (such as the sea palm *Postelsia*) whose survival depends on a sufficient level of disruption. Thus changes in the parameters of disturbance may cause qualitative shifts in fast-time scale behavior. Although in this case the parameter variation is on the slowest of time scales, similar examples exist at the next level.

For example, the principal dynamic variable on the slow-time scale for the model (9) is $n(a, \xi, t)$, the distribution of patches with respect to age $a$ and size $\xi$ at time $t$. On the scale explicitly treated in (9), this distribution will shift. A secondary consequence to such shifts, however, will be shifts in the regional densities of various species; for example, acorn barnacles and mussels. A number of tertiary consequences will then result. For one, the level of carnivorous snails will be affected as the resource base shifts, and this will have implications on colonization and succession patterns. Part of this effect will be felt on the fast-time scale, and it is possible that the system may shift from one regional steady state pattern to another as the result of shifts in predation. Clearly, understanding the organization of such complex systems involves an understanding of how things change on various time scales, and of the interactions between scales.

Naturally the situation becomes even more interesting when patches are reindexed by their spatial coordinates. Whether this added complexity is justified will depend upon the particular context, but it is clearly essential when the details of pattern are of interest. Consider, for example, the phenomenon just discussed, in which local disturbance initiates successional transients. The approach that Paine and I have taken is highly aggregative. It takes the viewpoint of the species, which disperse widely to buffer itself against the unpredictabilities of local environments. Hence we have emphasized only the mean characteristics of the environment as those parameters which will be of primary importance concerning the regional survival of species. The details of pattern are lost in such a formulation, however. When those details are of interest, one needs to graduate to the kinds of formulations discussed earlier. In those more detailed models, local changes in parameters (e.g. via disturbance) will result in localized successional sequences and (especially if qualitatively different successional sequences are feasible) possible bifurcations.

Consideration of the details of pattern becomes even more important when there is some synchrony between local environments, as with the travelling wave patterns in balsam fir forests (Sprugel, 1976). The variety of ways models can be constructed is tremendous, and the modeler must acquire a sensitivity to what amount of detail is essential to an understanding of the system.

## SUMMARY

Field ecologists and evolutionary theorists have long recognized the importance of spatial heterogeneity in the structure of ecological communities. Hutchinson (1951) introduced the notion of a fugitive species as one which exploits ephemeral resources, but the concept has much broader generality than is usually admitted. Most species are to some extent locally transient, dependent upon dispersal abilities to seek out new opportunities. Not only does this increase the potential for survival of individual species and hence the overall diversity of the system (Levin, 1974), it also can have stabilizing effects on ecosystem dynamics (Smith, 1972; May, 1974; T. Platt, 1975; Levin, 1976b; Sale, 1977).

Spatial pattern is especially apparent in plankton communities, and marine biologists have been interested in pattern and modelling approaches to it for over 30 years (Riley, 1946, 1947; Kierstead and Slobodkin, 1953). Pattern formation is a problem of general interest throughout biology, and the kinds of questions ecologists face in dealing with pattern are analogous, for example, to ones in developmental biology (Turing, 1952; Waddington, 1957; Keller and Segel, 1970; Gierer and Meinhardt, 1974; Othmer, 1977). It is therefore not surprising that the mathematical descriptions of these and other similar problems often take the same basic forms. In each of these cases one is dealing with a system in which the main forces are of interaction and transport. In each case, the system may assume one of an infinity of spatial designs. Quite naturally, one wishes to classify these designs, to study the ways in which they succeed one another in time, and to characterize their relative stabilities. Given the similarities of the basic forces and of the fundamental issues, it is hardly surprising that mathematical abstractions which preserve the key features appear similar for superficially different problems.

One mathematical entrée to these questions is through the study of diffusion-reaction systems, for which a tremendous theoretical literature has developed (e.g. Aronson, 1977a,b; Aronson and

Weinberger, 1975; Conway and Smoller, 1978; Conway, Hoff and Smoller, 1978; Fife, 1976, 1977, 1978a,b,c; Kopell and Howard, 1974; Segel and Levin, 1976). There are other possible places one might start, especially considering (as Clark points out elsewhere in this volume) that passive diffusion rarely applies to animal movement. But the diffusion assumption has a number of advantages as a point of departure. Even when movement is not diffusive, a diffusion approximation often provides a reasonable description at longer spatial scales. Moreover, by relaxing the pure diffusion assumption slightly, one can easily include effects such as advection, taxis, and dispersal keyed to local factors. With each modification comes increasing mathematical difficulty, but the most important classes of solutions remain those already present for the simplest forms. This increases the motivation for our in-depth treatment of the latter, and adds to their value as metaphors.

In this paper, I have introduced the mathematical approach to problems of pattern formation, with special reference to diffusion-reaction systems. A classification of solutions according to asymptotic behaviors is suggested, and this parallels the common approach in the classification of vegetational communities.

This classification is most easily done if parameters may be regarded as varying slowly over time. In this case, one may first direct attention to possible long-term behaviors (uniform and non-uniform steady states, periodic spatio-temporal patterns, and more complicated, perhaps chaotic, attractors) with some consideration of causes (characteristics of reaction schemes, interaction of biological and physical factors). Parameters may then be varied slowly, and the various threshold and bifurcation problems studied when critical values of parameters are reached. Such studies focus both on the transients (e.g. pulses, blooms, fronts) and on the longer term behavior.

When parameters vary on the faster time scale, somewhat different methods are necessary. One approach to this is the intertidal model of Levin and Paine (1974), and this was discussed in some detail in the next-to-last section.

The importance of spatial heterogeneity for adaptive strategies is clear. Moreover, only when the mechanisms of pattern formation are understood can the evolutionary problem be approached. This paper is an attempt to survey the status of the mathematical theory of pattern formation.

ACKNOWLEDGEMENTS

I gratefully acknowledge the support of NSF Grant MCS77-01-076. I am indebted to Bill Clark, Paul Fife, John Steele and Robert Whittaker for extremely useful comments on the manuscript.

REFERENCES

Albrecht, F., H. Gatzke, A. Haddad and N. Wax, 1974: The dynamics of two interacting populations. J. Math. Anal. and Applications, 46, 658-670.

Albrecht, F., H. Gatzke and N. Wax, 1973: Stable limit cycles in prey-predator populations. Science, 181, 1073-1074.

Aronson, D. G. and H. F. Weinberger, 1975: Nonlinear diffusion in population genetics, combustion, and nerve propagation. In Partial Differential Equations and Related Topics, J. A. Goldstein (ed.), Lecture Notes in Mathematics 446, Springer-Verlag, Heidelberg.

Aronson, D. G., 1977a: Topics in nonlinear diffusion. CBMS/NSF Lecture Notes, SIAM. To appear.

Aronson, D. G., 1977b: The asymptotic speed of propagation of a simple epidemic. In Nonlinear Diffusion. Research Notes in Mathematics, Pitman, London. To appear.

Bernstein, R. A., 1975: Foraging strategies of ants in response to variable food density. Ecology, 56, 213-219.

Brownlee, J., 1911: The mathematical theory of random migration and epidemic distribution. Proc. Roy. Soc. Edin., 31, 262-289.

Clements, F. E., 1916: Plant succession: an analysis of the development of vegetation. Carnegie Inst. Washington Publ. 242, 1-512.

Connell, J. H. and R. O. Slatyer, 1977: Mechanisms of succession in natural communities and their role in community stability and organization. Amer. Natur., 111, 1119-1144.

Conway, E. and J. Smoller, 1978: Diffusion and the classical ecological interactions. In Nonlinear Diffusion. Research Notes in Mathematics, Pitman, London. To appear.

Conway, E., D. Hoff and J. Smoller, 1978: Large time behavior of solutions of systems of nonlinear reaction-diffusion equations. SIAM J. Appl. Math., 35, 1-16.

Davis, M. B., 1976: Pleistocene biogeography of temperate deciduous forests. Geosci. and Man, 13, 13-36.

denBoer, P. J., 1968: Spreading of risk and stabilization of animal numbers. Acta Biotheor. Leiden, 18, 165-194.

denBoer, P. J., 1971: Stabilization of animal numbers and the heterogeneity of the environment: the problem of persistence of sparse populations. Proc. Adv. Study Inst. Dyn. Numbers Popul. Oosterbeek, 1970, pp. 77-97.

Drury, W. H. and I.C.T. Nisbet, 1973. Succession. J. Arnold Arboretum, 54, 331-368.

Fife, P. C., 1976: Pattern formation in reacting and diffusing systems. J. Chem. Phys., 64, 554-564.

Fife, P. C., 1977: Stationary patterns for reaction-diffusion equations. In Nonlinear Diffusion. Research Notes in Mathematics, Pitman, London. To appear.

Fife, P. C., 1978a: Asymptotic states for equations of reaction and diffusion. (Manuscript).

Fife, P. C., 1978b: Fisher's nonlinear diffusion equation and selection-migration models. (Manuscript).

Fife, P. C., 1978c: Results and open questions in the asymptotic theory of reaction-diffusion equations. (Manuscript).

Fisher, R. A., 1937: The wave of advance of advantageous genes. Ann. of Eugenics, 7, 355-369.

Gadgil, M., 1971: Dispersal: population consequences and evolution. Ecology, 52, 253-260.

Gierer, A. and H. Meinhardt, 1974: Biological pattern formation involving lateral inhibition. In Some Mathematical Questions in Biology, VI, S. A. Levin (ed.), pp. 163-184, Lectures on Mathematics in the Life Sciences, Vol. 7, American Mathematical Society, Providence, Rhode Island.

Gobber, F. and F. F. Seelig, 1975: Conditions for the applications of the steady-state approximation to systems of differential equations. J. Math. Biol.,2, 79-86.

Guckenheimer, J., 1978a: Catastrophe and chaos. In Lectures on Mathematics in the Life Sciences, Vol. 10, S. A. Levin (ed.), American Mathematical Society, Providence, Rhode Island. To appear.

Guckenheimer, J., 1978b: The catastrophe controversy. The Mathematical Intelligencer, 111, 15-20.

Haldane, J.B.S., 1948: The theory of a cline. J. Genet., 48, 277-284.

Hall, C.A.S., R. Howarth and B. Moore, 1978: Coastal zone siting of energy facilities. Ann. Rev. of Energy, in press.

Heinselman, M. L. and Wright, H. E., (eds.), 1973: The ecological role of fire in natural conifer forests of western and northern North America. Quat. Res., 3, 317-513.

Heinselman, M. L., 1974: Fire in the virgin forests of the Boundary Waters Canoe Area, Minnesota. J. Quat. Res., 3, 329-382.

Holling, C. S., 1965: The functional response of predators to prey density and its role in mimicry and population regulation. Mem. Entom. Soc. Can., 45, 1-60.

Holling, C. S., 1973: Resilience and stability of ecological systems. Ann. Rev. Ecol. Syst., 4, 1-24.

Hoppensteadt, F. C. and J. M. Hyman, 1977: Periodic solutions of a logistic difference equation. SIAM J. Appl. Math., 32, 73-81.

Horn, H. S. and MacArthur, R. H., 1972: Competition among fugitive species in a harlequin environment. Ecology, 53, 749-752.

Horn, H. S., 1974: The ecology of secondary succession. Ann. Rev. Ecol. Syst., 5, 25-37.

Horn, H. S., 1975: Markovian properties of forest succession. In Ecology of Species and Communities, J. Diamond and M. L. Cody (eds.), Harvard University, Cambridge.

Huffaker, C. B., 1958: Experimental studies on predation: dispersion factors and predator-prey oscillations. Hilgardia, 27, 343-383.

Hutchinson, G. E., 1951: Copepodology for the ornithologist. Ecology, 32, 571-577.

Hutchinson, G. E., 1961: The paradox of the plankton. Amer. Natur., 95, 137-145.

Hutchinson, G. E., 1965: The Ecological Theater and the Evolutionary Play. Yale University Press, New Haven, Conn., 139 pp.

Janzen, D. H., 1968: Host plants as islands in evolutionary and contemporary time. Amer. Natur., 102, 592-595.

Johnson, C. G., 1969: Migration and Dispersal of Insects by Flight. Methuen, London.

Keller, E. F. and L. A. Segel, 1970: Initiation of slime mold aggregation viewed as an instability. J. Theor. Biol., 26, 399-415.

Kennedy, J. S., 1975: Insect dispersal. In Insects, Science and Society, D. Pimentel (ed.), pp. 103-119, Academic Press, N.Y.

Kierstead, H. and L. B. Slobodkin, 1953: The size of water masses containing plankton blooms. J. Mar. Res., 12, 141-147.

Kolmogorov, A. N., 1936: Sulla teoria di Volterra della lotta per l'esisttenza. Giorn. Istituto Ital. Attuari., 7, 74-80.

Kopell, N. and L. N. Howard, 1973: Plane wave solutions to reaction-diffusion equations. Stud. Appl. Math., 52, 291-328.

Kopell, N. and L. N. Howard, 1974: Pattern formation in the Belousov reaction. In Some Mathematical Questions in Biology, VI, S. A. Levin (ed.), pp. 201-217, Lectures on Mathematics in the Life Sciences, Vol. 7, American Mathematical Society, Providence, Rhode Island.

Levin, S. A., 1974: Dispersion and population interactions. Amer. Natur., 108, 207-228.

Levin, S. A. 1976a: Spatial patterning and the structure of ecological communities. In Some Mathematical Questions in Biology, VII, S. A. Levin (ed.), pp. 1-36, Lectures on Mathematics in the Life Sciences, Vol. 8, American Mathematical Society, Providence, Rhode Island.

Levin, S. A., 1976b: Population dynamic models in heterogeneous environments. Ann. Rev. Ecol. Syst., 7, 287-310.

Levin, S. A., 1978: Population models and community structure in heterogeneous environments. In Mathematical Association of America Study in Mathematical Biology, Vol. II, S. A. Levin (ed.), Populations and Communities.

Levin, S. A. and R. T. Paine, 1974: Disturbance, patch formation, and community structure. Proc. Natl. Acad. Sci., USA, 71, 2744-2747.

Levin, S. A. and R. T. Paine, 1975: The role of disturbance in models of community structure. In Ecosystem Analysis and Prediction, S. A. Levin (ed.), pp. 56-67, SIAM, Philadelphia, Pennsylvania.

Levin, S. A. and L. A. Segel, 1976: Hypothesis for origin of planktonic patchiness. Nature, 259, 659.

Levins, R. and D. Culver, 1971: Regional coexistence of species and competition between rare species. Proc. Natl. Acad. Sci., USA, 68, 1246-1248.

Lewontin, R. C., 1969: The meaning of stability. In Diversity and Stability in Codogical Systems, G. W. Woodwell and H. H. Smith (eds.), pp. 13-24, Brookhaven Symp. Biol. 22.

Lorenz, E. N., 1964: The problem of deducing the climate from the governing equations. Tellus, 16, 1-11.

Ludwig, D., D. D. Jones and C. S. Holling, 1978: Qualitative analysis of insect outbreak systems: the spruce budworm and forest. J. Animal Ecol., 47, 315-332.

MacArthur, R. H. and E. O. Wilson, 1967: The Theory of Island Biogeography. Princeton Univ. Press, Princeton, 203 pp.

Marks, P. L., 1974: The role of pin cherry (Prunus pensylvancia L.) in the maintenance of stability in northern hardwood ecosystems. Ecol. Monogr., 44, 73-88.

Marsden, J. and M. McCracken, 1976: The Hopf Bifurcation. Springer-Verlay, Heidelberg.

May, R. M., 1972: Limit cycles in predator-prey communities. Science, 177, 900-902.

May, R. M., 1974: Ecosystem patterns in randomly fluctuating environments. In Progress in Theoretical Biology, R. Rosen and F. Snell (eds.), pp. 1-50, Academic Press, New York.

May, R. M., 1975: Biological populations obeying difference equations: stable points, stable cycles, and chaos. J. Theor. Bio., 49, 511-524.

May, R. M., 1977: Thresholds and breakpoints in ecosystems with a multiplicity of stable states. Nature, 269, 471-477.

May, R. M. and G. F. Oster, 1976: Bifurcations and dynamic complexity in simple ecological models. Amer. Natur., 110, 573-599.

Mech, L. D., 1977: Wolf-pack buffer zones as prey reservoirs. Science, 198, 320-321.

Nyblade, C., 1974: Coexistence in sympatric hermit crabs. Ph.D. Thesis, University of Washington.

Okubo, A., 1974: Diffusion-induced instability in model ecosystems: another possible explanation of patchiness. Tech. Rep. 86. Chesapeake Bay Inst., Johns Hopkins Univ., Baltimore.

Okubo, A., 1975: Ecology and Diffusion. Tsukiji Shokan, Tokyo 220 + vii pp.

Okubo, A., 1976: The critical size of water masses containing plankton blooms under scale-dependent diffusion. (Manuscript).

Oster G., 1978: The dynamics of nonlinear models with age structure. In Mathematical Association of America Study in Mathematical Biology, Vol. II, S. A. Levin (ed.), Populations and Communities.

Oster, G. and Y. Takahashi, 1974: Models for age-specific interactions in a periodic environment. Ecol. Monogr., 44, 483-501.

Othmer, H. G., 1977: Current problems in pattern formation. In Some Mathematical Questions in Biology, VIII, S. A. Levin (ed.), Lectures on Mathematics in the Life Sciences, Vol. 9, American Mathematical Society, Providence, Rhode Island, in press.

Paine, R. T. and S. A. Levin, 1978: Responses to perturbation in the intertidal zone. In The Biological Significance of Species Transients in Natural Resource Systems, R. Dugdale and O. Loucks (eds.), pp. 32-50, The Institute of Ecology, Madison, Wisconsin, in press.

Park, T., 1962: Beetles, competition, and populations. Science, 138, 1369-1375.

Pickett, S.T.A., 1976: Succession: an evolutionary interpretation. Amer. Natur., 110, 107-119.

Platt, T., 1975: The physical environment and spatial structure of phytoplankton populations. Mem. Soc. Roy. des Sci. de Liege, serie 6e, Vol. 7, 9-17.

Platt, T. and K. L. Denman, 1975: A general equation for the mesoscale distribution of phytoplankton in the sea. Mem. Soc. Roy. Sci. Liege, 7, 31-42.

Platt, W. J., 1975: The colonization and formation of equilibrium plant species associations on badger disturbances in a tall-grass prairie. Ecol. Monogr., 45, 285-305.

Reddingius, J., 1971: Gambling for existence: a discussion of some theoretical problem in animal population ecology. In Acta Theor. Suppl. Climun, added to Acta Biotheor. 20, Brill, Leiden, 208 pp.

Reddingius, J. and P. J. denBoer, 1970: Simulation experiments illustrating stabilization of animal numbers by spreading of risk. Oecologia, 5, 240-248.

Richerson, P., R. Armstrong and C. R. Goldman, 1970. Contemporaneous disequilibrium, a new hypothesis to explain the "paradox of the plankton." Proc. Nat. Acad. Sci., USA, 67, 1710-1714.

Riley, G., 1946: Factors controlling phytoplankton populations on Georges Bank. J. Mar. Res., 6, 54-73.

Riley, G., 1947: A theoretical analysis of the zooplankton populations of Georges Bank. J. Mar. Res., 6, 104-113.

Rinzel, J., 1978: Integration and propagation of neuroelectric signals. In MAA Study in Mathematical Biology, Vol. I, S. A. Levin (ed.), Mathematical Association of America, Washington, D.C.

Roff, D. A., 1974: Spatial heterogeneity and the persistence of populations. Oecologia, 15, 245-258.

Root, R. B., 1973: Organization of a plant-arthropod association in simple and diverse habitats: the fauna of collards (Brassica oleracea). Ecol. Monogr., 43, 95-124.

Rubinow, S. I., 1973: Mathematical problems in the Biological Sciences. SIAM, Philadelphia, 90 pp.

Rubinow, S. I., 1978: Age-structured equations in the theory of cell populations. In MAA Study in Mathematical Biology.

Ruelle, D. and F. Takens, 1971: On the nature of turbulence. Comm. Math. Phys., 20, 167-192.

Sale, P. F., 1976: Reef fish lottery. Nat. Hist., 85, 60-65.

Sale, P. F., 1977: Maintenance of high diversity in coral reef fish communities. Amer. Natur., 111, 337-359.

Sarkovskii, A. N., 1964: Coexistence of cycles of a continuous map of a line into itself. Ukr. Mat. Z., 16, 61-71 (in Russian).

Segel, L. A. and J. Jackson, 1972: Dissipative structure: an explanation and an ecological example. J. Theor. Biol., 37, 545-559.

Segel, L. A. and S. A. Levin, 1976: Application of nonlinear stability theory to the study of the effects of diffusion on predator-prey interactions. In Topics in Statistical Mechanics and Biophysics: A Memorial to Julius L. Jackson, R. A. Piccirelli (ed.), Proc. AIP Conf., 27, 123-152.

Shure, D. J. and H. J. Ragsdale, 1977: Patterns of primary succession on granite outcrop surfaces. Ecology, 58, 993-1006.

Simberloff, D. S., 1976: Experimental zoogeography of islands: effects of island size. Ecology, 57, 629-648.

Simberloff, D. S. and E. O. Wilson, 1969: Experimental zoogeography of islands: the colonization of empty islands. Ecology, 50, 278-296.

Simberloff, D. S. and E. O. Wilson, 1970: Experimental zoogeography of islands: a two-year record of colonization. Ecology, 51, 934-937.

Skellam, J. G., 1951: Random dispersal in theoretical populations. Biometrika, 38, 196-218.

Slatkin, M., 1974: Competition and regional coexistence. Ecology, 55, 128-134.

Smale, S., 1976: On the differential equations of species in competition. J. Math. Bio., 3, 5-8.

Smith, F. E., 1972: Spatial heterogeneity, stability, and diversity in ecosystems. In Growth by Intussusception: Ecological Essays in Honor of G. Evelyn Hutchinson. Trans. Conn. Acad. Arts Sci., 44, 309-335.

Sousa, W., 1977: Ph.D. Thesis, University of California, San Diego.

Southwood, T.R.E., 1962a: Migration of terrestrial anthropods in relation to habitat. Biol. Rev., 37, 171-214.

Southwood, T.R.E., 1962b: Migration - an evolutionary necessity for denizens of temporary habitats. Int. Kongr. Ent., 3, 55-58.

Sprugel, D., 1976: Dynamic structure of wave-regenerated Abies balsamea forests in the northeastern United States. J. Ecol., 64, 889-911.

Steele, J., 1973: Patchiness. Coastal Upwelling and Ecosystems Analysis Newsl., 2, 3-7.

Steele, J. H., 1974a: Spatial heterogeneity and population stability. Nature, 248, 83.

Steele, J. H., 1974b: Stability of plankton ecosystems. In Ecological Stability, M. B. Usher and M. H. Williamson (eds.), pp. 179-191, Chapman & Hall, London.

Stefan, P., 1976: A theorem of Sarkovskii on the existence of periodic orbits of continuous endomorphising of the real line. Ms. Bures sur Yvette IHES.

Strathman, R., 1974: The spread of sibling larvae of sedentary marine invertebrates. Amer. Natur., 108, 29-44.

Sutherland, J. P., 1974: Multiple stable points in natural communities. Amer. Natur., 108, 859-873.

Thom, R., 1969: Topological models in biology. Topology, 8, 313-335.

Tikhonov, A. N., 1948: On the dependence of the solutions of differential equations on a small parameter. Mat. Sb., 22, 193-204.

Tikhonov, A. N., 1952: Systems of differential equations with small parameters at the derivatives. Mat. Sb., 31, 575-586.

Turing, A., 1952: The chemical basis of morphogenesis. Phil. Trans. Roy. Soc. Lond. Ser. B, 237, 37-72.

Tyson, J. J., 1976: The Belousov-Zhabotinskii Reaction. Lecture Notes in Biomathematics 10. Springer-Verlag, Heidelberg.

von Foerster, H., 1959: Some remarks on changing populations. In The Kinetics of Cellular Proliferation, F. Stohlman (ed.), pp. 283-407, Stratton, New York.

Waddington, C. H., 1957: The Strategy of the Genes. Allen and Unwin, London.

Waltman, P., 1974: Deterministic Threshold Models in the Theory of Epidemics. Lecture Notes in Biomathematics 1. Springer-Verlag, Berlin-Heidelberg-New York.

Whittaker, R. H., 1953: A consideration of climax theory: the climax as a population and pattern. Ecol. Monogr., 23, 41-78.

Whittaker, R. H., 1967: Gradient analysis of vegetation. Biol. Rev., 42, 207-264.

Whittaker, R. H., 1974: Climax concepts and recognition. In Vegetation Dynamics, R. Knapp (ed.). Handb. Veget. Sci., 8, 137-154.

Whittaker, R. H., 1975: Communities and Ecosystems (2nd edition). MacMillan, New York.

Whittaker, R. H. and S. A. Levin, 1977: The role of mosaic phenomena in natural communities. Theor. Pop. Biol., in press.

Whittaker, R. H., S. A. Levin and R. B. Root, 1973: Niche, habitat, and ecotope. Amer. Natur., 107, 321-338.

Whittaker, R. H., S. A. Levin and R. B. Root, 1973: On the reasons for distinguishing 'niche, habitat and ecotope.' Amer. Natur., 109, 479-482.

Whittaker, R. H. and G. M. Woodwell, 1972: Evolution of natural communities. In Ecosystem Structure and Function, J. A. Wiens (ed.), pp. 137-159, Oregon State University Press.

Wright, H. E., Jr., 1974: Landscape development, forest fires, and wilderness management. Science, 186, 487-495.

Wynne-Edwards, V. C., 1962: Animal Dispersion in Relation to Social Behavior. Oliver and Boyd, Edinburgh.

Zaikin, A. N. and A. M. Zhabotinskii, 1970: Concentration wave propagation in two-dimensional liquid-phase self-oscillating system. Nature, 225, 535-537.

INDEX